Application and Characterization of Polymer Composites

Application and Characterization of Polymer Composites

Guest Editor

Fei Han

MDPI

Basel • Beijing • Wuhan • Barcelona • Belgrade • Novi Sad • Cluj • Manchester

Guest Editor
Fei Han
School of Life Sciences
Xi'an Jiaotong University
Xi'an
China

Editorial Office
MDPI AG
Grosspeteranlage 5
4052 Basel, Switzerland

This is a reprint of the Special Issue, published open access by the journal *Polymers* (ISSN 2073-4360), freely accessible at: https://www.mdpi.com/journal/polymers/special_issues/A639G5KY7R.

For citation purposes, cite each article independently as indicated on the article page online and as indicated below:

Lastname, A.A.; Lastname, B.B. Article Title. *Journal Name* **Year**, *Volume Number*, Page Range.

ISBN 978-3-7258-6175-0 (Hbk)
ISBN 978-3-7258-6176-7 (PDF)
https://doi.org/10.3390/books978-3-7258-6176-7

Contents

About the Editor

Fei Han

Fei Han, School of Life Sciences, Xi'an Jiaotong University, Xi'an 710049, China. My primary research interests focus on smart hydrogels, polymer composites, wearable colorimetric sensors, and unclonable anti-counterfeiting labels. To date, I have published over 40 high-quality papers, including more than 20 first-author or corresponding-author articles in internationally renowned journals such as *Advanced Materials, Nature Communications, Advanced Science, Chemical Engineering Journal, Biosensors and Bioelectronics, Small, Advanced Healthcare Materials*, and *ACS Macro Letters*. This includes 5 ESI Highly Cited Papers (one of which is a Hot Paper). My research findings have been cited and positively evaluated in top-tier international journals such as *Nature Nanotechnology; Chemical Reviews; Nature Biomedical Engineering; Nature Communications; Advanced Materials*; and *Journal of the American Chemical Society*. My published work has been cited over 3000 times (Google Scholar) and has an H-index of 24.

Preface

Polymer-based gels, occupying a unique niche between solids and liquids, represent one of the most dynamic and rapidly advancing frontiers in materials science. These three-dimensional networks of crosslinked polymers, capable of imbibing large quantities of solvent, exhibit a remarkable suite of properties, including stimuli-responsiveness, tunable mechanical strength, and high biocompatibility. This versatility positions them as critical materials for addressing some of the most pressing challenges across fields such as biomedicine, soft robotics, environmental science, and energy storage.

The motivation for this Special Issue stems from the need to consolidate and highlight the latest breakthroughs that bridge the fundamental design of these gels with their practical, real-world applications. While the understanding of gelation mechanisms and network structures has deepened, a significant and exciting challenge remains: the rational design of gels with precisely tailored functionalities for specific technological purposes. This involves intricate control over their chemical composition, physical architecture, and multi-scale responsiveness. The aim of this collection is to provide a platform for sharing cutting-edge research that demonstrates this journey—from innovative synthesis strategies and advanced characterization to proof-of-concept demonstrations and the scalable manufacturing of functional gel-based devices.

We extend our sincere gratitude to all the authors who have contributed their high-quality research to this collection and to the numerous reviewers whose diligent work and insightful comments have ensured the scientific rigor and quality of this Special Issue.

Fei Han
Guest Editor

polymers

MDPI

Article

Flexible $Ti_3C_2T_x$-Polyurethane Electrodes for Versatile Wearable Applications

Qiaohang Guo [1], Kepei Chen [1], Wei Yu [1], Man Peng [1], Nuozhou Yi [1], Zhen Wang [1], Peidi Zhou [2], Kaihuai Yang [3], Fei Han [4,5,*] and Mingcen Weng [1,6,*]

1 School of Materials Science and Engineering, Fujian University of Technology, Fuzhou 350118, China; guoqh@fjut.edu.cn (Q.G.); nuozhouyi@smail.fjut.edu.cn (N.Y.)
2 Institute of Smart Marine and Engineering, Fujian University of Technology, Fuzhou 350118, China
3 School of Mechanical and Intelligent Manufacturing, Fujian Chuanzheng Communications College, Fuzhou 350007, China
4 The Key Laboratory of Biomedical Information Engineering of Ministry of Education, School of Life Science and Technology, Xi'an Jiaotong University, Xi'an 710049, China
5 Bioinspired Engineering and Biomechanics Center (BEBC), Xi'an Jiaotong University, Xi'an 710049, China
6 Institute of Biology and Chemistry, Fujian University of Technology, Fuzhou 350118, China
* Correspondence: feihan@xjtu.edu.cn (F.H.); wengmc@fjut.edu.cn (M.W.)

Abstract: With the development of science and technology, wearable electronics are increasingly widely used in medical, environmental monitoring, and other fields. Thus, the demand for flexible electrodes is increasing. The two-dimensional material $Ti_3C_2T_x$ has attracted much attention in the manufacture of flexible electrodes due to its excellent mechanical and electrical properties. However, the brittleness of pure $Ti_3C_2T_x$ films has become a major obstacle for their use as flexible electrodes in wearable devices. Therefore, solving the brittleness problem of flexible electrodes based on $Ti_3C_2T_x$ while maintaining the excellent performance of $Ti_3C_2T_x$ has become an urgent problem. To solve this problem, $Ti_3C_2T_x$ was compounded with waterborne polyurethane (WPU), and a $Ti_3C_2T_x$-WPU composite film with a hierarchical structure was constructed by evaporation-assisted self-assembly. The $Ti_3C_2T_x$-WPU composite film not only retains the excellent electrical conductivity of $Ti_3C_2T_x$ (100 S m^{-1}) but also has flexibility (20 MJ m^{-3}). Furthermore, the $Ti_3C_2T_x$-WPU composite film is applied to functional devices such as contact pressure sensors and non-contact proximity sensors. Finally, the $Ti_3C_2T_x$-WPU composite film wearable device demonstrates its practical application potential in the field of wearable devices.

Keywords: wearable electronics; $Ti_3C_2T_x$; waterborne polyurethane; pressure sensor; proximity sensor

1. Introduction

With advancements in technology, flexible electronic devices have become a popular topic in the field of electronics [1–3]. Flexible electronic devices are increasingly applied across a wide range of fields, including healthcare [4], the aerospace industry, and the military, due to their superior flexibility and adaptability [5–7]. By optimizing the properties of materials [8], the development of deep learning systems [9] further extends the functionality and potential of flexible electronic devices, enabling them to meet ergonomic demands [10,11] while providing a new direction for the development of future smart wearable technologies. For example, OU et al. successfully prepared a dual-network-structured conductive hydrogel that can be used as a wearable strain sensor for real-time monitoring of human motion [12]. Wang et al. prepared a silk nanofiber self-powered humidity sensor that can be used for wireless monitoring systems and non-contact human-computer interaction [13]. Among these, waterborne polyurethane (WPU)-based flexible sensing materials have seen widespread application in wearable electronics, soft robotics, and biomedicine in recent years [14–16]. Polyurethane is an artificial polymer material with extensive applications, offering characteristics such as controllable cost, good stability, and adjustable performance compared to other polymer materials like polydimethylsiloxane, polyethylene terephthalate, and polyacrylic acid, making it an ideal substrate for flexible sensing devices. Therefore, the development of flexible sensing materials based on WPU is currently a hot topic of research.

Conductive materials in WPU-based flexible composite sensing materials include carbon-based conductive materials, conductive metals, and conductive polymers [17–19]. Yao et al. proposed a skin-like strain sensor that can be used in the fields of human sound signal detection, human motion detection, and electronic skin of bionic robots using multiwalled carbon nanotubes as the conductive material [20]. After compounding with the WPU substrate, flexible sensing materials with excellent electrical conductivity and mechanical properties can be prepared. However, the aforementioned conductive fillers have some drawbacks, such as carbon-based fillers being susceptible to environmental influences, leading to poor performance and short lifespans, and metal fillers, though stable, are costly [21]. Moreover, despite extensive research on WPU-based flexible sensing materials, their sensing functions and application scenarios remain relatively singular, presenting numerous challenges in practical applications. Therefore, there is a need to develop new multifunctional WPU-based flexible sensors to expand their applications to the field of sensing and beyond. As a two-dimensional transition metal carbide, nitride, and carbonitride, $Ti_3C_2T_x$ exhibits exceptional electrical conductivity, a high specific surface area, and excellent hydrophilicity, demonstrating significant potential for application in the realm of sensing technology [22–24]. The surface of $Ti_3C_2T_x$ is enriched with functional groups such as -OH, =O, or -F, which not only facilitates its dispersion in common solvents like ethanol and water but also renders the functionalization of the $Ti_3C_2T_x$ surface more accessible without markedly altering its conductivity [25]. The applications of $Ti_3C_2T_x$ in the field of sensors include strain, pressure, temperature, and gas sensors, as well as the detection of various biological markers and environmental pollutants [26–28]. The high electrical conductivity of $Ti_3C_2T_x$ significantly accelerates the rate of electron transfer, and the presence of surface functional groups aids in establishing stable electrode-to-target material contact in electrochemical sensors [29]. However, there are several drawbacks associated with the application of $Ti_3C_2T_x$ in the field of sensors. Typically, the relatively poor mechanical properties and high brittleness of $Ti_3C_2T_x$ render it challenging to withstand a certain degree of processing deformation, thereby limiting the expansion of its application domains.

Herein, WPU is selected as a toughening phase to fabricate a toughened $Ti_3C_2T_x$-WPU composite film, thereby enhancing its mechanical processability and further expanding its application in the field of wearable devices. $Ti_3C_2T_x$ was compounded with WPU, and a hierarchically structured $Ti_3C_2T_x$-WPU composite film was constructed through evaporation-assisted self-assembly. The composite film retains the excellent electrical conductivity of $Ti_3C_2T_x$ and also possesses toughness. As a well-known elastomer, WPU interacts with the soft polymer phase of $Ti_3C_2T_x$, promoting interlayer sliding of the $Ti_3C_2T_x$ nanosheets, thereby enhancing the toughness of the $Ti_3C_2T_x$-WPU composite film. Furthermore, the $Ti_3C_2T_x$-WPU composite film has been applied to functional devices such as contact pressure sensors and non-contact proximity sensors. Experimental results indicate that the contact pressure sensor based on the $Ti_3C_2T_x$-WPU composite film exhibits a sensitivity of up to 177 kPa^{-1} and a response time of only 50 ms, demonstrating its capability to respond sensitively and rapidly to external pressure changes. In addition, the $Ti_3C_2T_x$-WPU composite film has been applied to a non-contact proximity sensor, achieving a sensitivity of 321.4 mm^{-1}, which signifies its ability to swiftly detect approaching objects. The $Ti_3C_2T_x$-WPU composite film can be applied to soft and wearable devices, demonstrating its potential for practical applications in the domain of wearable devices.

2. Materials and Methods

2.1. Materials

Ti_3AlC_2 powder (200 mesh) was purchased from 11 Technology Co., Ltd. (Jilin, China). Hydrochloric acid (HCl), dimethyl sulfoxide (DMSO), lithium fluoride (LiF), polyvinyl alcohol (PVA), glycerol, and sulfuric acid (H_2SO_4) were purchased from Sinopharm Chemical Reagent Co., Ltd. (Shanghai, China). WPU was purchased from Shenzhen Jitian Chemical Co., Ltd. (Shenzhen, China).

2.2. Fabrication of the $Ti_3C_2T_x$ Suspension

Typically, HCl, LiF, and Ti_3AlC_2 powders were first added into a polytetrafluoroethylene beaker under stirring and further stirred. After repeatedly washing the resultant mixture with deionized water, DMSO was added to the mixture by sonification (sonication time was 180 min, power 660 W, room temperature). The mixture was centrifuged (10,000 rpm, 15 min) using a high-speed centrifuge to remove the supernatant and collect the sediment. Finally, deionized water was added to the centrifugal sediment with shaking to obtain the $Ti_3C_2T_x$ suspension (5 mg mL^{-1}).

2.3. Fabrication of $Ti_3C_2T_x$-WPU Composites

A pipette was used to remove 2 mL of the WPU and place it in a beaker. Subsequently, 20 mL of the $Ti_3C_2T_x$ suspension was added to the beaker. Next, the mixture was stirred with magnets on a magnetic mixer for 30 min (room temperature). When stirring was complete, the mixed solution was poured from the beaker into a petri dish. The $Ti_3C_2T_x$-WPU mixture was placed in an oven at 60 °C and dried for 18 h. At the end of drying, the $Ti_3C_2T_x$-WPU composites were peeled off the dish (200 μm thickness) for subsequent testing or application.

2.4. Fabrication of PVA-H_2SO_4 Hydrogel

First, 50 mL of H_2SO_4 (1 M) was placed in a beaker. Secondly, weigh 1 g of PVA was weighed added to the H_2SO_4 solution, and the above step was repeated five times. That is, a total of 5 g of PVA was added. After the fifth addition of PVA, another 50 mL of glycerol was added. After the glycerol was completely dissolved, it was poured into a petri dish and cooled to room temperature. Finally, three freeze-thaw cycles were are carried out to obtain the PVA-H_2SO_4 hydrogel.

2.5. Fabrication of Pressure Sensors Based on the $Ti_3C_2T_x$-WPU Composite Film

First, both the PVA-H_2SO_4 hydrogel (5 mm thickness, thicknesses less than 5 mm are not favorable for LRC bridge recognition and greater than 5 mm the capacitance rate of change decreases, resulting in a loss of sensitivity) and the $Ti_3C_2T_x$-WPU composite film were cut into 1 × 1 cm sizes. Second, utilizing the hydrogel's adhesive properties, two $Ti_3C_2T_x$-WPU composite films were pasted on the upper and lower surfaces to form a double-electrode layer structure. Finally, fine copper foils were pasted onto the upper and lower electrode surfaces.

2.6. Characterization and Measurements

The morphologies and microstructures of the materials were characterized using a transmission electron microscope (TEM) (JEM-2100) and a scanning electron microscope (SEM) (SU8000, JPN). The molecular compositions and phase structures were identified using a Fourier transform infrared spectrometer (FTIR) (Thermo Fisher, Nicolet 6700, USA) and an X-ray diffractometer (XRD) (Bruker Corporation, D8 advance, GER). The tensile properties of the samples were measured using a universal testing machine (Instron, 3343). The optical photos were captured by a digital camera (HUAWEI P40 Pro Plus). The electrical conductivity was measured using a Four-probe Tester (Helpass Electronic Technologies, Inc. HPS2526). The electrical signals of the pressure sensor were captured by a digital bridge (Tonghui 28320).

3. Results

3.1. Fabrication and Characterization of the $Ti_3C_2T_x$-WPU Composite Film

The process of fabricating a $Ti_3C_2T_x$-WPU composite film via an evaporation-assisted self-assembly technique is depicted in Figure 1a. Initially, the selective etching of the Al layer in the Ti_3AlC_2 MAX phase is conducted in a mixture of HCl/LiF, yielding $Ti_3C_2T_x$ flakes with high electrical conductivity. The synthesis of the $Ti_3C_2T_x$-WPU composite film utilizes an easily processable aqueous solution. After thoroughly stirring the $Ti_3C_2T_x$ flakes and WPU, the mixture is poured into a petri dish for drying. Upon evaporation of the water, a large-sized, flexible, and conductive thin film is formed, as shown in Figure 1a. The functional groups of WPU, including the amino, carbamate, and carboxyl groups, can form multiple hydrogen bonds with the terminal groups on the surface of $Ti_3C_2T_x$, thereby enhancing the mechanical properties of the $Ti_3C_2T_x$-WPU composite film. The microscopic analyses using SEM are conducted to observe the morphology of the $Ti_3C_2T_x$-WPU composite film. In Figure 1b,c, the upper and lower surface SEM images of the $Ti_3C_2T_x$-WPU composite film reveal a defect-free and uniform film. Figure 1d shows the cross-sectional SEM image of the $Ti_3C_2T_x$-WPU composite film; the thickness of the $Ti_3C_2T_x$-WPU composite film is about 8 μm. In order to further verify the composition of $Ti_3C_2T_x$-WPU composite films, detailed XRD analysis is carried out for $Ti_3C_2T_x$, WPU, and $Ti_3C_2T_x$-WPU composite films. As shown in Figure 1e, the analysis results show that the (002) diffraction peak of $Ti_3C_2T_x$ is sharp and well-defined, located at $2\theta = 6.32°$. However, after the successful recombination of $Ti_3C_2T_x$ and WPU, it is found that the characteristic peaks of $Ti_3C_2T_x$ shifted significantly to the left to the position of $2\theta = 5.24°$. The significant shift indicates that the WPU molecule has been successfully inserted into the molecular layer of $Ti_3C_2T_x$, resulting in a significant increase in the layer spacing [30]. The XRD patterns of WPU are then observed. It is found that it does not show the typical crystalline diffraction peak, but shows a broad peak at $2\theta = 20°$ [31]. Interestingly, a similar peak is also observed in the XRD pattern of the $Ti_3C_2T_x$-WPU composite film, occurring at $2\theta = 20°$. These results strengthen the conclusion that $Ti_3C_2T_x$ and WPU have successfully compounded and formed a $Ti_3C_2T_x$-WPU composite film. To investigate the interactions between $Ti_3C_2T_x$ and WPU, FTIR spectroscopy is employed to characterize the chemical structure [30,32–34]. Firstly, the FTIR spectrum of the WPU film is analyzed, as shown in Figure 1f. A distinct absorption peak is observed at 2920 cm^{-1}, which is attributed to the stretching vibrations of the -CH groups within the WPU molecules. A series of absorption peaks are identified in the range of 1000 to 1300 cm^{-1}, corresponding to the vibrations of the C-O-C bonds in the WPU molecules. Subsequently, the FTIR spectrum of $Ti_3C_2T_x$ has a significant absorption peak at 1630 cm^{-1} due to the -C=O groups. The stretching vibrations of the Ti-O bonds appear at 700 cm^{-1}, and the -OH absorption peak is observed at 3436 cm^{-1}. Finally, in the FTIR spectrum of the $Ti_3C_2T_x$-WPU composite film, an absorption peak corresponding to the stretching vibrations of the -CH groups in WPU is identified at 2920 cm^{-1}. Similarly, absorption peaks consistent with the vibrations of the C-O-C bonds in the WPU molecules are observed in the range of 1000 to 1300 cm^{-1}. The characteristic stretching vibrations of the -OH, -C=O, and Ti-O bonds in $Ti_3C_2T_x$ are also present at 3436 cm^{-1}, 1630 cm^{-1}, and 700 cm^{-1}, respectively. Finally, mechanical tests are conducted on the $Ti_3C_2T_x$-WPU composite film, WPU, and $Ti_3C_2T_x$. The stress-strain curves are presented in Figure 1g. The $Ti_3C_2T_x$-WPU composite film can endure a maximum stress of up to 12.12 MPa, a value significantly surpassing that of WPU (3.08 MPa). This indicates that the integration of $Ti_3C_2T_x$ has enhanced the maximum stress of WPU. Furthermore, as depicted in Figure 1g, WPU exhibits a fracture strain of 200%, whereas $Ti_3C_2T_x$ only has a fracture strain of 1.27%, indicating that $Ti_3C_2T_x$ is prone to fracturing under external forces. The incorporation of WPU has significantly increased the maximum deformation of the $Ti_3C_2T_x$-WPU composite film to 190%, demonstrating that the addition of WPU has substantially elevated the deformability of $Ti_3C_2T_x$. Fracture toughness, a pivotal indicator of mechanical capability, of the $Ti_3C_2T_x$-WPU composite film has demonstrated remarkable performance. Its fracture toughness reaches an impressive 20.73 MJ m^{-3}, vastly exceeding

WPU (5.01 MJ m^{-3}) and the $Ti_3C_2T_x$ (0.16 MJ m^{-3}). These results unequivocally illustrate the exceptional performance of the $Ti_3C_2T_x$-WPU composite film in resisting fractures and maintaining its structural integrity, thereby endowing it with enhanced reliability and durability for practical applications.

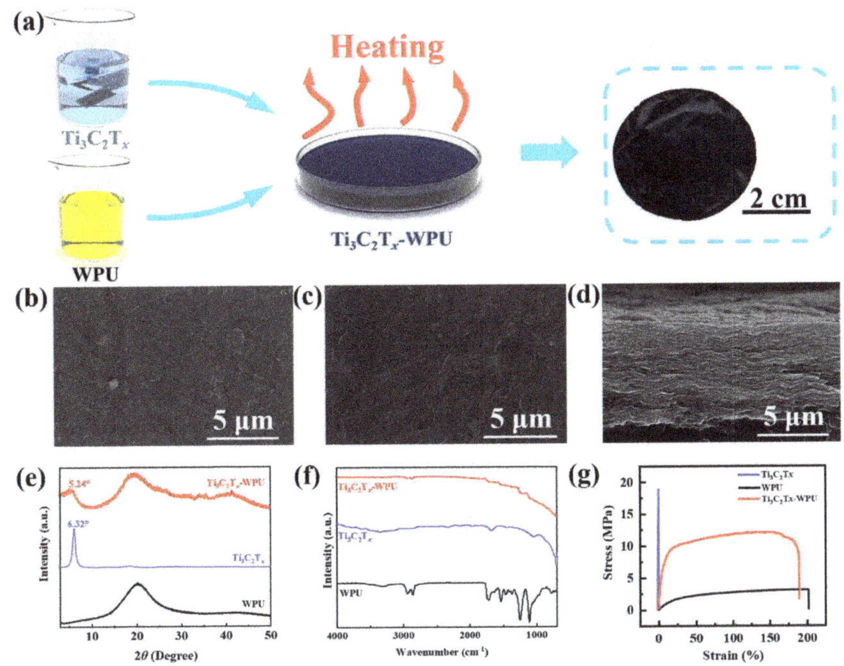

Figure 1. Fabrication and characterization of $Ti_3C_2T_x$-WPU composite film. (**a**) Schematic showing the fabrication process of the $Ti_3C_2T_x$-WPU composite film. (**b–d**) SEM images of the $Ti_3C_2T_x$-WPU composite film. (**e**) XRD patterns, (**f**) FTIR spectra, and (**g**) strain-stress curves of the WPU, $Ti_3C_2T_x$, and $Ti_3C_2T_x$-WPU composite films.

3.2. The Pressure Sensor Based on the $Ti_3C_2T_x$-WPU Composite Film

A capacitive pressure sensor with a sandwich structure is constructed using the $Ti_3C_2T_x$-WPU composite film as the flexible electrode and H_2SO_4-PVA as the ionic gel, as shown in Figure 2a. The sensing mechanism of the pressure sensor is as follows [35]. The electrons in the $Ti_3C_2T_x$-WPU composite film and the oppositely charged ions in the H_2SO_4-PVA ionic gel attract each other and accumulate at their contact interface to form an electric double-layer (EDL) capacitance with high capacitance. The capacitance changes with the change in the area of the interface formed between the ions and the electrons. When no pressure is applied to its surface, the $Ti_3C_2T_x$-WPU composite film and H_2SO_4-PVA ionic gel remain relatively loose. $Ti_3C_2T_x$-WPU composite film is in contact with the H_2SO_4-PVA ionic gel only in a small area. Thus, the capacitance of the pressure sensor is small. The capacitance of a pressure sensor is mainly related to its electrode/electrolyte contact area. The larger the contact area, the higher the capacitance of the pressure sensor. After applying pressure to the pressure sensor, the $Ti_3C_2T_x$-WPU composite film with a microstructure comes into contact with the H_2SO_4-PVA ionic gel. Thus, more EDL capacitances appear at the interface between the $Ti_3C_2T_x$-WPU composite film and H_2SO_4-PVA ionic gel. The capacitance of the pressure sensor increases dramatically, resulting in high sensitivity. The sensing performance of the pressure sensor at different pressures is shown in Figure 2b. Sensitivity is defined as the ratio of the rate of capacitance change to pressure. $\Delta C/C_0$ of the pressure sensor increases with a gradual increase in pressure. The sensitivity of a pressure sensor is typically defined as $S_P = \delta(\Delta C/C_0)/\delta P$. In particular, the sensitivity is as high as 177 kPa^{-1} in the pressure range of 0–1 kPa and 97 kPa^{-1} in the pressure range of 1–4 kPa. Subsequently, the response of the pressure sensor to the dynamic pressure signal is tested and shown in Figure 2c,d. It can be seen that the pressure sensor exhibits a relatively stable and repeatable response to the same dynamic pressure. The rate of change of capacitance increases with increasing pressure. The capacitance change under different dynamic pressures is basically the same as the capacitance change under static pressures in Figure 2d, indicating that the pressure sensor also has a stable sensing ability for dynamic pressures. In addition, the response/recovery time and detection limit of the pressure sensor are also evaluated and are presented in Figure 2e. Loading 1 Pa on the pressure sensor can obtain a clear and stable response and recovery curve, demonstrating a low detection limit. However, when the pressure is 1 Pa, the response and recovery times are only 50 ms, showing a fast response/recovery time. To demonstrate the low detection limit of the pressure sensor more intuitively, two interesting demonstrations are proposed. First of all, the pressure sensor can detect the airflow generated by the ear wash ball. As presented in Figure 2f, the capacitive signal quickly returns to the initial value when the airflow disappears, which intuitively shows that the pressure sensor can respond quickly and accurately to small pressures generated by the airflow. Secondly, objects with different masses are placed on the pressure sensor in turn, and the capacitive signal of the pressure sensor is recorded and shown in Figure 2g. It is found that the pressure sensor can quickly respond to small pressures. Finally, the pressure sensor is subjected to a stability test at a pressure of 4 kPa, as shown in Figure 2h. The capacitance change rate of the pressure sensor drifts slightly in the stability test. In summary, the pressure sensor has a stable pressure sensing performance with high sensitivity, wide sensing range, fast response time, and low detection limit.

3.3. The Proximity Sensor Based on the $Ti_3C_2T_x$-WPU Composite Film

The mechanism of proximity sensors is mainly derived from the perturbation of the fringing electric field in the parallel-plate capacitor. During the process, two types of capacitances are defined: mutual capacitance (C_M) between the upper and lower electrodes and fringing capacitance (C_F) [36]. As shown in Figure 3a, when the distance between the approaching object and the proximity sensor decreases, a portion of the fringe electric field is truncated and shunted to the ground, which reduces the strength of the electric field associated with the capacitive electrodes and decreases the charge stored in the capacitor. With further proximity, charge flows out of both electrodes, resulting in an increase in C_F and a decrease in C_M [37]. Thus, the capacitance response of the proximity sensor is more significant. We specify that a target object (Al block with a base area of 1 cm^2) approaches the proximity sensor from a distance of 100 mm from the upper surface of the proximity sensor and decelerates and stops at a distance of approximately 0.1 mm. The relationship between the $\Delta C/C_0$ and the proximity distance of the proximity sensor is shown in Figure 3b. The capacitance response of the proximity sensor is up to 2000 in the near range region (0–15 mm). The sensitivity of the proximity sensor is typically defined as $S_D = \delta(\Delta C/C_0)/\delta D$, where D represents the vertical distance between the proximity sensor and the object. The sensitivity is as high as 321.43 mm^{-1}. As shown in Figure 3c, the proximity sensor maintains a stable capacitance response at different distances, indicating that the proximity sensor can clearly identify different response signals at different distances. In Figure 3d, we plot the process of approaching and moving away from the proximity sensor into two segments (the red curve represents the object proximity sensor and the blue curve represents the object leaving the sensor) to facilitate the analysis of the hysteresis of the proximity sensor. The capacitance response of the proximity sensor is nearly coincident when the object is approaching and leaving, indicating that the proximity sensor has good reversibility and minimal hysteresis. As shown in Figure 3e, the proximity sensor exhibits a stable capacitance response when objects are approaching at different speeds. In addition, we tested the response/recovery times of the proximity sensor, as shown in Figure 3f. The response/recovery time of the proximity sensor is only 76 ms, indicating that the proximity sensor has a fast response ability. As can be seen in Figure 3g, the stability test of the proximity sensor at the same frequency is performed for over 800 cycles at a distance of 100 mm. As can be seen in the magnified image, the capacitance amplitude fluctuates only slightly over time. This shows that the proximity sensor has good stability.

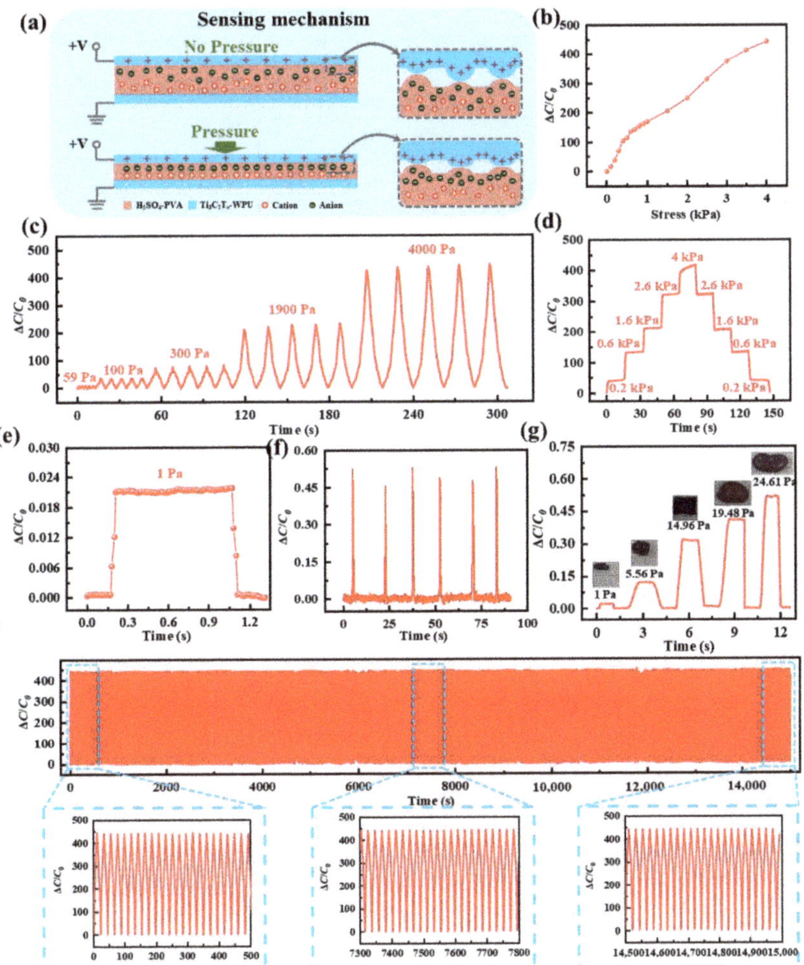

Figure 2. The pressure sensor based on the $Ti_3C_2T_x$-WPU composite film. (**a**) Schematic diagram of the sensing mechanism of the $Ti_3C_2T_x$-WPU composite film-based pressure sensor. (**b**) Sensitivity of the pressure sensor. (**c**) Dynamic step test curve of the pressure sensor. (**d**) Static step test curve of the pressure sensor. (**e**) Response and recovery curves of the pressure sensor. (**f**) The pressure sensor responds to airflow. (**g**) The pressure sensor responds to objects with different masses. (**h**) Stability of the pressure sensor.

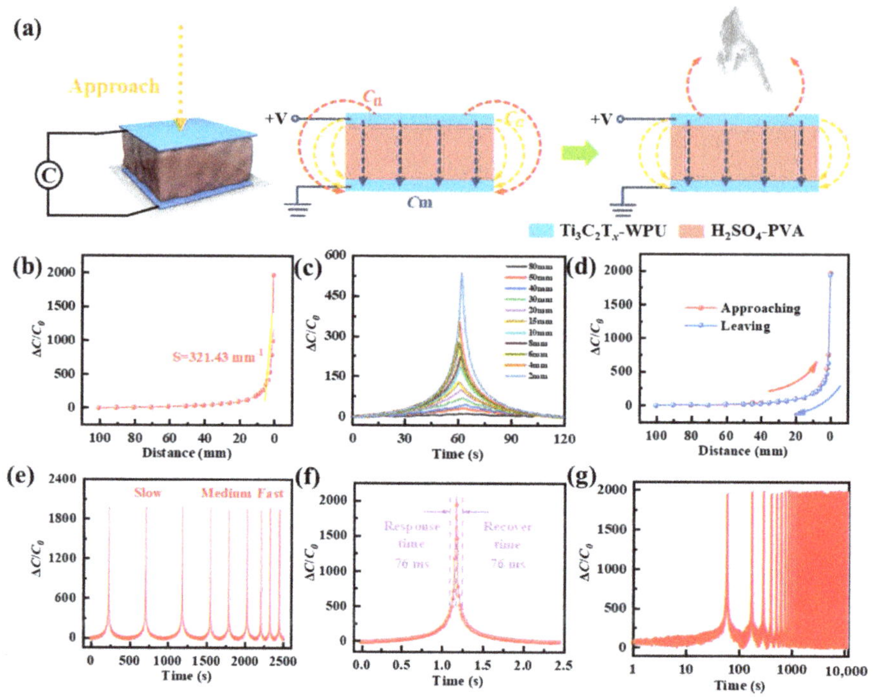

Figure 3. The proximity sensor based on the $Ti_3C_2T_x$-WPU composite film. (**a**) Schematic diagram of the proximity sensing mechanism. (**b**) Distance-dependent capacitance response of the proximity sensor with different distances of aluminum (Al) block. (**c**) The time-dependent change in the capacitance of the proximity sensor at different distances between the Al block and the proximity sensor. (**d**) The capacitance response of the proximity sensor during the approaching-leaving cycle is within 100 mm. (**e**) The capacitance response of the proximity sensor during the approaching-leaving cycle at different moving speeds. (**f**) The response/recovery time of the proximity sensor. (**g**) Stability test of the proximity sensor over 800 cycles.

3.4. The Applications Based on the $Ti_3C_2T_x$-WPU Composite Film

The monitoring of vocal cord health is critically important for maintaining human health. Unlike the bending motion of human joints, the vibration of the vocal cords is much more insignificant. Therefore, monitoring with higher accuracy and sensitivity sensors is required. Hence, the $Ti_3C_2T_x$-WPU composite film-based pressure sensors were mounted on the skin of the larynx of a volunteer (Figure 4a), and the capacitance change curve was used to monitor the vocal folds as they emitted the "Pressure", "Film", "Nano", "Sensor" and "Water" vibrations during the pronunciation of words (Figure 4b). When the volunteer spoke the word, the sensor immediately made a capacitance response to vocal cord vibration. This shows that different words correspond to different capacitance change curves. There are also differences in the capacitance change curves for multiple pronunciations of the same word (Figure 4c), which are due to the different amplitudes of vocal fold vibration during each pronunciation. To enable accurate recognition of word pronunciation, a fully connected multilayer perceptron (MLP) neural network model with five input layers, 100 hidden layers, and five output layers is used to classify the collected word pronunciation data. The algorithmic model achieves high classification accuracy after only 100 iterations, which proves its accuracy and feasibility (Figure 4e). In the end, the accuracy of classifying and recognizing the pronunciation of the five words in the test set is as high as 98.92%, which is slightly higher than the 98.66% accuracy of the training set

(Figure 4f). The confusion matrix also demonstrates the excellent classification ability of the constructed MLP neural network model for word pronunciation recognition (Figure 4g).

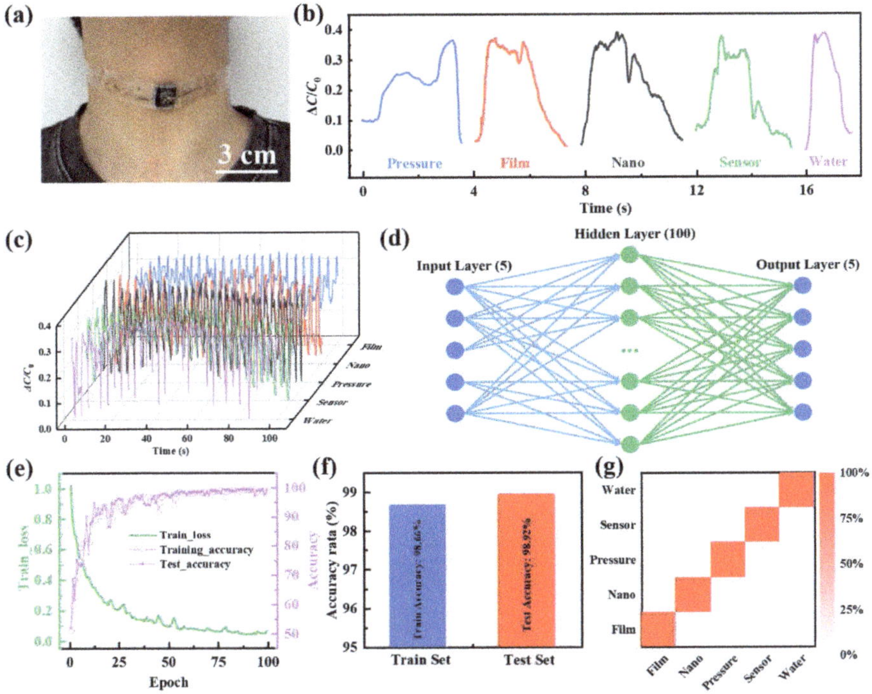

Figure 4. The application of pressure sensors based on the $Ti_3C_2T_x$-WPU composite film. (**a**) The $Ti_3C_2T_x$-WPU composite film-based pressure sensors are used in audible sound detection. (**b**) Capacitance rate of change curves for different word pronunciations. (**c**) Real-time capacitance rate of change signals from different word pronunciations. (**d**) Model structure of neural network algorithms. (**e**) Variation chart of loss and accuracy with the training cycle. (**f**) Accuracy comparison between the train and test set. (**g**) Confusion matrix for recognizing five different word pronunciations.

Human hands can recognize the shapes and sizes of different objects by touching and grasping them. However, the skin cannot distinguish the shape of an object without touching it. By forming a 6 × 6 array of the $Ti_3C_2T_x$-WPU composite film-based pressure sensors (Figure 5a), the shape and size of an object can be sensed as it approaches the sensing array by detecting the capacitance response of each sensing unit, and thus exhibiting sensing capabilities that exceed the human sense of touch. When a metal ball with a diameter of 40 mm approaches the sensing array, a spherical projection corresponding to the sensing unit produces a different capacitive response, which is able to show the hemispherical in a complete 3-dimensional contour map (Figure 5b). Three different metal shapes were designed for the measurement: a cylinder, circular cone, and cube. The mapping image of the metal cylinder is a cylinder with a circular base (Figure 5c), the metal circular cone has a protruding vertex that produces a circular cone-like mapping image (Figure 5d), and the metal cube mapping image is similar to that of the metal cylinder with a square base (Figure 5e).

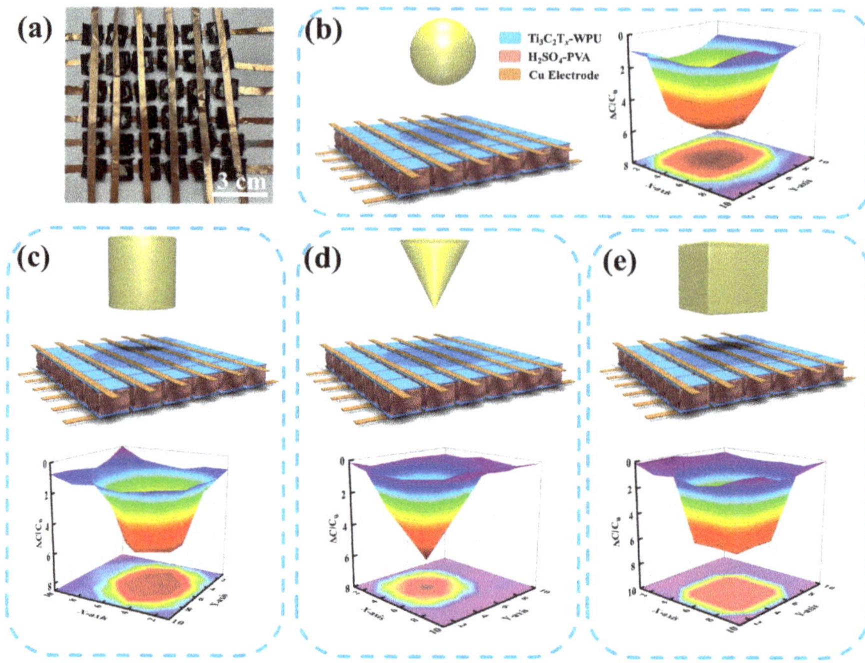

Figure 5. The application of proximity sensor based on the $Ti_3C_2T_x$-WPU composite film. (**a**) The optical image of a proximity sensing array consisting of 36 sensing units. Detect various shapes such as (**b**) a sphere, (**c**) a cylinder, (**d**) a circular cone, and (**e**) a cube.

4. Conclusions

Herein, we introduce a novel $Ti_3C_2T_x$-WPU composite film that has been meticulously engineered to exhibit enhanced mechanical processability and broadened applicability within the domain of soft sensors. The incorporation of WPU as a toughening phase within $Ti_3C_2T_x$ is a strategic choice that imparts the composite with superior mechanical robustness. The fabrication process of the $Ti_3C_2T_x$-WPU composite film is innovatively executed through evaporation-assisted self-assembly, culminating in a hierarchically structured material. This method ensures that the composite retains the intrinsic electrical conductivity of $Ti_3C_2T_x$, exceeding 100 S m^{-1}, while also endowing it with remarkable toughness, quantified at over 20 MJ m^{-3}. The interaction between WPU and $Ti_3C_2T_x$ is pivotal because it facilitates the interlayer sliding of $Ti_3C_2T_x$ nanosheets, thereby augmenting the toughness of the $Ti_3C_2T_x$-WPU composite film. The synergistic effect is a testament to the material design, which leverages the inherent properties of WPU as a well-established elastomer to enhance the mechanical integrity of the $Ti_3C_2T_x$-WPU composite film. The practical utility of the $Ti_3C_2T_x$-WPU composite film is underscored by its successful application in the development of functional devices, such as contact pressure sensors and non-contact proximity sensors. When integrated with an ionic gel, these pressure sensors harness the EDL effect, manifesting a comprehensive pressure sensing range from 0 to 4 kPa, coupled with high sensitivity (177 kPa^{-1}), expedited response and recovery times (50 ms), and an impressively low detection threshold (1 Pa). The efficacy of the pressure sensor is exemplified by its capability to capture and discern vocal cord sounds with a high degree of precision, achieving an accuracy rate of 98.9%. Additionally, the application of the $Ti_3C_2T_x$-WPU composite film in a non-contact proximity sensor has demonstrated a sensitivity of 321.4 mm^{-1}, which underscores its proficiency in the rapid detection of the proximity and shape of approaching objects. The material design and proof-of-concept demonstrations of the $Ti_3C_2T_x$-WPU composite film in various device components not only validate its efficacy but also illuminate the prospective opportunities for its integration into a spectrum of WPU-based soft sensors. The material design and proof-of-concept demonstrations of $Ti_3C_2T_x$-WPU composite film in various device components highlight the future opportunity in all WPU-based soft sensors.

Author Contributions: Conceptualization, Q.G. and K.C.; methodology, Q.G., K.C. and W.Y.; investigation, M.P., N.Y., Z.W., P.Z., K.Y., F.H. and M.W.; resources, Q.G., P.Z. and M.W.; data curation, Q.G. and K.C.; writing—original draft preparation, Q.G.; writing—review and editing, Q.G. and M.W.; funding acquisition, Q.G., P.Z. and M.W. All authors have read and agreed to the published version of the manuscript.

Funding: This work was supported by the National Natural Science Foundation of China [Grant number 52103138], the Natural Science Foundation of Fujian Province [Grant numbers 2022J01945, 2023J01159], the STS Project of Fujian-CAS [2023T3043], Educational Department of Fujian Province (No. JAT220223), and Fujian University of Technology (GY-Z220199, GY-Z23083).

Institutional Review Board Statement: Not applicable.

Data Availability Statement: Data are contained within the article.

Acknowledgments: The authors express gratitude to the editors and the reviewers for their constructive and helpful review comments.

Conflicts of Interest: The authors declare no conflicts of interest.

References

1. Han, F.; Ge, P.; Wang, F.; Yang, Y.; Chen, S.; Kang, J.; Ren, Y.; Liu, H.; Wei, Z.; He, Y.; et al. Smart Contact Lenses: From Rational Design Strategies to Wearable Health Monitoring. *Chem. Eng. J.* **2024**, *497*, 154823.
2. Han, F.; Xie, X.; Wang, T.; Cao, C.; Li, J.; Sun, T.; Liu, H.; Geng, S.; Wei, Z.; Li, J.; et al. Wearable Hydrogel-Based Epidermal Sensor with Thermal Compatibility and Long Term Stability for Smart Colorimetric Multi-Signals Monitoring. *Adv. Healthc. Mater.* **2023**, *12*, e2201730. [PubMed]
3. Torres Alonso, E.; Rodrigues, D.P.; Khetani, M.; Shin, D.; De Sanctis, A.; Joulie, H.; de Schrijver, I.; Baldycheva, A.; Alves, H.; Neves, A.I.S.; et al. Graphene Electronic Fibres with Touch-Sensing and Light-Emitting Functionalities for Smart Textiles. *NPJ Flex. Electron.* **2018**, *2*, 25.
4. Ray, T.R.; Choi, J.; Bandodkar, A.J.; Krishnan, S.; Gutruf, P.; Tian, L.; Ghaffari, R.; Rogers, J.A. Bio-Integrated Wearable Systems: A Comprehensive Review. *Chem. Rev.* **2019**, *119*, 5461–5533.
5. Chen, J.; Liu, F.; Abdiryim, T.; Liu, X. An Overview of Conductive Composite Hydrogels for Flexible Electronic Devices. *Adv. Compos. Hybrid Mater.* **2024**, *7*, 35.
6. Hou, S.; Chen, C.; Bai, L.; Yu, J.; Cheng, Y.; Huang, W. Stretchable Electronics with Strain-Resistive Performance. *Small* **2024**, *20*, 2306749.
7. Chen, S.; Wu, Z.; Chu, C.; Ni, Y.; Neisiany, R.E.; You, Z. Biodegradable Elastomers and Gels for Elastic Electronics. *Adv. Sci.* **2022**, *9*, e2105146.
8. Xu, J.; Chang, L.; Chen, T.; Ren, T.; Zhang, Y.; Cai, Z. Study of the Bending Properties of Variable Stiffness Chain Mail Fabrics. *Compos. Struct.* **2023**, *322*, 117369.
9. Zhou, Y.; Xie, J.; Zhang, X.; Wu, W.; Kwong, S. Energy-Efficient and Interpretable Multisensor Human Activity Recognition via Deep Fused Lasso Net. *IEEE Trans. Emerg. Top. Comput. Intell.* **2024**, 1–13, *early access*.
10. Wang, X.; Zhang, R.; Miao, Y.; An, M.; Wang, S.; Zhang, Y. PI2-Based Adaptive Impedance Control for Gait Adaption of Lower Limb Exoskeleton. *IEEE/ASME Trans. Mechatron.* **2024**, 1–11, *early access*.
11. Wang, K.; Boonpratatong, A.; Chen, W.; Ren, L.; Wei, G.; Qian, Z.; Lu, X.; Zhao, D. The Fundamental Property of Human Leg During Walking: Linearity and Nonlinearity. *IEEE Trans. Neural Syst. Rehabil. Eng.* **2023**, *31*, 4871–4881. [PubMed]
12. Ou, K.; Wang, M.; Meng, C.; Guo, K.; Shariar Emon, N.; Li, J.; Qi, K.; Dai, Y.; Wang, B. Enhanced Mechanical Strength and Stretchable Ionic Conductive Hydrogel with Double-Network Structure for Wearable Strain Sensing and Energy Harvesting. *Compos. Sci. Technol.* **2024**, *255*, 110732.
13. Wang, Z.; Yi, N.; Zheng, Z.; Zhou, J.; Zhou, P.; Zheng, C.; Chen, H.; Shen, G.; Weng, M. Self-Powered and Degradable Humidity Sensors Based On Silk Nanofibers and its Wearable and Human–Machine Interaction Applications. *Chem. Eng. J.* **2024**, *497*, 154443.
14. Wang, D.; Yu, H.; Jiang, L.; Qi, D.; Zhang, X.; Chen, L.; Lv, W.; Xu, W.; Tam, K.C. Flexible, Anti-Damage, and Non-Contact Sensing Electronic Skin Implanted with MWCNT to Block Public Pathogens Contact Infection. *Nano Res.* **2021**, *15*, 2616–2625.
15. Wang, Y.; Shu, R.; Zhang, X. Strong, Supertough and Self-Healing Biomimetic Layered Nanocomposites Enabled by Reversible Interfacial Polymer Chain Sliding. *Angew. Chem. Int. Ed.* **2023**, *62*, e202303446.

16. Yu, C.; Li, X.; Yang, X.; Qiu, X.; Zhang, X.; Chen, Z.; Luo, Y. Dynamic Covalent Bonded Gradient Structured Actuators with Mechanical Robustness and Self-Healing Ability. *Small* **2024**, *20*, 2311656.
17. Burmistrov, I.; Gorshkov, N.; Ilinykh, I.; Muratov, D.; Kolesnikov, E.; Anshin, S.; Mazov, I.; Issi, J.P.; Kusnezov, D. Improvement of Carbon Black Based Polymer Composite Electrical Conductivity with Additions of MWCNT. *Compos. Sci. Technol.* **2016**, *129*, 79–85.
18. Kang, H.; Buchman, J.T.; Rodriguez, R.S.; Ring, H.L.; He, J.; Bantz, K.C.; Haynes, C.L. Stabilization of Silver and Gold Nanoparticles: Preservation and Improvement of Plasmonic Functionalities. *Chem. Rev.* **2019**, *119*, 664–699. [PubMed]
19. Li, Y.; Shi, L.; Cheng, Y.; Wang, R.; Sun, J. Development of Conductive Materials and Conductive Networks for Flexible Force Sensors. *Chem. Eng. J.* **2023**, *455*, 140763.
20. Yao, Z.; Feng, H.; Shang, K.; Deng, X.; Yang, T. Skin-Like Strain Sensors Based On Multiwalled Carbon Nanotube/Polydimethylsiloxane Composite Films. *ACS Appl. Nano Mater.* **2023**, *6*, 6550–6558.
21. Hofmann, A.I.; Cloutet, E.; Hadziioannou, G. Materials for Transparent Electrodes: From Metal Oxides to Organic Alternatives. *Adv. Electron. Mater.* **2018**, *4*, 1700412.
22. Qin, R.; Nong, J.; Wang, K.; Liu, Y.; Zhou, S.; Hu, M.; Zhao, H.; Shan, G. Recent Advances in Flexible Pressure Sensors Based on MXene Materials. *Adv. Mater.* **2024**, *36*, e2312761. [PubMed]
23. Qin, R.; Shan, G.; Hu, M.; Huang, W. Two-Dimensional Transition Metal Carbides And/Or Nitrides (MXenes) and their Applications in Sensors. *Mater. Today Phys.* **2021**, *21*, 100527.
24. Wang, Y.; Yue, Y.; Cheng, F.; Cheng, Y.; Ge, B.; Liu, N.; Gao, Y. Ti_3C_2 T_x MXene-Based Flexible Piezoresistive Physical Sensors. *ACS Nano* **2022**, *16*, 1734–1758. [PubMed]
25. Li, G.; Lian, S.; Wang, J.; Xie, G.; Zhang, N.; Xie, X. Surface Chemistry Engineering and the Applications of MXenes. *J. Mater.* **2023**, *9*, 1160–1184.
26. Echols, I.J.; An, H.; Zhao, X.; Prehn, E.M.; Tan, Z.; Radovic, M.; Green, M.J.; Lutkenhaus, J.L. PH-Response of polycation/$Ti_3C_2T_x$ MXene Layer-By-Layer Assemblies for Use as Resistive Sensors. *Mol. Syst. Des. Eng.* **2020**, *5*, 366–375.
27. Guo, W.; Ma, Z.; Chen, Z.; Hua, H.; Wang, D.; Elhousseini Hilal, M.; Fu, Y.; Lu, P.; Lu, J.; Zhang, Y.; et al. Thin and Soft $Ti_3C_2T_x$ MXene Sponge Structure for Highly Sensitive Pressure Sensor Assisted by Deep Learning. *Chem. Eng. J.* **2024**, *485*, 149659.
28. Waheed, W.; Anwer, S.; Khan, M.U.; Sajjad, M.; Alazzam, A. 2D $Ti_3C_2T_x$-MXene Nanosheets and Graphene Oxide Based Highly Sensitive Humidity Sensor for Wearable and Flexible Electronics. *Chem. Eng. J.* **2024**, *480*, 147981.
29. Zhang, S.; Zahed, M.A.; Sharifuzzaman, M.; Yoon, S.; Hui, X.; Chandra Barman, S.; Sharma, S.; Yoon, H.S.; Park, C.; Park, J.Y. A Wearable Battery-Free Wireless and Skin-Interfaced Microfluidics Integrated Electrochemical Sensing Patch for On-Site Biomarkers Monitoring in Human Perspiration. *Biosens. Bioelectron.* **2021**, *175*, 112844.
30. Jiao, C.; Deng, Z.; Min, P.; Lai, J.; Gou, Q.; Gao, R.; Yu, Z.; Zhang, H. Photothermal Healable, Stretchable, and Conductive MXene Composite Films for Efficient Electromagnetic Interference Shielding. *Carbon* **2022**, *198*, 179–187.
31. Weng, M.; Zhou, J.; Ye, Y.; Qiu, H.; Zhou, P.; Luo, Z.; Guo, Q. Self-Chargeable Supercapacitor Made with MXene-bacterial Cellulose Nanofiber Composite for Wearable Devices. *J. Colloid Interface Sci.* **2023**, *647*, 277–286. [PubMed]
32. Zhao, J.; Wang, Z.; Xu, S.; Wang, H.; Li, Y.; Fang, C. Flexible Bilayer $Ti_3C_2T_x$ Mxene/Cellulose Nanocrystals/Waterborne Polyurethane Composite Film with Excellent Mechanical Properties for Electromagnetic Interference Shielding. *Colloids Surf. A Physicochem. Eng. Asp.* **2023**, *669*, 131556.
33. Cheng, H.; Zuo, T.; Chen, Y.; Yu, D.; Wang, W. High Sensitive, Stretchable and Weavable Fiber-Based PVA/WPU/MXene Materials Prepared by Wet Spinning for Strain Sensors. *J. Mater. Sci.* **2023**, *58*, 13875–13887.
34. Wang, Z.; Wang, L.; Chang, R.; Shi, M.; Sun, D. Construction of Alternating Multilayer MXene/WPU Thin Films with Excellent EMI Shielding Performance and Mechanical Properties. *J. Alloys Compd.* **2023**, *956*, 170367.
35. Tang, Y.; Wang, P.; Li, G.; Wang, G.; Yu, W.; Meng, C.; Guo, S. Flexible and Ultra-Sensitive Planar Supercapacitive Pressure Sensor Based on Porous Ionic Foam. *Adv. Eng. Mater.* **2023**, *25*, 2200814.
36. Niu, H.; Li, H.; Zhang, Q.; Kim, E.S.; Kim, N.Y.; Li, Y. Intuition-and-Tactile Bimodal Sensing Based on Artificial-Intelligence-Motivated All-Fabric Bionic Electronic Skin for Intelligent Material Perception. *Small* **2024**, *20*, 2308127.
37. Niu, H.; Li, H.; Li, N.; Niu, H.; Li, Y.; Gao, S.; Shen, G. Fringing-Effect-Based Capacitive Proximity Sensors. *Adv. Funct. Mater.* **2024**, *34*, 2409820.

polymers

MDPI

Article

Exploring the Impact of Reinforcing Filler Systems on Devulcanizate Composites

Rounak Ghosh [1], Christian Mani [2], Roland Krafczyk [2], Rupert Schnell [2], Auke Talma [3], Anke Blume [3] and Wilma K. Dierkes [1,*]

1 Sustainable Elastomer Systems, Elastomer Technology and Engineering, Department of Solids, Surfaces and Systems, Faculty of Engineering Technology, University of Twente, P.O. Box 217, 7500 AE Enschede, The Netherlands; ronforiit@gmail.com

2 Evonik Operations GmbH, Paul-Baumann-Straße 1, 45772 Marl, Germany; christian.mani@evonik.com (C.M.); roland.krafczyk@evonik.com (R.K.); rupert.schnell@evonik.com (R.S.)

3 Elastomer Technology and Engineering, Department of Solids, Surfaces and Systems, Faculty of Engineering Technology, University of Twente, P.O. Box 217, 7500 AE Enschede, The Netherlands; a.g.talma-1@utwente.nl (A.T.)

* Correspondence: w.k.dierkes@utwente.nl

Abstract: Composites revolutionize material performance, fostering innovation and efficiency in diverse sectors. Elastomer-based polymeric composites are crucial for applications requiring superior mechanical strength and durability. Widely applied in automotives, aerospace, construction, and consumer goods, they excel under extreme conditions. Composites based on recycled rubber, fortified with reinforcing fillers, represent a sustainable material innovation by repurposing discarded rubber. The integration of reinforcing agents enhances the strength and resilience of this composite, and the recycled polymeric matrix offers an eco-friendly alternative to virgin elastomers, reducing their environmental impact. Devulcanized rubber, with inherently lower mechanical properties than virgin rubber, requires enhancement of its quality for reuse in a circular economy: considerable amounts of recycled tire rubber can only be applied in new tires if the property profile comes close to the one of the virgin rubber. To achieve this, model passenger car tire and whole tire rubber granulates were transformed into elastomeric composites through optimized devulcanization and blending with additional fillers like carbon black and silica–silane. These fillers were chosen as they are commonly used in tire compounding, but they lose their reactivity during their service life and the devulcanization process. Incorporation of 20% (*w*/*w*) additional filler enhanced the strength of the devulcanizate composites by up to 15%. Additionally, increased silane concentration significantly further improved the tensile strength, Payne effect, and dispersion by enhancing the polymer–filler interaction through improved silanization. Higher silane concentrations reduced elongation at break and increased crosslink density, as it leads to a stable filler–polymer network. The optimal concentration of a silica–silane filler system for a devulcanizate was found to be 20% silica with 3% silane, showing the best property profile.

Keywords: elastomer; devulcanizate; filler; composite; silane; silica; carbon black; sustainability; recycling

1. Introduction

The widespread use of elastomers across various applications raises environmental concerns related to waste management. Effectively addressing this challenge requires the efficient recycling and reuse of rubber [1]. However, a substantial obstacle to successful recycling is the crosslinked nature of elastomeric products. Vulcanization involves crosslinking polymer chains, creating a robust three-dimensional network. This alteration renders elastomeric material insoluble and infusible, presenting a challenge for effective recycling [2].

Devulcanization, the reversion of the vulcanization process, stands out as a sustainable advancement in elastomer recycling, distinguishing itself from conventional techniques such as grinding or regeneration, the mere replasticization of vulcanized elastomers [3,4]. Unlike these methods, which impose limitations on both quality and quantity for application, devulcanization introduces a more selective breakdown of the network. In contrast to the random breakdown of polymer networks in regeneration, devulcanization focuses on selectively dismantling crosslinks while preserving the polymer chains. This results in a higher tensile strength compared to reclaimed rubber obtained through alternative methods like grinding or regeneration [4–10].

Polymeric composites, particularly those involving elastomers, play a pivotal role in various industries due to their unique properties and versatility. Elastomeric composites show enhanced inherent qualities compared to their non-reinforced counterparts, providing improved mechanical strength, durability, and resistance to wear and tear [11–14]. These composites find extensive applications in automotive components, aerospace technology, construction materials, and consumer goods. The ability of elastomeric composites to withstand extreme temperatures, harsh chemicals, and dynamic stresses makes them indispensable in engineering solutions. Their lightweight nature contributes to energy efficiency in transportation, while their damping characteristics make them valuable for reducing vibrations. In essence, the significance of polymeric composites, especially elastomeric composites, lies in their capacity to revolutionize material performance across diverse sectors, fostering innovation and efficiency [15–17].

Recycled elastomeric composites, enriched with reinforcing fillers, epitomize sustainable material innovation. Harnessing the benefits of recycled rubber and integrating reinforcing agents like carbon black or silica enhances a composite's strength and resilience. However, during the service life of a product and in the devulcanization process, fillers lose a part of their reinforcing strength. A way to mitigate this effect is to add fillers to the recycled rubber.

This sustainable solution not only repurposes discarded elastomers, but also contributes to reduced environmental impact. Widely applicable in diverse industries, these recycled elastomeric composites containing reinforcing fillers offer a sustainable alternative to fossil-based materials, aligning with the growing emphasis on circular economy practices.

The utilization of devulcanized elastomers is hampered by their inferior properties compared to virgin elastomers. To address this limitation, the study aimed to narrow this gap by reinforcing the devulcanizate, transforming it into a composite based on an elastomeric devulcanizate. Model passenger car and whole tire rubber granulate were subjected to devulcanization through an optimized reference process. The quality of the devulcanizates was subsequently enhanced via the incorporation of reinforcing fillers. This investigation focuses on two key aspects: evaluating the influence of additional fillers on the devulcanizate and examining the impact of a coupling agent forming a polymer-filler bond on the devulcanizate composite properties.

2. Effect of Reinforcing Fillers on Devulcanizate Composites

In this study, both a model passenger car tire compound and whole tire (WT) rubber granulates were subjected to devulcanization under optimized process conditions and subsequently compounded using two different filler systems:

1. A highly dispersible silica (ULTRASIL®7000GR, Evonik Industries AG, Essen, Germany) together with a bifunctional silane (TESPD, bis[3-(triethoxysilyl)propyl]-disulfide) as a coupling agent between the filler and polymer was used. TESPD is a commonly used coupling agent for silica reinforcement of rubber, as it is able to split at the disulfide moiety. This results in two molecules comprising a 3-(triethoxysilyl)propyl] moiety, which can each attach to the silica. The sulfur moiety couples to the polymer.

2. A reinforcing-grade carbon black (N330).

These fillers were individually incorporated into the devulcanizate at four varying concentrations, while maintaining a consistent revulcanization formulation. The primary objective of this part of the study was to assess the impact of fillers on the resulting devulcanizate composite, particularly in terms of mechanical properties as well as filler–filler and filler–polymer interactions.

2.1. Material and Methods

This section provides detailed information regarding the preparation of the feed materials, sampling plan, devulcanization, compounding with additional fillers, and the characterization processes. An overview of the experimental process is given in Figure 1.

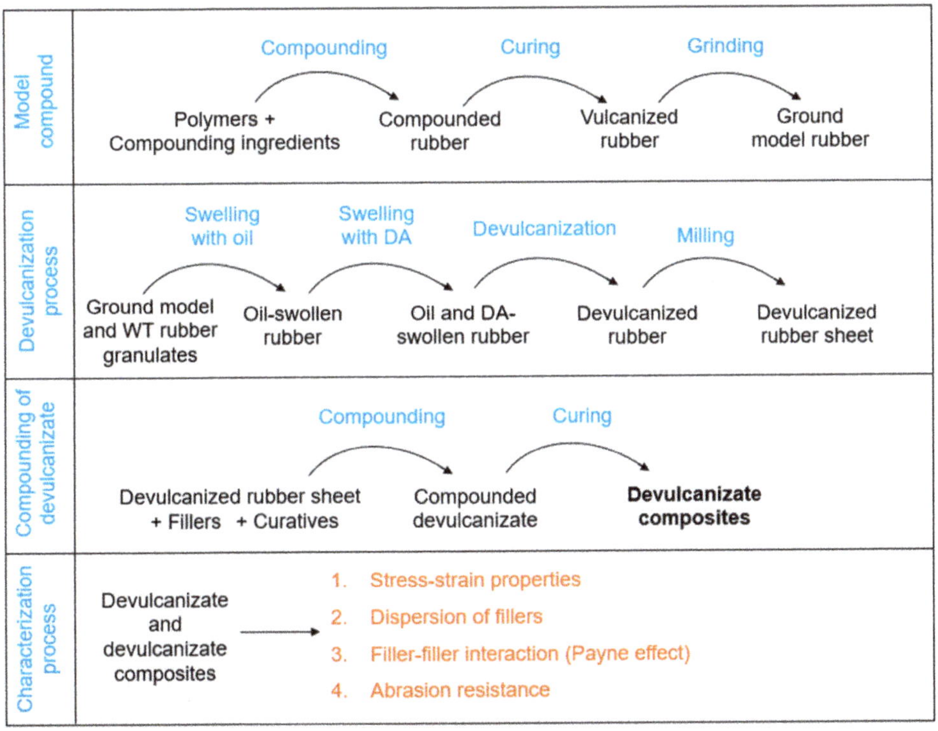

Figure 1. Experimental process of filler optimization for devulcanizate composites.

2.1.1. Model Compound Preparation

The formulation of a silica-filled model tire tread compound based on SBR/BR was developed for utilization as an input material in the devulcanization procedure. The various phases involved in the preparation of the model compound are described as follows.

Materials

The model tire tread formulation is given in Table 1.

Table 1. Compounding formulation of the model tire tread compound.

Function	Ingredient	Trade Name	Supplier	Quantity (phr)
Polymer	SSBR [1]	Sprintan 4601	Trinseo (Wayne, PA, USA)	70
	BR [2]	CB 24	Arlanxeo (Hague, The Netherlands)	30
Filler system	Silica	ULTRASIL® 7000 GR	Evonik (Essen, Germany)	80
	Silane	Si 266®	Evonik (Essen, Germany)	5.8 *
Activators	Zinc oxide	Merck Zinc Oxide	Sigma-Aldrich (St. Louis, MI, USA)	3
	Stearic acid	Merck Stearic Acid	Sigma-Aldrich (St. Louis, MI, USA)	2
Plasticizer	TDAE oil [3]	Vivatec	H&R (Houston, TX, USA)	25
Curatives	Sulphur	Merck Sulphur	Sigma-Aldrich (St. Louis, MI, USA)	1.5
	Primary accelerator	Santocure CBS	Flexsys (Drive Akron, OH, USA)	1.7
	Secondary accelerator	Perkacit DPG	Flexsys (Drive Akron, OH, USA)	2.5

[1]. Solution-polymerized styrene butadiene rubber; [2]. Polybutadiene rubber; [3]. Treated distillated aromatic extract.
* Silane amount was calculated according to the formula of Guy et al. [18].

Compounding

Compounding was carried out in two phases utilizing a 390 mL internal mixer (Model 350S) manufactured by Brabender GmbH & Co., Duisburg, Germany. In the initial stage, mixing commenced with a fill factor of 70%, an initial temperature of 80 °C, and a rotor speed set at 70 rpm. This initial mixing phase aimed to masticate the polymer and disperse the filler effectively, and enable the integration of other compounding ingredients. As the temperature rose, polymer viscosity decreased, while filler addition contributed to viscosity elevation. To optimize the silica–silane interaction, a temperature of 145 °C, known as the optimal reaction temperature, was chosen [19]. Consequently, adjustments were made to the initial temperature and rotor speed of the internal mixer to attain and uphold the desired temperature throughout the compounding process. The introduction of polymer, filler, and other compounding ingredients within the internal mixer led to a temperature rise from 80 °C to 145 °C due to high shear forces. Isothermal mixing and silanization were carried out at 145 °C for 5 min via rotor speed adjustments. The details of the mixing process are outlined in Table 2.

Table 2. Mixing process of the model tire tread compound.

Masterbatch (First) Step		Final (Second) Step	
Action	**Time [mm:ss]**	**Action**	**Time [mm:ss]**
Polymer Mastication	00:00–00:30 00:30–01:30	Masterbatch	-
$\frac{1}{2}$ (Silica + silane) Mixing	01:30–02:00 02:00–03:00	Mixing	00:00–02:00
$\frac{1}{2}$ (Silica + silane) + additives Mixing (140–150 °C)	03:00–03:30 03:30–04:30	Curatives	02:00–02:30
Ram sweep Mixing (target 145 °C)	04:30–05:00 05:00–09:00	Mixing	02:30–09:00
Discharge and sheeting	-	Discharge	-

The masterbatch, which resulted from the initial mixing step, was allowed to rest at room temperature for one day before the final mixing process. Final mixing was conducted using a laboratory-scale, two-roll mill with a diameter of 9 cm, manufactured by Schwabenthan GmbH & Co., Berlin, Germany, and the curatives were incorporated at room temperature. The specific procedure is mentioned in Table 2. Following the final mixing, the compound was left to stand at room temperature for one day before the curing process.

Curing

The optimal cure time was determined using an RPA Elite, manufactured by TA Instruments, based in Delaware, OH, USA. Samples underwent curing at 160 °C following the T_{95} specification. Compression molding was conducted using an automatic press produced by Wickert Maschinenbau GmbH, Landau, Germany, employing molds sized at either 200 mm × 200 mm × 4 mm or 80 mm × 80 mm × 2 mm. Post-curing, the vulcanized sheets exhibited a tensile strength averaging around 16 ± 1 MPa, with an elongation at break of approximately 310 ± 30%.

Chopping and Grinding

The vulcanized sheets were chopped using a bale cutter. Pre-treatment for cryogenic grinding involved immersing the chopped samples in liquid nitrogen for 4–5 min to reach a temperature below the glass transition point, followed by grinding at room temperature. Grinding was executed utilizing a mechanical grinder manufactured by Fritsch, Germany, equipped with a 0.7 mm mesh screen.

2.1.2. Preparation of WT Rubber Granulates

The WT rubber granulate used in this study was provided by Genan GmbH, Dorsten, Germany, and used without any further alteration.

2.1.3. Swelling of Rubber Granulates

The ground model compound and WT rubber granulate underwent initial mixing with processing oil (TDAE), followed by subsequent mixing with the DA at room temperature. Each addition was followed by a one-day room temperature incubation period to allow for swelling. The kinetic viscosity of the TDAE oil was measured at 331 mm^2/s at 40 °C and 18.4 mm^2/s at 100 °C [20]. Due to the high viscosity of the TDAE oil, manual stirring was employed to mix the rubber granulate with the oil. Vinyl silane with peroxide (VP) was selected as the best DA based on the literature comparisons with sulphidic, amino, and mercapto silanes; thus, VP was utilized in this study [2]. The oil-swollen sample was subsequently re-swollen with VP and left for one day to allow for DA migration into the particles.

2.1.4. Devulcanization of Rubber Granulates

The thermo-mechanical-chemical devulcanization process was conducted in a Plastograph EC internal mixer manufactured by Brabender GmbH & Co., Duisburg, Germany, featuring a mixing chamber volume of 50 cc; non-intermeshing, counter-rotating rotors; and a telescopic ram. To prevent oxidation at elevated temperatures, the cavity was sealed with paraffin wax to restrict air access. Devulcanization was carried out using the internal mixer set at a temperature of 180 °C, a rotor speed of 150 RPM, and a fill factor of 80%.

The same two-roll mill utilized in Section "Compounding" was employed for milling the devulcanized rubber at room temperature, employing a speed ratio of 1.25 and operating at 30 RPM. The nip gap was gradually reduced from 1 to 0.1 mm until the devulcanized rubber formed a band.

2.1.5. Sampling Plan

The devulcanizates were prepared by using the optimum devulcanization conditions based on an earlier study [2]. The two types of devulcanizate were compounded separately with two filler systems with four different concentrations. The quantities of silane used as coupling agent were calculated according to Guy's formula [18]. Considering the surface area of silica and functionality of the silane, 7.5% TESPT relative to the silica is sufficient for silanization under standard processing conditions. The sampling details of the filler optimization trials are given in Table 3.

Table 3. Sampling plan for filler optimization trials.

Devulcanizate	Abbreviations	Silica + Silane (*w/w* %)	Carbon Black (CB)
Model—VP	M-VP	1. 10% Silica + 0.75% Silane	1. 10% CB
WT granulates—VP	WT-VP	2. 20% Silica + 1.50% Silane 3. 30% Silica + 2.25% Silane 4. 40% Silica + 3.00% Silane	2. 20% CB 3. 30% CB 4. 40% CB

It is important to note that the feedstock, the model compound as well as the whole tire granulate, already contain reinforcing fillers.

2.1.6. Filler Compounding

The fillers were compounded with the devulcanizate using a Brabender Plastograph EC, Germany, with a 50 CC volume and 70% fill factor. The mixer was set to a temperature of 70 °C, which was increased to around 130 °C for silica compounding. The silica–silane compound was mixed for 3 min at an isothermally at 145 °C. For carbon black samples, the initial set temperature was 70 °C, but during compounding the temperature increased to around 120 °C; the compounding process, including mastication time, took a total of 6 min.

2.1.7. Revulcanization

The devulcanized rubber samples were compounded using the formulation shown in Table 4 and was subsequently revulcanized.

Table 4. Revulcanization formulation.

Function	Component	Weight (%)
Base polymer	Devulcanized rubber sample	100
Activators	Zinc oxide Stearic acid	4 2
Curing aid	Sulphur	2
Accelerator	CBS	1

The compounded devulcanized rubber underwent testing using the RPA 2000 Elite manufactured by TA Instruments, based in Delaware, USA at a temperature of 160 °C for a duration of 30 min, following the ASTM D7750-12 standard [21], to ascertain the optimal cure time. Sheets with a thickness of 2 mm were molded at 160 °C in accordance with the T_{95} specification using an automatic compression molding machine produced by Wickert Maschinenbau GmbH, Landau, Germany.

2.1.8. Characterization Process

The characterization process contains four parts:

Stress–Strain Analysis

The stress–strain properties of the revulcanized rubber were assessed using a Z010 tensile tester manufactured by Zwick Roell GmbH & Co., Ulm, Germany, following the ASTM D412 standard [22]. Seven tensile dumbbells were tested for each sample, with the highest and lowest results omitted. The average of the remaining five values, along with error bars, are reported.

Dispersion

The macro-dispersion analysis of the silica within the compounded sample was conducted using a Dispergrader (optical microscope equipped with software to measure dispersion) manufactured by Alpha Technologies, based in the USA, under full-top light conditions at room temperature. Cylindrical samples measuring 25 cm in diameter and 12 mm in thickness were cut using a sharp knife, and the exposed cross-sections were examined. The filler distribution within the devulcanizate composites was analyzed based on the filler concentration.

Payne Effect

The Payne effect gives an indication of the micro-dispersion, the filler–filler interactions on a small scale, in the compounds. The analysis of the silica-filled devulcanizate composites was performed according to ASTM D8059-19 [23] in a rheometer manufactured by TA Instruments, USA. To measure the Payne effect of cured samples, the samples were revulcanized at 160 °C up to T_{95}, and the Payne effect was measured at 60 °C in two strain sweeps, from low to high strains (from 0.33% to 14.1%) and from high to low strains (from 14.1% to 0.33%), for each sample. The curves from both sweeps were plotted, and the difference between the shear modulus at low strain and high strain of the low to high strain sweep was reported as the Payne effect.

Abrasion Resistance

The abrasion resistance of the rubber compounds was determined by the weight loss according to ASTM D5963 [24] on a rotary drum DIN abrasion tester manufactured by Montech, Switzerland.

2.2. Results

In a polymeric matrix, the filler system can create either a physical or a chemical interaction, depending on the type of filler: carbon fillers generate a physical interaction, while a silica–silane system forms a chemical bond with the polymer. As the concentration of filler increases, the mechanical strength of the elastomeric composite generally improves up to a certain point. However, if the polymer concentration is insufficient to interact with the further increased filler, it leads to a reduction in the overall mechanical properties of the system. This phenomenon is seen in Figure 2: The tensile strength of the de- and revulcanized elastomers increased with increasing filler concentration up to a certain level; above that limit reversal took place. For the model compound, the tensile strength reached around 10.3 MPa with 20% additional filler, and for WT granulate the maximum was around 9.7 MPa at 30% filler concentration. The samples filled with silica–silane demonstrated slightly higher values in comparison to carbon black. However, the average values of tensile strength follow a trend, which allows us to conclude that the silica-filled material performed, in general, better. Comparing the original and 20% additionally silica-filled model devulcanizate, a significant increase in tensile strength can be observed.

According to Figure 3, for both model compound and WT granulates, elongation at break decreased gradually with increasing filler concentration. For WT granulates, elongation at break was around 165% for the unfilled sample, and this was reduced to 118% for the 40% silica-filled samples and to 137% for the 40% carbon-black-filled samples. For the model compound, elongation at break was around 116% for original samples, which was reduced to 80% for 40% silica-filled samples and 83% for 40% carbon-black-filled samples.

The dispersion analysis of the WT devulcanizate composites with increasing silica concentration is shown in Figure 4. With the increase in added filler concentration, an elevation in viscosity of the devulcanizate was observed, consequently leading to an increase in shear forces during the mixing process. These higher shear forces contribute to more effective mixing and, thus, enhance the dispersion of filler particles within the devulcanized matrix.

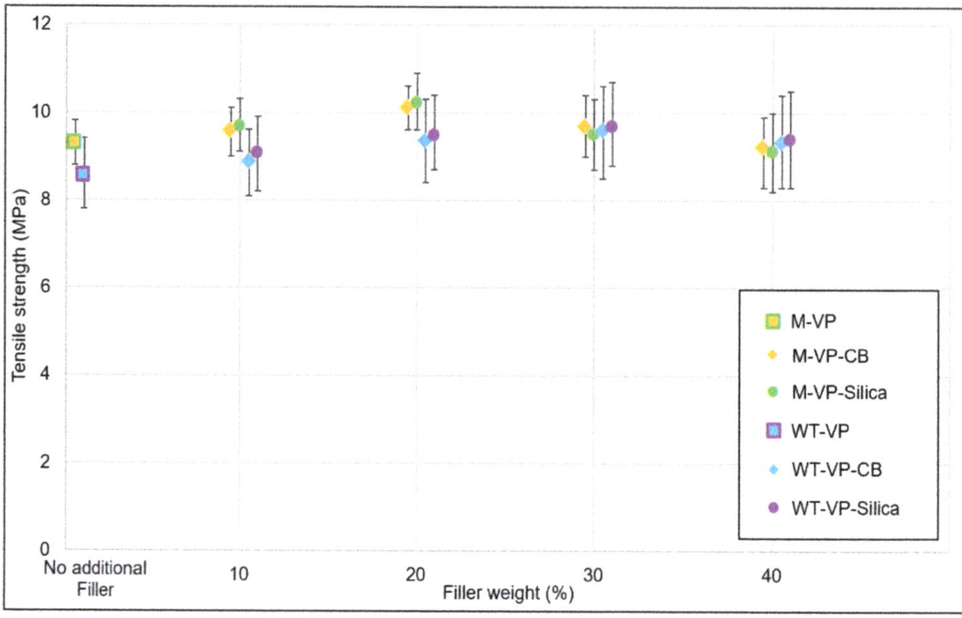

Figure 2. Filler optimization trials: tensile strength.

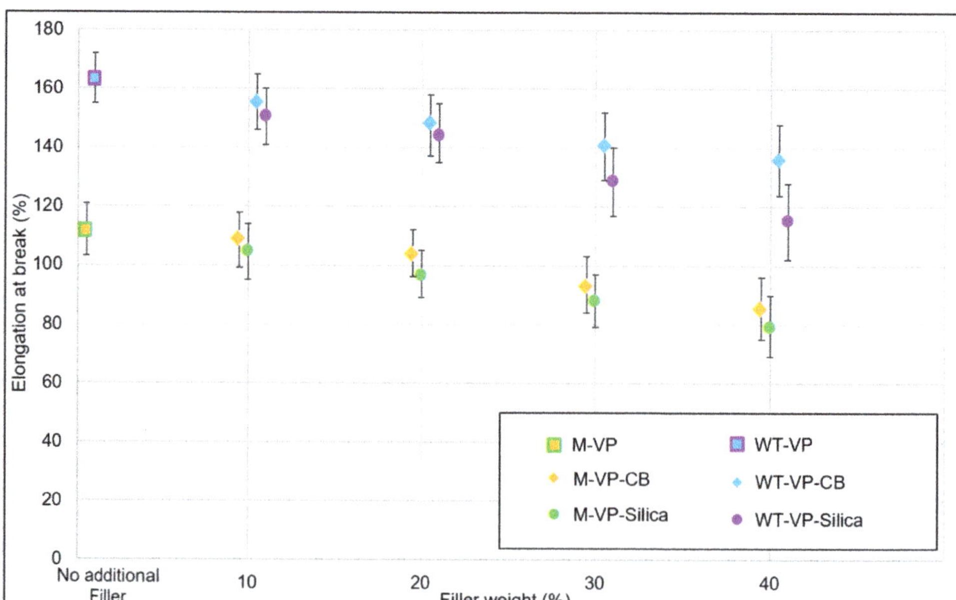

Figure 3. Filler optimization trials: elongation at break.

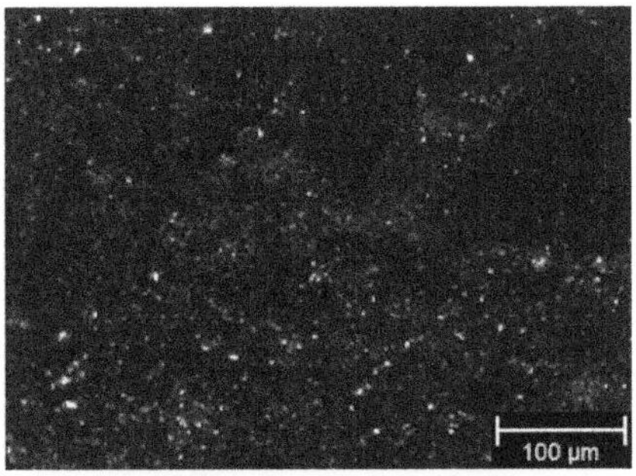

WT devulcanizate—Unfilled

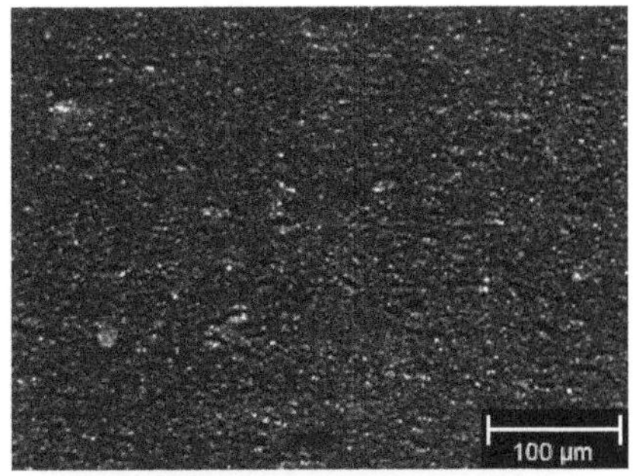

WT devulcanizate + 10% Silica

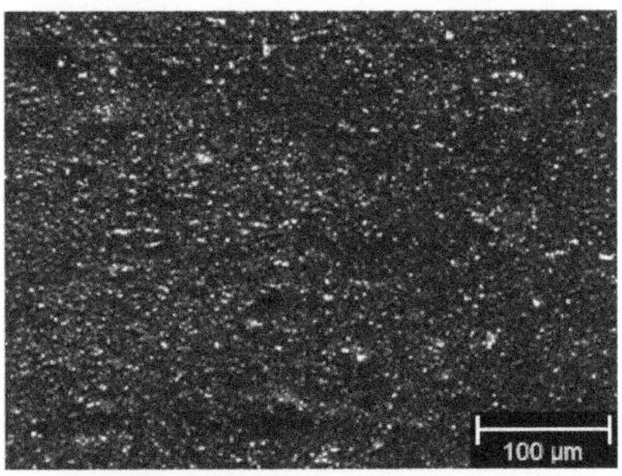

WT devulcanizate + 20% Silica

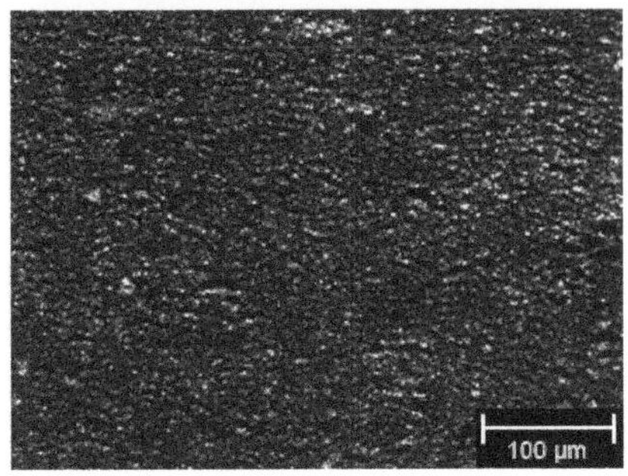

WT devulcanizate + 30% Silica

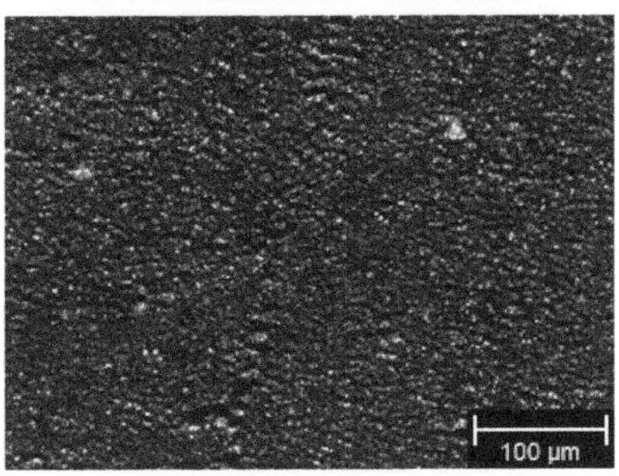

WT devulcanizate + 40% Silica

Figure 4. Filler optimization trials: dispersion measured using optical microscopy.

Due to instrumental limitations (very high shear modulus), it was not possible to measure the Payne effect of the 40% filled samples. The notation L-H stands for the strain sweep from low to high, and H-L is the high to low sweep, as mentioned in the legend of Figure 5. With increasing silica concentration, the Payne effect increased gradually, which indicates higher filler interactions. Similarly, the difference between the shear modulus values of the two sweeps at low strains also increased. This can be explained by a change in the elastic behavior due to less matrix material with increasing concentrations of added filler and a stiffer matrix, resulting in a delay of the recombination of the filler at higher loadings, as shown in Table 5.

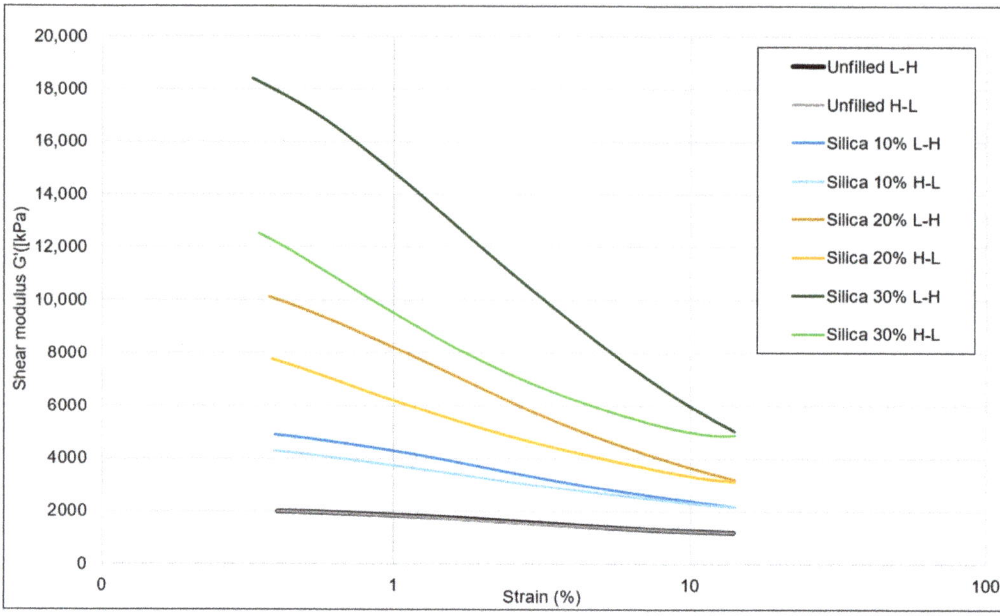

Figure 5. Filler optimization trials: Payne effect of the silica-filled samples.

Table 5. Payne effect values for filler optimization trials.

Filler Type	Filler Quantity (*w*/*w* %)	Low to High Strain Sweep: Shear Modulus G′ (KPa)		High to Low Strain Sweep: Shear Modulus G′ (KPa)		Payne Effect (KPa) #
		Max	Min	Min	Max	
Unfilled	-	1997	1212	1211	1987	785
Silica–silane	10	4867	2163	2147	4286	2704
	20	10,131	3191	3146	7745	6940
	30	18,392	5012	4875	12,512	13,380
Carbon black	10	6193	2369	2357	5465	3824
	20	19,400	5794	5724	14,954	13,606
	30	28,724	7878	7789	22,440	20,846

Payne effect = difference between the maximum and minimum shear modulus of the low to high strain sweep.

The devulcanizates without additional filler exhibited a minimal Payne effect, suggesting that the filler–filler interaction was already disrupted during devulcanization. The strain sweep conducted from low to high resulted in breaking of the remaining filler–filler network.

As shown in Figure 6 and Table 5, the carbon-black-filled samples showed the same trend in shear modulus as the silica-filled samples. The shear modulus values were higher for the carbon black samples compared to those of silica when compounded at the same weight percentage. This discrepancy in shear modulus is attributed to the higher volume of carbon black, as its density (1.8 g/cc) is lower than that of silica (2.5 g/cc). Consequently, the carbon black samples exhibited a comparatively elevated shear modulus.

Typically, abrasion resistance is quantified in terms of volume loss, but due to the density variations in the devulcanizate resulting from WT granulates, the abrasion resistance values were expressed in weight loss. The density of the model devulcanizate is constant within the sample as the feed material was uniform. However, for the whole tire (WT) devulcanizate, variations in density were observed, attributed to the diverse origins of granules from different tires and tire parts.

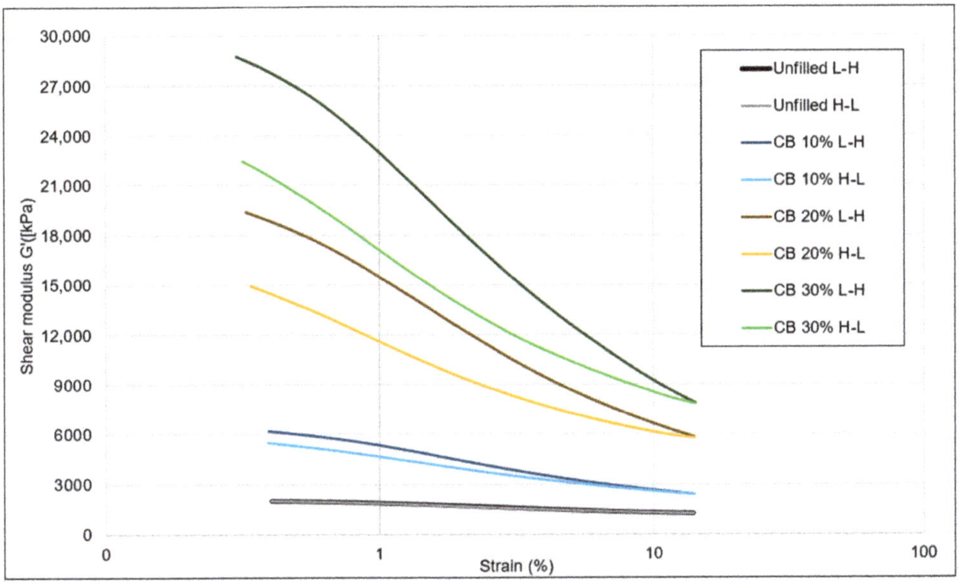

Figure 6. Filler optimization trials: Payne effect of the carbon-black-filled samples.

According to Figure 7, the abrasion resistance of the sample with 10% additional carbon black was similar to the wear of the sample without any additional filler. With further increasing additional filler concentration, abrasion resistance decreased gradually, resulting in higher losses.

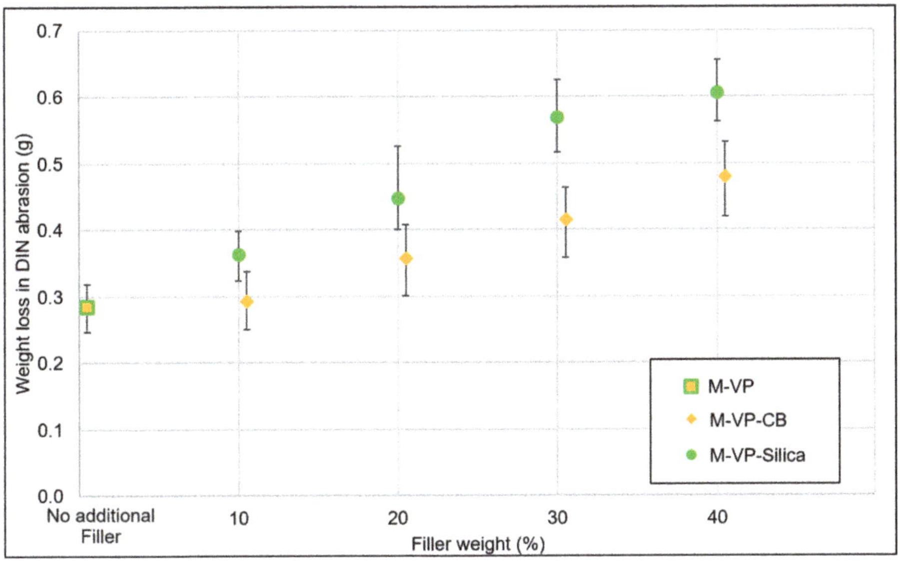

Figure 7. Filler optimization trials: abrasion.

2.3. Discussion

The model compound for devulcanization contained 80 phr silica, a commonly used filler concentration; therefore, it does not have much capacity for a further filler loading. With increasing additional filler decreasing polymer content, the filler–polymer interaction was reduced due to insufficient polymer. The limit for this sample was 20%: above this additional filler loading, the tensile strength decreased for the model devulcanizates. WT rubber devulcanizates were also filled with additional silica or carbon black, and they showed a maximum tensile strength for 30% additional filler loading. Above this concentration the trend reversed. Compared to the model devulcanizate, higher elongation at break values were observed for the WT devulcanizates, probably due to the presence of natural rubber and carbon black in the feedstock.

With increasing filler concentration, the macro-dispersion, as measured using optical microscopy, improved due to increasing shear forces during blending of the compound with the additional filler.

During the Payne effect measurement within the first sweep, the shear modulus decreases with increasing strain, which indicates the breakdown of the filler network. During the second sweep from the high to low strain, the recombination of the filler network takes place resulting in an increase in shear modulus. With increasing filler concentration, the mobility of the filler in the matrix reduced, indicating less recombination for higher filler loadings. In the case of inadequate dispersion of fillers, the abrasion index is adversely affected leading to increased material loss during abrasion resistance testing.

The primary objective of these experiments was enhancing the mechanical strength of the devulcanizate composite through the inclusion of reinforcing fillers; this part of the study did not assess the network breakdown and miscibility analysis.

2.4. Conclusions

Compounding of around 20% of additional filler can improve the tensile strength of the devulcanizate composites by up to 15%. In addition to the filler cost, a sacrifice in elongation at break and abrasion resistance was noticed.

In comparison to samples filled with silica, those filled with carbon black did exhibit significant improvements in mechanical or viscoelastic properties. Therefore, further optimization on the silica–silane filler system was performed in order to elaborate the best filler-coupling agent ratio. In addition, silica is a widely used in passenger car tires due to the enhancement in fuel efficiency of a vehicle.

3. Effect of Silanes on Devulcanizate Composite Properties

In this investigation, the silane concentration relative to the silica concentration was varied. In general, a fixed ratio of silica to silane is used; however, in the devulcanizate, carbon black as well as silica were already present in the feed material. Carbon black might interfere with the silanization of the additional filler by adsorbing and, thus, deactivating the silane, and the originally contained silica might become activated and, Thus, consume some of the added silane.

The devulcanizate was compounded with additional silica and varying amounts of silane. Highly dispersible silica (ULTRASIL® 7000GR) and TESPD were compounded in six different combinations, keeping the same revulcanizate formulation.

3.1. Experimental Process

The preparation of the feed devulcanizate, sampling plan, and characterization processes are described in this section and illustrated in Figure 8.

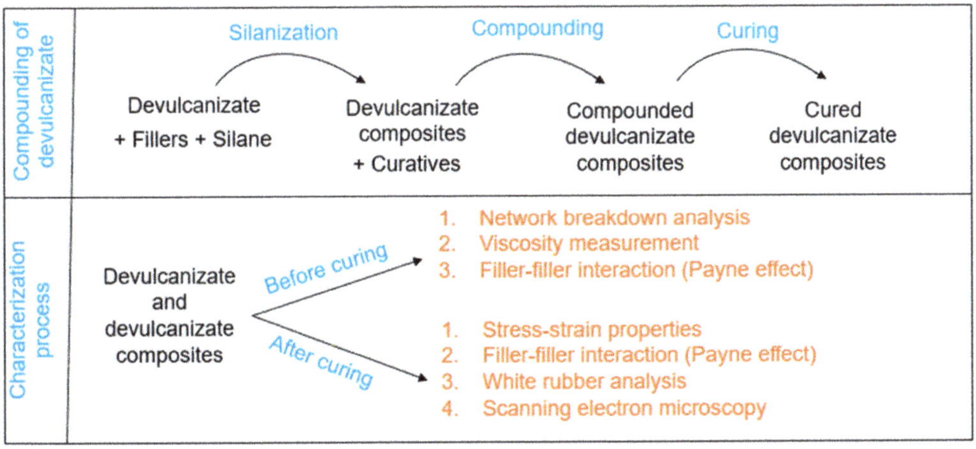

Figure 8. Experimental flowchart for silanization reactions on devulcanizates and devulcanizate composites.

3.1.1. Preparation of WT Devulcanizate

Preparation of the WT granulates was performed according to the same procedure described in Sections 2.1.2–2.1.4 and 2.1.6.

The mixing curve, illustrated in Figure 9, shows the mastication of the devulcanizate for 1 min, addition of silica for 1 min, ram cleaning and replacement of overflown silica for 30–45 s, and finally the silanization reaction during isothermal mixing for 220 to 250 s, depending on the time needed for returning escaped silica into the mixing chamber [19].

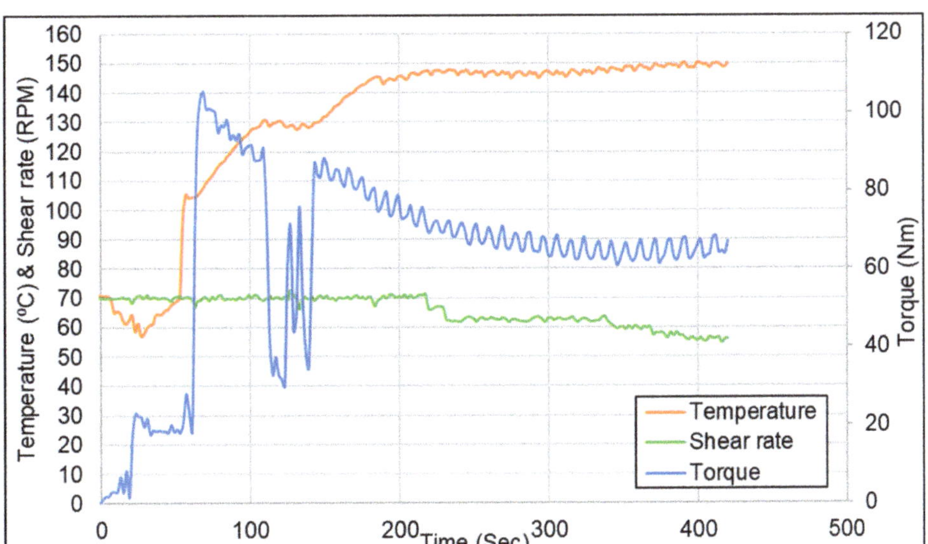

Figure 9. Compounding curve for the silanization reaction.

3.1.2. Sampling for Silanization Trials

A total of 20% silica with 1.5% silane was reported as the optimal filler concentration in Section 2.3; the quantity of silane coupling agent was calculated according to Guy et al. [18]. To evaluate the effect of silanization, samples without silica and silane were used as references, and the silane concentration was varied while keeping the silica concentration constant. The sample without additional silica but with additional silane was tested to assess the reactivation of silica present in the feed material in the devulcanizate composite. The sampling details of the silanization trials are given in Table 6.

Table 6. Sampling plan of silanization trials.

Sample	Compounding Details	
	Silica (*w*/*w*%)	Silane (*w*/*w*%)
1	0	0
2	0	1.5
3	20	0
4	20	1.5
5	20	3.0
6	20	4.5

Feed material: WT granulates; temperature: 180 °C; residence time: 6 min; shear rate: 150 RPM; fill factor: 80%; VP concentration: 5%.

All samples were devulcanized while maintaining the same procedure as described in Section 2.1.4. They were revulcanized using the same formulation mentioned in Section 2.1.7.

3.1.3. Characterization Process

The characterization process consists of six parts:

Stress–Strain Properties in Tensile Mode—Described in Section "Stress–Strain Analysis".

Filler Interaction by Payne Effect Analysis—Described in Section "Stress–Strain Analysis".

Degree of Network Breakdown by the Horikx Verbruggen Plot

The evaluation of crosslink density, as outlined in ASTM D 6814-02 [25], employs the equilibrium volume swelling method and the Flory–Rehner equation [26]. Initially, samples undergo extraction in acetone to eliminate polar components, followed by drying. Subsequently, extraction with tetrahydrofuran removes non-polar parts. The percentage of network breakdown was calculated using Equation (1) [27,28]:

$$\text{Network breakdown percentage}(\%) = \frac{V_{c1} - V_{c2}}{V_{c2}}. \quad (1)$$

In this equation, V_{c1} and V_{c2} represent the crosslink densities of the samples before and after devulcanization, respectively [26–28]. It is worth noting that the crosslink density determined using the Flory–Rehner equation may not reflect the true value in a filled compound. To ascertain the precise crosslink density, the Kraus correction was employed.

$$V_{actual} = \frac{V_{apparent}}{1 + k \times \Phi} \quad (2)$$

$$\Phi = \frac{\text{Weight fraction of the filler} \times \text{density of the compound} \times W_b}{\text{Density of the filler} \times W_a} \quad (3)$$

Here, $V_{apparent}$ represents the measured crosslink density as determined using the Flory–Rehner equation, while V_{actual} denotes the actual crosslink density after adjustment for the filler. K stands as a constant specific to the filler employed, Φ signifies the volume fraction of the filler in the specimen, W_b indicates the weight of the specimen prior to extraction, and W_a represents the weight of the specimen post-extraction of all soluble components, including the polymer sol fraction, oil, and soluble chemical residues [27,28].

The thresholds for random scission and crosslink scission were determined by assessing the sol content of the feed material utilizing the Horikx–Verbruggen method [28]. By plotting the sol fraction and network breakdown values of the devulcanizates on a graph, the nature of network breakdown can be inferred. For each data point of network breakdown versus sol content, the average outcome of five samples was computed.

In Figure 10, the highlighted green zone represents the desired range for the devulcanized rubber, aiming for the optimal balance between the minimal sol fraction and maximum devulcanization percentage. Attaining higher values necessitates the breaking of monosulphidic bonds, which correlates with an increased degree of random scission. Additionally, it is important to note that bound rubber cannot be dissolved, thereby restricting the sol content.

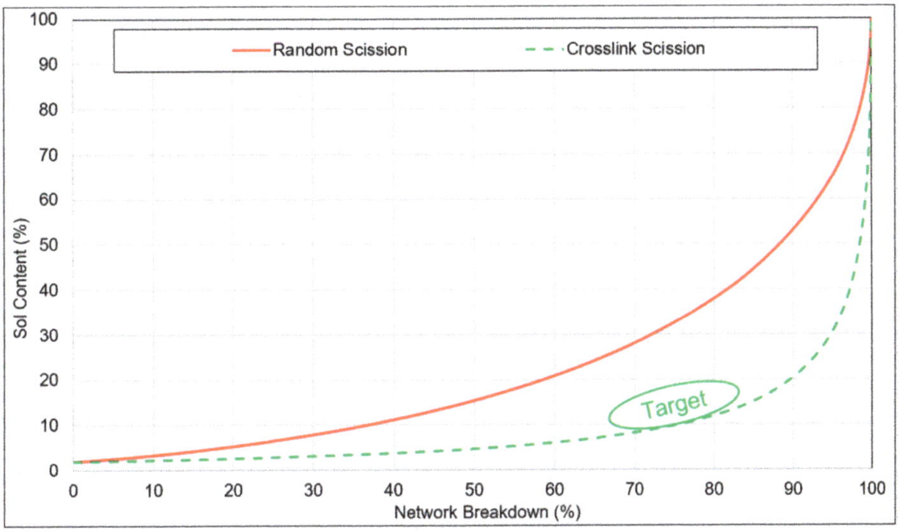

Figure 10. Sol content versus network breakdown percentage for random and crosslink scission according to the Horikx–Verbruggen plot [2].

Homogeneity Evaluation by White Rubber Analysis

Following the devulcanization process, residual undevulcanized particle cores may persist within the devulcanizate. While devulcanized polymer chains exhibit homogenous miscibility with compatible polymers or compounds, undevulcanized particle cores do not share the same property. Consequently, it becomes crucial to evaluate the quantity, dimensions, and overall area of these undevulcanized particle cores. To facilitate this analysis, the white rubber analysis (WRA) method was devised.

In this quantitative analytical approach, the devulcanizate was blended with a bright-white, polybutadiene-based compound incorporating titanium dioxide as a colorant. The selection of a white colorant was deliberate, enhancing the contrast between the background and devulcanizate, thus facilitating the quantitative characterization process. Samples were prepared by incorporating 10% devulcanizate into this white rubber compound, resulting in a gray compound where any remaining undevulcanized particles were discernible as brownish spots. Digital analysis of the particles and their size distribution was conducted using a VHX 5000 digital microscope manufactured by Keyence.

Processability of the Rubber in Terms of Viscosity

The viscosities samples were determined using an MV 2000 VS viscometer supplied by Alpha Technologies GmbH, Bellingham, WA, USA, following the guidelines specified in the ASTM D1646-19a standard [29]. In this testing procedure, a large rotor (ML) with a diameter of 38.1 mm was employed, and the test temperature was maintained at 100 °C. A fixed sample weight of 20 g was pre-heated for one minute, and the viscosity, recorded in Mooney units (MU), was obtained after another 4 min duration.

Scanning Electron Microscopy (SEM)

SEM analysis was performed using a TableTop SEM PhenomXL manufactured by ThermoFisher, Waltham, MA, USA. The cross-section and fracture surfaces of all the samples were analyzed to evaluate the dispersion of the fillers with variable silane concentrations. The experimental conditions were kept the same for all measurements: voltage: 5 kV, beam intensity: image, detector: BSD full, vacuum: low (60 Pa), averaging: high, and scan size: 3840 × 2400.

3.2. Results and Discussion

Figure 11 displays the stress–strain characteristics. The sample without additional silica and silane exhibited the lowest tensile strength and the highest elongation at break. When only 1.5% silane was added, no significant increase in tensile strength was noticed but a slight reduction in elongation at break was noticed. In the case of the additionally filled samples, the silica concentration remained constant at 20% (w/w), while the silane concentration was varied from no silane to 4.5% (w/w); the commonly used concentration is 1.5% (w/w). In this series, the tensile strength improved up to a silane concentration of 3%, beyond which no further enhancement was observed. Elongation at break gradually decreased as the silane concentration increased due to the formation of an additional filler–polymer and polymer–polymer network during the revulcanization process. This trend was expected, as stronger materials, in general, show reduced strain.

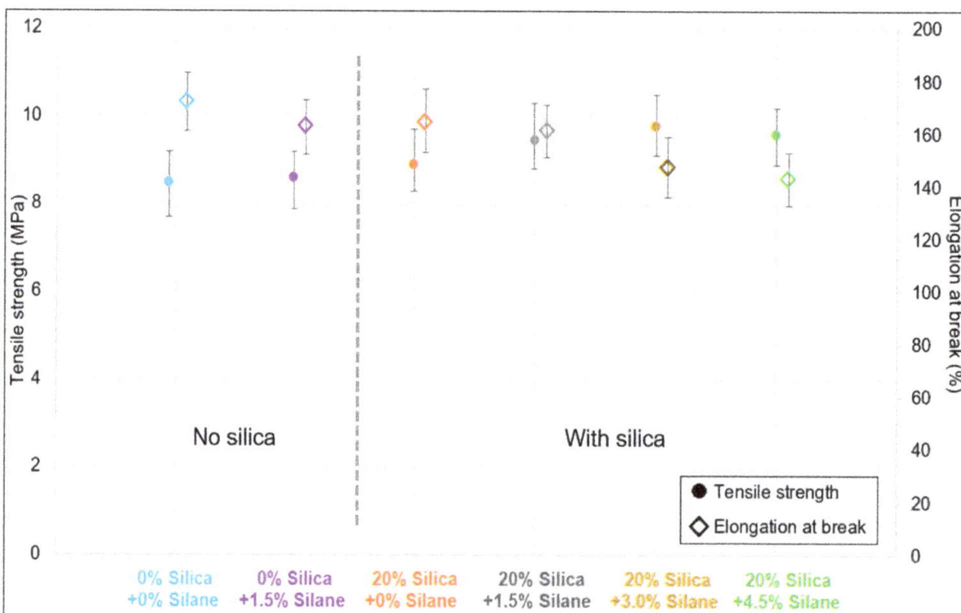

Figure 11. Silanization trials: stress–strain properties.

Figure 12 presents a comparative analysis of the Payne effect for not filled and silica-filled samples with varying silane concentrations. The strain sweep curves represented by dotted lines depict the sweep from low to high strain, during which the filler network is broken, leading to a decrease in shear modulus. Conversely, the solid curves represent the high to low strain sweep, during which recombination of the filler network occurs, resulting in an increase in the shear modulus.

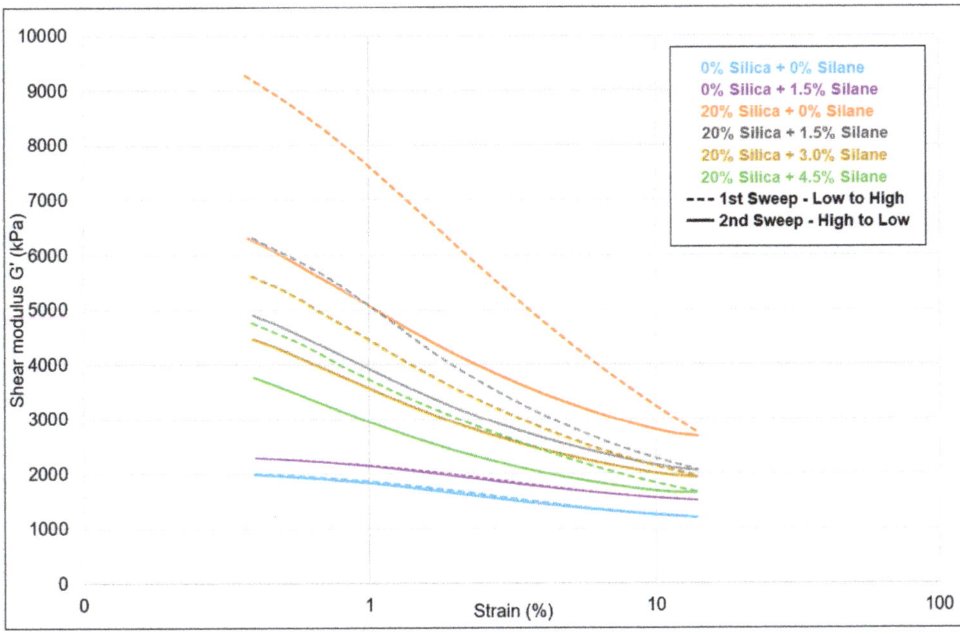

Figure 12. Silanization trials: Payne effect.

For the samples without additional silica (blue and violet), both sweeps show low shear moduli (Table 7), indicating effective dispersion of the filler. Upon the addition of silica, in the difference between the shear moduli in the first sweep of low to high strain, the Payne effect increased. This indicates stronger and more frequent interactions between the fillers. This is a commonly seen effect: as the number of silica particles increases, their distance decreases and filler–filler interaction becomes more probable. As the silane concentration increased, the Payne effect gradually decreased, indicating better dispersion of the filler with less filler–filler interactions: with increasing silane concentration, the degree of silanization increases; see Table 7. This increases the compatibility of the silica with the rather apolar polymers and, thus, enhances the filler–polymer interaction.

Table 7. Silanization trial: Payne effect data—shear modus.

Sample Details	Low to High Strain Sweep: Shear Modulus G′ (KPa)		High to Low Strain Sweep: Shear Modulus G′ (KPa)		Payne Effect # (KPa)
	Max	Min	Max	Min	
0% Silica + 0% Silane	1997	1212	1211	1987	785
0% Silica + 1.5% Silane	2297	1512	1511	2287	785
20% Silica + 0% Silane	9290	2747	2674	6317	6543
20% Silica + 1.5% Silane	6308	2073	2056	4898	4235
20% Silica + 3% Silane	5612	1953	1937	4469	3659
20% Silica + 4.5% Silane	4752	1675	1658	3776	3077

Payne effect = difference between the maximum and minimum shear modulus of the low to high strain sweep.

The Payne effect curves are a product of four variable effects, as illustrated in Figure 13: filler–filler interaction, filler–polymer interaction, polymer network, and the hydrodynamic effect. For samples without additional silica but containing a certain amount of silica from the original compounding, the only parameter that changes is the filler–polymer interaction: filler–polymer bonds might be broken in the mixing process these samples undergo. The other three parameters remain constant in this case. Conversely, in samples filled with additional silica, an increase in silane concentration results in an increase in filler–polymer interaction and a gradual decrease in filler–filler interaction due to the compatibilization and coupling of the silica by the silane. The polymer–polymer network remains unchanged. There might be a slight reduction in the hydrodynamic effect due to the addition of an excess of liquid silane, which did not react with silica.

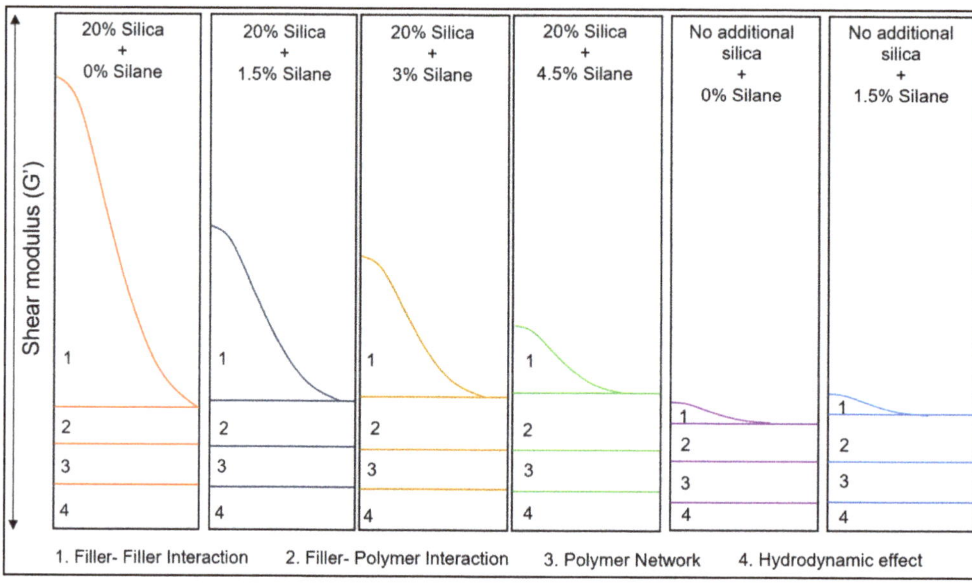

Figure 13. Silanization trials: schematic of different components of the Payne effect.

Based on the findings from the Payne effect data, no reduction was observed when silane was introduced to the samples that were not additionally filled. The consistent Payne effect values indicate the absence of silica reactivation: If silane had reacted with silica, a lower Payne effect would have been anticipated. The lack of any alteration in the Payne effect values confirms the absence of silica reactivation.

Figure 14 illustrates a comparative analysis of the viscosities of the samples with and without additional silica and different silane concentrations. The viscosity of the samples without additional silica was observed to be in the range of 75 MU to 78 MU. Upon adding 20% (w/w) silica, the viscosity increased to approximately 115 MU. The viscosity gradually decreased with increasing silane concentration, reaching around 110 MU. As the silica concentration was constant, an excess of silane contributed to the plasticization effect.

Figure 15 depicts a comparative Horikx–Verbruggen analysis of samples with and without additional silica with varying silane concentrations. The red line represents the limit of random scission, while the dotted green line represents the limit of crosslink scission. The average values of five samples are plotted in terms of network breakdown and sol content. All samples underwent the same devulcanization process, followed by separate compounding with silica and silane.

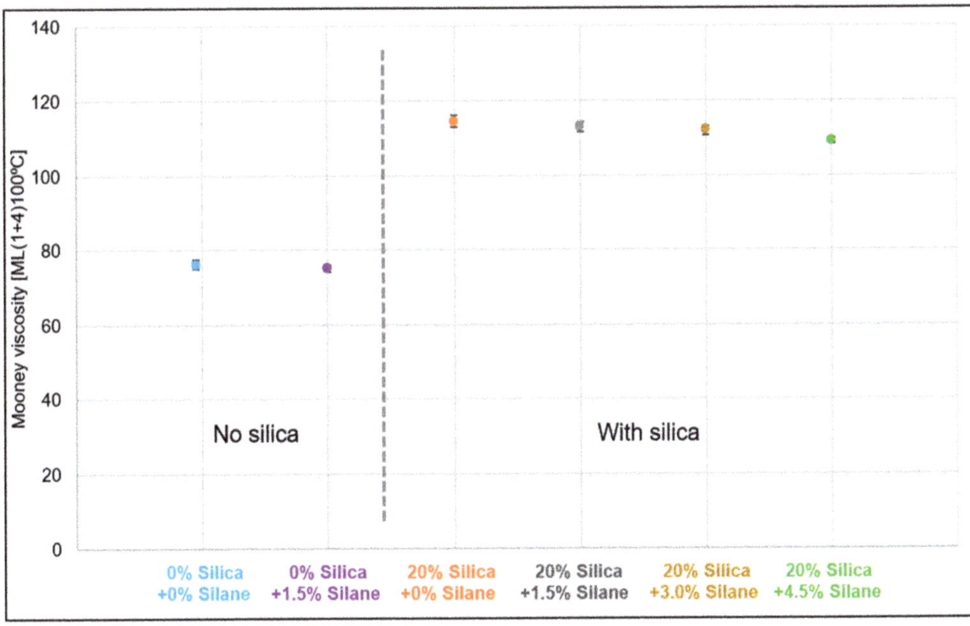

Figure 14. Silanization trials: viscosity.

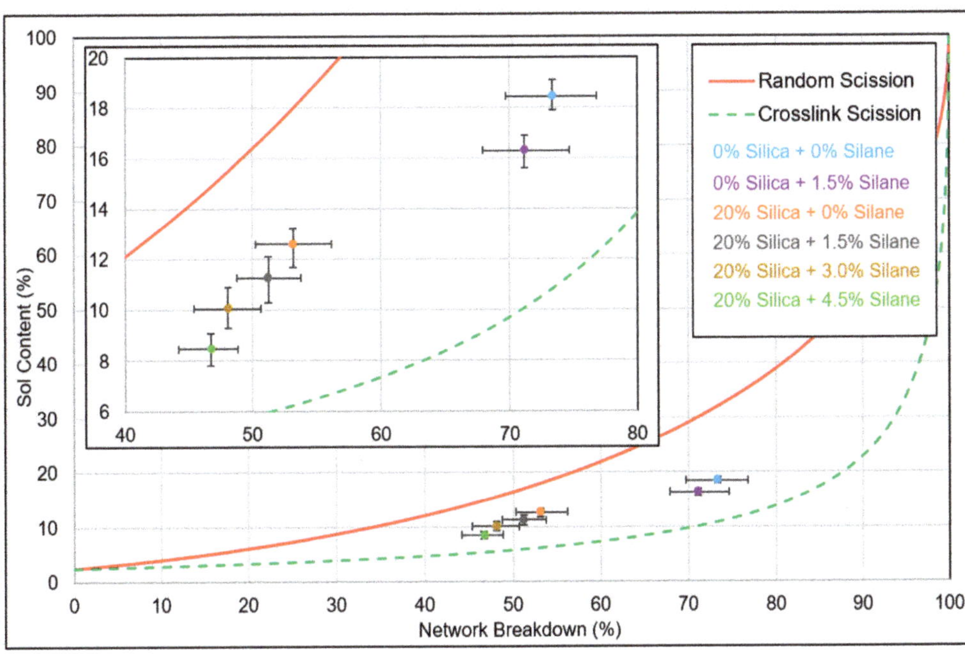

Figure 15. Silanization trials: network breakdown.

The unfilled samples exhibited a network breakdown ranging from 70% to 75%. Upon the addition of 20% (w/w) silica, when the silane concentration was varied from 0% to 4.5%, the network breakdown percentage decreased from 55% to 45%. The variation in network breakdown was attributed to the increase in silane concentration, as this resulted in an additional polymer–filler network, which counterbalances the reduction in crosslink density in the devulcanization process. Though a Krauss correction for the presence of fillers was performed, the reduction in network breakdown percentage could be due to the fact that the additional 20% silica forms a different type of reinforcement for which the Krauss correction has limited applicability. With increasing silane concentration, the interaction between the filler and the polymer improved due to the silanization reaction, leading to an additional network and, consequently, lower net network breakdown.

According to Figure 16, samples without silica exhibited a total number of visible undevulcanized particles in the range of 1290–1300. However, with the addition of 20% (w/w) silica together with varying amounts of silane, the total number of visible particles decreased to approximately 920–960. The presence of silica in the additionally filled composite samples led to a decrease in the total number of immiscible particles compared to not-filled devulcanizate, as shown in Figure 17. Due to the contribution of additional silica, the relative volume of undevulcanized particles was reduced, resulting in a lower number of visible particles.

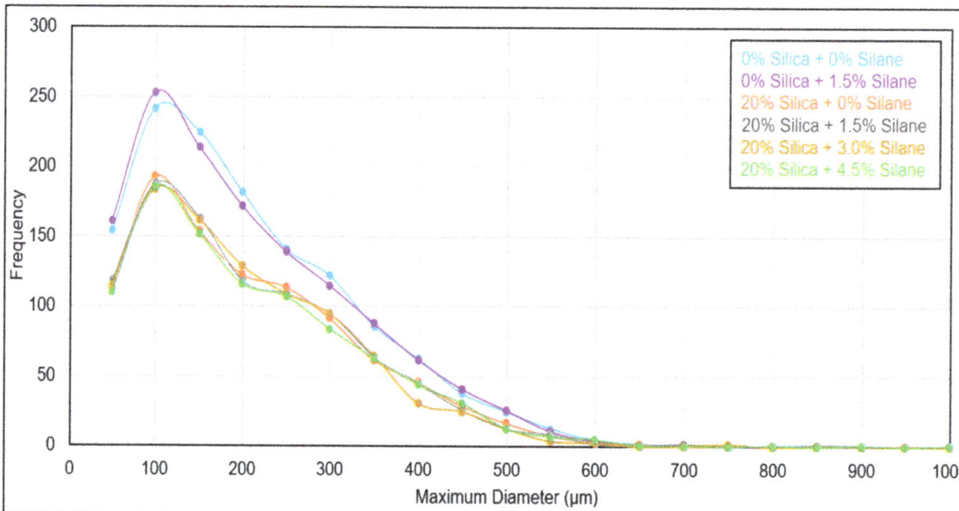

	50	100	150	200	250	300	350	400	450	500	550	600	650	700	750	800	850	900	950	1000	1000+	Total	TS MPa
0% Silica+0% Silane	154	242	225	182	141	122	86	63	38	25	13	5	2	1	0	1	1	1	0	1	0	1303	8.5
0% Silica+1.5% Silane	161	253	214	172	139	115	88	62	41	26	11	4	1	2	0	0	1	0	0	1	0	1291	8.6
20% Silica+0% Silane	113	193	154	123	114	92	62	46	29	17	8	3	2	0	1	1	0	0	1	0	0	959	8.9
20% Silica+1.5% Silane	119	188	163	118	109	95	64	45	26	13	9	2	1	1	1	1	0	0	0	1	0	956	9.5
20% Silica+3.0% Silane	115	184	162	129	109	95	65	31	25	13	4	3	0	1	2	0	1	1	0	0	0	940	9.8
20% Silica+4.5% Silane	110	186	152	116	107	84	63	44	31	13	7	5	1	1	0	1	0	1	0	1	0	923	9.6

Figure 16. Silanization trials: white rubber analysis.

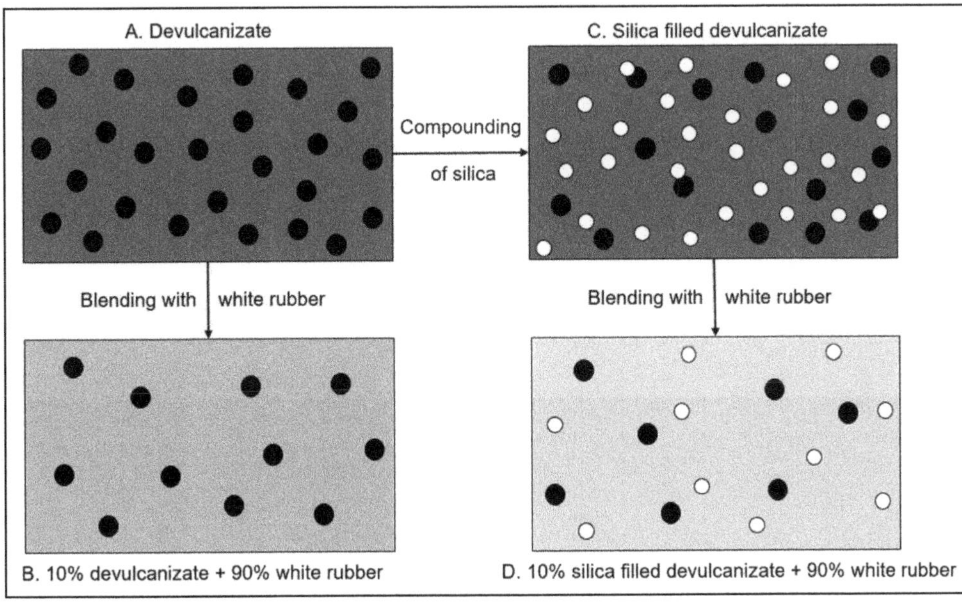

Figure 17. Schematic diagram of white rubber analysis: white rubber samples containing vulcanizates with and without additional silica.

According to the observations in Figure 18, an increase in silane concentration resulted in improved dispersion of silica: Figure 18C,D showed comparatively poor dispersion compared to Figure 18E,F, which correlates with the Payne effect values.

Based on the findings in Figure 19, in which fracture surfaces are illustrated, the material exhibited smoother fracture surfaces as the silane concentration increased compared to no-silane samples. With increasing silane concentration, the silica dispersion became more homogeneous. Fractures are generated from defective spots; therefore, a more homogeneous and uniform distribution will cause less defects, leading to a smoother fracture surface and resulting in better tensile properties, as seen in Figure 11.

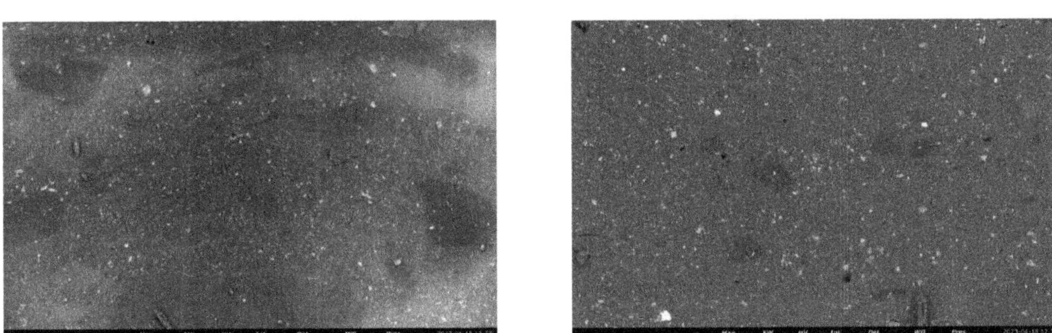

(**A**) 0% Silica + 0% Silane

(**B**) 0% Silica + 1.5% Silane

Figure 18. *Cont.*

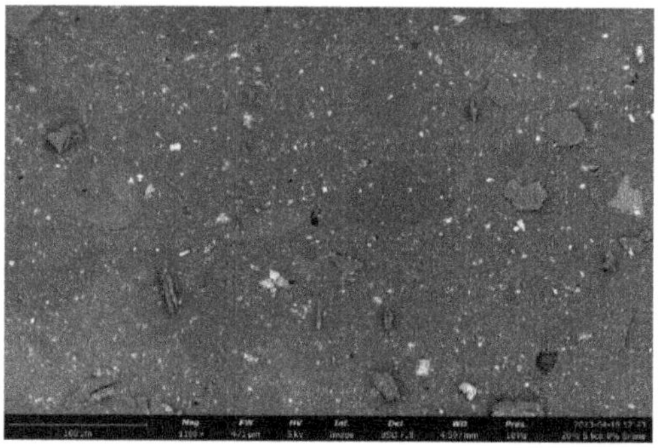

(**C**) 20% Silica + 0% Silane

(**D**) 20% Silica + 1.5% Silane

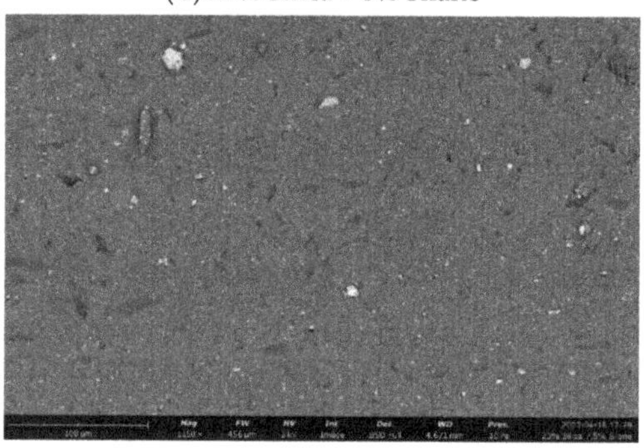

(**E**) 20% Silica + 3% Silane

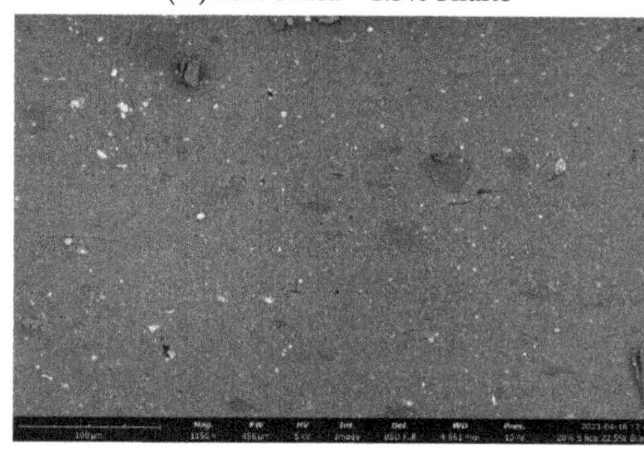

(**F**) 20% Silica + 4.5% Silane

Figure 18. Silanization trials: macro-dispersion (cross section analysis by SEM).

(**A**) 0% Silica + 0% Silane

(**B**) 0% Silica + 1.5% Silane

Figure 19. *Cont.*

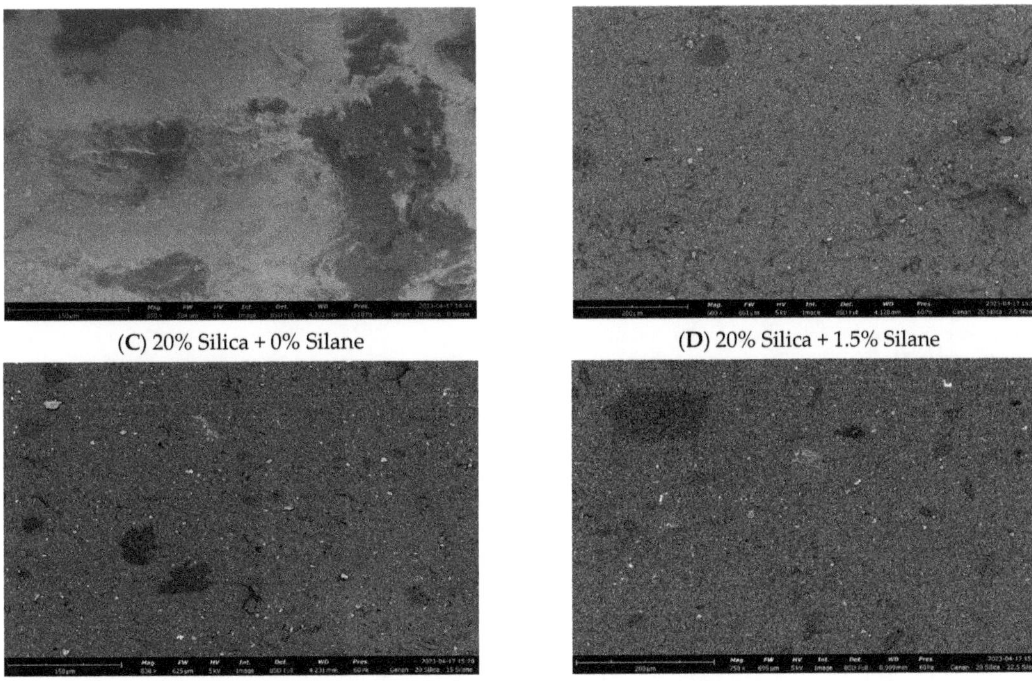

(C) 20% Silica + 0% Silane

(D) 20% Silica + 1.5% Silane

(E) 20% Silica + 3% Silane

(F) 20% Silica + 4.5% Silane

Figure 19. Silanization trials: fracture surface analysis via SEM.

3.3. Conclusions

With additional 20% (w/w) silica and increasing silane concentration, the tensile strength, Payne effect, and dispersion improved due to better polymer–filler interactions following a higher degree of silanization. With increasing silane concentration, elongation at break and net network breakdown were reduced due an increase in crosslink density by the formation of short, stable filler–polymer bonds during the silanization reaction. Viscosity was reduced due to the plasticization effect of silanes.

A total of 20% (w/w) silica with 3% (w/w) silane showed the best properties in terms of mechanical strength; beyond this concentration, no further improvement was noticed. This can be considered as the optimum concentration for the addition of a silica–silane filler system to the devulcanizate composite.

This high-quality devulcanized composite can be partially integrated into new high-performance rubber products, such as tires and conveyor belts. It can also be used to create lower tensile strength products like road underpads and acoustic protection materials, which can be made entirely from the devulcanized composite.

Author Contributions: R.G.: Methodology; Investigation; Data curation; Formal analysis; Writing—original draft; W.K.D.: Conceptualization; Formal analysis; Validation; Writing—review and editing; Supervision; Project administration; C.M.: Supervision; Project administration; Validation; R.K., R.S., A.T. and A.B.: Supervision; Validation. All authors have read and agreed to the published version of the manuscript.

Funding: This project was fully funded by the company Evonik Operations GmbH, Paul-Baumann-Straße 1, 45772 Marl, Germany.

Institutional Review Board Statement: Not applicable.

Data Availability Statement: Data are contained within the article.

Acknowledgments: The research work was supported by Evonik Operations GmbH and the University of Twente.

Conflicts of Interest: Authors Christian Mani, Roland Krafczyk and Rupert Schnell were employed by the company Evonik Operations GmbH. The remaining authors declare that the research was conducted in the absence of any commercial or financial relationships that could be construed as a potential conflict of interest.

References

1. Ghosh, R. Design of a Devulcanization Process for Tire Material in a Twin-Screw Extruder. Ph.D. Thesis, University of Twente, Enschede, The Netherlands, 2021.
2. Ghosh, R.; Mani, C.; Krafczyk, R.; Schnell, R.; Paasche, A.; Talma, A.; Blume, A.; Dierkes, W.K. New route of tire rubber devulcanization using silanes. *Polymers* **2023**, *15*, 2848. [CrossRef]
3. van Hoek, H. Closing the Loop: Reuse of Devulcanized Rubber in New Tires. Ph.D. Thesis, University of Twente, Enschede, The Netherlands, 2022.
4. Ghosh, R.; Blume, A.; Dierkes, W.K. GTR devulcanization review. *Tire Technol. Int. Nov.* **2023**, 84.
5. Formela, K. Sustainable development of waste tires recycling technologies–recent advances, challenges and future trends. A review. *Adv. Ind. Eng. Polym. Res.* **2021**, *4*, 209.
6. Saputra, R.; Walvekar, R.; Khalid, M.; Mubarak, N.M.; Sillanpää, M. Current progress in waste tire rubber devulcanization. *Chemosphere* **2021**, *265*, 129033. [CrossRef] [PubMed]
7. Wintersteller, R.; Markl, E.; Lackner, M. Feedstock recycling of rubber—A review on devulcanization technologies. *Materials* **2020**, *1*, 1246.
8. Valentini, F.; Pegoretti, A. End-of-life options of tyres. *Adv. Ind. Eng. Polym. Res.* **2022**, *5*, 203.
9. Markl, E.; Lackner, M. Devulcanization technologies for recycling of tire-derived rubber: A review. *Materials* **2020**, *13*, 1246. [CrossRef]
10. Ghorai, S.; Hait, S.; Mondal, D.; Wießner, S.; Das, A.; De, D. Fill two needs with one deed: Simultaneous devulcanization and silica reinforcement of waste rubber for green tyre tread compound. *Mater. Today Commun.* **2023**, *35*, 106065. [CrossRef]
11. Sienkiewicz, M.; Janik, H.; Borzędowska-Labuda, K.; Kucińska-Lipka, J. Environmentally friendly polymer-rubber composites obtained from waste tyres: A review. *J. Clean. Prod.* **2017**, *147*, 560. [CrossRef]
12. Colom, X.; Carrillo-Navarrete, F.; Saeb, M.R.; Marin, M.; Formela, K.; Canavate, J. Evaluation and rationale of the performance of several elastomeric composites incorporating devulcanized EPDM. *Polym. Test.* **2023**, *121*, 107976. [CrossRef]
13. Zhang, T.; Asaro, L.; Gratton, M.; Hocine, N.A. An overview on waste rubber recycling by microwave devulcanization. *J. Environ. Manag.* **2024**, *353*, 120122. [CrossRef] [PubMed]
14. Hittini, W.; Mourad, A.H.I.; Abu-Jdayil, B. Utilization of devulcanized waste rubber tire in development of heat insulation composite. *J. Clean. Prod.* **2021**, *280*, 124492. [CrossRef]
15. Robertson, C.G.; Hardman, N.J. Nature of carbon black reinforcement of rubber: Perspective on the original polymer nanocomposite. *Polymers* **2021**, *13*, 538. [CrossRef] [PubMed]
16. Duncan, A.M.; Akutagawa, K.; Ramier, J.L.; Busfield, J.J. Improved Dual Network Model for Aging of Rubber Composites under Set Strains. *Macromolecules* **2023**, *56*, 6704. [CrossRef] [PubMed]
17. Görbe, Á.; Kohári, A.; Bárány, T. Rubber Compounds from Devulcanized Ground Tire Rubber: Recipe Formulation and Characterization. *Polymers* **2024**, *16*, 455. [CrossRef] [PubMed]
18. Guy, L.; Daudey, S.; Cochet, P.; Bomal, Y. New insights in the dynamic properties of precipitated silica filled rubber using a new high surface silica. *Kautsch. Gummi Kunststoffe* **2009**, *62*, 383–391.
19. Blume, A.; Gatti, L.; Luginsland, H.D.; Maschke, D.; Moser, R.; Nian, J.C.; Röben, C.; Wehmeier, A. Silica and silanes. In *Rubber Compounding: Chemistry and Applications*; CRC Press: Boca Raton, FL, USA, 2015; p. 251.
20. Product Safety Data Sheet; Vivatech 500, H&R Chem Pharm. Available online: https://www.hur.com/en/products/process-oils (accessed on 26 March 2024).
21. *ASTM D7750-12*; Standard Test Method for Cure Behavior of Thermosetting Resins by Dynamic Mechanical Procedures Using an Encapsulated Specimen Rheometer. ASTM International: West Conshohocken, PA, USA, 2017.
22. *ASTM D412-16*; Standard Test Methods for Vulcanized Rubber and Thermoplastic Elastomers–Tension. ASTM International: West Conshohocken, PA, USA, 2021.
23. *ASTM D8059-19*; Standard Test Method for Rubber Compounds—Measurement of Unvulcanized Dynamic Strain Softening (Payne Effect) Using Sealed Cavity Rotorless Shear Rheometers. ASTM International: West Conshohocken, PA, USA, 2022.
24. *ASTM D5963-22*; Standard Test Method for Rubber Property—Abrasion Resistance—Rotary Drum Abrader. ASTM International: West Conshohocken, PA, USA, 2022.
25. *ASTM D6814*; Standard Test Method for Determination of Percent Devulcanization of Crumb Rubber Based on Crosslink Density. ASTM International: West Conshohocken, PA, USA, 2018.
26. Flory, P.J.; Rehner, J., Jr. Statistical mechanics of cross-linked polymer networks, I.I. Swelling. *J. Chem. Phys.* **1943**, *11*, 521. [CrossRef]

27. Marzocca, A.J.; Garraza, A.R.; Mansilla, M.A. Evaluation of the polymer–solvent interaction parameter χ for the system cured polybutadiene rubber and toluene. *Polym. Test.* **2010**, *29*, 119. [CrossRef]
28. Verbruggen, M. Devulcanization of EPDM Rubber: A Mechanistic Study into a Successful Method. Ph.D. Thesis, University of Twente, Enschede, The Netherlands, 2007.
29. *ASTM D1646-19a*; Standard Test Methods for Rubber—Viscosity, Stress Relaxation, and Pre-Vulcanization Characteristics—Mooney Viscometer. ASTM International: West Conshohocken, PA, USA, 2019.

polymers

MDPI

Article

Preparation of an Antibacterial Branched Polyamide 6 via Hydrolytic Ring-Opening Co-Polymerization of ε-Caprolactam and Lysine Derivative

Xiaoyu Mao [1], Wei Liu [1], Zeyang Li [1], Shan Mei [2] and Baoning Zong [3,*]

1 Research Center of Renewable Energy, Research Institute of Petroleum Progressing, SINOPEC, Beijing 100083, China; maoxiaoyu.ripp@sinopec.com (X.M.); liuwei2.ripp@sinopec.com (W.L.); lizeyang.ripp@sinopec.com (Z.L.)

2 NO. 22 Research Department, Research Institute of Petroleum Progressing, SINOPEC, Beijing 100083, China; meishan.ripp@sinopec.com

3 State Key Laboratory of catalytic Material and Reaction Engineering, Research Institute of Petroleum Progressing, SINOPEC, 18th Xueyuan Road, Haidian District, Beijing 100083, China

* Correspondence: zongbn.ripp@sinopec.com

Abstract: In this study, we successfully realized the hydrolytic ring-opening co-polymerization of ε-caprolactam (CPL) and lysine derivative. A novel antibacterial modified polyamide 6 with a branched structure was obtained after the quaternization of the co-polymers. The co-polymers exhibited a significant increase in zero shear viscosity, melt index and storage modulus at the low frequency region. The quaternized co-polymers displayed thermal properties different from pure PA6 and good mechanical (tensile) properties. The antibacterial activity of the quaternized co-polymers depends on the quaternary ammonium groups' incorporated content. At 6.2 mol% incorporation of quaternary ammonium groups, the strong antibacterial activity has been introduced to the co-polymers. As the quaternary ammonium groups approached 10.1 mol%, the antibacterial polymers demonstrated nearly complete killing of *Staphylococcus aureus* (Gram positive) and *Escherichia coli* (Gram negative). The above research results provided a new approach for the study of high-performance nylon.

Keywords: antibacterial PA6; branched PA6; hydrolysis open-ring polymerization; thermal properties; rheological properties; mechanical properties; antibacterial activity

1. Introduction

Polyamide 6 (PA6), also known as Nylon 6, is a kind of versatile polymer with a wide range of applications. It is primarily used in fibers (80–90%), engineering plastics, adhesives, and packing materials [1–4]. Since first obtained, PA6 as well as its monomer ε-caprolactam (CPL) has resulted in remarkable development in the synthesis technology, with PA6 now accounting for over 80% of the total nylon production [5]. Given the continuous expansion of PA6 production capacity and the growing demand for high-performance materials, the functional modification of PA6 has become a key development focus. Especially in recent years, the COVID-19 pandemic has heightened public awareness of hygiene and health, making the study of antibacterial materials a current research hot spot [6,7]. With biocompatibility, excellent mechanical properties and high resistance to organic solvents, oils, and bodily fluids, PA6 is suitable for the materials of medical equipment [8,9]. However, it is inherently non-antibacterial. Thus, the development of antibacterial PA6 through the incorporation of functional components is crucial.

Antibacterial PA6 can be prepared via modification techniques and post-treatment techniques [10]. Adding antibacterial agents during the polymerization or spinning process is considered the main method of modification techniques, while coating or impregnating PA6 with antibacterial components is involved in post-treatment techniques. Antibacterial PA6 prepared through modification techniques exhibits long-lasting antibacterial effects, but the other properties of the polymer may be impacted with the introduction of antibacterial components.

Erem et al. [11] prepared antibacterial PA6 fibers with different silver nanoparticle (AgNP) contents using the melt intercalation method. Antibacterial activity tests showed that the fibers exhibited high bactericidal rates against *Staphylococcus aureus* (Gram positive) and *Klebsiella pneumoniae* (Gram negative), and the bactericidal rates against both were enhanced with the increasing silver content. However, the partial aggregation of AgNP components led to the generation of defects in PA6, resulting in a decrease in the tensile properties of antibacterial PA6 fibers prepared by this method. In addition, the high thermal conductivity of AgNP components accelerate the cooling of the polymer melt, resulting in a decrease in the crystallinity. Buchenska et al. [12] grafted acrylic acid (AA) onto PA6 and combined antibiotics with the introduced carboxyl groups to create antibacterial PA6 fibers, which demonstrated good antibacterial properties against both Gram-positive and Gram-negative bacteria. Shi et al. [13] grafted chitosan onto PA6 molecular chains under the initiation of potassium persulfate. Although the grafting rate was low (1.58 mol%), the grafted PA6 showed a bactericidal rate of 89% against Gram *Staphylococcus aureus* and over 90% against *Escherichia coli* and *Candida albicans*. In addition, there was no significant decrease in the bactericidal rate of the modified antibacterial PA6 against the aforementioned bacteria after 50 washes. Zhang et al. [14] employed co-irradiation to graft poly(methacrylic acid-2-(dimethylamino) ethyl ester) (PDMAEMA) onto PA66, followed by protonation and anion exchange, resulting in antibacterial PA66 fibers. These fibers exhibited strong antibacterial efficacy against *Candida albicans* and *Staphylococcus aureus*, with a slight decrease in mechanical properties. Lian et al. [15] obtained the co-polymer of CPL and dimethyl-protected cyclic lysine (DMCL) mediated by t-BuP_4, followed by quaternization with 1-bromoethane. When the quaternized ammonium groups content reached 17 mol%, the product exhibited a 99% bactericidal rate against *Staphylococcus aureus* and *Escherichia coli*, albeit with some reduction in thermal and mechanical properties.

Based on the above research results, the introduction of antibacterial groups may have an impact on the thermal and mechanical properties of polymers. However, the above two properties play an important role in the processing and use of polymers. Therefore, it is necessary to prepare an antibacterial polymer with excellent thermal and mechanical properties. In the previous research of our research group, a branched co-polymer was synthesized by introducing α-Amino-ε-caprolactam (ACL) into the CPL polymerization process [16]. When the amount of ACL added was appropriate, the thermal properties of the co-polymer had little change compared to pure PA6, and the tensile properties were significantly improved, and the rheological properties showed obvious change. In addition, our research group has previously achieved hydrolytic ring-opening co-polymerization of CPL and DMCL. Based on the above achievements, ACL was introduced to prepare a co-polymer with a branched chain structure during the co-polymerization of CPL and DMCL (Figure 1a), and the resulting product was subjected to rheological properties and tensile properties tests. Then the co-polymer underwent quaternization with 1-Bromohexane (Figure 1b). The thermal properties as well as the antibacterial activities against *Staphylococcus aureus* and *Escherichia coli* of the co-polymer were tested.

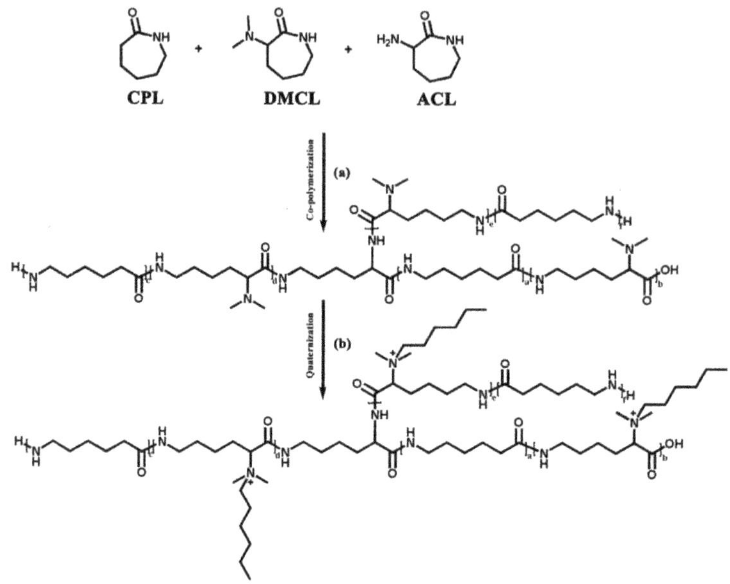

Figure 1. The main reaction of this system.

2. Materials and Methods

2.1. Materials

The CPL used in this study was supplied by Baling Petrochemical Company, SINOPEC (Yueyang, China). ACL as well as DMCL was synthesized in our earlier work. Sulfuric acid and 2,2,2-Trifluoroethanol was purchased from Acros (Fukuoka, Japan). 1-Bromohexane provided by Innochem (Beijing, China) was used to be the quaternization reagent in our research.

2.2. Co-Polymerization

The reactant mixture consisted of CPL, DMCL and ACL, with the initiator, deionized water, and these were added to a reactor. Vacuuming was applied to remove the O_2 component in the reactor and then N_2 gas was inflated to form the N_2 atmosphere. The pre-polymerization was carried out at 225 °C, lasting for 1 h, followed by 6 h polymerization at 250 °C under pressure. Then, the reactant mixture was co-polymerized under atmospheric pressure for 1 h, followed by another 2 h reaction below −0.05 MPa. The obtained co-polymers were discharged and cut into pellets. The residual monomers and low molecular weight polymers contained in theco-polymers pellets were removed through the water bath extraction at 90 °C.

2.3. Quaternization of the Co-Polymers

The mixed liquid consisted of the co-polymers 2,2,2-Trifluoroethanol solution and the quaternized reagent 1-romohexane and was kept at 70 °C for a 24 h reaction. Then the solvent and unreacted quaternized reagent were removed through the rotary evaporation. The quaternized co-polymer samples were obtained after stoving.

2.4. Characterization and Test of Properties

The capillary viscometer was used to determine the relative viscosity (η_r), intrinsic viscosity ([η]), and viscosity-average molecular weight (M_W) of the co-polymers. The co-polymers were dissolved in a 96 wt% sulfuric acid solution (solute:solvent, 1 g:100 mL). The time it took for the solvent (t_0) and the co-polymers solution (t) to pass through the capillary at 25 °C was measured. The above-mentioned feature can be calculated through the following equations.

$$\text{Relative viscosity } (\eta_r) = \frac{t}{t_0}$$

$$\text{Specific viscosity } (\eta_{sp}) = \eta_r - 1$$

$$\text{Intrinsic viscosity } ([\eta]) = \frac{\sqrt{2(\eta_{sp} - \ln\eta_r)}}{c}$$

the M_w can be obtained through the Mark–Houwink equation:

$$[\eta] = k\,[M_w]^{a}.$$

The Mark–Houwink constants for PA6 in a 25 °C, 96% sulfuric acid solution are as follows [17]:

$$k = 6.3 \times 10^{-4}\ a = 0.764$$

The structures of the co-polymers were investigated by ^{1}H NMR (AVANCE NEO 500 M, Bruker, Billerica, MA, USA), using formic acid/trifluoroacetic acid-d (1:1, v:v) as the solvent. To check the rheological properties of co-polymers, a rotational rheometer (P25CSL, HAAKE MARS, Waltham, MA, USA) was used. The samples were prepared by injection molding at 240 °C in a mold with a 20 mm diameter and 1 mm thickness. Complex viscosity (η), storage modulus (G′) and loss modulus (G″) of the co-polymers were tested through a frequency sweep ranging from 0.1 to 500 rad/s under oscillatory mode with a strain of 1%. The melt index was tested by the melt index instrument (MI40, GOETTFRT, Essen, Germany). The test temperature was 230 °C, and the load was 2.16 kg. The mass of the sample passing through the capillary column for a certain period of time was recorded.

The thermal properties of the co-polymers were determined by differential scanning calorimetry (DSC3, METLER TOLEDO, Zurich, Switzerland) and thermal gravimetric analysis (RT-800, METLER TOLEDO, Zurich, Switzerland) under N_2 atmosphere. In DSC, the samples were first heated from 25 °C to 300 °C to remove the thermal history, with the rate of 10 °C/min. Then, the samples were cooled to 25 °C and heated to 300 °C again at the sample rate. The degree of crystallinity (Xc) can be obtained by the following equation.

$$X_c = \Delta H_m / \Delta H_0 \times 100\%$$

where ΔH_m is the specific enthalpy of melting, and ΔH_0 is the enthalpy of melting with 100% crystalline PA6 (188 J/g) [18].

The tensile mechanical properties were checked by the universal tensile machine (CMT2000, MTS/SANS, Eden Prairie, MN, USA) according to ASTM D638 [19] with the strain rate of 10 mm/min. All tested co-polymer samples were dried for 24 h in a desiccator under vacuum.

2.5. Antimicrobial Activities Test of Co-Polymers after Quaternization

The surface antibacterial activities of quaternized co-polymers were tested according to GB/T31402-2015 [20]. Culture medium, liquid culture medium, and test samples were prepared before the test as well as the activation of *Staphylococcus aureus* (Gram positive) and *Escherichia coli* (Gram negative). The activated two types of bacteria were inoculated onto liquid culture medium, followed by incubation at 36 °C for 24 h, and then, the bacterial suspension can be obtained after diluting. An appropriate amount of bacterial suspension was added onto the surface of PA6 samples and the quaternized co-polymers samples to be tested. Then the bacterial suspension was covered with polypropylene film and squeezed evenly. After a 24 h incubation at 36 °C, the bacterial suspension was retrieved. Then it was applied to the surface of the culture dish after dilution, followed by incubation at 36 °C for 24 h. The number of colonies in the culture dish was calculated by the plate counting method. The antibacterial activities were determined by comparing the number of colonies in the PA6 samples and the quaternized co-polymers samples.

3. Results

3.1. The Reaction Mechanism of Co-Polymerization

The reaction mechanism of our co-polymerization system is similar to the synthesis of PA6 via hydrolytic ring-opening polymerization of CPL. Under high temperature and with water existence, the amide bonds of the monomers (CPL and DMCL) break, resulting in the formation of 6-aminocaproic acid and linear DMCL (Figure 2a,b). Low molecular weight co-polymers are generated through poly-condensation between linear monomers (Figure 2c). The terminal amino groups of the low weight co-polymers attack the protonated ACL, which leads the branching agent added to the co-polymers chain (Figure 2d). The incorporation of ACL contributes to the two chain growth sites in the co-polymers. The two amino groups in a co-polymer chain can attack the protonated monomers, through which the growth of the main chain and the side chain can be realized (Figure 2e). A kind of branched co-polymer is obtained through the above reaction steps.

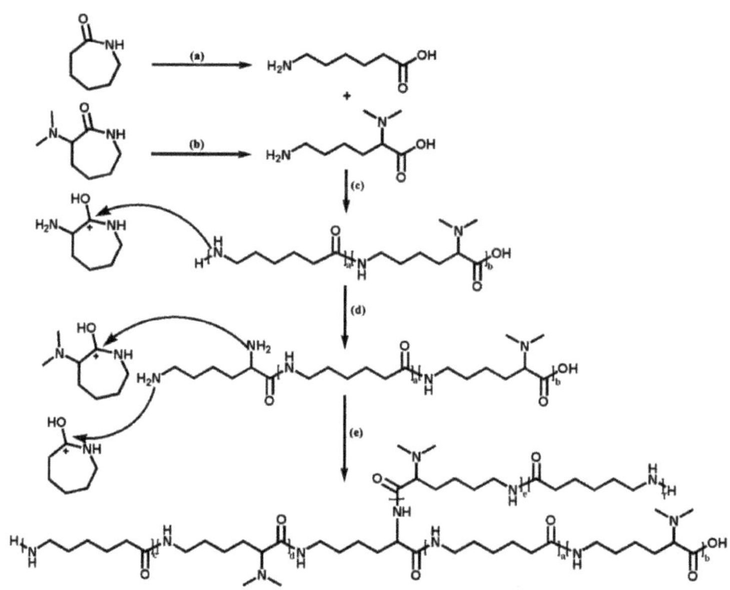

Figure 2. The reaction mechanism of CPL and ACL: (**a**) The ring-opening of CPL. (**b**) The ring-opening of DMCL. (**c**) The poly-condensation of linear monomers. (**d**) The incorporation of two chain growth sites. (**e**) The growth of the main chain and the side chain.

3.2. The Structure Characterization of the Co-Polymers

The structure of the co-polymers and the quaternized co-polymers were determined by ^{1}H NMR. The structural formula and ^{1}H NMR spectrum of the co-polymers are shown in Figure 3a, with the chemical shifts of H atoms as follows: δ4.67 (*l*, branching site); δ4.18 (a, dimethylamino site); δ3.48, 2.68, 1.42–1.75 (main chain and side chain); δ3.12 and δ3.05 (f, methyl H). The structural formula and ^{1}H NMR spectrum of the quaternized co-polymers are given in Figure 3b, with the chemical shifts of H atoms as follows: δ4.67 (*l*, branching site); δ4.07 (a, dimethylamino site); δ3.48, 2.68, 1.42–1.75 (main chain, side chain and quaternized groups); δ3.25 and δ3.18 (f, methyl H); δ3.33 and δ0.82 (quaternized groups).

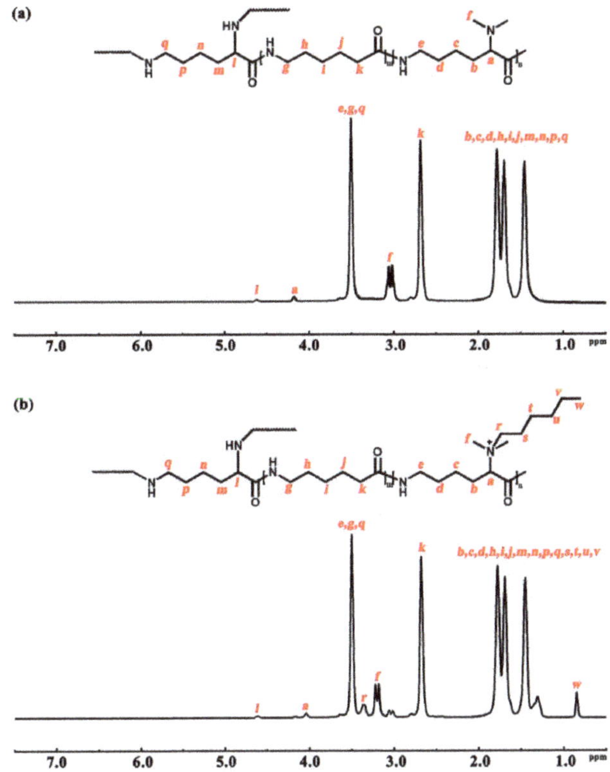

Figure 3. The ^{1}H NMR of the co-polymers (**a**) and the quaternized co-polymers (**b**).

3.3. Monomer Conversion, Configuration and Incorporation of the Co-Polymers

The co-polymers obtained under different feed ratios are shown in Figure 4a–c. When the proportion of DMCL in the feed was below 22 mol%, the co-polymers can be drawn and pelletized after extrusion. The PA6 pellets, shown in Figure 4a, indicated that the melt strength of the co-polymers was sufficiently high to maintain continuity during drawing. When the DMCL content in the feed was 30 mol%, the melt strength of the co-polymers failed to meet the requirements for drawing and pelletizing, and the pellets were obtained by simple crushing (Figure 4b). When the DMCL content in the feed reached 50 mol%, the co-polymers obtained after cooling exhibited properties similar to elastomers, suggesting that the co-polymers had converted to an amorphous state (Figure 4c). When the DMCL content in the feed reached 70 mol%, the co-polymer appeared as a gel rather than a solid at room temperature, and it had a certain solubility in hot water. Table 1 shows that with a constant feed amount of ACL, the relative viscosity and molecular weight of the co-polymers decreased with the increase in the DMCL proportion in the feed. This corresponded with the aforementioned morphological changes of the co-polymers. The monomers in the feed cannot be completely co-polymerized into the molecular chain. As shown in Table 1, there was little change in the conversion of ACL and CPL, while the conversion of DMCL was significantly affected by its content in the feed. It is due to the notably lower polymerization activity of DMCL compared to CPL, that the polymerization rate is influenced by the reactivity and monomers concentration. When the concentration of DMCL was low, its polymerization rate was considerably smaller than that of CPL, leaving a substantial amount of un-polymerized DMCL at the end of the reaction. In the feed ratio shown in entry 11, the proportion of DMCL to CPL in the co-polymer molecular chains approaches 1:1, indicating that the sample was just like an alternating co-polymer.

Figure 4. (**a**–**c**) The co-polymers with the feed shown in entry 2, 5 and 7 in Table 1, respectively. (**d**) The quaternized co-polymer with the feed shown in entry 5 in Table 1. (**e**) The samples after molding. (**f**) The co-polymers solution.

Table 1. Monomer conversion and incorporation of the co-polymers.

Entry	Feed ACL/DMCL/CPL	η_r	M_w	Conversion(%) ACL/DMCL/CPL	Incorporation ACL/DMCL/CPL
1	0/0/100	2.63	17010	0/91.2	-
2	1/4/95	2.51	15850	35.7/33.4/92.1	0.4/1.6/98
3	1/8/91	2.49	15640	35.3/36.2/91.1	0.4/3.4/96.2
4	1/13/86	2.42	14910	35.9/41.2/91.2	0.5/7.4/92.1
5	1/22/77	2.20	12610	34.8/47.3/90.9	0.4/12.8/86.8
6	1/30/69	1.98	10290	33.7/54.1/90.2	0.4/20.7/78.9
7	1/50/49	1.89	9340	34.1/63.2/91.4	0.4/37.4/62.2
8	1/70/29	1.42	4410	33.1/75.1/90.2	0.3/49.9/49.8

3.4. *Rheological Properties Test of Co-Polymers with Different ACL Feed*

It has been proven that the rheological properties of polymers are closely related to the molecular chain structure [21]. During the co-polymerization process, incorporating different amounts of ACL can generate different amounts of branched chains, thereby altering the rheological properties of the co-polymers. In this section, co-polymers with different ACL feed content were prepared while keeping the DMCL feed content constant. The feed ratio and rheological characteristic parameters of the co-polymers are shown in Table 2. Figure 5 exhibits the complex viscosity (η) of co-polymers with different AACL feed melt as a function of angular frequency (ω). The Newtonian liquid behavior was observed through the η-ω curve of linear P(DMCL-*co*-CPL). A Newtonian plateau of low viscosity appeared at low frequency. As the ACL was incorporated, the complex viscosity of the melt increased substantially higher than that of linear P(DMCL-*co*-CPL). Meanwhile, the Newtonian plateau of these samples disappeared, and the strong shear-thinning behavior was observed. As the incorporated DMCL increased, the shear thinning behavior became more obvious. The above phenomenon was caused by the formation of the branched structure and the entanglement effect of molecular chains. According to the mechanism of co-polymerization mentioned in Figure 2, the generation of the branched structure was formed with the incorporation of ACL, followed by the intensified entanglement effect between molecular chains. The movement of the chain was hindered by the increasing entanglement effect, which contributed to shearing and deformation not being instantaneous, manifested externally as the increase in melt viscosity at low frequency and exhibiting non-Newtonian fluid behavior [22]. The zero-shear viscosity is the complex viscosity of the melt when the shear rate tends to zero and the system approaches an equilibrium state. It can be calculated through the simple Carreau equation with the Cox–Merz rule [23]

$$\eta(\gamma)/\eta_0 = (1 + (\gamma\tau_n)^2)^{(n-1)/2}$$

where η_0 is the zero-shear viscosity, γ is the shear rate, τ_n is the characteristic time, and n is a parameter. It is notable in Table 2 that as the is ACL incorporated, the zero-shear viscosity increased more than 50 times. For linear chain polymers, the increase in M_w can also enhance the zero-shear viscosity. The 3.4 power-law is used to describe the relationship of zero shear viscosity and M_w [24]. However, the increase in molecular weight of Sample 1 and Sample 2 can only result in a five-fold increase in the zero-shear viscosity, assuming that they follow the 3.4 power-law. For Sample 3, the zero-shear viscosity increased as well as a decreasing M_w, which was unable to follow the law. The above phenomena indicate that changes in molecular weight are not the main factor affecting zero-shear viscosity. The huge increase in zero-shear viscosity can be also explained with the generation of gel structures after chemical cross-linking [22]. Cross-linked polymers can only swell in a solution without dissolving. Furthermore, as thermosetting materials, cross-linked polymers cannot be hot worked after molding [25]. However, the co-polymers we obtained can easily dissolve in 2,2,2-Trifluoroethanol (Figure 4f) and be transformed into a test sample through a molding injection (Figure 4e). Thus, we believed that there are no cross-linked structures in our co-polymers.

Table 2. Rheological properties of the co-polymers with different ACL feed.

Entry	Feed ACL/DMCL/CPL	η_r	M_w	Zero Shear Viscosity (Pa·s)	Melt Index
1	0/13/87	2.31	13780	405	71.4
2	0.5/13/86.5	2.40	14470	20542	251
3	1/13/86	2.42	14910	31454	279
4	2/13/85	2.38	14510	49523	386

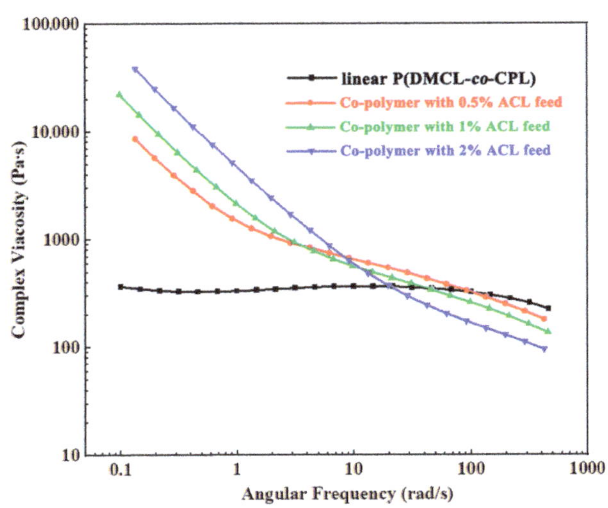

Figure 5. The complex viscosity of the co-polymers.

It can be also observed that as the ACL was incorporated, the melt index exhibits an increase. During the testing process of the melt index, the shear rate is high and the entanglement between the molecular chains is released, leading to the disappearance of the effect of the viscosity increase caused by branched chains disappearance. As the amount of ACL incorporated increases, the branched chain content increases, and the hydrogen bonds between the co-polymer molecular chains are disrupted. The macroscopic manifestation is that the resistance to internal flow of the co-polymer melt is weakened, the flow performance of the co-polymer improves, and the melt index increases. This is consistent with the variation pattern of the complex viscosity of the co-polymers at high frequency in Figure 5.

Figure 6a,b show the variation of storage modulus (G′) and loss modulus (G″) with the angular frequency (ω), respectively. The storage modulus is the elastic, solid-like behavior, while the loss modulus is the viscous response [26]. The melt of linear polymers exhibits typical terminal behaviors at the low frequency region. The G′–ω curve and G″–ω are close to straight lines, whose slopes are 2 and 1, respectively, in logarithmic coordinates [27]. The G′–ω curve of P(DMCL-*co*-CPL) exhibited like a straight line with a slope close to 2 in the low-frequency region, agrees with the terminal behaviors. As the introduction of ACL, the G′–ω curve deviated significantly from the terminal behavior. More branched chains strengthened the entanglement of the molecular chains, with enhancement of the solid-like behavior of the co-polymer melt and prolongation of the relaxation time. At the low frequency region, the decrease in G′ with ω is not significant. Meanwhile, the enhancement of solid-like behavior means that the energy stored in the melt increases during an alternating stress cycle, corresponding to a nearly 1000-fold increase in the storage modulus of ACL-incorporated co-polymers over P(DMCL-*co*-CPL) at the low frequency region.

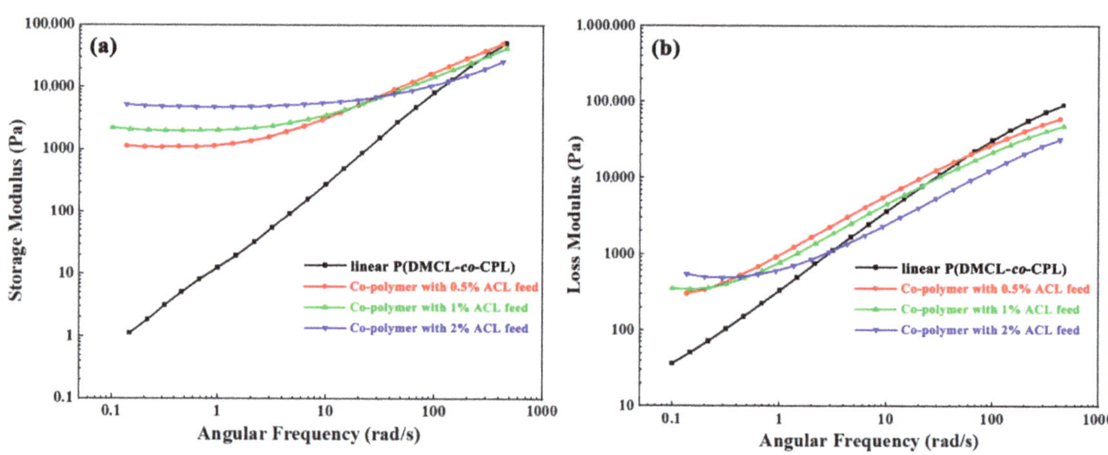

Figure 6. The storage modulus (**a**) and loss modulus (**b**) of the co-polymers.

3.5. Mechanical Properties of the Co-Polymers

The mechanical properties of the PA6 and the co-polymers(Entry 3–5 in Table 1) are shown in Figure 7. Linear DMCL/CPL co-polymers with the same content of DMCL (Sample A-C) were the control. Samples A-C have the same DMCL incorporated as Enrty 3–5, respectively. The linear co-polymers exhibit a significant decrease in tensile strength compared to pure PA6; this is because the introduction of DMCL significantly reduces the crystallinity of the polymer. For the co-polymers of Entry 3 and Entry 4, despite the reduction of crystallinity, the tensile strength was improved with the ACL incorporated because the molecular weight as well as the branches had increased, and more chain entanglement was proceeded. During the tensile process, each entanglement site of the chain needs to be disentangled after reaching the yield, resulting in an increase in the total stress required, which is reflected in the enhancement of tensile strength. For the co-polymer of Entry 5, due to a significant decrease in crystallinity, its tensile strength is lower than PA6.

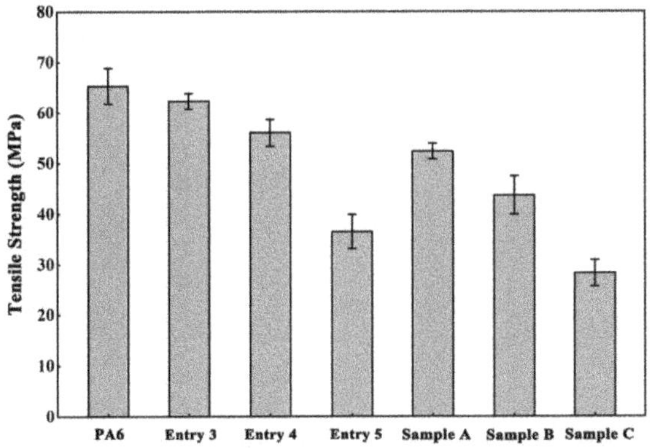

Figure 7. The mechanical properties of the PA6 and co-polymers.

3.6. Quaternization of the Co-Polymers

The solid co-polymers (Entry 2–7 in Table 1) were quaternized by reacting with 1-bromohexane, and the quaternized products were sequentially named as Samples 1–6. The quaternization rate was determined by the ^{1}H NMR, and the content of quaternary ammonium groups incorporated was calculated, as shown in Table 3. It can be seen that in the quaternization reaction of the co-polymer with 1-bromobutane, the quaternization rate remained around 80%.

Table 3. Quaternization rate and quaternary ammonium group content.

Sample	Feed ACL/DMCL/CPL	Incorporation DMCL Content	Quaternization Rate (%)	Quaternary Ammonium Groups Incorporated (%)
1	1/4/95	1.6	80.2	1.3
2	1/8/91	3.4	80.9	2.7
3	1/15/84	7.4	81.9	6.2
4	1/22/77	12.8	79.7	10.1
5	1/30/69	20.7	81.2	16.8
6	1/50/49	37.4	80.1	30.2

3.7. Thermal Properties Test of Co-Polymers with Different Quaternary Ammonium Groups Incorporated

Figure 8a shows the DSC scanning curve obtained from the second heat progress of PA6 and Samples 1–6 with different incorporated quaternary ammonium groups. The thermal properties parameters are shown in Table 4. PA6 exhibited a sharp melting peak at 221.2 °C. As the quaternary ammonium groups incorporated increased, the melting peak became wide and moved to a low-temperature area, which was reflected in polymer properties as a decrease in melting point and crystallinity. When the incorporated quaternary ammonium groups reached 16.8 mol% (Sample 5), the melting peak became particularly wide, and the glass transition temperature was observed on the curve. At the highest level of quaternary ammonium group incorporation (30.2 mol%, Sample 6), the melting peak disappeared, which meant that the co-polymer was amorphous and there was no crystallization. Short-branched chains, formed during the branching process, as well as the quaternary ammonium groups affects, the arrangement of the main chain and long branch chains, hindering their formation of ordered and regular structures [27]. As the main chain gradually forms an ordered arrangement, the areas where short chains gather form amorphous regions. The appearance of amorphous regions affects the overall structural regularity of the polymer, leading to a decrease in its melting point and crystallinity. Compared to pure PA6, the thermal stability of co-polymers decreases. All of the co-monomers, branched chains, and quaternary ammonium groups had an impact on the regularity of the molecular chains, disrupting the regularity, resulting in abatement in hydrogen bonding and intermolecular forces, leading to the decrease in thermal stability [28].

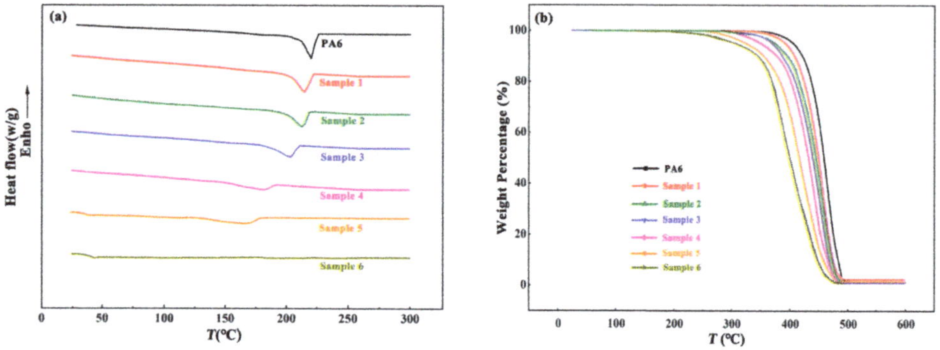

Figure 8. (**a**) DSC scanning curve and (**b**) TG curve of PA6 and Samples 1–6.

Table 4. The thermal properties of PA6 and Samples 1–6.

The Polymer	Quaternary Ammonium Groups Incorporated (%)	Melting Point (°C)	Crystallinity (%)	Initial Thermal Degradation Temperature (°C)	Maximum-Rate Thermal Degradation Temperature (°C)
PA6	0	221.2	27.38	424.4	458.1
Sample 1	1.3	211.4	25.59	414.2	456.2
Sample 2	2.7	209.8	24.26	401.2	453.1
Sample 3	6.2	201.4	22.32	398.4	451.9
Sample 4	10.1	189.2	20.19	382.2	450.7
Sample 5	16.8	168.4	18.20	361.1	449.2
Sample 6	30.2	-	-	283.4	398.6

3.8. Antibacterial Activities Test of Co-Polymers

Quaternized co-polymers with crystallinity and capable of being molded into sample pieces were selected for the antibacterial activities test. The results are shown in Figure 9. The small yellow or white dots on the culture dish represent the surviving bacterial colonies. Due to the absence of antibacterial activities, the bacterial content in the bacterial suspension did not decrease after a 24 h culture on the PA6 sample piece. Therefore, when the bacterial suspension is recovered, diluted, and transferred to a culture dish for another 24 h, dense colonies of bacteria can be obviously seen. It has been proved that long chain PA6 containing quaternary ammonium groups has antibacterial activities. Quaternary ammonium groups have a positive charge, while bacterial cell membranes composed of phospholipid bilayers have a negative charge. When the two come into contact, the quaternary amine groups will adsorb onto the bacterial cell membrane due to the electrostatic interactions, inhibiting bacterial growth and division. In addition, the long chains of aliphatic polyamides have good compatibility with phospholipids. Long molecular chains can penetrate the cell membrane, destroy the cell membrane skeleton, and cause bacterial cell membrane rupture [29]. After the bacterial suspension underwent a 24 h culture on the piece of the co-polymer with 6.2 mol% quaternary ammonium group incorporation (Sample 3), a significant decrease in colonies appeared on the culture dish. The antibacterial rate of this co-polymer sample against *Staphylococcus aureus* and *Escherichia coli* was calculated by the plate counting method to be around 85%. When the incorporated quaternary ammonium groups reached 10.1 mol% (Sample 4 and 5), the antibacterial rate of the co-polymer sample against the above two bacteria reached 99.9%, indicating that the co-polymer demonstrated nearly complete killing of *Staphylococcus aureus* and *Escherichia coli*.

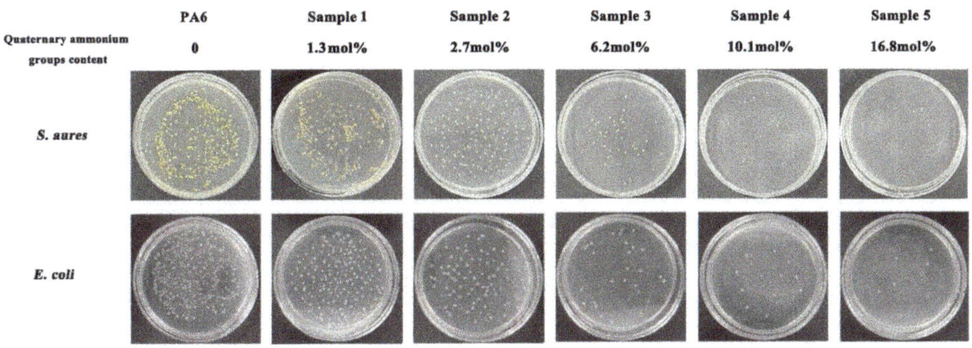

Figure 9. The results of antimicrobial activities test.

4. Conclusions

In this study, we successfully synthesized an antibacterial branched PA6 with CPL, ACL and DMCL via hydrolytic ring-opening co-polymerization, followed by the quaternization with 1-Bromohexane. The met index (MFR), zero-shear rate viscosity and storage modulus at the low frequency region of the co-polymers have a remarkable increase as well as the shear thinning phenomenon becoming more obvious. The melting point, degree of crystallinity and thermal stability of the quaternized co-polymers decreases with the increase in functional monomers incorporated. The quaternized co-polymers with 6.2 mol% quaternary ammonium groups incorporated (Sample 3) exhibit excellent antimicrobial activities against *Staphylococcus aureus* (Gram positive) and *Escherichia coli* (Gram negative) and when the quaternary ammonium groups incorporated reach 10.1% (Sample 4), the co-polymers demonstrate complete killing of the two above bacterium. This research provides a new method for the preparation of high performance modified PA6.

Author Contributions: Software, W.L.; Validation, X.M., Z.L. and B.Z.; Formal analysis, X.M. and S.M.; Investigation, B.Z.; Resources, W.L. and Z.L.; Data curation, Z.L. and S.M.; Writing—original draft, X.M.; Writing—review & editing, X.M., W.L., S.M. and B.Z.; Supervision, W.L. and B.Z. All authors have read and agreed to the published version of the manuscript.

Funding: This research received no external funding.

Institutional Review Board Statement: Not applicable.

Data Availability Statement: Data are contained within the article.

Acknowledgments: The authors would like to thank the support from the Major Program Foundation of SINOPEC.

Conflicts of Interest: All authors were employed by the Research Institute of Petroleum Progressing, SINOPEC. The authors declare that the research was conducted in the absence of any commercial or financial relationships that could be construed as a potential conflict of interest.

References

1. Kudo, H.; Buya, K. Mechanistic Study of Ring-Opening Copolymerization of ε-caprolactam with epoxide: Development of novel thermosetting epoxy resin system. *J. Polym. Sci. Part A Polym. Chem.* **2016**, *54*, 2220–2228. [CrossRef]
2. Gong, Y.; Liu, A.; Yang, G. Polyamide single polymer composites prepared via insitu anionic polymerization of ε-caprolactam. *Compos. Part. A Appl. Sci. Manuf.* **2010**, *41*, 1006–1011. [CrossRef]

3. Naumann, S.; Epple, S.; Bonten, C.; Buchmeiser, M.R. Polymerization of ε-Caprolactam by Latent Precatalysts Based on Protected N-Heterocyclic Carbenes. *ACS Macro Lett.* **2013**, *2*, 609–612. [CrossRef] [PubMed]
4. Hashimoto, K. Ring-opening polymerization of lactams. Living anionic polymerization and its applications. *Prog. Polym. Sci.* **2000**, *25*, 1411–1462. [CrossRef]
5. Wang, J.; Kuai, P.; Liu, H.; Feng, Z. Development Status of Nylon 6 and Nylon 66 in China. *Synth. Fiber China* **2021**, *50*, 8–11.
6. Kumaran, S.; Han, S.; Choi, H. Photopolymerizable, Universal Antimicrobial Coating to Produce High-Performing, Multifunctional Face Masks. *Nano Lett.* **2021**, *21*, 5422–5429. [CrossRef] [PubMed]
7. Zhou, J.; Zabihi, F.; Chen, Z.; Zhu, M. Progress and Perspective of Antiviral Protective Material. *Adv. Fiber. Mater.* **2020**, *2*, 123–139. [CrossRef] [PubMed]
8. Ulery, B.D.; Nair, L.S.; Laurencin, C.T. Biomedical applications of biodegradable polymers. *J. Polym. Sci. Part B Polym. Phys.* **2011**, *49*, 832–864. [CrossRef]
9. Hernández-Escolano, M.; Juan-Díaz, M.; Martínez-Ibáñez, M.; Jimenez-Morales, A.; Goñi, I.; Gurruchaga, M. The design and characterisation of sol gel coatings for the controlled-release of active molecules. *J. Sol-Gel Sci. Technol.* **2012**, *64*, 442–451. [CrossRef]
10. Tang, W.; Xiao, W.; Ye, E.; Liao, J. Discussion on the anti-microbe fabrics. *Chem. Fiber Text. Technol.* **2003**, *004*, 28–35.
11. Erem, A.D.; Ozcan, G.; Skrifvars, M.; Cakmak, M. In vitro assesment of antimicrobial activity and characteristics of polyamide 6/silver nanocomposite fibers. *Fibers Polym.* **2013**, *14*, 1415–1421. [CrossRef]
12. Bucheńska, J. Polyamide fibers (PA6) with antibacterial propertie. *J. Appl. Polym. Sci.* **1996**, *61*, 567–576. [CrossRef]
13. Shi, Z. Grafting chitosan oxidized by potassium persulfate onto Nylon 6 fiber, and characterizing the antibacterial property of the graft. *J. Polym. Res.* **2014**, *21*, 534–539. [CrossRef]
14. Tao, Z.; Kang, H. Enhancement of the Processability and Properties of Nylon 6 by Blending with Polyketone. *Polymers* **2021**, *13*, 3403. [CrossRef]
15. Lian, J.; Chen, J.; Luan, S.; Liu, W.; Zong, B.; Tao, Y.; Wang, X. Organocatalytic Copolymerization of Cyclic Lysine Derivative and ε-Caprolactam toward Antibacterial Nylon-6 Polymers. *ACS Macro Lett.* **2022**, *11*, 46–52. [CrossRef]
16. Liu, K.; Shao, B.; Liu, H.; He, J.; Zheng, A.; Li, G.; Zheng, B.; Li, Z.; Liu, W.; Li, X.; et al. Sustainable production of dimethyl-protected cyclic lysine over Pd/m-Al_2O_3-Si catalysts and its application in synthesis of antibacterial nylon6 copolymers. *Chem. Eng. J.* **2023**, *463*, 142504. [CrossRef]
17. Kohan, M.I. *Nylon Plastics Handbook*; Hanser: New York, NY, USA, 1995; pp. 112–118.
18. Galleny, P.G. Nylon 6. In *Polymer Data Handbook*; Mark, J.E., Ed.; Oxford University Press: New York, NY, USA, 1999; pp. 180–185.
19. *ASTM D638*; Standard Test Method for Tensile Properties of Plastics. ASTM International: West Conshohocken, PA, USA, 2022.
20. *GB/T 31402-2015*; Plastics—Measurement of antibacterial activity on plastics surfaces. Standardization Administration of China: Beijing, China, 2015.
21. Sugimoto, M.; Tanaka, T.; Masubuch, Y. Effect of chain structure on the melt rheology of modified polypropylene. *J. Appl. Polym. Sci.* **1999**, *73*, 1493–1500. [CrossRef]
22. Kima, H.; Oha, K.; Seo, Y. Rheological and mechanical properties of a novel polyamide 6 synthesized by anionic polymerization of ε-caprolactam in a twin-screw extruder. *Polymer* **2019**, *177*, 196–201. [CrossRef]
23. Seo, Y.P. Effect of molecular structure change on the melt rheological properties of a polyamide (nylon 6). *ACS Omega* **2018**, *12*, 16549–16555. [CrossRef]
24. Doi, M. Explanation for the 3.4-power law for viscosity of polymeric liquids on the basis of the tube model. *J. Polym. Sci. B* **1983**, *21*, 667–684. [CrossRef]
25. Young, R.J.; Lovell, P.A. *Introduction to Polymers*; CRC Press: Boca Raton, FL, USA, 2011; p. 263.
26. Gotsis, A.D.; Zeevenhoven, B.L.F.; Tsenoglou, C. Effect of long branches on the rheology of polypropylene. *J. Rheol.* **2004**, *48*, 895–914. [CrossRef]
27. Ruymbeke, E.; Stephenne, V.; Daoust, D. A sensitive method to detect very low levels of long chain branching from the molar mass distribution and linear viscoelastic response. *J. Rheol.* **2005**, *49*, 1503–1520. [CrossRef]
28. Núñez, E.; Ferrando, C.; Malmström, E.; Claesson, H.; Gedde, U.W. Crystallization behavior and morphology of star polyesters with poly (ε-caprolactone) arms. *J. Macromol. Sci. B* **2004**, *43*, 1143–1160. [CrossRef]
29. Timofeeva, L.; Kleshcheva, N. Antimicrobial polymers: Mechanism of action, factors of activity, and applications. *Appl. Microbiol. Biotechnol.* **2011**, *89*, 475–492. [CrossRef]

polymers

MDPI

Article

Self-Floating Polydopamine/Polystyrene Composite Porous Structure via a NaCl Template Method for Solar-Driven Interfacial Water Evaporation

Xiao Wang [1,2], Zhen Li [3], Xiaojing Wu [1,*], Bingjie Liu [4], Tian Tian [2], Yi Ding [2,3], Haibo Zhang [2], Yuanli Li [2], Ye Liu [2] and Chunai Dai [2,*]

1 School of Undergraduate Education, Shenzhen Polytechnic University, Shenzhen 518055, China; 23126690@bjtu.edu.cn
2 School of Physical Science and Engineering, Beijing Jiaotong University, Beijing 100044, China; ttian@bjtu.edu.cn (T.T.); 23121824@bjtu.edu.cn (Y.D.); 22341149@bjtu.edu.cn (H.Z.); 23342012@bjtu.edu.cn (Y.L.); 22341062@bjtu.edu.cn (Y.L.)
3 Advanced Materials and Energy Center, Academy of Aerospace Science and Innovation, Beijing 100088, China; chemlizhen@gmail.com
4 Beijing Institute of Space Mechanics & Electricity, Beijing 100094, China; bingjie1118@outlook.com
* Correspondence: wuxj0308@szpu.edu.cn (X.W.); chadai@bjtu.edu.cn (C.D.)

Abstract: Solar energy, as a clean and renewable energy source, holds significant promise for addressing water shortages. Utilizing solar energy for water evaporation is seen as an effective solution in this regard. While many existing interfacial photothermal water evaporation systems rely on nanoparticles or graphene as photothermal or support materials, this study introduced polydopamine (PDA) as a photothermal material due to its environmental friendliness and excellent photon absorption characteristics that closely match the solar spectrum. Polystyrene (PS) was also introduced as a support material for its porous structure and density similar to water, enabling it to float on water. The resulting PS-PDA composite porous structure solar evaporator exhibited a photothermal conversion efficiency comparable to nanoparticles (over 75%), yet with lower production costs and minimal environmental impact. This innovative approach offers a scalable solution for water-scarce regions, providing a cost-effective and efficient means to address water scarcity. The use of PDA and PS in this context highlights the potential for utilizing common materials in novel ways to meet pressing environmental challenges.

Keywords: polydopamine; polystyrene; photothermal materials; porous structure; NaCl template method; solar-driven interfacial water evaporation

1. Introduction

The global water shortage issue is rapidly escalating, impacting approximately one-third of the world's population. This crisis has emerged as one of the most pressing challenges of the 21st century. The lack of access to safe and clean water not only affects human health but also poses significant challenges to agriculture, industry, and ecosystems [1–4]. Solar-driven water evaporation—the extraction of vapor from liquid water using solar energy—provides the basis for the development of freshwater production. Traditional solar water evaporation systems use their surfaces or cavities to absorb solar radiation and then transfer heat to the water. This method produces a large amount of energy loss due to the influence of thermal radiation and convection, and the photothermal conversion efficiency is not high.

To this end, an interfacial evaporation system floating on the water was proposed, which can lock thermal energy at the interface to minimize heat loss and improve conversion efficiency [5,6]. The materials that make up the interfacial evaporation system should have excellent photothermal conversion properties and good water transmission properties.

In recent years, interfacial evaporation materials with different structural designs and excellent properties have been proposed, including porous graphene oxide [7,8], gold, aluminum, copper, silicon, and nanoparticles such as manganese dioxide, titanium dioxide, molybdenum disulfide, iron oxide, cobalt oxide, copper sulfide, and so on [9–19]. However, these are non-biodegradable photothermal materials and support materials. Once performance degradations occur, they will not only affect the photothermal conversion efficiency but also have certain impacts on the environment. The extent of the impact of these materials currently used in large quantities on interface evaporator systems entering the environment remains a matter of uncertainty and requires further discussion [20,21]. On this basis, the choice of interface evaporator material must take into account not only the photothermal conversion efficiency and water transport properties but also the environmental impact of the material itself.

Polydopamine (PDA) is a black biopolymer formed by the oxidative self-polymerization of dopamine (DA) and can be used as a coating to adhere to a variety of materials and as a platform for surface reactions [22]. PDA has good photothermal properties and can absorb the vast majority of photon energy in the solar spectrum and convert it into heat [23,24]. Moreover, PDA has good biodegradability [25] and PDA is harmless in mammalian cells [26–28]. Therefore, PDA can be used as a multifunctional surface treatment material with huge application prospects in biology, energy, and industry [29–31].Polystyrene (PS) is a widely used thermoplastic polymer known for its cost-effectiveness, stable physical and chemical properties, and excellent biocompatibility [32,33]. Its recyclability further adds to its appeal, making it a primary material in everyday packaging, disposable containers, and insulation products [34,35]. One of its key advantages is its compatibility with PDA, making it an optimal support material for PDA loading [36–39]. Its compatibility and versatility make PS a compelling choice as a support material in various studies.

In this study, we investigated the composite formation of PS with PDA at varying concentrations. Based on the composite strategy, we developed a novel PS-PDA composite porous structure solar-driven interfacial water evaporator. The design exhibited outstanding water vapor collection efficiency under specific solar irradiation intensities. Additionally, the PS-PDA composite porous structure solar evaporator demonstrated low raw material costs and a simple, environmentally friendly preparation process, which shows significant potential for large-scale production and utilization. The characteristics of low cost and high-water vapor collection efficiency make the PS-PDA evaporator a promising solution for water-scarce regions. This advancement not only provides a sustainable method for water evaporation but also addresses the global challenge of water scarcity in a practical and scalable manner.

2. Materials and Methods

2.1. Materials and Characterization

In this study, dopamine hydrochloride (DA·HCl, 98%, TCI Chemicals, Shanghai, China), PS (98%, Aladdin's reagent, Shanghai, China), sodium chloride (NaCl, AR, Aladdin's reagent, Shanghai, China), toluene (AR, Aladdin's reagent, Shanghai, China), ethanol (AR, Beijing Chemical Factory, Beijing, China) and ammonia (28%, AR, Beijing Chemical Plant, Beijing, China) were used. All materials were used directly without processing.

A scanning electron microscope (SEM, Hitachi SU8000, Tokyo, Japan) was employed to photograph the samples. Absorbance or reflectance of PS-PDA was tested using UV–VIS–NIR diffuse reflection spectroscopy (DRS, Hitachi, U-3010, Tokyo, Japan). The water contact angle was measured using an OCA 15Pro instrument (Dataphysics, Stuttgart, Germany), and an infrared thermal imager (FLIR E40, Wilsonville, OR, USA) was used for measuring the temperature of the object.

2.2. Synthesis of PDA Nanoparticles

PDA nanoparticles were synthesized following previous literature reports [27,40,41]. The specific synthesis method is as follows.

A mixed solution of 180 mL deionized water and 80 mL ethanol was prepared, to which 5 mL of 28–30% ammonia water was added, and the mixed solution was stirred at room temperature for 10 min.

Subsequently, 1 g of DA·HCl was dissolved in 20 mL of deionized water and added dropwise to the mixed solution above; after stirring for 30 min, the mixture was then left at room temperature for 24 h to allow for the polymerization of DA.

The PDA nanoparticles were obtained by centrifuging the reaction mixture at 8000 rpm multiple times, washing the precipitate with water, and freeze-drying.

2.3. Fabrication of PS-PDA Material

The production of the PS-PDA composite porous structure solar evaporator for photo-evaporation of water was modified based on the literature [42]. It was divided into three steps as shown in Figure 1. The first step was to make a toluene solution of PS-PDA(PS-PDA/PhMe), which was prepared for later use by placing 1.32 g of PS, 4 mL of toluene, and 1, 3, 5, or 7% PDA nanoparticles in a small glass bottle and stirring for 24 h. The reason for choosing toluene is that it is a common organic solvent in the laboratory and has good solubility and low toxicity. The second step was to fabricate a PS-PDA gel. First, 200-mesh NaCl was ground to a Teflon template with a depth of 0.5 mm and placed in an atmosphere of saturated NaCl solution to solidify. The above PS-PDA/PhMe solution with different PDA concentrations was then dropwise added to the hardened NaCl template and placed in a toluene atmosphere for percolation until no bubbles escaped. The third step was curing and demolding. After scraping off the liquid on the surface with a scraper, the PS-PDA gel on the NaCl template was sprinkled with NaCl particles and put in a fume hood to solidify in the air. This step aims to create more porous channels in the surface. The cured PS-PDA was then de-molded, ultrasonically treated, and soaked in deionized water for 24 h to remove NaCl. For PS-PDA materials prepared with this template method, the surface formed on the air interface side is named the top and the surface formed on the Teflon interface side is named the bottom. The PS template followed the same preparation method, except that the PDA was not added.

Figure 1. Schematic diagram of the preparation of PS-PDA composite porous structure material.

2.4. Solar-Driven Interfacial Water Evaporation Measurements

Solar-driven interfacial water evaporation measurements were conducted indoors and outdoors. In the case of indoors, a solar simulator composed of a xenon light source (Beijing Zhongjiao Jinyuan Technology Corporation CEL-S500, Beijing, China) was used, surrounded by thermal insulation glass to avoid heat loss. A 3 cm × 3 cm × 10 cm sample stage was placed in the center on which the solar water evaporator was placed. The weight of the lost water was measured by a precision balance (Mettler Toledo ME204, Zurich, Switzerland) placed under the sample stage. Before testing, a solar standard cell (Oriel Newport/VLSI 91150V, Newport, RI, USA) was used for solar intensity calibration. While testing, the solar water evaporator of the PS-PDA material was placed and floated on the water in a quartz cup. The water evaporated from the interface of the solar evaporator due to the heat generated by the xenon. When the evaporation time reached 15 min, it was considered to be balanced. In the case of outdoors, the measurements were carried out in sunlight. Our outdoor experiments took place in Beijing during the summer, with an average maximum weather temperature of 36 °C throughout the week. We conducted the tests in the afternoon when the sun was at its strongest.

3. Results

3.1. Morphology of PS-PDA Materials

Figure 2 shows the photos and SEM images of the prepared PS-PDA materials. The actual photos of the PS-PDA composite porous structure materials with different PDA contents are shown in Figure 2A. As the PDA content increases, the color of the PS-PDA material gradually deepens. Figure 2B is an SEM image of a PS support material. It can be seen from the picture that the PS has a highly open porous structure. The average aperture of the PS measured using the Nano Measurer software (V1.2.0) is 76 μm. Figure 2C shows an SEM image of the PDA particles. The PDA particles are spherical, with an average diameter of approximately 300 nm measured using Nano Measurer software. Figure 2D shows the SEM image of the PS-PDA with 5% PDA doped. It can be observed that the PDA particles (dotted) were loaded into the PS support material. Based on these morphology studies, it can be estimated that highly dense PDA particles can be accommodated in the pores of PS, and the porous nature of PS also facilitates the transmission of water from the liquid to the evaporation surface, greatly improving the water transmission efficiency.

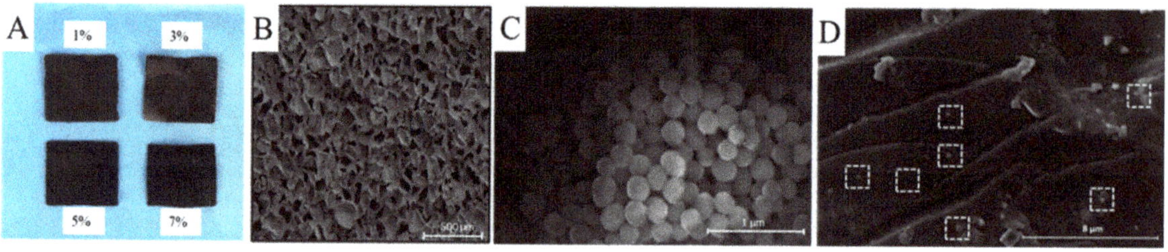

Figure 2. (**A**) The actual photos of the PS-PDA composite porous structure materials with different PDA contents. SEM images of (**B**) PS, (**C**) PDA, and (**D**) PS-PDA with 5% PDA doped.

3.2. Wettability of PS-PDA Materials

We tested the wettability of the PS-PDA material with 5% PDA doped; the results are listed in Figure 3. As shown in Figure 3A, the evaporator can self-float well at the air–water interface. One of the reasons for choosing PS as the support material is that the density of PS is close to that of water. This property will help the evaporator float on the water, maximizing the surface exposed to sunlight. Figure 3B shows the water contact angle test results of PS and PS-PDA materials. The water contact angle of 108° indicates that PS is a hydrophobic material. The picture below shows that PS loaded with 5% PDA particles has a decreased water contact angle of 82°, indicating that the surface of PS-PDA is more hydrophilic. Because the loaded PDA particles have exposed hydrophilic polar groups such as amines and hydroxyl groups [43], which increases the number of hydrophilic sites on the PS-PDA surface and improves the surface wettability, the PS-PDA evaporator has a stronger ability to transport water to the heat-absorbing surface. Moreover, the enhanced surface hydrophilicity of the PS-PDA evaporator can promote the adsorption of pollutants in water, indicating its potential as a material for water pollutant treatment. These findings highlight the multifaceted capabilities and promising applications of the PS-PDA composite material in water treatment and solar evaporation systems. The composite's ability to adsorb pollutants, coupled with its potential for use in solar evaporation systems, underscores its versatility and potential impact in addressing water treatment challenges [12].

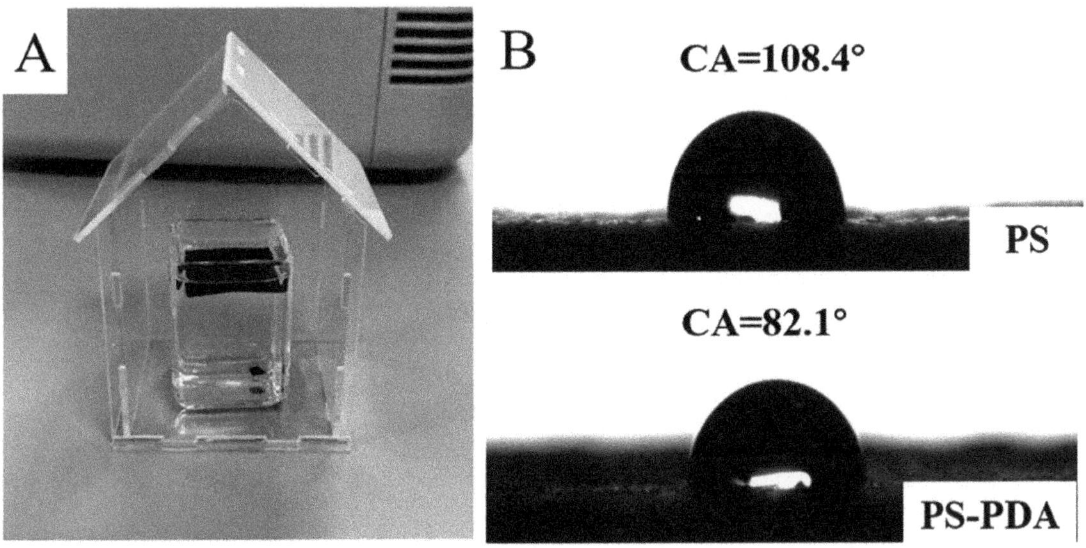

Figure 3. (**A**) Picture of the PS-PDA composite evaporator floating on the water. (**B**) Water contact angle test: the picture above shows the water droplet (4 μL) on PS and the picture below shows the water droplet (4 μL) on PS-PDA.

3.3. *Light Absorption Property of PS-PDA Materials*

The UV–VIS–NIR DRS spectra of PS-PDA material with 5% PDA doped are shown in Figure 4. Figure 4A shows the spectra of the dry and wet PS-PDA materials. The reflectivity of the wet PS-PDA significantly decreased, indicating that the material had better solar light absorption performance in water. As shown in Figure 4B, the wet PS-PDA had a relatively low reflectivity (R); since the PS-PDA material had almost no transmitted light (T), the absorption rate A can be calculated using the formula $A = 1 - T - R$ [44]. The calculated average absorption value over the entire solar spectral range reached 90%, indicating that the wet PS-PDA material has excellent solar light absorption performance and is therefore a candidate material for solar water evaporation applications.

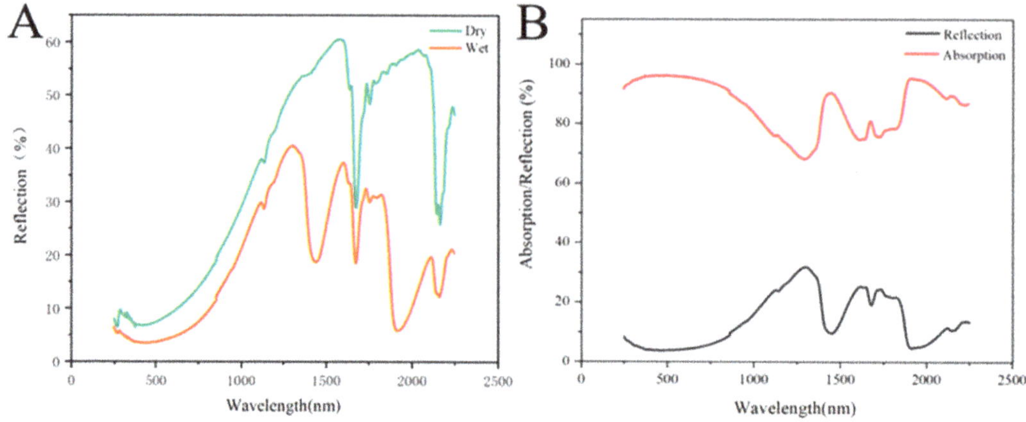

Figure 4. (**A**) UV–VIS–NIR DRS spectra of dry and wet PS-PDA membranes. (**B**) UV–VIS–NIR DRS spectra of wet PS-PDA membrane.

3.4. Water Evaporation Test under Simulated Solar Radiation Indoors

Figure 5A shows the photothermal conversion water vapor collection test device, in which the enlarged part is a top-view photo. The device was surrounded by insulated glass and equipped with a sample stage that could measure weight. This device simulated sunlight evaporating water and reflected the quality of the evaporation by measuring the amount of water reduced in the cup. Figure 5B represents the infrared imaging images of the photothermal conversion water vapor collection test device after 15 min of simulated solar irradiation, The cross head represents the highest temperature position in the diagram. The left image shows the temperature of the PS-PDA material and the maximum temperature has exceeded 150 °C. The right image depicts the temperature of the test water, with the highest recorded temperature being approximately 38.4 °C. The infrared imaging images indicate that the heat was focused within the material rather than in the water. This is due to the good photothermal properties of the PDA, which absorbs most of the photons of the solar spectrum and converts them into thermal energy [23,24]. During the photothermal conversion water vapor collection test, PS-PDA materials containing varying amounts of PDA were tested with their top and bottom surfaces facing the simulated sun to avoid uneven distribution of PDA. The results are shown in Figure 5C. For PS-PDA doped with 1% to 7% PDA, as the doped PDA concentration increased, the water evaporation rate first increased and then decreased. The evaporation rate peaked at 5% PDA doping. Because PDA has favorable photo-thermal characteristics, the water evaporation rate increases with the increase in PDA content. However, once the PDA content exceeds a certain level, the PDA will absorb too much light and heat, thus preventing the heat from reaching the water layer below. As a result, the water layer lacks the necessary heat from the light source to reach the required temperature for evaporation, leading to a decrease in the evaporation.

The water vapor collection efficiency can be calculated by the following commonly used formula in this field [6,44]:

$$\eta = \frac{mH_{LV}}{E_i} \quad (1)$$

In this equation, m is the mass of water that decreases per unit time and unit area, H_{LV} is the evaporation enthalpy, and E_i is the light intensity.

According to the given formula, the water vapor collection efficiency of PS-PDA with PDA content ranging from 1% to 7% has been calculated and is illustrated in Table 1 and Figure 5D. Remarkably, for PS-PDA with 5% PDA doping, the efficiency under one sun has exceeded 75%, higher than that of metal nanoparticles and graphene oxide. These efficiencies highlight the competitive performance of PS-PDA photothermal materials, showcasing the potential of PDA-PDA materials in solar-driven applications for water treatment and evaporation systems. This is because PDA has an absorption spectrum that closely matches the solar spectrum, and the porous nature of PS-PDA helps water transport from the liquid to the surface for evaporation.

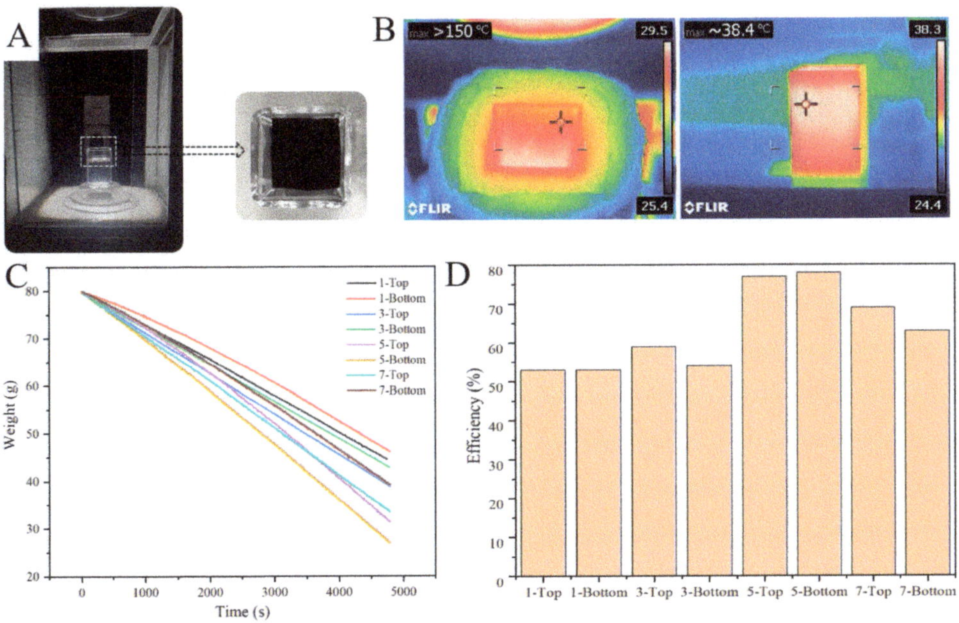

Figure 5. Water evaporation test under simulated solar radiation indoors. (**A**) Physical diagram of the photothermal conversion water vapor collection test device. (**B**) Infrared imaging of the photothermal conversion water vapor collection test device, where the left is for the PS-PDA material and the right is for the water. (**C**) Photothermal conversion water vapor collection test results of PS-PDA evaporators with different proportions of PDA doped. (**D**) Water vapor collection efficiency of PS-PDA with different proportions of PDA doped.

Table 1. Evaporation efficiency of PS-PDA doped with different PDA concentrations.

Group	1-Top	1-Bottom	3-Top	3-Bottom	5-Top	5-Bottom	7-Top	7-Bottom
Efficiency	53%	53%	59%	54%	77%	78%	69%	63%

3.5. Solar-Driven Interfacial Water Evaporation Test Outdoors

To explore the practical application possibilities of the PS-PDA evaporators, we further performed water vapor collection tests on the 5% PDA-doped PS-PDA interface evaporator under outdoor solar radiation, as shown in Figure 6A. For comparison, contrast tests were conducted under indoor conditions of 25 °C and 52% air humidity, as well as outdoor conditions of 36 °C and 56% air humidity. The results are plotted in Figure 6B. During the indoor test, the mass of the test water decreased from 80 g to 79.76 g. The relatively weak indoor sunlight led to a lower photon energy absorption by the PS-PDA interface evaporator. In contrast, during the outdoor testing, the mass of the test water decreased from 79.8 g to 78.44 g. The efficiency observed in the outdoor test is in close agreement with the results of other photothermal conversion water vapor harvesting devices. These findings support the practical applicability of the PS-PDA evaporators in real-world settings.

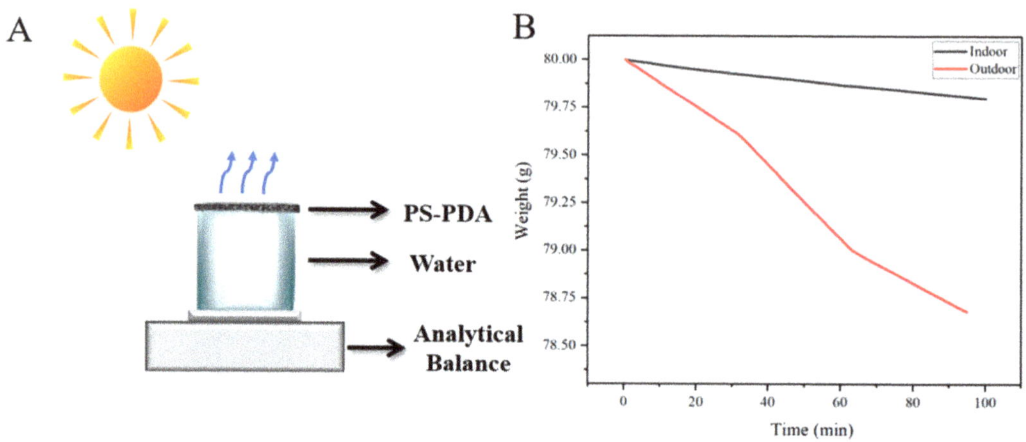

Figure 6. Outdoor test of water vapor collection. (**A**) Schematic diagram of the outdoor photothermal conversion water vapor collection test device. (**B**) Comparison of indoor and outdoor water vapor collection.

4. Conclusions

In this study, we present a novel solar evaporator design with exceptional performance metrics, including high photothermal conversion efficiency, cost-effectiveness, and environment-friendly. The solar evaporator is a PS-PDA composite porous structure prepared through a NaCl template method. The absorption spectrum of PDA particles closely matches the photon absorption characteristics of the solar spectrum, while PS's porous structure and density that are similar to water enable its self-floating manner. This composite porous structure exhibits not only high photothermal conversion efficiency but also efficient water transmission. Our study demonstrates that a 5% PDA doping level in PS achieves the optimal water vapor collection efficiency, surpassing 75%. One of the key advantages of this solar evaporator is its use of readily available and environmentally friendly materials. The raw materials, including food-grade PS and biomass-derived PDA, are common and inexpensive. Additionally, the preparation process is simple and scalable, making it suitable for large-scale applications. Overall, the PS-PDA solar evaporator represents a significant advancement in solar evaporation technology, offering a sustainable solution for water-scarce regions. Its high efficiency, low cost, and eco-friendly nature make it a promising candidate for addressing the global water crisis.

Author Contributions: Conceptualization, Z.L., X.W. (Xiaojing Wu) and C.D.; methodology, Z.L.; software, Y.L. (Yuanli Li); validation, Y.D., H.Z. and T.T.; formal analysis, B.L.; investigation, Y.L. (Ye Liu); resources, Z.L.; data curation, B.L.; writing—original draft preparation, X.W. (Xiao Wang); writing—review and editing, X.W. (Xiaojing Wu); visualization, B.L.; supervision, C.D.; project administration, Z.L.; funding acquisition, X.W. (Xiaojing Wu). All authors have read and agreed to the published version of the manuscript.

Funding: This research was funded by The High-Level Talent Initiation Project of Shenzhen Polytechnic University (6024330003K). The funder is Xiaojing Wu.

Institutional Review Board Statement: Not applicable.

Data Availability Statement: Data are contained within the article.

Conflicts of Interest: The authors declare no conflicts of interest.

References

1. Ahmed, F.E.; Hashaikeh, R.; Hilal, N. Solar powered desalination—Technology, energy and future outlook. *Desalination* **2019**, *453*, 54–76. [CrossRef]

2. Jiang, Q.S.; Derami, H.G.; Ghim, D.; Cao, S.S.; Jun, Y.S.; Singamaneni, S. Polydopamine-filled bacterial nanocellulose as a biodegradable interfacial photothermal evaporator for highly efficient solar steam generation. *J. Mater. Chem. A* **2017**, *5*, 18397–18402. [CrossRef]
3. Shannon, M.A.; Bohn, P.W.; Elimelech, M.; Georgiadis, J.G.; Marinas, B.J.; Mayes, A.M. Science and technology for water purification in the coming decades. *Nature* **2008**, *452*, 301–310. [CrossRef]
4. Elimelech, M. The global challenge for adequate and safe water. *J. Water Supply Res. Technol.* **2006**, *55*, 3–10. [CrossRef]
5. Shang, W.; Deng, T. Solar steam generation: Steam by thermal concentration. *Nat. Energy* **2016**, *1*, 16133. [CrossRef]
6. Ghasemi, H.; Ni, G.; Marconnet, A.M.; Loomis, J.; Yerci, S.; Miljkovic, N.; Chen, G. Solar steam generation by heat localization. *Nat. Commun.* **2014**, *5*, 4449. [CrossRef]
7. Li, X.; Xu, W.; Tang, M.; Zhou, L.; Zhu, B.; Zhu, S.; Zhu, J. Graphene oxide-based efficient and scalable solar desalination under one sun with a confined 2D water path. *Proc. Natl. Acad. Sci. USA* **2016**, *113*, 13953–13958. [CrossRef] [PubMed]
8. Ito, Y.; Tanabe, Y.; Han, J.; Fujita, T.; Tanigaki, K.; Chen, M. Multifunctional Porous Graphene for High-Efficiency Steam Generation by Heat Localization. *Adv. Mater.* **2015**, *27*, 4302–4307. [CrossRef]
9. Zhang, J.T.; Qin, Y.J.; Pan, W.; Wang, Z.B.; Qi, Y.; He, J.X.; Zhang, H.Q. Dual-Driven Functional Fabric with High Electrothermal and Photothermal Conversion Efficiency Modified by CuS Nanoparticles. *ACS Appl. Polym. Mater.* **2023**, *5*, 5747–5757. [CrossRef]
10. Simayee, M.; Zad, A.I.; Esfandiar, A. Green synthesize of copper nanoparticles on the cotton fabric as a self-regenerating and high-efficient plasmonic solar evaporator. *Sci. Rep.* **2023**, *13*, 12762. [CrossRef]
11. Simayee, M.; Zad, A.I.; Esfandiar, A. Boosting-photothermal properties of Cu/Black TiO_2 nanoparticles on biomimetics texture structure as high-performance and self-regenerating solar-evaporator. *Sol. Energy* **2023**, *265*, 112097. [CrossRef]
12. Shafaee, M.; Niazi, Z.; Asarnia, M.; Goharshadi, E.K.; Dehghani, R. Modified pine cone with MnO_2 nanoparticles as a photoabsorber for highly efficient seawater desalination and wastewater treatment. *Appl. Phys. A* **2023**, *129*, 656. [CrossRef]
13. Mitra, D.; Chanda, K.; Bhattacharjee, S.; Bairi, P.; Chattopadhyay, K.K.; Chattopadhyay, P. Enhanced interfacial evaporation and desalination by solar heat localisation using nitrogenated graphitic carbon and Co_3O_4 nanorods. *Sol. Energy Mater. Sol. Cells* **2023**, *257*, 112361. [CrossRef]
14. Jiang, G.; Fang, X.; Yu, W.; Xie, H.; Lei, H. Magnetic recyclable Fe_3O_4@$Ti_3C_2T_X$ nanoparticles for high-efficiency solar membrane distillation. *Desalination* **2023**, *564*, 116784. [CrossRef]
15. Joo, B.S.; Kim, I.S.; Han, I.K.; Ko, H.; Kang, J.G.; Kang, G. Plasmonic silicon nanowires for enhanced heat localization and interfacial solar steam generation. *Appl. Surf. Sci.* **2022**, *583*, 152563. [CrossRef]
16. Chen, R.; Wang, X.; Gan, Q.M.; Zhang, T.Q.; Zhu, K.H.; Ye, M.M. A bifunctional MoS_2-based solar evaporator for both efficient water evaporation and clean freshwater collection. *J. Mater. Chem. A* **2019**, *7*, 11177–11185. [CrossRef]
17. Zhou, L.; Tan, Y.; Wang, J.; Xu, W.; Yuan, Y.; Cai, W.; Zhu, S.; Zhu, J. 3D self-assembly of aluminium nanoparticles for plasmon-enhanced solar desalination. *Nat. Photonics* **2016**, *10*, 393–398. [CrossRef]
18. Liu, Y.; Yu, S.; Feng, R.; Bernard, A.; Liu, Y.; Zhang, Y.; Duan, H.; Shang, W.; Tao, P.; Song, C.; et al. A Bioinspired, Reusable, Paper-Based System for High-Performance Large-Scale Evaporation. *Adv. Mater.* **2015**, *27*, 2768. [CrossRef] [PubMed]
19. Wang, Z.; Liu, Y.; Tao, P.; Shen, Q.; Yi, N.; Zhang, F.; Liu, Q.; Song, C.; Zhang, D.; Shang, W.; et al. Bio-Inspired Evaporation Through Plasmonic Film of Nanoparticles at the Air-Water Interface. *Small* **2014**, *10*, 3234–3239. [CrossRef]
20. Lead, J.R.; Batley, G.E.; Alvarez, P.J.J.; Croteau, M.-N.; Handy, R.D.; McLaughlin, M.J.; Judy, J.D.; Schirmer, K. Nanomaterials in the environment: Behavior, fate, bioavailability, and effectsAn updated review. *Environ. Toxicol. Chem.* **2018**, *37*, 2029–2063. [CrossRef]
21. Selck, H.; Handy, R.D.; Fernandes, T.F.; Klaine, S.J.; Petersen, E.J. Nanomaterials in the Aquatic Environment: A European Union-United States Perspective on the Status of Ecotoxicity Testing, Research Priorities, and Challenges Ahead. *Environ. Toxicol. Chem.* **2016**, *35*, 1055–1067. [CrossRef]
22. Ryu, J.H.; Lee, Y.; Kong, W.H.; Kim, T.G.; Park, T.G.; Lee, H. Catechol-Functionalized Chitosan/Pluronic Hydrogels for Tissue Adhesives and Hemostatic Materials. *Biomacromolecules* **2011**, *12*, 2653–2659. [CrossRef] [PubMed]
23. Li, Y.; Xu, H.; Li, H.; Zhong, S. Controlled preparation and photothermal properties of polydopamine submicrospheres. *Inorg. Chem. Commun.* **2021**, *124*, 108395. [CrossRef]
24. Liu, Y.; Ai, K.; Lu, L. Polydopamine and Its Derivative Materials: Synthesis and Promising Applications in Energy, Environmental, and Biomedical Fields. *Chem. Rev.* **2014**, *114*, 5057–5115. [CrossRef]
25. Amin, D.R.; Higginson, C.J.; Korpusik, A.B.; Gonthier, A.R.; Messersmith, P.B. Untemplated Resveratrol-Mediated Polydopamine Nanocapsule Formation. *ACS Appl. Mater. Interfaces* **2018**, *10*, 34792–34801. [CrossRef] [PubMed]
26. Amin, D.R.; Sugnaux, C.; Lau, K.H.A.; Messersmith, P.B. Size Control and Fluorescence Labeling of Polydopamine Melanin-Mimetic Nanoparticles for Intracellular Imaging. *Biomimetics* **2017**, *2*, 17. [CrossRef] [PubMed]
27. Liu, Y.; Ai, K.; Liu, J.; Deng, M.; He, Y.; Lu, L. Dopamine-Melanin Colloidal Nanospheres: An Efficient Near-Infrared Photothermal Therapeutic Agent for In Vivo Cancer Therapy. *Adv. Mater.* **2013**, *25*, 1353–1359. [CrossRef]
28. Ju, K.-Y.; Lee, Y.; Lee, S.; Park, S.B.; Lee, J.-K. Bioinspired Polymerization of Dopamine to Generate Melanin-Like Nanoparticles Having an Excellent Free-Radical-Scavenging Property. *Biomacromolecules* **2011**, *12*, 625–632. [CrossRef]
29. Yazdi, M.K.; Zare, M.; Khodadadi, A.; Seidi, F.; Sajadi, S.M.; Zarrintaj, P.; Arefi, A.; Saeb, M.R.; Mozafari, M. Polydopamine Biomaterials for Skin Regeneration. *ACS Biomater. Sci. Eng.* **2022**, *8*, 2196–2219. [CrossRef] [PubMed]

30. Sun, F.; Lu, J.; Wang, Y.; Xiong, J.; Gao, C.; Xu, J. Reductant-assisted polydopamine-modified membranes for efficient water purification. *Front. Chem. Sci. Eng.* **2021**, *15*, 109–117. [CrossRef]
31. Postma, A.; Yan, Y.; Wang, Y.; Zelikin, A.N.; Tjipto, E.; Caruso, F. Self-Polymerization of Dopamine as a Versatile and Robust Technique to Prepare Polymer Capsules. *Chem. Mater.* **2009**, *21*, 3042–3044. [CrossRef]
32. Berthier, E.; Young, E.W.K.; Beebe, D. Engineers are from PDMS-land, Biologists are from Polystyrenia. *Lab Chip* **2012**, *12*, 1224–1237. [CrossRef] [PubMed]
33. An, S.; Lim, J.; Choi, D.; Hong, H.; Kim, H.W.; Park, S.M.; Rhie, J.W.; Kim, D.S. Fabrication of polystyrene-based multi-well screening platform for micrometer-scale surface topographies promoting stem cell functions. *Microelectron. Eng.* **2017**, *174*, 28–34. [CrossRef]
34. Marquez, C.; Martin, C.; Linares, N.; De Vos, D. Catalytic routes towards polystyrene recycling. *Mater. Horiz.* **2023**, *10*, 1625–1640. [CrossRef] [PubMed]
35. Ahmed, D.S.; El-Hiti, G.A.; Yousif, E.; Hameed, A.S. Polyphosphates as Inhibitors for Poly(vinyl Chloride) Photodegradation. *Molecules* **2017**, *22*, 1849. [CrossRef] [PubMed]
36. Qiu, J.; Shi, Y.; Xia, Y. Polydopamine Nanobottles with Photothermal Capability for Controlled Release and Related Applications. *Adv. Mater.* **2021**, *33*, 2104729. [CrossRef] [PubMed]
37. Zhu, Y.; Cheng, Z.; Weng, W.; Cheng, K. A facile synthesis of polydopamine/TiO_2 composite films for cell sheet harvest application. *Colloids Surf. B* **2018**, *172*, 355–361. [CrossRef] [PubMed]
38. Wang, R.; Long, Y.H.; Zhu, T.; Guo, J.; Cai, C.; Zhao, N.; Xu, J. Fabrication of oriented wrinkles on polydopamine/polystyrene bilayer films. *J. Colloid Interface Sci.* **2017**, *498*, 123–127. [CrossRef]
39. Moon, S.; Lee, W.; Ahn, Y. Fabrication of Superhydrophobic Surface on Polydopamine-coated Al Plate by Using Modified SiO_2 Nanoparticles/Polystyrene Nano-Composite Coating. *Bull. Korean Chem. Soc.* **2016**, *37*, 600–603. [CrossRef]
40. Li, Z.; Yang, Y.; Wang, Z.; Zhang, X.; Chen, Q.; Qian, X.; Liu, N.; Wei, Y.; Ji, Y. Polydopamine nanoparticles doped in liquid crystal elastomers for producing dynamic 3D structures. *J. Mater. Chem. A* **2017**, *5*, 6740–6746. [CrossRef]
41. Li, Z. Polydopamine Based Photo-Responsive Shape Memory Polymer and the Achievement of Specific 3D Structures. Ph.D. Thesis, Tsinghua University, Beijing, China, 2019.
42. Salt Fusion: An Approach to Improve Pore Interconnectivity within Tissue Engineering Scaffolds. *Tissue Eng.* **2002**, *8*, 43–52. [CrossRef] [PubMed]
43. Xi, Z.Y.; Xu, Y.Y.; Zhu, L.P.; Wang, Y.; Zhu, B.K. A facile method of surface modification for hydrophobic polymer membranes based on the adhesive behavior of poly(DOPA) and poly(dopamine). *J. Membr. Sci.* **2009**, *327*, 244–253. [CrossRef]
44. Chen, Q.M.; Pei, Z.Q.; Xu, Y.S.; Li, Z.; Yang, Y.; Wei, Y.; Ji, Y. A durable monolithic polymer foam for efficient solar steam generation. *Chem. Sci.* **2018**, *9*, 1392. [CrossRef]

polymers

MDPI

Article

Molecular Dynamics Simulation of Hydrogen Barrier Performance of Modified Polyamide 6 Lining of IV Hydrogen Storage Tank with Graphene

Jin Li [1,2], Xiaokou Zhao [1], Jianguo Liang [3], Chunjiang Zhao [1,4,*], Ning Feng [1], Guanyu Guo [1] and Zhengze Zhou [1]

1 College of Mechanical Engineering, Taiyuan University of Science and Technology, Taiyuan 033024, China; s202112210646@stu.tyust.edu.cn (G.G.)
2 Department of Resource and Mechanical Engineering, Lyuliang University, Lvliang 033001, China
3 College of Mechanical and Vehicle Engineering, Taiyuan University of Technology, Taiyuan 030024, China
4 College of Intelligent Manufacturing Industry, Shanxi University of Electronic Science and Technology, Linfen 041000, China
* Correspondence: zhaochj75@163.com

Abstract: The polymer liner of the hydrogen storage cylinder was studied to investigate better hydrogen storage capacity in Type-IV cylinders. Molecular dynamics methods were used to simulate the adsorption and diffusion processes of hydrogen in a graphene-filled polyamide 6 (PA6) system. The solubility and diffusion characteristics of hydrogen in PA6 systems filled with different filler ratios (3 wt%, 4 wt%, 5 wt%, 6 wt%, and 7 wt%) were studied under working pressures (0.1 MPa, 35 MPa, 52 MPa, and 70 MPa). The effects of filler ratio, temperature, and pressure on hydrogen diffusion were analyzed. The results show that at atmospheric pressure when the graphene content reaches 5 wt%, its permeability coefficient is as low as 2.44×10^{-13} $cm^3 \cdot cm/(cm^2 \cdot s \cdot Pa)$, which is a 54.6% reduction compared to PA6. At 358 K and 70 MPa, the diffusion coefficient of the 5 wt% graphene/PA6 composite system is 138% higher than that at 298 K and 70 MPa. With increasing pressure, the diffusion coefficients of all materials generally decrease linearly. Among them, pure PA6 has the largest diffusion coefficient, while the 4 wt% graphene/PA6 composite system has the smallest diffusion coefficient. Additionally, the impact of FFV (free volume fraction) on the barrier properties of the material was studied, and the movement trajectory of H_2 in the composite system was analyzed.

Keywords: polyamide 6; graphene; hydrogen diffusion; molecular dynamics; free volume

1. Introduction

Hydrogen is a crucial component of global decarbonization strategies due to its carbon-free, efficient, and renewable nature. It will play a central role in the ongoing development and successful transition of traditional energy systems, effectively mitigating the negative impacts of carbon dioxide emissions, such as global warming [1,2]. In the transportation sector, hydrogen fuel cell vehicles (HFCVs) have been extensively researched. Many companies, including Toyota, Honda, and Hyundai, are dedicated to developing high-performance hydrogen fuel cell vehicles [3].

Hydrogen, as fuel for hydrogen fuel cell vehicles, is typically compressed and stored in pressure vessels, which is currently the mainstream technology for onboard hydrogen storage. Among these, Type III and Type IV pressure vessels are the most widely used [4,5]. Compared to Type III hydrogen storage cylinders with metal liners, Type IV hydrogen storage cylinders use polymer liners [6], including high-density polyethylene (HDPE), polyamide (PA), polyethylene terephthalate (PET), and various polyether materials. These Type IV cylinders offer numerous advantages, such as high hydrogen storage density, lightweight, corrosion resistance, and fatigue resistance [7,8].

Polyamide (PA6) has gradually become a potential choice for Type IV hydrogen storage vessel liners due to its strong molecular polarity and hydrogen bonding interactions [9,10]. However, Type IV hydrogen storage vessels face challenges where the polymer liner directly contacts hydrogen while being bonded to an external fiber-wound layer. In high-pressure hydrogen environments, hydrogen permeates into the material, diffuses to the plastic-composite interface, and creates internal pressure. Upon depressurization, hydrogen escapes from the material to the exterior, creating a pressure differential at the interface, which leads to liner deformation and bulging [11,12]. Understanding the hydrogen transport properties of polymers under harsh operating conditions (233–358 K, 0–87 MPa) is crucial [13,14]. Some researchers have studied polymer permeability to hydrogen, focusing on external conditions (including temperature [15,16] and pressure [17,18]), gas-material interactions [19,20], material properties (including crystallinity [21–23], polymer molecular weight [20,24], filler structure [23,25,26], filler content [23,27,28]), and others. However, further research is needed to understand the hydrogen permeation characteristics of PA6 fully.

Using the Einstein relation [29], the molecular dynamics (MD) software Materials Studio (2019) is employed to simulate the structure-property relationships and study the diffusion coefficients of small gas molecules in polymer material models. This method effectively predicts the hydrogen permeability of polymer liners in hydrogen storage cylinders [20,30]. Wu [31] investigated the adsorption and diffusion processes of hydrogen in PA6 systems filled with modified montmorillonite (OMMT) under different filler contents (3–7%), temperatures (288–328 K), and pressures (0–60 MPa). The results indicated that at a filler content of 5%, the material's permeability coefficient was less than 2×10^{-13} $cm^3{\cdot}cm/(cm^2{\cdot}s{\cdot}Pa)$. Additionally, as the pressure increased, the permeability coefficient of the modified material first decreased and then increased. Su [14] comprehensively explored the dissolution and diffusion behavior of H_2 in PA6 under service conditions (233–358 K, 0–87 MPa), finding that the diffusion coefficient and permeability coefficient were positively correlated with temperature, while the solubility coefficient was also positively correlated with temperature. MURARU [28] evaluated the gas permeability of PSF-cnt-g membranes and compared them with three other membranes (PSF, PSF-cnt, and PSF-g), discovering that the addition of carbon nanotubes and graphene to polysulfone membranes significantly increased the diffusion coefficients of gases such as CH_4, CO_2, H_2, N_2, and O_2. Yi [23] studied the diffusion of gases, including hydrogen and its isotopes, at room temperature and pressure, noting that the diffusion decreased with increasing polystyrene molecular weight. Zhang [32] analyzed the diffusion characteristics of hydrogen in HDPE under temperature (room temperature to 80 °C) and pressure (2.5–10 MPa) conditions. The results showed that when the temperature increased from 30 °C to 80 °C, hydrogen's solubility, diffusion coefficient, and permeability coefficient in HDPE increased by 18.7%, 92.9%, and 129.0%, respectively. Fang [12] simulated the diffusion and adsorption processes of hydrogen molecules in polyethylene (PE) and PA6 at temperatures ranging from 263 K to 353 K. The results indicated that under the same conditions, the solubility, diffusion, and permeability coefficients of hydrogen in PE were higher than those in PA6, suggesting that PA6 has better hydrogen barrier properties than PE. Hu [33] studied the diffusion characteristics and local structure of a mixed system composed of CH_4, CO_2, SO_2, and H_2O, finding that higher temperatures and lower pressures favored gas diffusion.

Previous studies have shown that the diffusion characteristics of polymers are related to additives, different temperatures, and pressures. Although the aforementioned simulations focus on gas diffusion in various polymers under different temperatures and pressures, the impact of graphene-modified PA6 on H_2 diffusion at extreme temperatures (233 K–358 K) and operating pressures of gas cylinders (0.1–70 MPa) has not been systematically investigated.

Materials Studio (2019) was used in this study to construct a molecular model of a composite material with PA6 as the polymer liner and graphene as the nanofiller. Analyzing the permeation behavior of H_2 in modified PA6 materials from a microscopic perspective to elucidate the characteristics and mechanisms of hydrogen permeation seems to be a very interesting endeavor. This approach can provide a method for evaluating the performance of liner materials for Type IV hydrogen storage cylinders. Our research aims to offer more evidence for the selection of liner materials for hydrogen storage cylinders.

2. Model and Simulation Methods

2.1. Establishment of Modified Liner Models

The MD simulation study utilized amorphous unit cells to construct polymer liners. Single-chain structures were built using the Builder module, and monomer structures were constructed using the Sketch toolbar, establishing 60 repeated units to form random single-chain molecules of PA6. The AC module was employed to generate models of five different systems, each containing 20 H_2 molecules and identical PA6 chains of 8 repeated units [28,31,34]. Hydrogen positions were randomly distributed to investigate their diffusion within the liner models.

In the composite model systems, graphene, initially composed of single sheets containing 48 carbon atoms, was introduced. Subsequently, the graphene mass fraction was varied by gradually increasing the number of graphene sheets to achieve mass fractions of 3 wt%, 4 wt%, 5 wt%, 6 wt%, and 7 wt%. The unit cell dimensions for different mass fractions were approximately 44.06 Å, 44.11 Å, 44.06 Å, 44.32 Å, and 44.40 Å, with identical lattice dimensions in the a, b, and c directions for all unit cells. Figure 1 illustrates the structures of PA6 chains and graphene. Figure 2 depicts the construction of mixed models of PA6 chains and graphene using the random copolymer option.

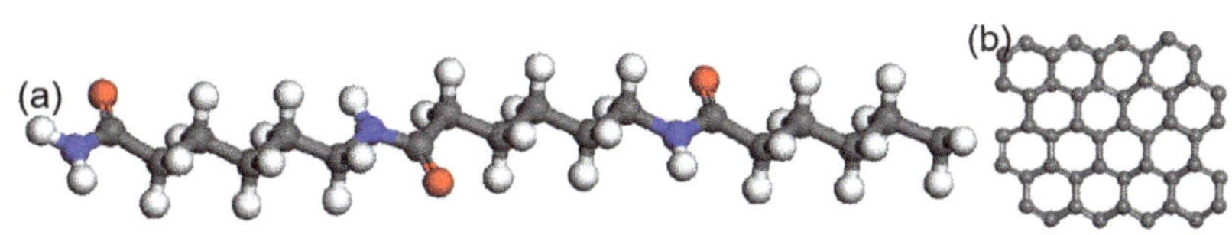

Figure 1. (**a**) PA6 repeating unit (gray atoms—carbon, white atoms—hydrogen, red atoms—oxygen, blue atoms—nitrogen). (**b**) Graphene sheet.

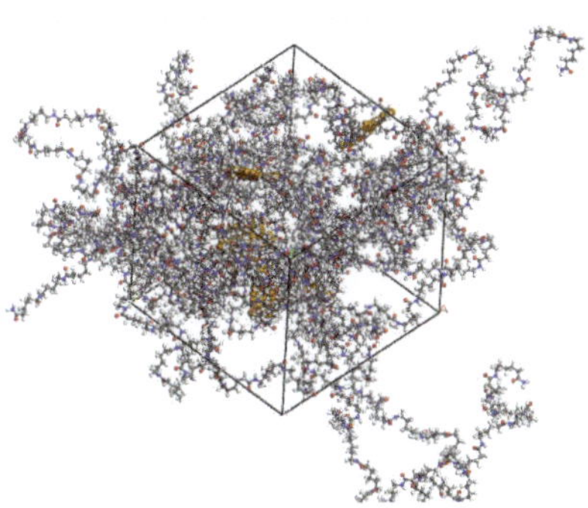

Figure 2. Model of graphene-modified PA6 (yellow atom-graphene sheet).

In the composite model systems, graphene, initially composed of single sheets containing 48 carbon atoms, was introduced. Subsequently, the graphene mass fraction was varied by gradually increasing the number of graphene sheets to achieve mass fractions of 3 wt%, 4 wt%, 5 wt%, 6 wt%, and 7 wt%. The unit cell dimensions for different mass fractions were approximately 44.06 Å, 44.11 Å, 44.06 Å, 44.32 Å, and 44.40 Å, with identical lattice dimensions in the a, b, and c directions for all unit cells. Figure 1 illustrates the structures of PA6 chains and graphene. Figure 2 depicts the construction of mixed models of PA6 chains and graphene using the random copolymer option.

2.2. Simulation Method

The initial models underwent geometric optimization using the Smart method with energy minimization applied to all monomers and models to eliminate local non-equilibrium states. The maximum iteration step was set to 5×10^4 to achieve stability across all models. Subsequently, a cyclic annealing process was applied to the polymer material models to obtain more realistic polymer structures. According to the simulated annealing method, starting from 300 K and ramping up to a midpoint temperature of 600 K, a total of 30 annealing cycles were conducted. This thermal annealing reduced internal stresses in the modified liner models and nearly eliminated any structurally unreasonable configurations generated during optimization.

The optimized models were subjected to molecular dynamics (MD) simulations using the COMPASS II force field [35]. The simulations were conducted in the NVT [12,14,20] ensemble at an initial temperature of 298 K for a relaxation period of 1 ns. The MD simulations aimed to stabilize the energy and density of the entire system. During MD simulations, the Andersen temperature control method and Berendsen pressure control were applied to maintain the models' constant temperature [36–38]. The Group-Based [39] method was utilized to calculate non-bonded interactions and Coulomb forces. Integration of the equations of motion was performed with a time step of 1 fs for all simulations [38–41].

Figure 3 illustrates the change in total energy during the dynamic processing. Over time, the total energy of the system stabilizes, oscillating within a narrow range around a fixed value. This indicates that the models have achieved full relaxation and obtained stable structures.

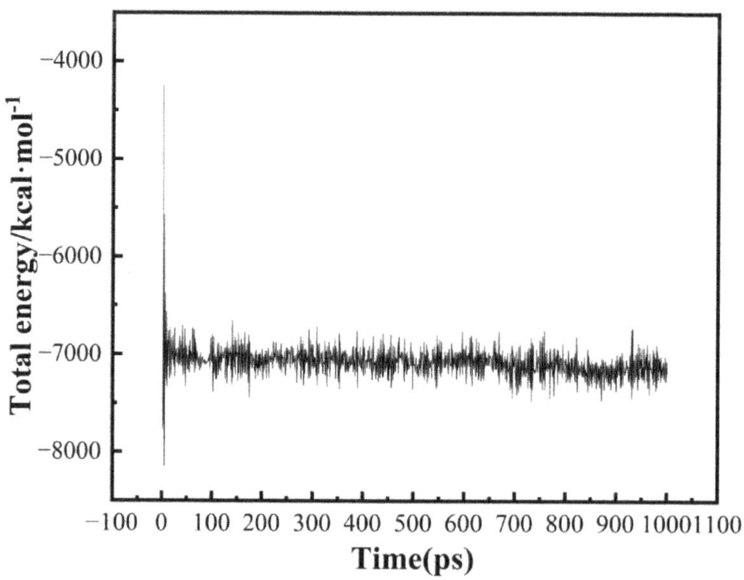

Figure 3. Total energy changes in the PA6 model during dynamic treatment.

2.3. Model Reliability Verification

Taking PA6 as an example, it exhibits a fully amorphous structure. Figure 4 depicts the density variations observed throughout the relaxation process. It can be observed that the final density of PA6 is 1.095 g/cm^3. Comparing this with the actual density of 1.13 g/cm^3 for PA6 in an amorphous state, the relative error is calculated to be 3.1%, indicating the validity of the models proposed in this study.

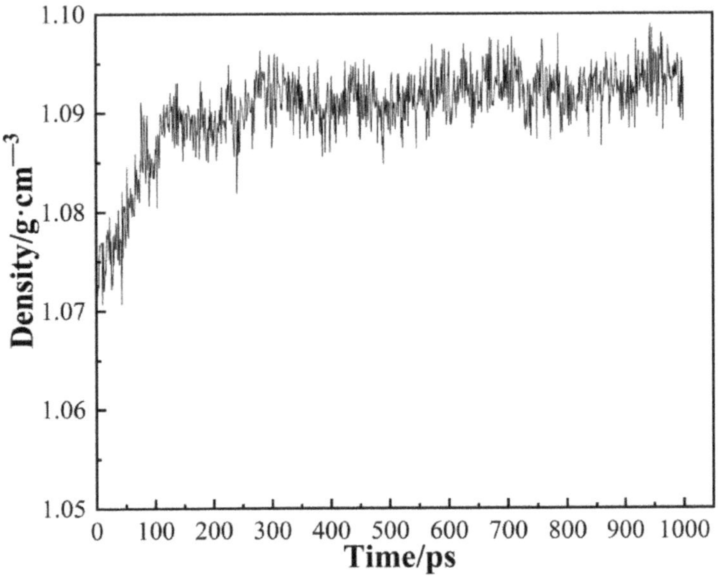

Figure 4. The density of PA6 during the relaxation process (Conditions: 298 K and 0.1 MPa).

The solubility coefficients of H_2 in the six systems were obtained using adsorption isotherms within the pressure range of 0.01 kPa to 10,000 kPa. According to the Chinese National Standard (GB/T 42612-2023) [42], Type IV hydrogen storage cylinders operate at a maximum working pressure of 70 MPa, with a working temperature range not lower than −40 °C and not higher than 85 °C. Exceeding this temperature range can cause irreversible damage to the carbon fibers in the gas cylinder. Therefore, diffusion analysis was conducted at three temperatures: 233 K, 298 K, and 358 K. Therefore, to determine the diffusion coefficients under different conditions, pressures of half the maximum working pressure of Type IV hydrogen storage cylinders (35 MPa) and an intermediate value of 52.5 MPa (between 35 MPa and 70 MPa) were selected. The model was subjected to four different pressures (0.1 MPa, 35 MPa, 52.5 MPa, and 70 MPa). One ns molecular dynamics simulations were performed under the NPT ensemble for six different systems. The mean square displacement (MSD) curves of hydrogen molecules were calculated, and the diffusion coefficients of H_2 were obtained.

3. Theoretical Basis of Permeability

3.1. Adsorption Concentration

The determination of the solubility coefficient (S) of gas molecules in polymers can be achieved through the application of adsorption isotherms. Under constant temperature conditions, adsorption isotherms are derived based on the concentration corresponding to different solubilities. Henry's law applies to systems with the smallest solubility and can be used to explain the dissolution process of small gas molecules such as hydrogen in polymers [41,43]. This relationship is expressed in Equation (1). It is noteworthy that when the fugacity approaches zero, the solubility coefficient can be determined as the limiting slope of the adsorption isotherm, as shown in Equation (2).

$$C = K_D P + C_H \frac{bP}{1 + bP} \tag{1}$$

$$S = \lim_{P \to 0} \frac{C}{P} = K_D + C_H P \tag{2}$$

where K_D is the Henry constant, C_H is the Langmuir capacity parameter, b is the Langmuir parameter, P is the pressure, and C is the adsorption of gas molecules.

3.2. Diffusion Coefficient

The diffusion coefficient (D) describes the Brownian motion of particles under concentration gradient conditions. It represents the dynamic characteristics of interaction between permeable gas molecules and polymers. The diffusion coefficient D can be determined by analyzing the MSD of molecular motion. In MD simulations, the correlation between MSD and molecular motion time is established by tracking the motion trajectory of permeating molecule centers. Subsequently, the diffusion coefficient is calculated using the Einstein formula [24], as shown in Equation (3):

$$D = \frac{1}{6N} \lim_{t \to \infty} \frac{d}{dt} \left\langle \sum_{i}^{N} [r_i(t) - r_i(0)]^2 \right\rangle = \frac{a}{6} \tag{3}$$

where D is the diffusion coefficient, $r_i(t)$ and $r_i(0)$ are the position vectors of molecule i at times t and 0, respectively. N represents the total number of gas particles, and $\left\langle [r_i(t) - r_i(0)]^2 \right\rangle$ denotes the ensemble average of the MSD of gas molecules. Gas molecules collide within a small pocket of free volume, jumping from one confined region to another; the repeated jumps of molecules constitute diffusion, which is characterized by the MSD, and a represents the gradient of the MSD curve obtained from molecular simulations.

3.3. Permeability Coefficient

The "dissolution-diffusion" theory can describe the permeation process of gas molecules in polymers. Gas molecules dissolve from the surface into the polymer and diffuse from one side of the polymer to the other. Finally, gas molecules desorb and escape from the polymer [44]. Therefore, the permeation process of H_2 in polymers can be divided into dissolution and diffusion processes. The permeability coefficient [45] is the product of the solubility and diffusion coefficients, as shown in Equation (4).

$$P = S \times D \tag{4}$$

where P is permeability coefficient, $cm^3 \cdot cm/(cm^2 \cdot s \cdot Pa)$, S is the solubility coefficient, $cm^3 \cdot cm^{-3} \cdot Pa^{-1}$, and D is the diffusion coefficient, cm^2/s.

3.4. Free Volume

To obtain the free fraction volume (FFV) of the polymer [12,31,46], an analysis of the five models was conducted using the Atom V volume and Surface tools. The Connolly surface of the system was calculated based on the known van der Waals radius of hydrogen atoms, with a Connolly radius of 1.4 Å, to determine the system's FFV.

4. Results and Discussion

4.1. Effect of Filler Ratio on Permeation Coefficient

Figures 5 and 6 illustrate the isothermal adsorption curves and MSD curves of H_2 in various graphene/PA6 systems under conditions of 298 K and 0.1 MPa. The simulated data were fitted, and solubility coefficient S and diffusion coefficient D were calculated using Equations (2) and (3), respectively. Permeability coefficient P was then derived using Equation (4), and the calculated results are presented in Table 1.

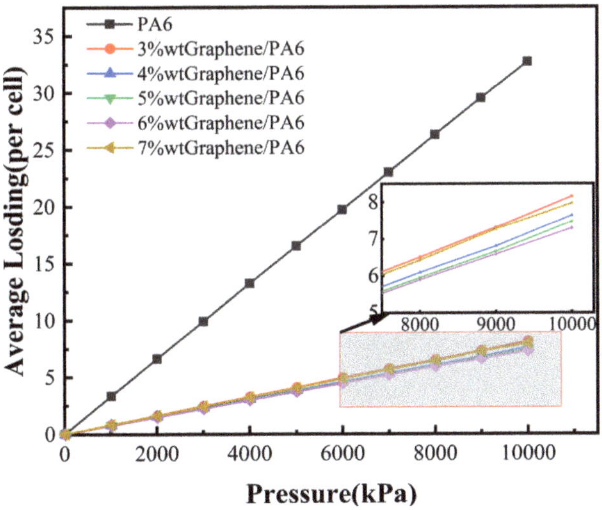

Figure 5. Isothermal Adsorption Curves of H_2 in Different Systems at 298 K.

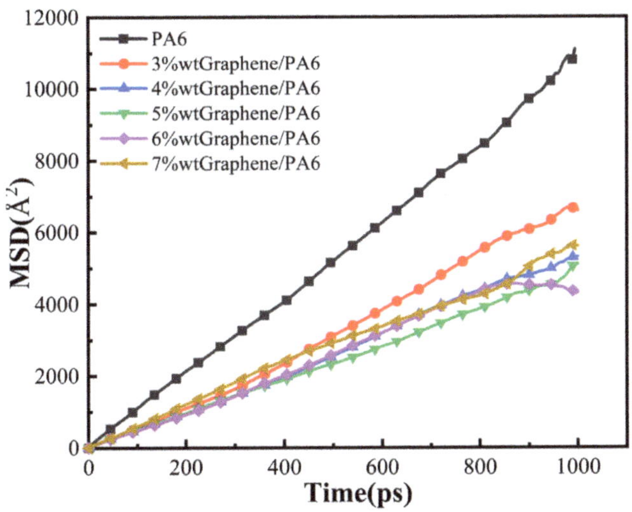

Figure 6. MSD Curves of H_2 in Different Systems (Conditions: 298 K and 0.1 MPa).

Table 1. Dissolution, diffusion coefficients, and permeability coefficients of H_2 in different material systems.

Material System	Solubility ($cm^3 \cdot cm^{-3} \cdot Pa^{-1}$)	Diffusion Coefficient (cm^2/s)	Permeation Coefficient ($cm^3 \cdot cm/(cm^2 \cdot s \cdot Pa)$)
PA6	1.32×10^{-6}	1.785×10^{-6}	2.356×10^{-12}
3%graphene/PA6	3.33×10^{-7}	9.34×10^{-7}	3.11×10^{-13}
4%graphene/PA6	3.22×10^{-7}	9.19×10^{-7}	2.95×10^{-13}
5%graphene/PA6	3.02×10^{-7}	8.09×10^{-7}	2.44×10^{-13}
6%graphene/PA6	3.02×10^{-7}	8.56×10^{-7}	2.58×10^{-13}
7%graphene/PA6	3.28×10^{-7}	8.99×10^{-7}	2.95×10^{-13}

Based on Figure 5, it is evident that the adsorption concentration of H_2 in the polymer shows an approximately linear relationship with pressure. Additionally, PA6 exhibits the highest adsorption concentration. Combined with Table 2, it is observed that the solubility coefficient varies with the addition of graphene in different graphene/PA6 composite systems. When the graphene content reaches 6 wt%, the adsorption of H_2 decreases to its minimum, with a solubility coefficient of 3.01×10^{-7} $cm^3 \cdot cm^{-3} \cdot Pa^{-1}$, representing a reduction of 77.2% compared to pure PA6. The solubility coefficient reflects the thermodynamic interactions between gas molecules and polymer chains. According to the free volume theory, permeating gas molecules generally occupy less dense regions within the polymer, known as free volume, through physical adsorption. With the addition of graphene filler, interactions between the filler and PA6 chains restrict the mobility of polymer chains, thereby reducing the formation of free volume within the system. Consequently, the available area for gas adsorption within the material decreases, impacting its permeability to leaking gases.

Table 2. Diffusion coefficients D (cm^2/s) of six systems at different test temperatures under 70 MPa.

Material System	PA6	3 wt%	4 wt%	5 wt%	6 wt%	7 wt%
233 K	3.14×10^{-7}	2.89×10^{-7}	1.97×10^{-7}	3.05×10^{-7}	3.33×10^{-7}	3.35×10^{-7}
298 K	7.35×10^{-7}	6.9×10^{-7}	6.09×10^{-7}	5.28×10^{-7}	6.32×10^{-7}	7.10×10^{-7}
358 K	2.28×10^{-6}	1.02×10^{-6}	1.08×10^{-6}	$1.26 \times 10-6$	1.36×10^{-6}	1.34×10^{-6}

At a content of 3 wt%, the solubility coefficient reaches its maximum at 3.33×10^{-7} $cm^3 \cdot cm^{-3} \cdot Pa^{-1}$, which is not significantly different from the solubility coefficient of the 6 wt% graphene/PA6 composite system. This suggests that when the graphene mass fraction exceeds 3 wt%, the solubility coefficient is not the primary factor influencing the permeation performance of H_2 in the composite system. This finding is consistent with the results reported by Zheng [47].

Figure 6 shows that the MSD curve exhibits a linear relationship with time. Before Einstein diffusion, anomalous diffusion may occur due to structural reasons within the system. It is essential to characterize the system diffusion properties to verify if the system has undergone normal diffusion.

The verification can be conducted by examining the slope of the logarithmic plot of MSD against time [48,49]. Figure 7 depicts the logarithmic plot log(MSD) vs. log(t) for determining diffusion coefficients using the Einstein relationship. Here, the log(MSD) vs. log(t) curve primarily assesses whether H_2 molecules in the graphene/PA6 composite systems transition from anomalous diffusion ($K < 1$) to normal diffusion ($K > 1$). It is observed that H_2 undergoes this transition in various graphene/PA6 composite systems with different filler ratios, allowing for the calculation of diffusion coefficients using the Einstein relationship.

The diffusion coefficient reflects the dynamic interactions between gas molecules and polymer chains. Figure 6 and Table 1 demonstrate that the introduction of graphene reduces the diffusion coefficients of H_2 in all systems. As the proportion of graphene increases, the diffusion coefficients initially decrease and then increase, which affects the permeability of the materials. Specifically, when graphene reaches 5 wt%, the composite system exhibits the lowest diffusion coefficient, leading to the minimum permeability. At this composition, the permeability of the 5 wt% graphene/PA6 composite for H_2 is as low as 2.44×10^{-13} $cm^3 \cdot cm/(cm^2 \cdot s \cdot Pa)$, indicating optimal barrier properties under these conditions. Compared to pure PA6, the permeability of the 5 wt% graphene/PA6 composite is reduced by 54.6%. The reasons behind these results may be attributed to several factors. Firstly, the introduction of a small amount of filler disrupts the continuity of "pores" within the system to some extent, forming barriers (or "walls"). When H_2 molecules move within the polymer matrix, they encounter these graphene filler barriers, increasing the energy required for diffusion and complicating the diffusion pathways. This restriction limits the diffusion of H_2 molecules. Secondly, the presence of graphene reduces the free volume within the system. However, when the filler content is too high (e.g., 6 wt%), the randomly arranged graphene interacts with the polymer to form new "pores," increasing the internal free volume and consequently increasing the diffusion coefficient of H_2 molecules.

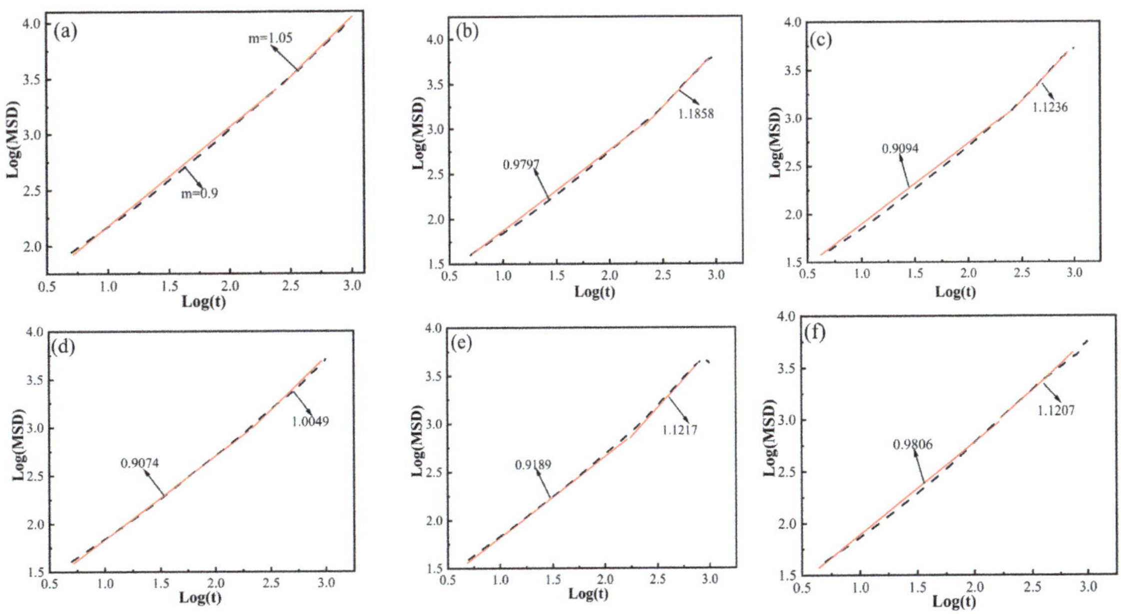

Figure 7. Logarithmic Plot of H_2 MSD vs. Simulation Time at 298 K and 0.1 MPa: (**a**) PA6, (**b**) 3 wt%, (**c**) 4 wt%, (**d**) 5 wt%, (**e**) 6 wt%, and (**f**) 7 wt%.

4.2. Impact of Temperature Diffusion Coefficients

Figure 8 and Table 2 present the diffusion coefficients of six systems at three different temperatures (233 K, 298 K, and 358 K). The results indicate that at 358 K and 70 MPa pressure, PA6 exhibits the highest diffusion coefficient of 2.28×10^{-6} cm^2/s. As shown in Figure 8, the diffusion coefficient of PA6 increases correspondingly with temperature. This phenomenon can be attributed to the energy required for hydrogen molecules to overcome inter-molecular interactions within the polymer matrix during their diffusion. At higher temperatures, the kinetic energy of polymer chains increases, facilitating greater mobility of hydrogen molecules within the polymer structure [10,50]. Therefore, the diffusion coefficient increases with temperature.

It is important to emphasize that Dong [8] experimentally measured the diffusion coefficient of PA6 at 358 K and 87.5 MPa to be 2.36×10^{-6} cm^2/s. Therefore, we specifically analyzed the diffusion coefficient of PA6 under these conditions, resulting in 2.11×10^{-6} cm^2/s. By comparing these two values, we found the error within 10.6%. In the context of gas diffusion coefficients at the small magnitude of 10^{-7}, errors within 20% are generally considered reasonable. Considering the crystallinity of PA6, these findings demonstrate that molecular dynamics simulations can effectively approximate the hydrogen barrier properties of the material. This also validates the reliability of the research methods and molecular dynamics simulation techniques used in this study.

Figure 8 and Table 2 show that the diffusion coefficients of all systems increase with temperature at 358 K compared to 233 K and 298 K. Specifically, the diffusion coefficient of 3 wt% graphene/PA6 at 358 K is 1.02×10^{-6} cm^2/s, which is 48% higher than at 298 K. Similarly, the diffusion coefficients of 4 wt% to 7 wt% graphene/PA6 increase by approximately 77%, 138%, 115%, and 89%, respectively, at 358 K compared to 298 K.

This illustrates the significant influence of temperature on the diffusion coefficients of the materials, a phenomenon supported by previous studies [41,42].

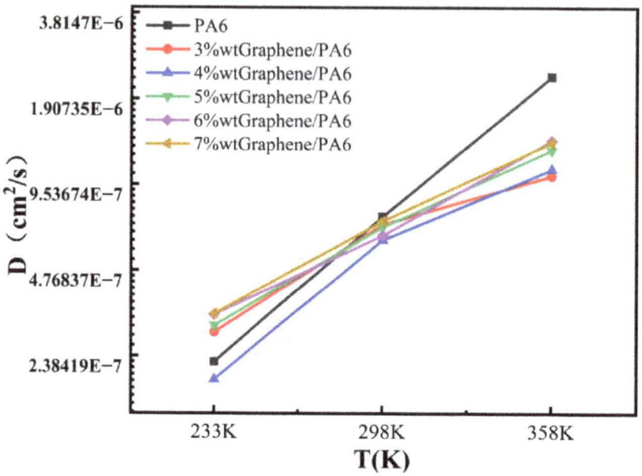

Figure 8. Diffusion coefficients of six systems at different temperatures.

4.3. Effect of Pressure on Diffusion Coefficients

Figure 5 demonstrates that at constant temperature, the adsorption of H_2 in the materials is approximately proportional to pressure. The solubility of H_2 in the materials shows almost no change with pressure; thus, pressure variations have minimal impact on the materials' solubility coefficient. Gas solubility is considered constant at different pressures because pressure does not affect the thermodynamic properties of gas molecules and polymers. Dong [8] analyzed the influence of different test temperatures and pressures on the solubility coefficient of PA6, noting a complex interaction between temperature and gas activity capacity affecting H_2 solubility in PA6. This results in limited effects of test temperature and pressure on the solubility coefficient. Fujiwara [50] conducted permeability coefficient experiments on HDPE under pressures ranging from 10 to 90 MPa, indicating that solubility coefficients do not significantly depend on pressure; changes in permeability coefficients are primarily diffusion-controlled.

Figure 9 illustrates the MSD curves of six material systems at 298 K under different pressures. It can be observed that as the pressure increases, the slope of the curves gradually decreases, indicating a significant influence of pressure on the diffusion characteristics of the materials. Figure 10 shows the calculated diffusion coefficients under different pressures, demonstrating a substantial impact of pressure on the diffusion of H_2 within the systems. As pressure increases, the diffusion coefficients of all materials generally decrease linearly. Specifically, pure PA6 exhibits the highest diffusion coefficient, while the 4 wt% graphene/PA6 composite system shows the lowest diffusion coefficient.

Table 3 presents the permeability coefficients of different systems at 298 K under various pressures. It can be observed that within the same system, permeability decreases with increasing pressure. At 0.1 MPa pressure, the permeability of 5 wt% graphene/PA6 is the lowest, at 2.44×10^{-13} $cm^3 \cdot cm/(cm^2 \cdot s \cdot Pa)$. At pressures above 35 MPa, the permeability of the 4 wt% graphene/PA6 is the lowest, with values of 2.15×10^{-13} $cm^3 \cdot cm/(cm^2 \cdot s \cdot Pa)$, 2.00×10^{-13} $cm^3 \cdot cm/(cm^2 \cdot s \cdot Pa)$, and 1.96×10^{-13} $cm^3 \cdot cm/(cm^2 \cdot s \cdot Pa)$, respectively. This is consistent with Dong et al.'s analysis of PA6 hydrogen permeation behavior under different pressures [8], which indicated a decreasing trend in permeability coefficient with increasing experimental pressure.

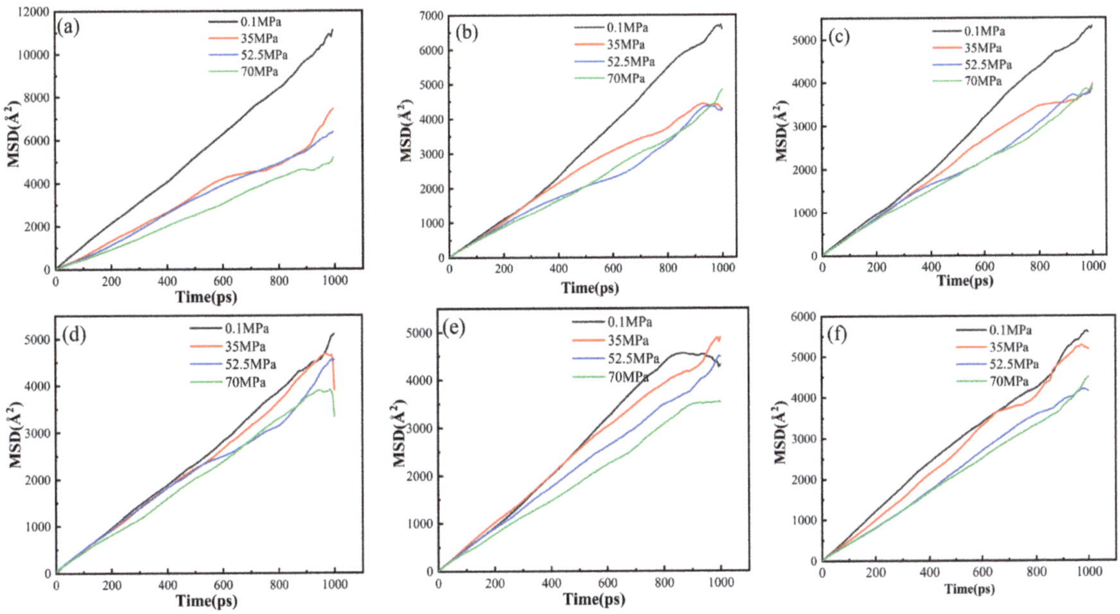

Figure 9. MSD Curves of Different Graphene Content Systems at 298 K: (**a**) PA6, (**b**) 3 wt%, (**c**) 4 wt%, (**d**) 5 wt%, (**e**) 6 wt%, and (**f**) 7 wt%.

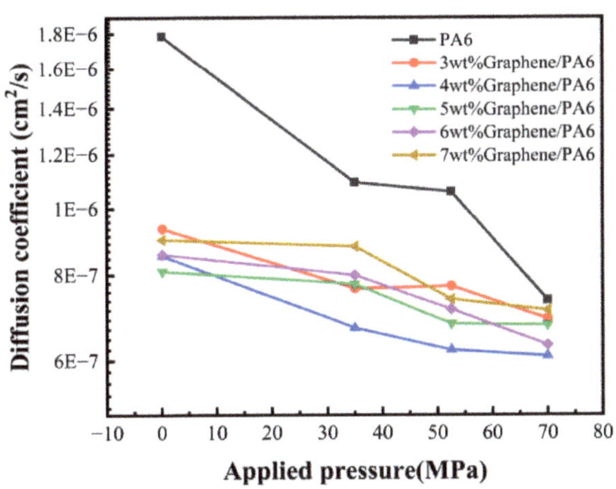

Figure 10. Diffusion Coefficients of 6 Systems at 298 K under Different Pressures.

Table 3. Permeability Coefficients of 6 Systems at 298 K under Different Pressures ($cm^3 \cdot cm/(cm^2 \cdot s \cdot Pa)$).

	PA6	3 wt%	4 wt%	5 wt%	6 wt%	7 wt%
0.1 MPa	2.35×10^{-12}	3.11×10^{-13}	2.74×10^{-13}	2.44×10^{-13}	2.58×10^{-13}	2.95×10^{-13}
35 MPa	1.44×10^{-12}	2.54×10^{-13}	2.15×10^{-13}	2.34×10^{-13}	2.41×10^{-13}	2.88×10^{-13}
52.5 MPa	1.39×10^{-12}	2.56×10^{-13}	2.00×10^{-13}	2.05×10^{-13}	2.15×10^{-13}	2.42×10^{-13}
70 MPa	9.70×10^{-13}	2.29×10^{-13}	1.96×10^{-13}	2.04×10^{-13}	1.90×10^{-13}	2.33×10^{-13}

For non-soluble gases like H_2, the permeability coefficient decreases with increasing test pressure [51]. This phenomenon is attributed to the compression of the polymer and the compaction of voids as the test pressure rises [20,51]. Consequently, the reduction in free volume within the material more significantly hinders the diffusion of H_2 molecules, decreasing the hydrogen permeability coefficient. Fumitoshi et al. [52] observed that under high-pressure hydrogen environments, the crystallinity of polymer materials is further enhanced, which impedes hydrogen diffusion. Therefore, as experimental pressure increases, the diffusion coefficient of the material decreases.

4.4. Analysis of H_2 Diffusion Mechanism in Graphene/PA6 Composite Systems

Distribution of Free Volume:

The formula is as follows (5):

$$FFV = \frac{V - V_o}{V} = \frac{V_f}{V} \tag{5}$$

In the equation, V represents the volume of the polymer unit cell, V_o denotes the volume occupied by the polymer within the unit cell, and V_f signifies the free volume within the polymer unit cell.

According to Fox and Flory's [53] free volume theory, the volume of a polymer can be divided into the occupied volume by polymer chains and the free volume not occupied by polymer chains. Hydrogen molecules (H_2) can only diffuse within the free volume of the polymer. Changes in the free volume provide more space for H_2 diffusion within the polymer, thereby affecting the diffusion coefficient. This study employed the hard probe method to investigate the free volume of graphene/PA6 composite systems at different proportions. The probe radius corresponds to the van der Waals radius of H_2 (1.40 Å), representing the spatial extent of H_2 diffusion within the polymer. The distribution of free volume for various graphene/PA6 composite systems at 298 K and 0.1 MPa is illustrated in Figure 11, where the blue regions indicate the distribution of free volume within the composite materials.

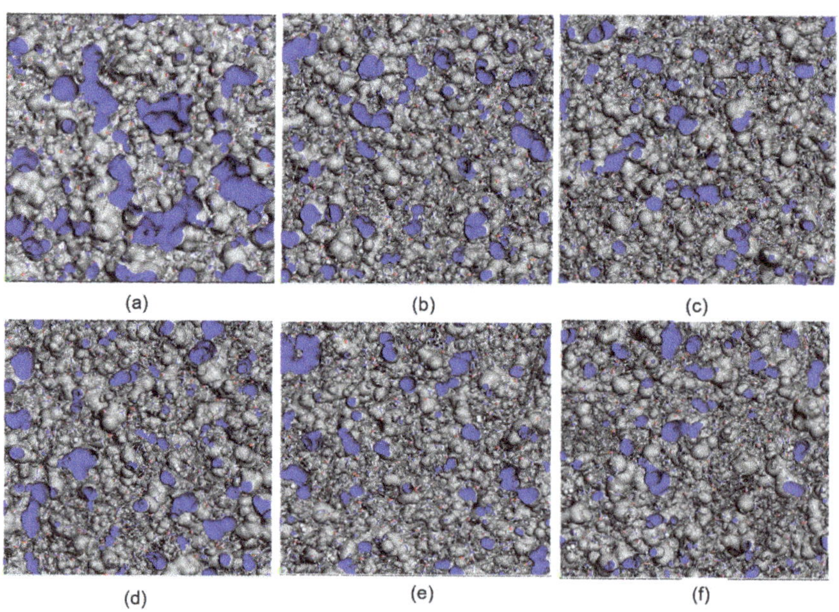

Figure 11. FFV of Different Systems at 298 K and 0.1 MPa: (**a**) PA6, (**b**) 3 wt%, (**c**) 4 wt%, (**d**) 5 wt%, (**e**) 6 wt%, and (**f**) 7 wt% (Blue area: free volume distribution of the composite).

It can be observed that the blue area representing the FFV is the largest for pure PA6. Upon adding the 3 wt% graphene, the blue area decreases and then fluctuates with increasing graphene content. The FFV initially decreases and then increases, reaching a minimum when the graphene content is 4 wt%. However, when graphene content exceeds 5%, the blue area begins to increase again. These results indicate that the addition of graphene restricts the movement of polymer chains within PA6, thereby disrupting the continuity of "voids" within PA6 to some extent and reducing the polymer's FFV. This limitation consequently restricts H_2 diffusion within the polymer matrix, enhancing the gas barrier properties of the polymer. As graphene content continues to increase, new "voids" form between the filler and PA6, resulting in an increase in free volume. This provides additional channels for gas diffusion, deteriorating the barrier properties and consequently increasing the diffusion coefficient while reducing the barrier performance.

Figure 12 shows the FFV calculated from simulation results to further analyze the impact of free volume fraction (FFV). Figure 12a depicts the FFV of different composite systems at 298 K and 0.1 MPa. The trend aligns with the distribution of the blue regions in Figure 11.

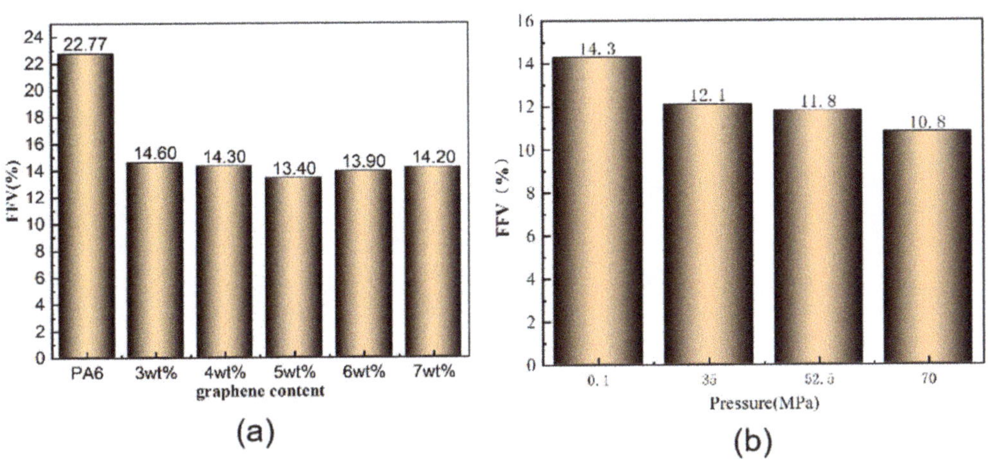

Figure 12. (**a**) FFV for different filler contents at 298 K and 0.1 MPa; (**b**) FFV of 4% Graphene/PA6 at four different pressure conditions.

Figure 12b illustrates the FFV of 4% graphene/PA6 under four different pressure conditions. It is evident from the figure that the FFV decreases gradually with increasing pressure. This phenomenon arises because pressure reduces the distance between molecules in the composite material, thereby slightly diminishing the size and number of pores within the system.

The trajectories of H_2 in various graphene/PA6 systems were studied at 298 K and 0.1 MPa to investigate the motion of H_2 molecules within polymers. Figure 13a illustrates that H_2 diffusion in the PA6 system follows a "hop and jump" mechanism, where H_2 molecules move between adjacent pores in a relatively short time. Over time, H_2 molecules move from their initial positions, increasing overall displacement. Upon adding 3% graphene, Figure 13b shows that H_2 molecules in the graphene/PA6 composite exhibit a distinct "hop and vibration" motion, indicating that the presence of graphene effectively hinders gas molecules. When hydrogen molecules encounter graphene during their motion, they cannot pass through directly, thus deviating from their original diffusion path; instead, they hop back and forth within the free volume. In Figure 13b, blue ellipse represent hopping, while red circle indicate vibration.

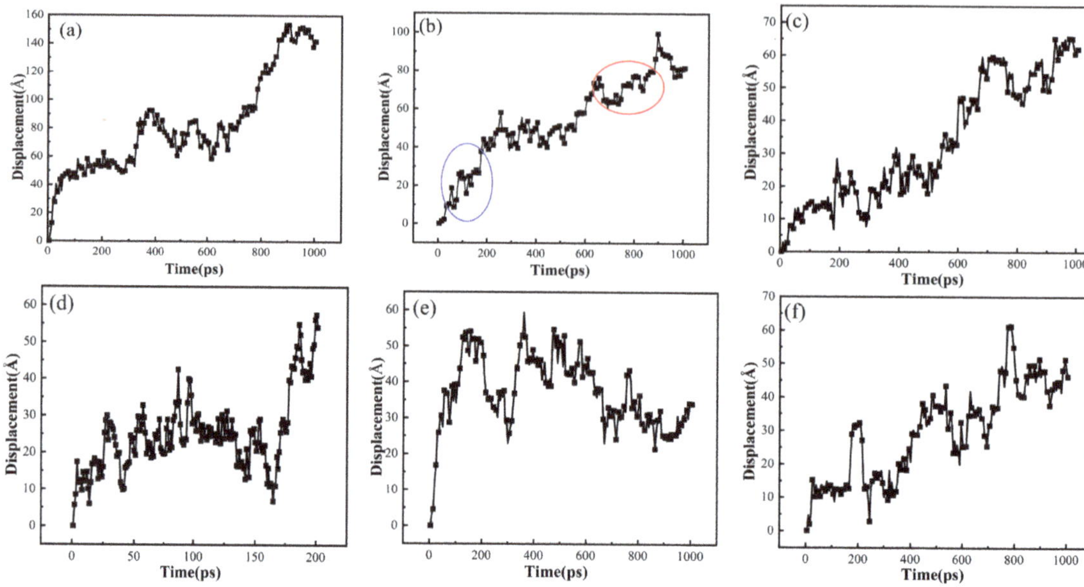

Figure 13. Trajectories of H_2 in different systems at 298 K and 0.01 MPa: (**a**) PA6, (**b**) 3 wt%, (**c**) 4 wt%, (**d**) 5 wt%, (**e**) 6 wt%, and (**f**) 7 wt% (blue ellipse: hopping, red circle: vibration).

With the increase of graphene content, the vibration of H_2 molecules within the polymer decreases, and there is a noticeable increase in hopping frequency, as shown in Figure 13d. Moreover, as the graphene content reaches 6 wt% and 7 wt%, the movement trajectory of H_2 exhibits a pattern of "forward hopping" + "backward hopping", with the backward hopping displacements being larger.

The movement trajectories of H_2 in the 4 wt% graphene/PA6 system under different pressures are shown in Figure 14. When the pressure reaches 35 MPa, a distinct pattern of forward "hopping" + backward "hopping" trajectories is observed, indicating that pressure enhances molecular motion while also affecting the distribution of free volume. As the pressure further increases to 70 MPa, the available space for hydrogen molecule movement decreases. Although pressure promotes molecular motion, the polymer chains become more distorted during movement. The more intense the local segment movements, the greater the probability of molecular transitions. Consequently, H_2 molecules exhibit longer and more frequent hopping displacements.

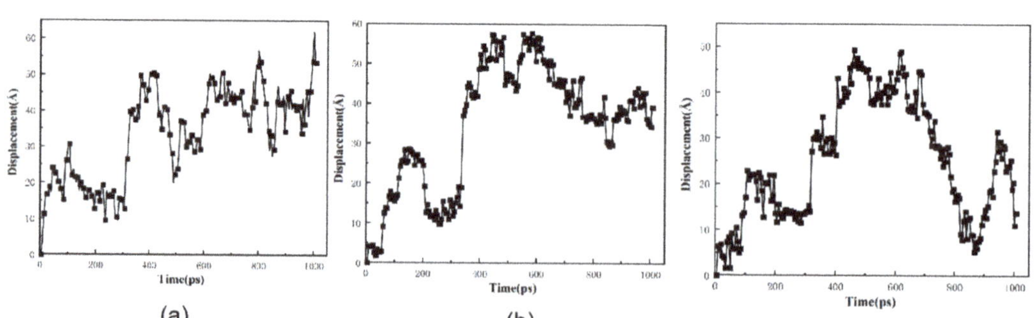

Figure 14. Displacement of H_2 in 5% graphene/PA6 at different pressures: (**a**) 35 MPa, (**b**) 52.5 MPa and (**c**) 70 MPa.

5. Conclusions

Using molecular dynamics (MD) simulations, the dissolution, diffusion, and permeability characteristics of H_2 in PA6 were studied across varying graphene contents (3 wt%–7 wt%), temperatures (233 K–358 K), and pressures ranging from 0.1 to 70 MPa. The following conclusions were drawn from this study.

(1) The addition of graphene restricts the movement of polymer chains and disrupts the continuity of "pores" within the polymer. When H_2 moves within the polymer composite system, it influences the formation of free volume to a certain extent. Compared to pure PA6, under conditions of 0.1 MPa and 298 K, the permeability coefficient of the 5 wt% graphene/PA6 composite system decreases by 54.6%.

(2) Temperature has no significant effect on the solubility coefficient of graphene-added composite systems. The diffusion coefficients of materials in each system increase with rising temperature, and the increase in diffusion coefficient becomes more pronounced at higher temperatures. The permeability coefficients of each system decrease with increasing pressure, and within the same system, the permeability coefficient decreases as pressure increases.

(3) Additionally, the FFV and movement trajectories of H_2 in each system were studied. FFV directly influences the diffusion coefficients of materials: the larger the FFV, the larger the diffusion coefficient. H_2 diffusion in graphene/PA6 composite systems follows a "vibration + leap" mechanism. In the 3 wt% graphene/PA6 composite system, H_2 exhibits noticeable vibration. In the 7 wt% graphene/PA6 composite system, H_2 displays both significant "forward leaps" and "backward leaps". With increasing pressure, H_2 in the 4 wt% graphene/PA6 system shows longer leap distances and higher frequencies.

Author Contributions: Methodology, data curation, and writing—original draft preparation, J.L. (Jin Li); funding and software, J.L. (Jianguo Liang); conceptualization, methodology, supervision, writing—review, and editing, C.Z.; formal analysis, N.F. and G.G.; investigation, X.Z. and Z.Z.; All authors have read and agreed to the published version of the manuscript.

Funding: This research were funded by the National Natural Science Foundation of China (grant number 52075361), the Shanxi Province Major Science and Technology Project (grant number 20201102003), the Key Research and Development Projects in Shanxi Province (grant number 201903D421030), the Startup Fund for Talent Introduction of Shanxi Electronic Science and Technology Institute (grant number 2023RKJ017), and the Key Research Project of School-Land Cooperation in Lvliang City (grant number 2022XDHZ08).

Institutional Review Board Statement: Not applicable.

Informed Consent Statement: Not applicable.

Data Availability Statement: The data presented in this study are contained within the article and are also available upon request from the corresponding author.

Conflicts of Interest: The authors declare no conflicts of interest.

References

1. Liu, G.; Yang, F.; Liu, W.; Han, C.; Jiao, W.; Wang, P. Ultra-high gas barrier composites with aligned graphene flakes and polyethylene molecules for high-pressure gas storage tanks. *J. Energy Storage* **2021**, *40*, 102692. [CrossRef]
2. Rivard, E.; Trudeau, M.; Zaghib, K. Hydrogen Storage for Mobility: A Review. *Materials* **2019**, *12*, 1973. [CrossRef] [PubMed]
3. Yamashita, A.; Kondo, M.; Goto, S.; Ogami, N. Development of High-Pressure Hydrogen Storage System for the Toyota "Mirai". *SAE Int. J. Engines* **2018**, *01*, 1169.
4. Jebeli, M.A.; Heidari-Rarani, M. Development of Abaqus WCM plugin for progressive failure analysis of type IV composite pressure vessels based on Puckfailure criterion. *Eng. Fail. Anal.* **2022**, *131*, 105851. [CrossRef]
5. Guillon, D.; Espinassou, D.; Pichon, P.; Carrillo, J.J.R.; Landry, C.; Clainchard, D.; Juras, L.; Brault, R. Manufacturing, burst test and modeling of high pressure thermoplastic composite overwrap pressure vessel. *Compos. Struct.* **2023**, *316*, 116965. [CrossRef]
6. Wang, X.L.; Tian, M.M.; Chen, X.D.; Xie, P.C.; Yang, J.N.; Chen, J.X.; Yang, W.M. Advances on materials design and manufacture technology of plastic liner of type IV hydrogen storage vessel. *Int. J. Hydrogen Energy* **2022**, *47*, 8382–8408. [CrossRef]
7. Wang, D.L.; Liao, B.B.; Zheng, J.Y.; Huang, G.; Hua, Z.L.; Gu, C.H.; Xu, P. Development of regulations, codes and standards on composite tanks for on-board gaseous hydrogen storage. *Int. J. Hydrogen Energy* **2019**, *44*, 22643–22653. [CrossRef]

8. Dong, C.F.; Liu, Y.T.; Li, J.P.; Bin, G.F.; Zhou, C.L.; Han, W.L.; Li, X. Hydrogen Permeability of Polyamide 6 Used as Liner Material for Type IV On-Board Hydrogen Storage Cylinders. *Polymers* **2023**, *15*, 3715. [CrossRef] [PubMed]
9. Dao, D.Q.; Luche, J.; Rogaume, T.; Richard, F.; Bustamante-Valencia, L.; Ruban, S. Polyamide 6 and Polyurethane Used as Liner for Hydrogen Composite Cylinder: An Estimation of Fire Behaviours. *Fire Technol.* **2016**, *52*, 397–420.
10. Sun, Y.; Lv, H.; Zhou, W.; Zhang, C.M. Research on hydrogen permeability of polyamide 6 as the liner material for type IV hydrogen storage tank. *Int. J. Hydrogen Energy* **2020**, *45*, 24980–24990. [CrossRef]
11. Chen, M.; Hu, Z.; Jia, X.; Yang, Q.; Shen, A.; Xu, K. Research progress on key technologies of Type IV vehicle-mounted hydrogen storage vessel, Press. *Vessel. Technol.* **2022**, *37*, 39–50.
12. Fang, Q.; Ji, D.M. Molecular simulation of hydrogen permeation behavior in liner polymer materials of Type IV hydrogen storage vessels. *Mater. Today Commun.* **2023**, *35*, 106302. [CrossRef]
13. Maus, S.; Hapke, J.; Ranong, C.N.; Wuchner, E.; Friedlmeier, G.; Wenger, D. Filling procedure for vehicles with compressed hydrogen tanks. *Int. J. Hydrogen Energy* **2008**, *33*, 4612–4621. [CrossRef]
14. Su, Y.; Lv, H.; Feng, C.; Zhang, C.M. Hydrogen permeability of polyamide 6 as the liner material of Type IV hydrogen storage tanks: A molecular dynamics investigation. *Int. J. Hydrogen Energy* **2024**, *50*, 1598–1606. [CrossRef]
15. Dutta, R.C.; Bhatia, S.K. Transport Diffusion of Light Gases in Polyethylene Using Atomistic Simulations. *Langmuir* **2017**, *23*, 936–946. [CrossRef] [PubMed]
16. Barth, R.R.; Simmons, K.L.; San Marchi, C. *Polymers for Hydrogen Infrastructure and Vehicle Fuel Systems: Applications, Properties, and Gap Analysis*; Pacific Northwest National Laboratory: Washington, DC, USA, 2013.
17. Naito, Y.; Mizoguchi, K.; Terada, K.; Kamiya, Y. The effect of pressure on gas permeation through semicrystalline polymers above the glass-transition temperature. *J. Polym. Sci. Polym. Phys. Ed.* **1991**, *29*, 457–462. [CrossRef]
18. Klopffer, M.H.; Flaconneche, B.; Odru, P. Transport properties of gas mixtures through polyethylene. *Plast. Rubber Compos.* **2007**, *36*, 184–189. [CrossRef]
19. Liao, L.Q.; Fu, Y.Z.; Liang, X.Y.; Mei, L.Y.; Liu, Y.Q. Diffusion of CO_2 Molecules in Polyethylene Terephthalate/Polylactide Blends Estimated by Molecular Dynamics Simulations. *Bull. Korean Chem. Soc.* **2013**, *34*, 753–758. [CrossRef]
20. Zhao, J.W.; Wang, X.W.; Yang, Q.Q.; Yin, H.; Zhang, B.; Zhang, S.J. Molecular dynamics simulation of H_2 in amorphous polyethylene system: H_2 diffusion in various PE matrices and bubbling during rapid. *Int. J. Hydrogen Energy* **2022**, *47*, 39572–39585. [CrossRef]
21. Takeuchi, K.; Kuo, A.T.; Hirai, T.; Miyajima, T.; Urata, S.; Okazaki, S.; Shinoda, W. Hydrogen permeation in hydrated perfluorosulfonic acid polymer membranes: Effect of polymer crystallinity and equivalent weight. *J. Phys. Chem. C* **2019**, *123*, 20628–20638. [CrossRef]
22. Kane, M. Permeability, Solubility, and Interaction of Hydrogen in Polymers—An Assessment of Materials for Hydrogen Transport. *Eng. Mater. Sci.* **2008**.
23. Yi, Y.; Bi, P.; Zhao, X.F.; Wang, L.L. Molecular dynamics simulation of diffusion of hydrogen and its isotopic molecule in polystyrene. *J. Polym. Res.* **2018**, *25*, 43. [CrossRef]
24. Alirezaie, A.H.H.; Navarchian, A.H.; Sabzyan, H. Molecular Dynamics Simulation of Gas Diffusion in Polyethylene-Clay Nanocomposites with Different Silicate Layers Configurations. *Polym. Sci. Ser. A* **2016**, *58*, 487–498. [CrossRef]
25. Gao, J.G.; Sun, W.F. Molecular dynamics simulation study of parallel orientation structure and gas transport in graphite-nanoplatelet/polyethylene composites. *Mater. Today Commun.* **2017**, *13*, 54–67. [CrossRef]
26. Yang, Q.; Achenie, L.E.K. Exploration of gas molecules transport in composite poly (4-methyl-2-pentyne) and carbon nanotubes employing molecular dynamics simulation. *Comp. Mater. Sci.* **2018**, *143*, 87–94. [CrossRef]
27. Cui, Y.B.; Kundalwal, S.I.; Kumar, S. Gas barrier performance of graphene/polymer nanocomposites. *Carbon* **2016**, *98*, 313–333. [CrossRef]
28. Muraru, S.; Ionita, M. A Molecular Dynamics Study of Small Gas Molecules Diffusion Through Composite Polysulfone/Carbon Nanotube and Polysulfone/Graphene Membranes. *Mater. Plast.* **2020**, *57*, 87–93. [CrossRef]
29. Einstein, A. Über die von der molekularkinetischen The der Wärme geforderte Bewegung von in ruhenden Flüssigkeiten suspendierten Teichen [AdP 17, 549 (1905)]. *Ann. Phys.* **1905**, *14*, 182–193.
30. Mozaffari, F.; Eslami, H.; Moghadasi, J. Molecular dynamics simulation of diffusion and permeation of gases in polystyrene. *Polymer* **2010**, *51*, 300–307. [CrossRef]
31. Wu, P.; Zhang, J.M.; Yang, Z.H.; Zhao, J.P. Molecular Dynamics Simulation of Helium Barrier Performance of Modified Polyamide 6 Lining of IV Hydrogen Storage Tank with Montmorillonite. *Molecules* **2023**, *28*, 3333. [CrossRef]
32. Zhang, X.M.; Zhai, L.Z.; Li, H.B.; Qi, G.Q.; Gao, X.; Yang, W.H. Molecular Simulation Study on the Hydrogen Permeation Behavior and Mechanism of Common Polymers. *Polymers* **2024**, *16*, 953. [CrossRef] [PubMed]
33. Hu, M.H.; Gao, W.; Zhang, L.S.; Wang, Y.Z.; Feng, H.J. Transformation of Diffusion and Local Structure of CH_4, CO_2, SO_2 and H_2O Mixtures in Graphene Under Wide Temperature and Pressure Range: A Molecular Dynamics Simulation Study. *Chemphyschem* **2024**, *25*, e202300851. [CrossRef] [PubMed]
34. Stalker, M.R.; Grant, J.; Yong, C.W.; Ohene-Yeboah, L.A.; Mays, T.J. Molecular simulation of hydrogen storage and transport in cellulose. *Mol. Simul.* **2021**, *47*, 170–179. [CrossRef]
35. Sun, H. COMPASS: An ab initio force-field optimized for condensed-phase applications overview with details on alkane and benzene compounds. *J. Chem. Phys.* **1998**, *102*, 7338–7364. [CrossRef]

36. Lippert, R.A.; Predescu, C.; Ierardi, D.J. Accurate and efficient integration for molecular dynamics simulations at constant temperature and pressure. *J. Chem. Phys.* **2013**, *139*, 164106. [CrossRef] [PubMed]
37. Andersen, H.C. Molecular dynamics simulations at constant pressure and/or temperature. *J. Chem. Phys.* **1980**, *72*, 2384–2393. [CrossRef]
38. Chen, L.J.; Qian, H.J.; Lu, Z.Y. An automatic coarse-graining and fine-graining simulation method: Application on polyethylene. *J. Phys. Chem. B* **2006**, *110*, 24093–24100. [CrossRef] [PubMed]
39. Ding, H.; Karasawa, N.; Goddard, W.A.G. Atomic level simulations on a million particles: The cell multipole method for Coulomb and London nonbond interactions. *J. Chem. Phys.* **1992**, *97*, 4309–4315. [CrossRef]
40. Khosravanian, A.; Dehghani, M.; Pazirofteh, M. Grand canonical Monte Carlo and moleculardynamics simulations of the structural properties, diffusion and adsorption of hydrogen moleculesthrough poly(benzimidazoles)/nanoparticle oxidescomposites. *Int. J. Engines* **2018**, *43*, 2803–2816.
41. Bian, L.; Shu, Y.; Xu, J.; Wang, L. Molecular dynamics study on permeability of gas molecules through amorphous PPX polymers. *Int. Polym. Proc.* **2013**, *28*, 24–33. [CrossRef]
42. Theodorou, D.N.; Suter, U.W. Atomistic modeling of mechanical properties of polymeric glasses. *Macromolecules* **1986**, *19*, 139–154. [CrossRef]
43. *GB/T 42612-2023*; Fully-Wrapped Carbon Fiber Reinforced Cylinders with a Plastic Liner for the On-Board Storage of Compressed Hydrogen as a Fuel for Land Vehicles. China Gas Cylinder Standardization Technical Committee (TC31SC8): Beijing, China, 2023.
44. Klopffer, M.H.; Flaconneche, B. Transport properties of gases in polymers: Bibliographic review, Oil. *Gas. Sci. Technol.* **2001**, *56*, 223–244. [CrossRef]
45. Li, Y.S.; Hu, H.R.; Xia, Y.; Li, G.Z.; Liu, Q.D. Molecular dynamics simulation of the microscopic properties of nano-MgO-doped polyethylene. *J. At. Mol. Phys.* **2020**, *39*, 52–60.
46. Xie, Z.; Liu, G.S.; Zhao, C.H.; Li, M.; Gan, C. Molecular simulation of water diffusion behavior in cellulose. *Build. Energy Environ.* **2020**, *39*, 18–24.
47. Zheng, D.K.; Li, J.F.; Liu, B.; Yu, B.; Yang, Y.F.; Han, D.X.; Li, J.L.; Huang, Z.Q. Molecular dynamics investigations into the hydrogen permeation mechanism of polyethylene pipeline material. *J. Mol. Liq.* **2022**, *368*, 120773. [CrossRef]
48. Hu, H.; Li, X.; Fang, Z.; Wei, N.; Li, Q. Small-molecule gas sorption and diffusion in coal: Molecular simulation. *Energy* **2010**, *35*, 2939–2944. [CrossRef]
49. Yu, S.; Fangkai, Q.; Junhong, Y. Diffusion of guest molecules in coal: Insights from simulation. *Fuel* **2022**, *323*, 124295. [CrossRef]
50. Fujiwara, H.; Ono, H.; Ohyama, K.; Kasai, M.; Kaneko, F.; Nishimura, S. Hydrogen permeation under high pressure conditions and the destruction of exposed polyethylene-property of polymeric materials for high-pressure hydrogen devices (2)-. *Int. J. Hydrogen Energy* **2021**, *46*, 11832–11848. [CrossRef]
51. Stern, S.; Fang, S.M.; Frisch, H. Effect of pressure on gas permeability coefficients. A new application of" free volume" theory. *J. Polym. Sci. Part A-2 Polym. Phys.* **1972**, *10*, 201–219. [CrossRef]
52. Fumitoshi, K.; Hirotada, F.; Shin, N. Influence of high-pressure hydrogen gas on crystalline polymers. In Proceedings of the 33rd Polymer Degradation Discussion Group Conference (PDDG), St Julians, Malta, 1–5 September 2019.
53. Fox, T.G.; Flory, P.J. Second-Order Transition Temperatures and Related Properties of Polystyrene. I. Influence of Molecular Weight. *J. Appl. Phys.* **1950**, *21*, 581–591. [CrossRef]

polymers

MDPI

Article

Toward the Production of Hydroxyapatite/Poly(Ether-Ether-Ketone) (PEEK) Biocomposites: Exploring the Physicochemical, Mechanical, Cytotoxic and Antimicrobial Properties

Meirilany Rozeno Costa [1,*], José Adeilton Carvalho Filho [1], Carlos Bruno Barreto Luna [2], Gleydis Manalig Pereira Dantas [1], Ana Cristina Figueiredo de Melo Costa [1] and Nadja Maria da Silva Oliveira [3]

[1] Ceramic Materials Synthesis Laboratory, Federal University of Campina Grande, Av. Aprígio Veloso, 882, Bodocongó, Campina Grande 58429-900, PB, Brazil; adeiltoncarvalho87@yahoo.com.br (J.A.C.F.); manaliggg@gmail.com (G.M.P.D.); ana.figueiredo@professor.ufcg.edu.br (A.C.F.d.M.C.)

[2] Polymer Processing Laboratory, Federal University of Campina Grande, Av. Aprígio Veloso, 882, Campina Grande 58429-140, PB, Brazil; brunobarretodemaufcg@hotmail.com

[3] Postgraduate Program in Health Science and Technology—PPGCTS, Dentistry Department, State University of Paraíba, R. Baraúnas, 351, Bodocongó, Campina Grande 58429-500, PB, Brazil; nadjamso@paqtc.org.br

* Correspondence: meirilany.rozeno@estudante.ufcg.edu.br

Abstract: The development of hydroxyapatite (HAp) and polyether ether ketone (PEEK) biocomposites has been extensively studied for bone repair applications due to the synergistic properties of the involved materials. In this study, we aimed to develop HAp/PEEK biocomposites using high-energy ball milling, with HAp concentrations (20%, 40%, and 60% *w/v*) in PEEK, to evaluate their physicochemical, mechanical, cytotoxicity, and antimicrobial properties for potential applications in Tissue Engineering (TE). The biocomposites were characterized by structure, morphology, apparent porosity, diametral compression strength, cytotoxicity, and antimicrobial activity. The study results demonstrated that the HAp/PEEK biocomposites were successfully synthesized. The C2 biocomposite, containing 40% HAp, stood out due to the optimal distribution of HAp particles in the PEEK matrix, resulting in higher compression strength (246 MPa) and a homogeneous microstructure. It exhibited antimicrobial activity against *Staphylococcus aureus*, *Pseudomonas aeruginosa*, and *Escherichia coli*, with no cytotoxicity observed. These properties make the C2 biocomposite promising for regenerative medicine applications, combining mechanical strength, bioactivity, and biocompatibility.

Keywords: biomaterials; hydroxyapatite; PEEK; high-energy ball milling; bone repair; mechanical properties; cytotoxicity

1. Introduction

According to the Global Burden of Diseases, Injuries, and Risk Factors Study, there were 178 million (95% uncertainty interval, 162–196) new fractures globally in 2019, representing a 33.4% increase since 1990, with 455 million prevalent cases of acute or long-term symptoms from a fracture [1]. In Brazil, injuries or other conditions caused by external factors have been among the top five leading causes of hospital admissions. It is important to note that among the morbidities categorized in the "injuries due to external causes" chapter of the eleventh revision of the International Classification of Diseases (ICD-11), bone fractures are the most frequent hospitalization cases, accounting for 43.1% of the total [2].

AO Trauma International and the Orthopaedic Trauma Association (AO/OTA) indicate that the repair process is complex, involving multiple stages that begin with immediate reactions to injury, including damage to soft tissues, blood vessel rupture, and bone necrosis, all co-occurring. The hematoma response triggers the release of inflammatory cells, activates osteoclasts and mesenchymal stem cells, and stimulates chondrocytes to form a rich, soft callus. Cycles of osteoblast and osteoclast activity subsequently restore bone morphology and mechanical properties [3–5].

The global orthopedic implants market is expected to reach over $64 billion by 2025, driven by factors such as an aging population, increased sports injuries, and the growth of minimally invasive surgeries [6]. Although orthopedic implants made from metals, ceramics, and synthetic polymers have been successful, each material class has limitations that impact long-term performance [7]. Metals are rigid and durable but can corrode and cause adverse tissue reactions over time [8]. Ceramics have high compressive strength but are brittle and prone to fractures. On the other hand, polymers better match the rigidity of bone but lack robustness and bioactivity [9]. This has driven the research into new biomaterials that can overcome the deficiencies of current implants. Choosing an implant to restore structural integrity is crucial for treatment, as it can avoid trauma to the donor site, reduce patient pain, and lower the overall treatment cost, providing a new strategy for the clinical repair of bone defects [10,11].

Bone substitutes should be biocompatible, bioresorbable, osteoconductive, osteointegrative, non-immunogenic, and non-inflammatory. Additionally, various technological strategies can be applied for the "customization" of treatment and effectiveness in bone regeneration, as these materials can enhance the irregularity of tissue defects and reduce inflammation and infection in the application area [12–15]. For example, a study conducted by Safavi et al. [16] evaluated the osteogenic activity of NiTi orthopedic implants through HAp-Nb_2O_5. composite coatings. The results demonstrated that the incorporation of HAp-Nb_2O_5 reinforcement particles leads to a significant increase in adhesion, proliferation, and osteogenic activity of SAOS-2 cells (osteosarcoma). In this context, hydroxyapatite (HAp) ($Ca_{10}(PO_4)_6\ (OH)_2$) has such characteristics, and additionally, its biological application is due to its chemical similarity to the inorganic components of human bones and teeth, being the main component in the form of calcium apatite [14]. This bioceramic calcium phosphate can be obtained from natural and synthetic sources and is hydrophilic and stable even at high temperatures [17]. However, it has some disadvantages when used in isolation, as it is highly brittle, with low tensile strength (18 MPa) and high compressive strength (917 MPa), making it insufficient to promote vascularization and osteoinductivity [18].

Thus, combining bioactive inorganic phases with polymers can produce composites with tailored biological and mechanical properties [19]. For example, in a study conducted by Ma et al. [20] using a composition and injection molding technique, they assessed the bioactivity of a polyether ether ketone (PEEK) biocomposite with HAp. The in vitro bioactivity evaluation demonstrated cellular attachment and proliferation of MC3T3-E1 cells (pre-osteoblast cell type) on the material's surface, showing enhanced spreading efficiency. Additionally, this composite induced the formation of an apatite layer, and the attracted osteoblasts proliferated and differentiated to produce collagen and protein, which mineralized to form a new bone interface. Another study conducted by Qiu et al. [21] examined the osteointegration and antimicrobial properties of PEEK implants by applying a double-layer coating of amorphous magnesium phosphate (AMP) and HAp, where the PEEK was sulfonated using an ultrasonic method and the AMP/HAp was processed using microwave technology. The results demonstrated that surface modification conferred antibacterial properties, highlighting the potential of these composite coatings in preventing infections.

It is important to note that PEEK has been approved by the U.S. Food and Drug Administration (FDA) as an implant material since the 1980s [22]. It belongs to a family of high-temperature thermoplastic polymers, semicrystalline with chemical stability and an elastic modulus around 3.6 GPa, close to bone tissue [23]. Despite its high macromolecular rigidity, its planar conformation allows the polymer chains to be organized into crystalline and amorphous domains [24]. Biocomposites made from this polymer or surface modifications have been prepared using various methodologies such as direct injection molding, three-dimensional (3D) printing technology [25], sol–gel processing [26], plasma spraying [27,28], and physical or chemical vapor deposition [21]. However, they have limitations such as multi-step preparation processes and high costs. Additionally, the high cost of implant-grade raw materials, such as PEEK, can lead companies to face large-scale process reproducibility challenges. Therefore, alternative methods to achieve higher productivity in the production of these implants should be considered, as the level of crystallinity and crystalline morphology can influence the physical properties of PEEK-containing composites [29].

Thus, high-energy ball milling can ensure optimized processing through a single-step methodology, such as physical mixing, allowing for high reproducibility. Moreover, it can lead to a more uniform distribution of HAp particles within the PEEK matrix, and this homogeneity is essential for achieving consistent mechanical and biological properties throughout the material. Therefore, this study aimed to evaluate the properties of HAp/PEEK biocomposites formed by high-energy milling.

2. Materials and Methods

2.1. Materials

The biocomposites, a significant advancement in materials science, were produced using the following materials: Victrex® polyether ether ketone (PEEK) under the commercial code Vicote 702. Hydroxyapatite (HAp) was synthesized by precipitation at the Laboratory of Ceramic Materials Synthesis (LabSMaC, Campina Grande, Brazil), and acetone was 99%.

2.2. Synthesis of Biocomposites by High-Energy Ball Milling

The synthesis of HAp followed the methodology proposed by Saeri et al.; Barandehfard et al.; and Sarkar et al. [30–32], using the precipitation method with a phosphorus/calcium ratio of 1.67. Initially, solutions of calcium hydroxide ($Ca(OH)_2$) and phosphoric acid (H_3PO_4) were prepared at 2 M. The calcium hydroxide solution was subjected to constant stirring for 30 min in an IKA® RH essential KT/C mixer/heater until it reached a temperature of 80 °C. After this, the H_3PO_4 solution was added dropwise while stirring at 100 rpm. The solution was left in an oven at 110 °C for 24 h. The product was sifted through an ABNT 100 mesh sieve (150 µm). The produced material was then characterized and used to prepare the biocomposite formulations.

The addition of HAp to PEEK, based on the mass/volume ratio, was carried out in proportions of 20% (C1), 40% (C2), and 60% (C3). The proportions of HAp were determined based on bone composition, which has approximately 70% inorganic phase [33]. A ball mill was used to mill a PEEK/HAp composite using alumina balls as the grinding medium. The ratio between the weight of the grinding balls and the weight of the material was optimized to maximize milling efficiency and ensure a homogeneous material distribution. A jar with a total volume of 225 mL was used, and alumina balls with a diameter of 5 mm were added. The total mass of the balls used was 270 g. In addition to the alumina balls, 23.33 g of PEEK/HAp composite powder was added to the jar. This configuration provides enough space for the balls to move during the milling process, promoting effective impacts and adequate milling of the PEEK/HAp powder. The milling time was set to 18 h, allowing the material to be sufficiently processed and ensuring uniformity without causing excessive milling. The rotation speed was adjusted to 120 rpm, which is ideal for avoiding material segregation and ensuring efficient mixing. Figure 1 shows the production flowchart of PEEK/HAp biocomposites.

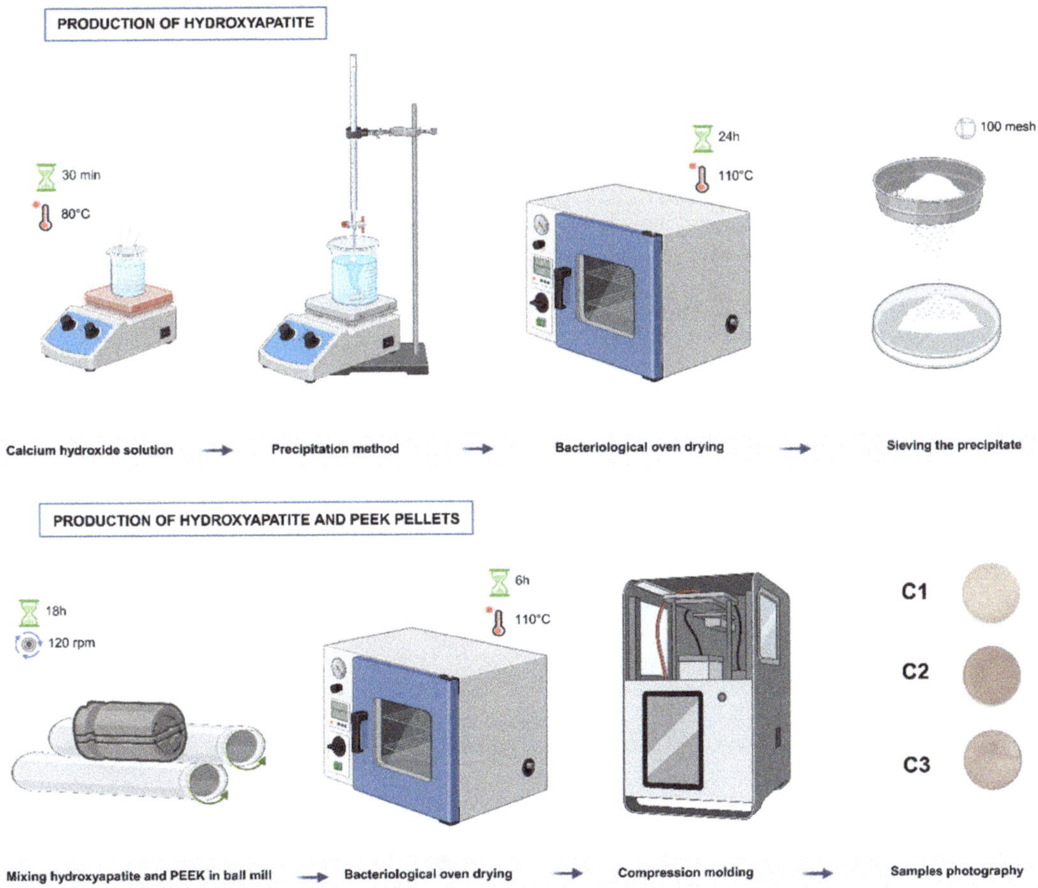

Figure 1. Manufacturing process of the biocomposites.

2.3. Characterization

2.3.1. X-ray Diffraction (XRD)

The phases present and crystallinity were determined from X-ray diffraction (XRD) data using a BRUKER diffractometer (Bruker, Billerica, MA, USA), model D2 Phaser, with copper radiation (CuKα1 = 1.54056 Å) at 40 kV and 30 mA. The scanning was performed in the range of $10^\circ \leq 2\theta \leq 80^\circ$, with a speed of 0.016°/min and a time of 5 s.

2.3.2. Fourier Transform Infrared Spectroscopy (FTIR)

The spectra obtained in the Fourier Transform Infrared (FTIR) region were recorded using a BRUKER VERTEX 70 FT-IR (Bruker, Billerica, MA, USA), model 660-IR, between 4000 and 500 cm^{-1}, with a resolution of 4 cm^{-1} and 32 scans.

2.3.3. Scanning Electron Microscopy (SEM/EDS)

The morphology and the presence of clusters were analyzed using Scanning Electron Microscopy (SEM) with a VEGA 4, TESCAN (TESCAN, Brun, Czech Republic), operating between 10 and 15 mA, using a magnification of 5000×. The samples were gold-coated.

2.3.4. Apparent Porosity Determination

For the apparent porosity (PA) test, samples with a thickness of 3.94 mm and a diameter of 22.6 mm were used. The evident porosity test was based on Archimedes' principle. The samples were prepared and dried at 345 °C for 30 min to obtain the dry weight (Ps). Subsequently, they were immersed in 20 mL of distilled water for 24 h, and the immersed weight (Pi) and the wet weight (Pu) were measured.

2.3.5. Compressive Strength Mechanical Testing

This was performed using an INSTRON 5582 universal testing machine (INSTRON, São Paulo, Brazil), with a load of 10 N and a speed of 1.3 mm/min (ASTM D 695-23) [34], with one grip fixed and the other mobile. Three samples were tested for each batch, with tests conducted at room temperature between 20 and 25 °C. The average values and standard deviations were calculated based on the three samples from each group.

2.3.6. Cytotoxicity Testing

Cytotoxicity was assessed using the Agar Diffusion method, according to ISO 10993-5:2009 [35], and adapted by Pina et al. and Wanderley et al. [36,37]. This method assesses the effects of the material on cells by using an agar layer that allows the diffusion of chemical substances from the sample to the cell layer [38,39]. The L929 fibroblast cell line (ATCC NCTC clone 929, Rio de Janeiro Cell Bank, Rio de Janeiro, Brazil) was cultured in RPMI 1640 medium (Gibco—Invitrogen Corporation, Grand Island, NE, USA) and incubated in a humidified incubator at 37 °C and 5% CO_2 until 80% confluence was reached. Subsequently, trypsinization with 0.25% trypsin (Gibco®, Life Technologies, Carlsbad, CA, USA) and cell counting was performed using an automated cell counter Interwoven—Thermo Fisher (Waltham, MA, USA). Suspensions of 1.0×10^5 cells/mL were distributed into 6-well plates, with 4 mL added to each well, and incubated under the same conditions described above for 24 h. Upon reaching uniform 80% confluence in the plates, a detailed examination was conducted using a microscope, the culture medium was aspirated, and 1 mL/well of medium prepared with MEM 2X concentrated (Gibco®—Invitrogen Corporation, Grand Island, NE, USA) and 1.8% agar solution with 0.01% neutral red (Sigma–Aldrich, St. Louis, MI, USA) was added. This was kept in the dark at room temperature for 10 minutes to solidify. Test samples of 1 cm^2 were placed in the center of the agar surfaces, along with positive (latex sheet) and negative (high-density polyethylene—HDPE) controls in triplicate. The plates were wrapped in aluminum foil and incubated upside down for 24 h in a humid chamber at 37 °C and 5% CO_2. In the end, following the guidelines of ISO 10993-5:2009, the areas of discoloration around the sample were measured, and an inverted microscope Nikon Eclipse TS100 (Minato, Tokyo, Japan) was used to analyze cell lysis, with images recorded after 24 h of incubation. The discolored areas and lysis were graded according to the ISO 10993-5:2009 guidelines, as shown in Table 1, as follows: 0 = no detectable cell lysis; 1 = less than 20% cell lysis; 2 = 20–40% cell lysis; 3 = 40–60% cell lysis; 4 = 60–80% cell lysis; 5 = more than 80% cell lysis.

Table 1. Degree of cytotoxicity.

Degree	Cytotoxicity	Description of the Cytotoxicity Zone
0	Absence	Absence of bleaching around or under the sample.
1	Light	Bleaching zone limited to the area under the sample
2	Soft	Sample bleaching zone size less than 0.45 cm.
3	Moderate	Sample bleaching zone size less than 0.45 cm to 1.0 cm.
4	Severe	Size of the sample bleaching zone greater than 1.0 cm, but not involving the entire plate.

2.3.7. Determination of Antimicrobial Activity from the Minimum Inhibitory Concentration (MIC)

Preparation of Bacterial Suspension and Inoculum Standardization

To determine antimicrobial activity based on the minimum inhibitory concentration (MIC), microbial strains from the American Type Culture Collection (ATCC) were used: *S. aureus* (ATCC 25923), *P. aeruginosa* (ATCC 27853), and *E. coli* (ATCC 25922). These strains were maintained at the Drug Development and Testing Laboratory of UEPB (Labdem-UEPB), stored in brain–heart infusion broth (BHIB) (DIFCO®) and 20% (*v/v*) glycerol. The inoculum was standardized according to Clinical Laboratory Standards Institute (CLSI) M07 guidelines, using a Mueller Hinton broth (MHB) (DIFCO®) culture incubated for 24 h at 35 ± 2 °C. The inoculum was standardized to achieve a concentration corresponding to a MacFarland 0.5 scale in MHB, and for the assays, the initial inocula were diluted to a concentration of 2.0 to 8.0 × 10^5 CFU/mL [40].

Broth Microdilution Method for Minimum Inhibitory Concentration Determination

The minimum inhibitory concentrations (MICs) were determined in 96-well microdilution plates following the methodologies outlined by the Clinical and Laboratory Standards Institute M07 and M27 (CLSI, 2018). To this end, 190 µL of Mueller Hinton broth was added to each well of sterile round-bottom plates, and then 10% (*w/v*) of the PEEK/HAp C2 powder from the test compounds was added. Serial dilutions of the composite were performed to obtain final concentrations ranging from 100 × 10^3 µg/mL to 390.62 µg/mL. Each well received 10 µL of a bacterial suspension with a final concentration of 1.5 × 10^8 CFU/mL. After treatment, the plates were incubated at 35 ± 2 °C for 24 h. The positive and negative controls involved bacterial suspension combined with broth and only the culture medium. Antibiotics such as ciprofloxacin hydrochloride (CIPRO), oxacillin (OXA), and ceftazidime (CFT) were used as positive controls at a concentration of 200 µg/mL. After the incubation period, 20 µL of 2% triphenyl tetrazolium chloride (TTC) in sterilized water was added, and after 1 hour of incubation, visual readings were performed.

3. Results and Discussion

Figure 2 presents the X-ray diffractograms of HAp, PEEK, and the composites with varying HAp contents. In Figure 2, it can be observed that calcium HAp was successfully obtained using the precipitation method, as evidenced by the presence of characteristic peaks corresponding to the crystal planes (2 1 3), (2 2 2), (3 1 0), (2 0 2), (1 1 2), (2 1 1), (2 1 0), and (0 0 4) at approximately 26°, 31.8°, 33°, 40°, 46°, and 50°, respectively. This behavior was also reported by Li et al. [41], who studied the 3D printing of PEEK biocomposites with $Ca_{10}(OH)(PO_4)_3$ for bone implants. For PEEK, as shown in Figure 2b, a broadband with four peaks corresponding to the diffraction planes (110), (111), (200), and (211) was observed at around 18.8°, 21°, 22.6°, and 28.7°, respectively. This is typical of a semicrystalline polymer. These characteristics were also reported by Asante et al. [42], who studied sulfonated carbon fiber-reinforced PEEK at room temperature with HAp coating.

The diffractograms in Figure 2c–e show that the biocomposites were formed by adding 20%, 40%, and 60% HAp to PEEK, respectively. It can be noted that the addition of HAp did not significantly alter the characteristics of the PEEK matrix, as the main diffraction peaks characteristic of both HAp and PEEK are displayed. This behavior demonstrates that the HAp/PEEK biocomposite was successfully synthesized. Such characteristics were also reported by Ma et al., Qi et al., and Yusong et al. [20,43,44], when evaluating the bioactivity of a HAp-incorporated PEEK biocomposite.

Figure 3 illustrates the FTIR spectra of HAp, PEEK, and the biocomposites, as a function of HAp content, in the range of 4000–500 cm^{-1}.

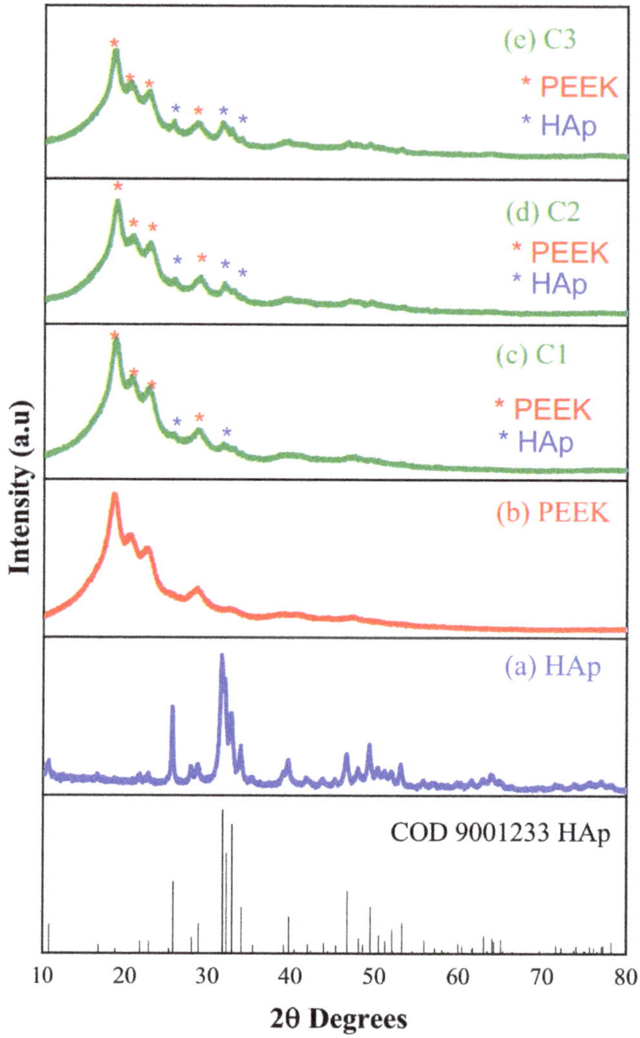

Figure 2. X-ray diffractograms: (**a**) HAp, (**b**) PEEK, (**c**) C1, (**d**) C2, and (**e**) C3.

When analyzing the FTIR spectrum of HAp, characteristic bands for HAp functional groups are observed around 1026 cm^{-1} and 1094 cm^{-1}, corresponding to the asymmetric stretching vibration of the PO_4^{3-} group, and another at 960 cm^{-1}, associated to the symmetric stretching vibrations of PO_4^{3-}. A band at 871 cm^{-1} is attributed to the stretching vibrations of CO_3^{2-}, indicating the presence of carbonate ions in phosphate units (type B substitution). Absorption bands appearing at 3569.59 $cm^{-1}$and 632.53 cm^{-1} correspond to the stretching and bending vibrations of structural OH. These characteristic absorption bands of HAp are consistent with results obtained by Xu et al., Nascimento et al., and Tabrizi et al. [45–47].

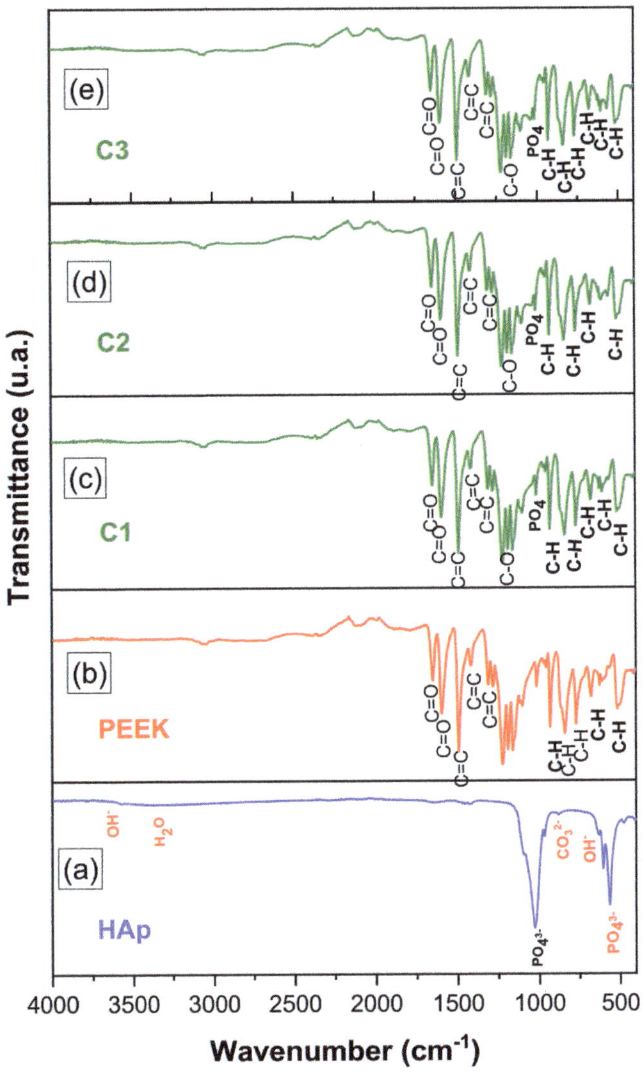

Figure 3. Infrared spectra: (**a**) HAp, (**b**) PEEK, (**c**) C1, (**d**) C2, and (**e**) C3.

The FTIR spectrum of PEEK showed absorption bands for aromatic CH groups at 671, 766, 830, and 929 cm^{-1}, suggesting angular deformation of the aromatic CH group. Bands at 1011, 1159, 1170, and 1213 cm^{-1} correspond to asymmetric stretching of the C-O group. Axial deformation bands at 1278, 1305, and 1488 cm^{-1} are associated with the C=C group, related to the aromatic ether C-O group. Bands at 1594 and 1645 cm^{-1} correspond to the primary and secondary stretching of the carbonyl C=O group, respectively. These bands were also reported by [48], who studied the characterization of the chemo-mechanical properties of polyether ether ketone (PEEK).

From the FTIR spectra of the HAp/PEEK biocomposites (Figure 3c–e), a band at 1000 cm^{-1} was observed, corresponding to the asymmetric stretching deformation of the $PO_4{}^3$ group. The bands at 671, 766, 830, and 929 cm^{-1} are associated with PEEK, suggesting angular deformation of the aromatic CH group. Bands at 1011, 1159, 1170, and 1213 cm^{-1} indicate asymmetric stretching of the C-O group. Bands at 1594 and 1645 cm^{-1} correspond to the primary and secondary stretching of the carbonyl C=O group, respectively. The results indicate that HAp is incorporated into the PEEK chain, which is essential for enhancing the functional behavior of the biocomposite.

Figure 4 shows the scanning electron microscopy (SEM) micrographs, EDS spectra, and EDS mapping of HAp, PEEK, and the biocomposites, at a magnification of 5000×. Figure 4a demonstrates that the HAp sample consists of irregular, rounded aggregates with a broad distribution. This morphology is typical of HAp synthesized by coprecipitation, as also reported by Afshar et al. [49]. According to Ponciano et al. [50], this synthesis route produces nanometric particles that enhance bone tissue bonding, cell proliferation, and better dispersion. The EDS spectrum confirmed the predominant presence of calcium and phosphate, the main elements in the chemical composition of HAp.

In Figure 4b, it can be seen that PEEK exhibits a morphology of rounded grains with a dense and spherical structure, characteristic of PEEK, as also reported by Manzoor et al. [51]. EDS mapping showed a homogeneous distribution of carbon and oxygen elements, while the EDS spectrum highlighted the prominent peaks of these two elements, confirming the predominant chemical composition of PEEK.

Regarding the HAp/PEEK biocomposites, the presence of distributed granules was observed, indicating that the increased amount of HAp favored more significant densification and interaction of the HAp particles, which tend to form agglomerates due to Van der Waals forces and hydrogen bonding [52]. In Figure 4e, the HAp particles are distributed heterogeneously on the surface, with a discrete increase in particle agglomerates of various sizes and shapes dispersed on the PEEK surface. In Figure 4d, the HAp particles are dispersed uniformly within the PEEK matrix. The addition of HAp makes the PEEK surface more hydrophilic, which can facilitate cell adhesion to the surface of an implant and, consequently, promote bone growth [20].

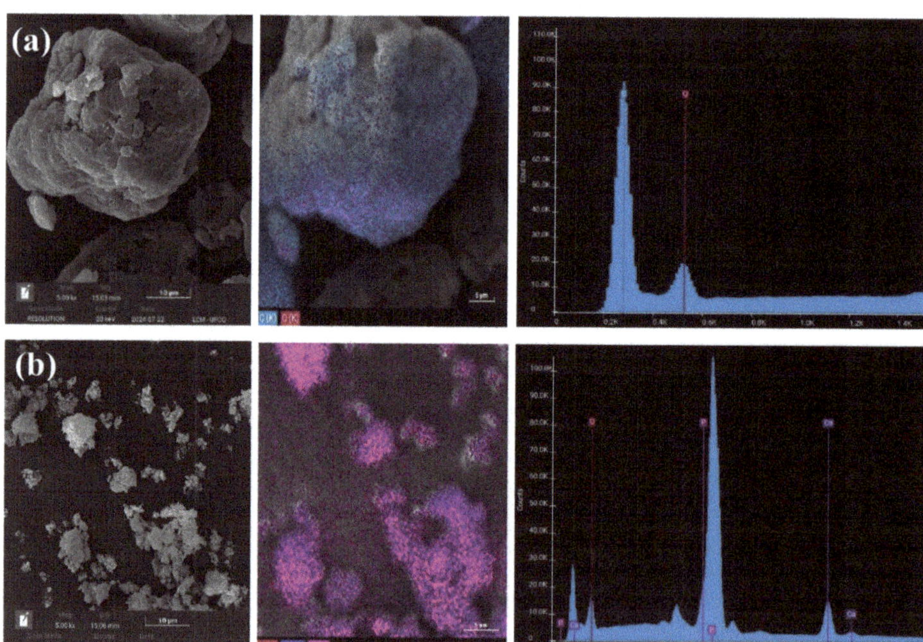

Figure 4. *Cont.*

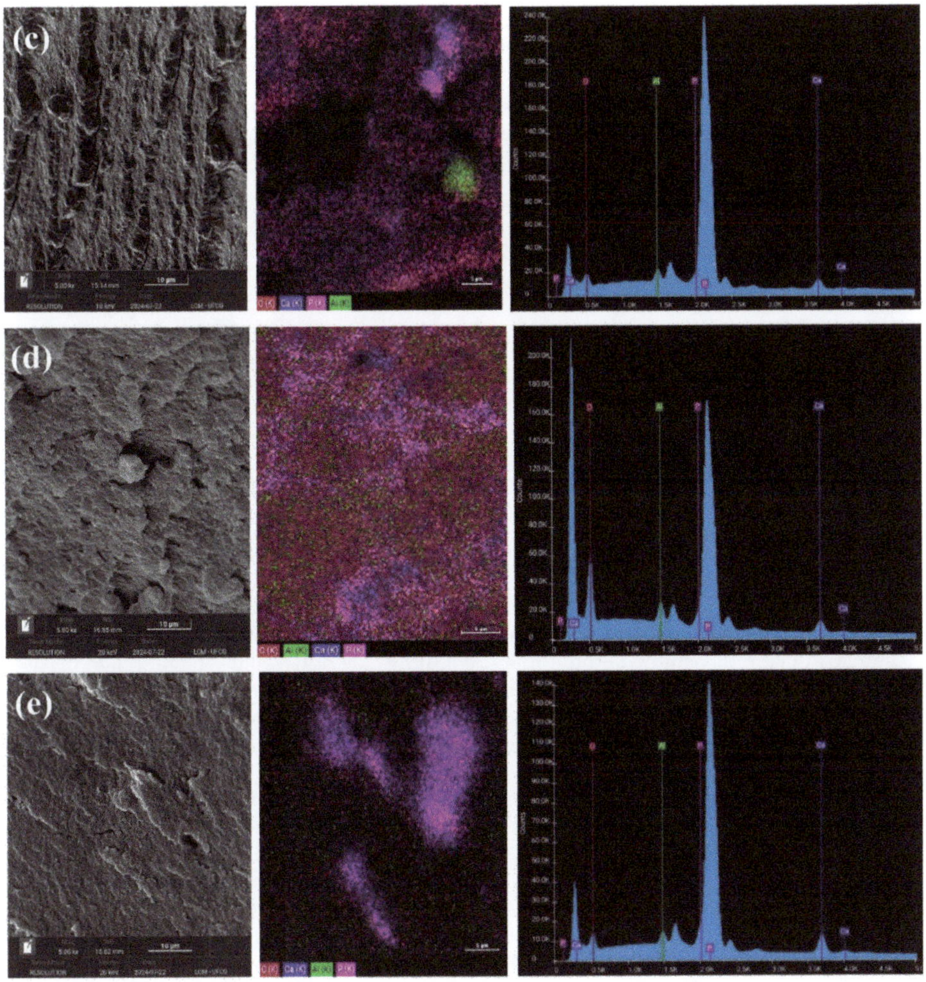

Figure 4. SEM images with chemical mapping by EDS, for: (**a**) HAp, (**b**) PEEK, (**c**) C1, (**d**) C2, and (**e**) C3.

Figure 5 illustrates the apparent porosity of the HAp/PEEK biocomposites with different HAp concentrations. It can be observed that all biocomposites exhibited an apparent porosity ranging from 0.10 to 0.16%.

When analyzing the porosity of the samples, it is noteworthy that the apparent porosity of the C3 biocomposite was higher than that of the C1 and C2 biocomposites. An increase in apparent porosity was observed for each of the biocomposites obtained with the incorporation of HAp. This behavior was expected due to the increased amount of HAp in PEEK. This phenomenon was also reported by Li et al. [53], who studied PEEK-based biocomposites for orthopedics. HAp, when processed at high concentrations, can hinder the complete sintering of grains. This occurs because the material's density may limit the diffusion of calcium and phosphate ions between particles. Sintering can result in a network of interconnected microchannels and pores, which increase the material's porosity. HAp granules may agglomerate during the molding process, creating voids between them. These voids remain as pores after sintering, contributing to the total porosity of the material.

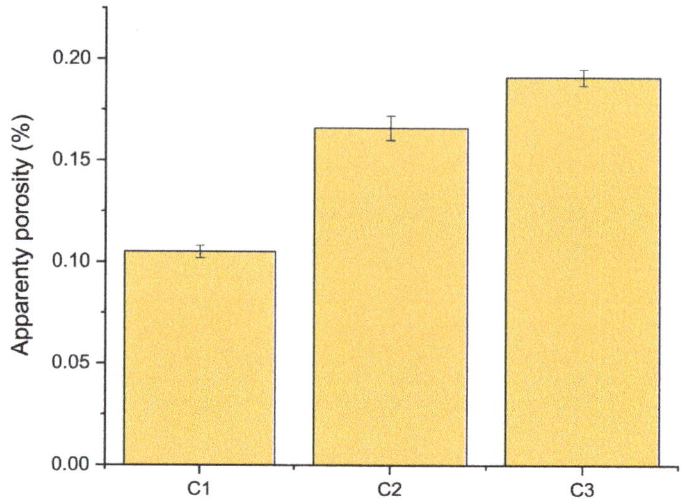

Figure 5. Histogram of apparent porosity of the C1, C2, and C3 biocomposites.

According to Bastan et al. [54], the PEEK particles spread over the substrate and acted as a continuous matrix due to the applied heating temperature (above the melting point). The PEEK matrix can maintain mechanical integrity while the HAp particles provide in vitro bioactivity to the composite coating. Previous studies have shown that porous surfaces can provide a larger surface area and enhance blood supply and oxygen delivery within bone tissues [55–58]. Additionally, HAp can promote bone cell growth and regeneration, improve adhesion between bone tissue and the implant, and regulate the rate and direction of bone remodeling [53,58].

Understanding the density and porosity of HAp/PEEK biocomposites is extremely important because the more significant the chemical, physical, and structural similarity between the prosthesis and bone tissue, the lower the likelihood of future issues with the implant. The porous surface is essential in a bone repair biomaterial, as it aids in new bone formation and neovascularization, produces more excellent wettability, and promotes the diffusion of biological fluids. This facilitates adherence and enhances cellular activity, characteristics that are essential in grafts. According to Dorozhkin [33], porous structures increase the surface area, improving space for cell attachment and favoring chemical bonding with the adjacent tissue. Additionally, porosity regulates bioactivity as it directly affects structural permeability, which controls the initial rate of tissue regeneration and influences mechanical properties.

Properly combining raw materials and processing techniques makes it possible to obtain biocomposites with high structural uniformity and mechanical strength values that make them suitable for various applications, including bone repair. Figure 6 illustrates the compressive stress vs deformation curves of the HAp/PEEK biocomposites with different HAp concentrations.

PEEK and hydroxyapatite-based biocomposites have shown characteristic behavior of semicrystalline materials in their compression strength curves, as observed in Figure 6. Despite its macromolecular rigidity, PEEK exhibits considerable deformation and can withstand significant levels of plastic deformation in both stress and compression. These materials are influenced by parameters such as temperature, test speed, and crystallinity, which can affect the penetration of HAp particles and consequently influence the morphology of the biocomposites, impacting their mechanical properties.

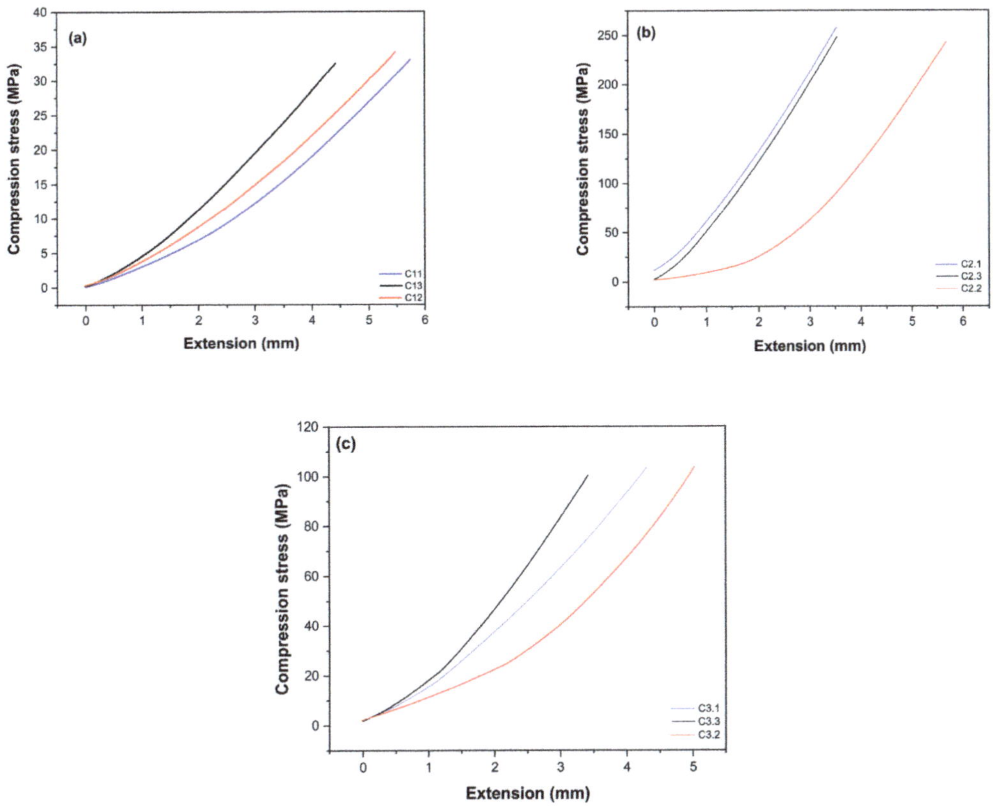

Figure 6. Compression strength curves as a function of extension for the biocomposites: (**a**) C1, (**b**) C2, and (**c**) C3.

The results of the compression tests indicate that the mechanical strength of the biocomposites does not follow a linear response as the concentration of HAp increases in PEEK. For example, the C2 biocomposite exhibited an average compressive strength of 246 MPa $\pm$ 0.9, significantly higher than the other samples (C1: 34 MPa $\pm$ 0.8 and C3: 96 MPa $\pm$ 0.7). This behavior can be attributed to the uniform and well-dispersed distribution of HAp within the PEEK matrix, as previously observed in the EDS results. This confirmed the homogeneity of the constituent mixture in the C2 sample, resulting in higher compressive strength. On the other hand, SEM revealed areas of porosity and HAp particle aggregation in the C3 biocomposite. These defects act as weak points under load, leading to a decline in mechanical properties. In a study conducted by Zheng et al., it was demonstrated that an excess of HAp reduced the mechanical properties of the HAp/PEEK composite [58].

The results suggest that a well-refined and homogeneous microstructure, with good interfacial adhesion between PEEK and HAp, is essential for achieving a composite with excellent mechanical performance. The C2 biocomposite exhibited higher mechanical strength, which is attractive for biomaterial applications requiring compressive stress. According to Silva [59], materials for bone repair should demonstrate the strength to withstand mechanical forces while regenerating. However, this strength does not necessarily need to be equivalent to that of bone, as over time, there is a significant increase in compression, ranging from 10 to 30 MPa, due to the growth of bone tissue in vivo.

The agar diffusion test was chosen to analyze the cytotoxicity of the HAp/PEEK biocomposites. This method involves evaluating the effects of the samples through an agar layer, which protects the cells from mechanical damage and allows the diffusion of substances from polymeric samples. The discoloration index was obtained by averaging the halo measurements in four sample quadrants, starting from its edge, as presented in Table 2. The positive control had an index of three; the average discoloration zone was 0.88 cm, with a moderate cytotoxicity level. The negative control, as expected, did not cause cell lysis and was classified as non-cytotoxic. The HAp/PEEK sample tested by the agar diffusion method did not show halo formation, a result similar to the negative control. This result indicates the absence of cytotoxicity, as shown in Table 2.

Table 2. Results of the agar diffusion test on HAp/PEEK biocomposites.

Tested Material	Degree Discoloration	Cell Lysis	Interpretation
Positive control	3	4	Moderate cytotoxicity
Negative control	0	0	Non-cytotoxic
white	0	0	Non-cytotoxic
C1	0	0	Non-cytotoxic
C2	0	0	Non-cytotoxic
C3	0	0	Non-cytotoxic

The images obtained from inverted digital microscopy were recorded after scanning the wells, prioritizing the sample's contour to verify the occurrence of cell lysis. The images presented in Figure 7 were qualitatively evaluated for the presence or absence of cell lysis, and the results were compared with the positive and negative controls. The positive control (Figure 7a) caused complete clearing, clearly reflecting cell lysis under and around the sample, confirming the cytotoxic nature indicated in Table 2. On the other hand, the negative and white controls (Figure 7b,c), as well as the samples (HAp/PEEK) (Figure 7d–f), exhibited a vital fibroblastic cell layer, where the uptake of neutral red dye occurred after 24 h of incubation. This indicates the absence of cytotoxicity in the tested (HAp/PEEK) samples and their viability for biological use in general biomaterial applications. The blank well (Figure 7c) should be treated as a parameter to observe the uptake of neutral red dye by living cells.

These results indicate that the HAp/PEEK samples were not toxic to the cells, making them viable for biomaterial applications, especially in regenerative medicine. These findings are consistent with reports in the literature that evaluated the cytotoxicity of HAp/PEEK [60]. Therefore, these biocomposites have promising potential for biomedical applications, considering their biocompatibility and safety of the tested cells.

The HAp/PEEK (C2—40/60%) biocomposite showed better mechanical behavior, which favors its application as a high-performance biomaterial. Therefore, it is essential to consider that exposed orthopedic wounds are highly susceptible to bacterial infection, the primary cause of implant surgery failure. Thus, implants with antimicrobial activity can help prevent infections at the site [61]. Given this, the antimicrobial activity of the HAp/PEEK (C2—40/60%) biocomposite was evaluated, and the results are presented in Table 3. According to the results, the biocomposite C2 demonstrated antibacterial activity, with minimum inhibitory concentration (MIC) against *S. aureus*, *P. aeruginosa*, and *E. coli*. The bactericidal effect of the biocomposite C2 was observed at twice the MIC concentration. It is important to note that isolated PEEK has low antimicrobial properties, which limits its application [62]. Thus, several strategies have been employed to enhance the antimicrobial activity of PEEK, such as surface modification or combining it with other materials [63]. In this regard, using coating materials like hydroxyapatite (HAp) is considered to improve the antibacterial properties of the implant material [64], as previously verified by the EDS results, which confirmed the homogeneity of the constituent mixture in this biocomposite.

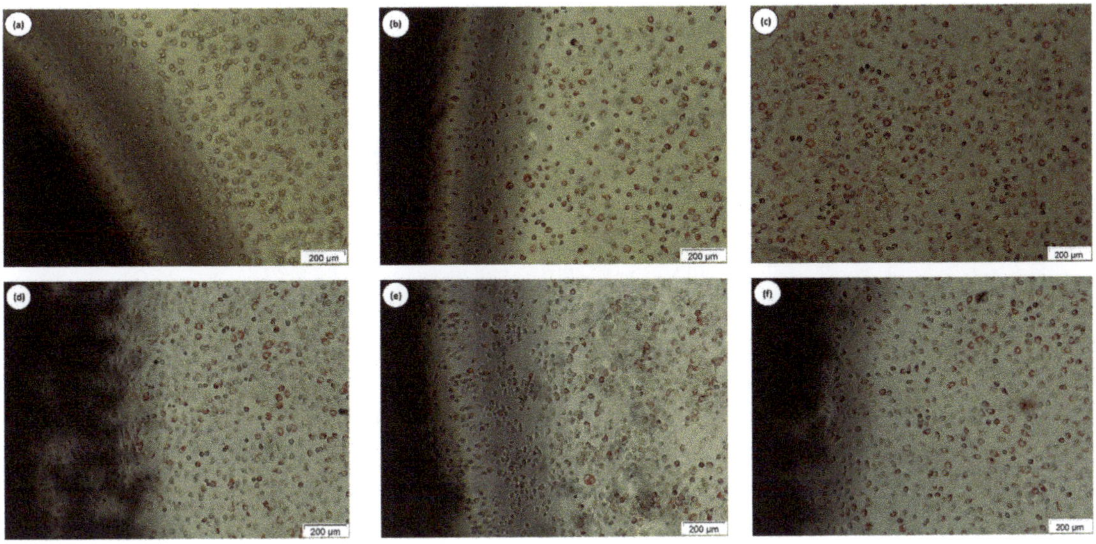

Figure 7. Images obtained by inverted digital microscopy at 100× magnification of the L929 cell line in the discoloration zones around the samples in the agar diffusion test: (**a**) positive control, (**b**) negative control, (**c**) blank, (**d**) C1, (**e**) C2, and (**f**) C3.

Table 3. MIC/MBC of Biocomposite C2.

Microrganisms	MIC/MBC (μg.mL^{-1})			
	C2	CIPRO	OXA	CFT
S. aureus ATCC (25923)	1562.5/3125	200	200	-
P. aeruginosa ATCC (27853)	1562.5/3125	200	-	200
E. coli ATCC (25922)	390.62/781.25	200	-	200

ATCC—American Type Culture Collection; MIC—Minimum Inhibitory Concentration; MBC—Minimum Bactericidal Concentration; CIPRO—Ciprofloxacin Hydrochloride; OXA—Oxacillin; CFT—Ceftazidime.

The study conducted by Wang et al. [65] demonstrated that bacterial biofilms formed and reached a thickness of about 90 μm with live bacteria visible on the surface of isolated PEEK. In contrast, bacterial cells were killed on the surface of the PEEK/nanofluorohydroxyapatite biocomposite, suggesting the antimicrobial activity of this composite. This bioceramic material affects both Gram-positive and Gram-negative bacteria due to its ability to penetrate bacterial cell walls through electrostatic interaction with the cell wall [66]. Additionally, other studies have shown activity through the production of reactive oxygen species (OH^-, H_2O_2 e $O_2{}^{-2}$) that can disrupt bacterial membranes, abrasive surface ordering due to surface effects, and aggregates that may contribute to mechanical damage to the bacterial cell membrane [67].

4. Conclusions

The synthesis of hydroxyapatite (HAp) and polyether ether ketone (PEEK) biocomposites using high-energy ball milling has proven to be a promising approach for creating materials for bone regeneration. The incorporation of HAp into PEEK resulted in significant changes in the material's morphology, particularly regarding porosity and mechanical strength. Among the analyzed biocomposites, C2, with 40% HAp, stood out for its effective antibacterial activity against *S. aureus*, *P. aeruginosa*, and *E. coli*, showing notable minimum inhibitory concentration (MIC) and bactericidal effect at concentrations above the MIC. Additionally, cytotoxicity tests revealed that C2 is non-toxic to cells, confirming its viability in biomaterials and safety in contact with biological tissues. The developed system offers significant potential for various medical applications, especially in bone restoration, due to its well-balanced combination of mechanical and biological properties.

Author Contributions: Conceptualization, J.A.C.F., C.B.B.L., M.R.C. and G.M.P.D.; Methodology, J.A.C.F., C.B.B.L., M.R.C. and G.M.P.D.; Validation, M.R.C., A.C.F.d.M.C. and N.M.d.S.O.; Formal analysis, J.A.C.F., C.B.B.L., M.R.C., G.M.P.D., A.C.F.d.M.C. and N.M.d.S.O.; Investigation, J.A.C.F., C.B.B.L., M.R.C. and G.M.P.D.; Resources, A.C.F.d.M.C. and N.M.d.S.O.; Data curation, A.C.F.d.M.C. and N.M.d.S.O.; Writing—original draft, J.A.C.F., C.B.B.L., M.R.C., G.M.P.D., A.C.F.d.M.C. and N.M.d.S.O., A.C.F.d.M.C. and N.M.d.S.O.; Writing—review & editing, M.R.C., A.C.F.d.M.C. and N.M.d.S.O.; Visualization, A.C.F.d.M.C. and N.M.d.S.O.; Supervision, A.C.F.d.M.C. and N.M.d.S.O.; Funding acquisition, A.C.F.d.M.C. and N.M.d.S.O. All authors have read and agreed to the published version of the manuscript.

Funding: The authors are grateful for the financial support from CAPES (001) and CNPq (Ana Cristina Figueiredo de Melo Costa—Process: 407850/2022-8; Carlos Bruno Barreto Luna—Process: 350025/2023-1).

Institutional Review Board Statement: Not applicable.

Data Availability Statement: The original contributions presented in the study are included in the article; further inquiries can be directed to the corresponding author.

Acknowledgments: The authors would like to thank the Coordination for the Improvement of Higher Education Personnel (CAPES—Brazil) and the National Council for Scientific and Technological Development (CNPq—Brazil) for the financial support and the Laboratory for the Evaluation and Development of Biomaterials of the Northeast—CERTBIO.

Conflicts of Interest: The authors declare no conflicts of interest.

References

1. Wu, A.-M.; Bisignano, C.; James, S.L.; Abady, G.G.; Abedi, A.; Abu-Gharbieh, E.; Alhassan, R.K.; Alipour, V.; Arabloo, J.; Asaad, M. Global, regional, and national burden of bone fractures in 204 countries and territories, 1990–2019: A systematic analysis from the Global Burden of Disease Study 2019. *Lancet Healthy Longev.* **2021**, *2*, e580–e592. [CrossRef]
2. Celiński, M.; Cybulski, M.; Fiłon, J.; Muszalik, M.; Goniewicz, M.; Krajewska-Kułak, E.; Ślifirczyk, A. Analysis of medical Management in Geriatric Patients in the hospital emergency department by example of selected cities with county status in Poland: A retrospective cohort study. *Int. J. Environ. Res. Public Health* **2021**, *19*, 48. [CrossRef]
3. Kodama, J.; Wilkinson, K.J.; Iwamoto, M.; Otsuru, S.; Enomoto-Iwamoto, M. The role of hypertrophic chondrocytes in regulation of the cartilage-to-bone transition in fracture healing. *Bone Rep.* **2022**, *17*, 101616. [CrossRef] [PubMed]
4. Park, S.; Rahaman, K.A.; Kim, Y.-C.; Jeon, H.; Han, H.-S. Fostering tissue engineering and regenerative medicine to treat musculoskeletal disorders in bone and muscle. *Bioact. Mater.* **2024**, *40*, 345–365. [CrossRef]
5. Tsang, K.Y.; Cheah, K.S.E. The extended chondrocyte lineage: Implications for skeletal homeostasis and disorders. *Curr. Opin. Cell Biol.* **2019**, *61*, 132–140. [CrossRef] [PubMed]
6. Zhang, J.; Zhang, L.; Li, C.; Chai, W.; Zhang, L.; Chen, H.; Zhang, W.; Hou, Z.; Chen, B.; Sun, T. Clinical guidelines for the diagnosis and treatment of fragility fractures of the pelvis. *Orthop. Surg.* **2023**, *15*, 2195–2212. [CrossRef] [PubMed]
7. Hu, C.; Ashok, D.; Nisbet, D.R.; Gautam, V. Bioinspired surface modification of orthopedic implants for bone tissue engineering. *Biomaterials* **2019**, *219*, 119366. [CrossRef]
8. Bartolomeu, F.; Costa, M.M.; Alves, N.; Miranda, G.; Silva, F.S. Additive manufacturing of NiTi-Ti6Al4V multi-material cellular structures targeting orthopedic implants. *Opt. Lasers Eng.* **2020**, *134*, 106208. [CrossRef]
9. Rodriguez-Contreras, A.; Punset, M.; Calero, J.A.; Gil, F.J.; Ruperez, E.; Manero, J.M. Powder metallurgy with space holder for porous titanium implants: A review. *J. Mater. Sci. Technol.* **2021**, *76*, 129–149. [CrossRef]
10. Jia, Z.; Xu, X.; Zhu, D.; Zheng, Y. Design, printing, and engineering of regenerative biomaterials for personalized bone healthcare. *Prog. Mater. Sci.* **2023**, *134*, 101072.
11. Ślusarczyk, K.; Flejszar, M.; Spilarewicz, K.; Wytrwal, M.; Awsiuk, K.; Wolski, K.; Raczkowska, J.; Janiszewska, N.; Chmielarz, P. On the way to increase osseointegration potential: Sequential SI-ATRP as promising tool for PEEK-based implant nano-engineering. *Eur. Polym. J.* **2024**, *210*, 112953. [CrossRef]
12. Adhikara, A.G.; Maharani, A.P.; Puspitasari, A.; Nuswantoro, N.F.; Juliadmi, D.; Maras, M.A.J.; Nugroho, D.B.; Saksono, B. Bovine hydroxyapatite for bone tissue engineering: Preparation, characterization, challenges, and future perspectives. *Eur. Polym. J.* **2024**, *214*, 113171. [CrossRef]

13. Daneshvar, A.; Farokhi, M.; Bonakdar, S.; Vossoughi, M. Synthesis and characterization of injectable thermosensitive hydrogel based on Pluronic-grafted silk fibroin copolymer containing hydroxyapatite nanoparticles as potential for bone tissue engineering. *Int. J. Biol. Macromol.* **2024**, *277*, 134412. [CrossRef]
14. Fendi, F.; Abdullah, B.; Suryani, S.; Usman, A.N.; Tahir, D. Development and application of hydroxyapatite-based scaffolds for bone tissue regeneration: A systematic literature review. *Bone* **2024**, *183*, 117075. [CrossRef] [PubMed]
15. Soleymani, S.; Naghib, S.M. 3D and 4D printing hydroxyapatite-based scaffolds for bone tissue engineering and regeneration. *Heliyon* **2023**, *9*, e19363. [CrossRef]
16. Safavi, M.S.; Khalil-Allafi, J.; Visai, L. Improved osteogenic activity of NiTi orthopedic implant by HAp-Nb2O5 composite coatings: Materials and biological points of view. *Biomater. Adv.* **2023**, *150*, 213435. [CrossRef]
17. Chopra, V.; Fuentes-Velasco, V.; Nacif-Lopez, S.R.; Melendez-Malpicca, J.; Mendez-Hernandez, A.S.; Ramos-Mendez-Iris, L.F.; Arroyo-Jimenez, D.A.; Reyes-Segura, D.G.; Gonzalez-Y-Mendoza, P.; Sanchez-Hernandez, K.A. Advancements in 3D-4D printing of hydroxyapatite composites for bone tissue engineering. *Ceram. Int.* **2024**, *in press*. [CrossRef]
18. Gani, M.A.; Budiatin, A.S.; Shinta, D.W.; Ardianto, C.; Khotib, J. Bovine hydroxyapatite-based scaffold accelerated the inflammatory phase and bone growth in rats with bone defect. *J. Appl. Biomater. Funct. Mater.* **2023**, *21*, 22808000221149193. [CrossRef]
19. El-Kady, A.M.; Mahmoud, E.M.; Sayed, M.; Kamel, S.M.; Naga, S.M. In-vitro and in-vivo evaluation for the bio-natural Alginate/nano-Hydroxyapatite (Alg/n-HA) injectable hydrogel for critical size bone substitution. *Int. J. Biol. Macromol.* **2023**, *253*, 126618. [CrossRef]
20. Ma, R.; Guo, D. Evaluating the bioactivity of a hydroxyapatite-incorporated polyetheretherketone biocomposite. *J. Orthop. Surg. Res.* **2019**, *14*, 32. [CrossRef]
21. Qiu, B.; Zhao, C.; Pan, J.; Zhou, Q.; Yao, W. Enhancing osteointegration and antibacterial properties of PEEK implants via AMP/HA dual-layer coatings. *Surf. Interfaces* **2024**, *51*, 104761. [CrossRef]
22. He, M.; Huang, Y.; Xu, H.; Feng, G.; Liu, L.; Li, Y.; Sun, D.; Zhang, L. Modification of polyetheretherketone implants: From enhancing bone integration to enabling multi-modal therapeutics. *Acta Biomater.* **2021**, *129*, 18–32. [CrossRef]
23. Han, X.; Gao, W.; Zhou, Z.; Yang, S.; Wang, J.; Shi, R.; Li, Y.; Jiao, J.; Qi, Y.; Zhao, J. Application of biomolecules modification strategies on PEEK and its composites for osteogenesis and antibacterial properties. *Colloids Surf. B Biointerfaces* **2022**, *215*, 112492.
24. Li, J.; Liu, H.; Wang, Y.; Wang, L.; Liu, G.; Chen, C.; Wei, L.; Li, H. Strength-plasticity synergetic CF/PEEK composites obtained by adjusting melt flow rate. *Polymer* **2024**, *305*, 127186.
25. Hu, Q.; Wang, Y.; Liu, S.; Liu, Q.; Zhang, H. 3D printed polyetheretherketone bone tissue substitute modified via amoxicillin-laden hydroxyapatite nanocoating. *J. Mater. Sci.* **2022**, *57*, 18601–18614.
26. Shimizu, T.; Fujibayashi, S.; Yamaguchi, S.; Yamamoto, K.; Otsuki, B.; Takemoto, M.; Tsukanaka, M.; Kizuki, T.; Matsushita, T.; Kokubo, T.; et al. Bioactivity of sol–gel-derived TiO2 coating on polyetheretherketone: In vitro and in vivo studies. *Acta Biomater.* **2016**, *35*, 305–317. [CrossRef] [PubMed]
27. Hu, F.; Fan, X.; Peng, F.; Yan, X.; Song, J.; Deng, C.; Liu, M.; Zeng, D.; Ning, C. Characterization of porous titanium-hydroxyapatite composite biological coating on polyetheretherketone (PEEK) by vacuum plasma spraying. *Coatings* **2022**, *12*, 433. [CrossRef]
28. Yabutsuka, T.; Fukushima, K.; Hiruta, T.; Takai, S.; Yao, T. Effect of pores formation process and oxygen plasma treatment to hydroxyapatite formation on bioactive PEEK prepared by incorporation of precursor of apatite. *Mater. Sci. Eng. C* **2017**, *81*, 349–358. [CrossRef]
29. Oladapo, B.I.; Zahedi, S.A.; Ismail, S.O.; Omigbodun, F.T. 3D printing of PEEK and its composite to increase biointerfaces as a biomedical material-A review. *Colloids Surf. B Biointerfaces* **2021**, *203*, 111726. [CrossRef]
30. Barandehfard, F.; Kianpour Rad, M.; Hosseinnia, A.; Rashidi, A.; Tahriri, M.; Tayebi, L. The evaluation of the mechanical characteristics of the synthesized glass-ionomer cements (GICs): The effect of hydroxyapatite and fluorapatite nanoparticles and glass powders. *J. Aust. Ceram. Soc.* **2019**, *55*, 507–517. [CrossRef]
31. Saeri, M.R.; Afshar, A.; Ghorbani, M.; Ehsani, N.; Sorrell, C.C. The wet precipitation process of hydroxyapatite. *Mater. Lett.* **2003**, *57*, 4064–4069. [CrossRef]
32. Sarkar, C.; Sahu, S.K.; Sinha, A.; Chakraborty, J.; Garai, S. Facile synthesis of carbon fiber reinforced polymer-hydroxyapatite ternary composite: A mechanically strong bioactive bone graft. *Mater. Sci. Eng. C* **2019**, *97*, 388–396. [CrossRef] [PubMed]
33. Dorozhkin, S.V. Bioceramics of calcium orthophosphates. *Biomaterials* **2010**, *31*, 1465–1485. [CrossRef] [PubMed]
34. *ASTM D695-23*; Standard Test Method for Compressive Properties of Rigid Plastics. ASTM: West Conshohocken, PA, USA, 2023. [CrossRef]
35. *ISO 10993-5:2009*; Biological Evaluation of Medical Devices—Part 5: Tests for In Vitro Cytotoxicity. ISO: Geneva, Switzerland, 2009.
36. Pina, H.d.V.; Farias, A.J.A.d.; Barbosa, F.C.; de Lima Souza, W.J.; de Sousa Barros, A.B.; Batista Cardoso, M.J.; Fook, M.V.L.; Wellen, R.M.R. Microbiological and cytotoxic perspectives of active PCL/ZnO film for food packaging. *Mater. Res. Express* **2020**, *7*, 025312. [CrossRef]
37. Wanderley, D.M.S.; Melo, D.F.; Silva, L.M.; Souza, J.W.L.; Pina, H.V.; Lima, D.B.; Amoah, S.K.S.; Borges, S.M.P.; Fook, M.V.L.; Moura, R.O. Biocompatibility and mechanical properties evaluation of chitosan films containing an N-acylhydrazonic derivative. *Eur. J. Pharm. Sci.* **2020**, *155*, 105547. [CrossRef]
38. Kim, D.-H.; Lee, S.-H.; Kim, K.-N.; Kim, K.-M.; Shim, I.-B.; Lee, Y.-K. Cytotoxicity of ferrite particles by MTT and agar diffusion methods for hyperthermic application. *J. Magn. Magn. Mater.* **2005**, *293*, 287–292. [CrossRef]

39. Puškar, T.; Trifković, B.; Koprivica, D.D.; Kojić, V.; Jevremović, A.; Mirković, S.; Eggbeer, D. In vitro cytotoxicity assessment of the 3D printed polymer based epoxy resin intended for use in dentistry. *Vojnosanit. Pregl.* **2019**, *76*. [CrossRef]
40. CLSI. *Methods for Dilution Antimicrobial Susceptibility Tests for Bacteria That Grow Aerobically*, 11th ed.; CLSI: Wayne, PA, USA, 2018.
41. Li, S.; Li, G.; Lian, X.; Hu, J.; Li, M.; Wang, B.; Zou, Y.; Zhou, Z. Integrated porous polyetheretherketone/hydroxyapatite scaffolds: Design, manufacturing and performance evaluation. *Compos. Part A Appl. Sci. Manuf.* **2023**, *173*. [CrossRef]
42. Asante, N.A.; Wang, Y.; Bakhet, S.; Kareem, S.; Owusu, K.A.; Hu, Y.; Appiah, M. Ambient temperature sulfonated carbon fiber reinforced PEEK with hydroxyapatite and reduced graphene oxide hydroxyapatite composite coating. *J. Biomed. Mater. Res. Part B Appl. Biomater.* **2021**, *109*, 2174–2183. [CrossRef]
43. Qi, D.; Wang, N.; Wang, S.; Liu, L.; Zhu, S.; She, P.; Yue, X. High-strength porous polyetheretherketone/hydroxyapatite composite for the treatment of bone defect. *Compos. Commun.* **2023**, *38*, 101473. [CrossRef]
44. Yusong, P.; Qianqian, S.; Yan, C. Fabrication and characterisation of functional gradient hydroxyapatite reinforced poly (ether ether ketone) biocomposites. *Micro Nano Lett.* **2013**, *8*, 357–361. [CrossRef]
45. Nascimento, M.; Dos Santos Almeida, A.R.; Hirata, M.C.; Elzubair, A.; Navarro da Rocha, D.; Prado da Silva, M.H. Biomineralization of calcium phosphates functionalized with hydroxyapatite-binding peptide. *J. Mech. Behav. Biomed. Mater.* **2023**, *146*, 106082. [CrossRef] [PubMed]
46. Nasiri-Tabrizi, B.; Basirun, W.J.; Yeong, C.H.; Walvekar, R.; Phang, S.W. Mechanosynthesis and microstructural characterization of F− and CO32− mono-and co-substituted hydroxyapatite. *J. Mol. Struct.* **2024**, *1315*, 138809. [CrossRef]
47. Xu, H.; Yang, L.; Wang, P.; Liu, Y.; Peng, M. Kinetic research on the sorption of aqueous lead by synthetic carbonate hydroxyapatite. *J. Environ. Manag.* **2008**, *86*, 319–328. [CrossRef] [PubMed]
48. Santos, F.S.F.; Ferreira, V.P.; Sá, M.D.; Fook, M.V.L. Modificação da superfície do poli (éter-éter-cetona). *Matéria (Rio de Janeiro)* **2017**, *22*, e-11883. [CrossRef]
49. Afshar, A.; Ghorbani, M.; Ehsani, N.; Saeri, M.R.; Sorrell, C.C. Some important factors in the wet precipitation process of hydroxyapatite. *Mater. Des.* **2003**, *24*, 197–202. [CrossRef]
50. Ponciano, R.C.d.O.; de Melo Costa, A.C.F.; Barbosa, R.C.; Fook, M.V.L.; Ponciano, J.J. Chitosan and hydroxyapatite scaffolds with amoxicillin for bone repair. *Res. Soc. Dev.* **2021**, *10*, e13410514790. [CrossRef]
51. Manzoor, F.; Golbang, A.; Jindal, S.; Dixon, D.; McIlhagger, A.; Harkin-Jones, E.; Crawford, D.; Mancuso, E. 3D printed PEEK/HA composites for bone tissue engineering applications: Effect of material formulation on mechanical performance and bioactive potential. *J. Mech. Behav. Biomed. Mater.* **2021**, *121*, 104601. [CrossRef]
52. Zhang, D.; George, O.J.; Petersen, K.M.; Jimenez-Vergara, A.C.; Hahn, M.S.; Grunlan, M.A. A bioactive "self-fitting" shape memory polymer scaffold with potential to treat cranio-maxillo facial bone defects. *Acta Biomater.* **2014**, *10*, 4597–4605. [CrossRef]
53. Li, S.; Li, M.; Hu, J.; Li, G.; Wang, B.; Zhou, Z. A new strategy for PEEK-based biocomposites to achieve porous surface for bioactivity and adjustable mechanical properties for orthopedic stress matching. *Compos. Part A Appl. Sci. Manuf.* **2024**, *177*, 107909. [CrossRef]
54. Bastan, F.E.; Atiq Ur Rehman, M.; Avcu, Y.Y.; Avcu, E.; Ustel, F.; Boccaccini, A.R. Electrophoretic co-deposition of PEEK-hydroxyapatite composite coatings for biomedical applications. *Colloids Surf. B Biointerfaces* **2018**, *169*, 176–182. [CrossRef] [PubMed]
55. Landy, B.C.; VanGordon, S.B.; McFetridge, P.S.; Sikavitsas, V.I.; Jarman-Smith, M. Mechanical and in vitro investigation of a porous PEEK foam for medical device implants. *J. Appl. Biomater. Funct. Mater.* **2013**, *11*, 35–44. [CrossRef] [PubMed]
56. Torstrick, F.B.; Lin, A.S.P.; Potter, D.; Safranski, D.L.; Sulchek, T.A.; Gall, K.; Guldberg, R.E. Porous PEEK improves the bone-implant interface compared to plasma-sprayed titanium coating on PEEK. *Biomaterials* **2018**, *185*, 106–116. [CrossRef]
57. Wang, L.; Zeng, X.; Chen, X.; Zeng, X.; Luo, K. Programmable, biodegradable composite scaffolds with variable pore morphology for minimal invasive bone repair. *Compos. Part A Appl. Sci. Manuf.* **2022**, *162*, 107130. [CrossRef]
58. Zheng, J.; Zhao, H.; Dong, E.; Kang, J.; Liu, C.; Sun, C.; Li, D.; Wang, L. Additively-manufactured PEEK/HA porous scaffolds with highly-controllable mechanical properties and excellent biocompatibility. *Mater. Sci. Eng. C* **2021**, *128*, 112333. [CrossRef]
59. Silva, L.A.J. Obtenção e Caracterização de Scaffolds de Hidroxiapatita Utilizando Amido de Milho Como Agente Porogênico. Master's Thesis, State University of Campinas, Faculty of Mechanical Engineering, Campinas, Brazylia, 2012.
60. Oladapo, B.I.; Ismail, S.O.; Bowoto, O.K.; Omigbodun, F.T.; Olawumi, M.A.; Muhammad, M.A. Lattice design and 3D-printing of PEEK with Ca10 (OH)(PO4) 3 and in-vitro bio-composite for bone implant. *Int. J. Biol. Macromol.* **2020**, *165*, 50–62. [CrossRef]
61. Wang, Z.; Yu, Z.; Wang, Z.; Li, S.; Song, L.; Xu, T.; Shen, G.; Wang, Y.; Huang, T.; Dong, X. Surface-activated 3D-printed PEEK implant enhances anti-infection and osteogenesis. *Compos. Part B Eng.* **2024**, *273*, 111258. [CrossRef]
62. Sun, A.A.; Lin, X.; Xue, Z.; Huang, J.; Bai, X.; Huang, L.; Lin, X.; Weng, S.; Chen, M. Facile surface functional polyetheretherketone with antibacterial and immunoregulatory activities for enhanced regeneration toward bacterium-infected bone destruction. *Drug Deliv.* **2021**, *28*, 1649–1663. [CrossRef] [PubMed]
63. Zheng, Z.; Liu, P.; Zhang, X.; Zou, X.; Mei, X.; Zhang, S.; Zhang, S. Strategies to improve bioactive and antibacterial properties of polyetheretherketone (PEEK) for use as orthopedic implants. *Mater. Today Bio* **2022**, *16*, 100402. [CrossRef]
64. He, X.; Deng, Y.; Yu, Y.; Lyu, H.; Liao, L. Drug-loaded/grafted peptide-modified porous PEEK to promote bone tissue repair and eliminate bacteria. *Colloids Surf. B Biointerfaces* **2019**, *181*, 767–777. [CrossRef]
65. Wang, L.; He, S.; Wu, X.; Liang, S.; Mu, Z.; Wei, J.; Deng, F.; Deng, Y.; Wei, S. Polyetheretherketone/nano-fluorohydroxyapatite composite with antimicrobial activity and osseointegration properties. *Biomaterials* **2014**, *35*, 6758–6775. [CrossRef] [PubMed]

66. Abdulkareem, M.H.; Abdalsalam, A.H.; Bohan, A.J. Influence of chitosan on the antibacterial activity of composite coating (PEEK/HAp) fabricated by electrophoretic deposition. *Prog. Org. Coat.* **2019**, *130*, 251–259. [CrossRef]
67. Resmim, C.M.; Dalpasquale, M.; Vielmo, N.I.C.; Mariani, F.Q.; Villalba, J.C.; Anaissi, F.J.; Caetano, M.M.; Tusi, M.M. Study of physico-chemical properties and in vitro antimicrobial activity of hydroxyapatites obtained from bone calcination. *Prog. Biomater.* **2019**, *8*, 1–9. [CrossRef] [PubMed]

polymers

MDPI

Article

Structural Similarity-Induced Inter-Component Interaction in Silicone Polymer-Based Composite Sunscreen Film for Enhanced UV Protection

Yuyan Chen [1], Hanwen Xu [1,*], Yuhang Liu [1], Qiuting Fu [1], Pingling Zhang [1], Jie Zhou [1], Hongyu Dong [1] and Xiaodong Yan [2,*]

1 Research & Innovation Center, Proya Cosmetics Co., Ltd., Hangzhou 310023, China; chenyuyan@proya.com (Y.C.); liuyuhang@proya.com (Y.L.); fuqiuting@proya.com (Q.F.); zhangpingling@proya.com (P.Z.); zhoujie74@proya.com (J.Z.); donghongyu@proya.com (H.D.)

2 Key Laboratory of Synthetic and Biological Colloids, Ministry of Education, School of Chemical & Material Engineering, Jiangnan University, Wuxi 214122, China

* Correspondence: xuhanwen48@proya.com (H.X.); xiaodong.yan@jiangnan.edu.cn (X.Y.)

Abstract: Film-forming agents are key ingredients in achieving long-lasting and effective sun protection by sunscreens. However, studies on the synergistic effects of film-forming agents with different properties as well as the interaction between film-forming agents and powders are scarce, restricting the development of sunscreens with strong ultraviolet (UV)-shielding effects. Herein, we innovatively adopt polysiloxane-15 as the soft film, trimethylsiloxysilicate as the hard film, and vinyl dimethicone/methicone silsesquioxane crosspolymer as the functional powder to construct a co-assembled sunscreen film, and we investigate the property-enhancing effects of the sunscreen film as well as the interaction between the silicone polymer-based film-forming agents and functional powder therein. The results show that structural similarity is essential to generating film-forming agent–powder interactions, which primarily enhance the Si−O bond binding energy, thereby enhancing the lasting protection effect of sunscreens. In addition, the inter-component interaction of the co-assembled sunscreen film inhibits the agglomeration of sunscreen paste to facilitate the formation of a homogeneous film, endowing the sunscreen with excellent UV protection abilities, with the sun protection factor (SPF) and protection factor of UVA (PFA) values increased by 61.58 and 43.84%, respectively. This work offers novel insights into the optimization of film-forming agent properties and the development of durable and efficient sunscreens.

Keywords: sunscreen; film-forming agent; polymer composites; ultraviolet protection

1. Introduction

The increased ultraviolet (UV) intensity caused by continuous destruction of the ozone layer has led to a series of problems, such as accelerated melanin synthesis, collagen degradation, and increased skin sensitivity [1–5], which has motivated consumers to increase their awareness of sun protection and development of the sunscreen market [6–11]. It is generally accepted that sunscreen products with high sun protection factor (SPF) provide better UV shielding [12–14], but this does not mean that a high level of protection can be achieved on a long-lasting basis. For instance, external friction and perspiration can disrupt the distribution of sunscreen agents, which in turn reduces the effectiveness of sunscreens [15–19]. Moreover, the film-forming behavior of the sunscreen film can also alter the distribution of sunscreen agents, thereby directly affecting the sun protection value of the sunscreen [20,21]. Therefore, an effective strategy urgently needs to be developed to enhance the long-lasting and efficient UV-shielding effect of sunscreens.

The sunscreen film is mainly shaped by the film-forming agent in the formula, and many studies have confirmed that the sunscreen film is the key component in achieving durable and efficient sun protection by sunscreen [22–31]. For example, Infante et al. found that the addition of film-forming agent PEG-75 lanolin was able to achieve better sun protection without increasing the amount of UV filters, and in vivo tests showed that the SPF value of a sample containing the film-forming agent was increased by 53.2% compared with a blank sample, which confirmed the sun protection enhancement by the film-forming agent [23]. Keshavarzi et al. evaluated the sweat resistance of sunscreens by varying the concentration of hydrophobic film-forming agents, which showed that medium levels of film-forming agents inhibited the redistribution of sunscreens when encountered with sweat and higher levels of film-forming agents reduced the amount of sunscreen washed off by sweat [26]. Although numerous efforts have been devoted to the study of film-forming agents, focusing only on a single component, the synergistic effects of film-forming agents with different properties are rarely investigated, especially the intrinsic interaction between film-forming agents and other components, resulting in limited long-lasting exertion of strong UV-shielding effects when applied in sunscreen formulations. The development of composite sunscreen films with excellent film-forming properties is required to meet the growing consumer demand for sunscreens.

Herein, we adopted a co-assembly strategy to combine rigid trimethylsiloxysilicate and flexible polysiloxane-15 film-forming agents to achieve a balance of film-forming strength and toughness, and we innovatively screened the specific functional powder vinyl dimethicone/methicone silsesquioxane crosspolymer (VDSC) to further reinforce the sunscreen film. The anti-stretching, anti-migration, anti-friction, sunscreen enhancement, and water resistance properties of different formulations were compared and tested. In addition, a combination of testing and characterization was employed to analyze the mechanisms of action of the film-forming agents and powder in the co-assembled sunscreen film.

2. Materials and Methods

2.1. Materials

All reagents were purchased from raw material vendors and used directly without further purification. The ingredients used in the experiment are listed according to the International Nomenclature of Cosmetic Ingredients: water (Wahaha, Hangzhou, China); propanediol (Meijingrong Chemistry, Suzhou, China); 1,2-hexanediol (Shinsung Materials, Seoul, Republic of Korea); citric acid (Seven Star Lemon Technology, Linyi, China); sodium citrate (Xilong Scientific, Shantou, China); polysiloxane-15 (DSM, Maastricht, The Netherlands); trimethylsiloxysilicate (Wacker, Munich, Germany); vinyl dimethicone/methicone silsesquioxane crosspolymer (ShinEtsu, Tokyo, Japan); ethylhexyl methoxycinnamate (DSM, Maastricht, The Netherlands); ethylhexyl salicylate (DSM, Maastricht, The Netherlands); octocrylene (Symrise, Holzminden, Germany); titanium dioxide (Hope-Tec Biotechnology, Shanghai, China); zinc oxide (Hope-Tec Biotechnology, Shanghai, China); cyclopentasiloxane (Momentive, Wilton, CT, USA); dimethicone (Momentive, Wilton, CT, USA); PEG-9 polydimethylsiloxyethyl dimethicone (ShinEtsu, Tokyo, Japan); PEG-10 dimethicone (ShinEtsu, Tokyo, Japan); tocopheryl acetate (NHU, Shaoxing, China); isododecane (Goodlight, Qingyuan, China); alcohol (Ante food, Suzhou, China); phenoxyethanol (Ashland, Covington, KY, USA); ethylhexylglycerin (Ashland, Covington, KY, USA); dextrin palmitate (Chiba Flour Milling, Chiba, Japan); VP/eicosene copolymer (Ashland, Covington, KY, USA); and talc (Ritian Nanotechnology, Dandong, China).

2.2. Preparation of Sunscreen Samples

The co-assembled sunscreen film was composed of three silicone polymer-based components: polysiloxane-15, trimethylsiloxysilicate, and VDSC, in which polysiloxane-15 served as the soft film, trimethylsiloxysilicate served as the hard film, and VDSC served as the functional powder. The three components possessed similar silicone-oxygen bonded structures, and the corresponding structural formulae are shown schematically in Figures S1–S3.

The formulation of the sunscreen sample containing the co-assembled sunscreen film is shown in Table 1. The specific process was as follows: Firstly, phase A and phase B were heated to 78~80 °C and homogenized using a blender (RW20, IKA, Staufen, Germany). Then, phase A was slowly transferred to phase B, stirred to homogeneity, and emulsified using a homogenizer (T25, IKA, Staufen, Germany) at a homogenization speed of 6000 rpm for 3 min. Next, the temperature was cooled down to below 40 °C, phase C was added, and the mixture was homogenized continuously for 2 min, ultimately obtaining the sunscreen containing the co-assembled sunscreen film, labeled as CA.

Preparation of control samples: Based on the above method and the formulation shown in Table 1, polysiloxane-15, trimethylsiloxysilicate, and VDSC were removed to obtain the blank control sample without the co-assembled sunscreen film, labeled as BC; polysiloxane-15 was removed to obtain the sunscreen sample with a non-soft film, labeled as NS; trimethylsiloxysilicate was removed to obtain the non-hard-film sample, labeled as NH; and VDSC was removed to obtain the sample (labeled as NP) without powder. In addition, polysiloxane-15 was replaced with dextrin palmitate to obtain the sunscreen sample with replacement of the soft film, labeled as RS; trimethylsiloxysilicate was replaced with VP/eicosene copolymer to obtain the sunscreen sample with replacement of the hard film, labeled as RH; and VDSC was replaced with talc to obtain the sunscreen sample with replacement of the functional powder, labeled as RP.

Table 1. The formulation of the sunscreen sample containing the co-assembled sunscreen film.

Phase	Component	Content (%)
A	Water	To 100
	Propanediol	5
	1,2-Hexanediol	0.4
	Citric acid	0.01
	Sodium citrate	0.15
B	Polysiloxane-15	2
	Trimethylsiloxysilicate	3
	Vinyl dimethicone/methicone silsesquioxane crosspolymer	2
	Ethylhexyl methoxycinnamate	6
	Ethylhexyl salicylate	3
	Octocrylene	3
	Titanium dioxide	4
	Zinc oxide	7
	Cyclopentasiloxane	9
	Dimethicone	8
	PEG-9 Polydimethylsiloxyethyl dimethicone	2.5
	PEG-10 Dimethicone	1.5
	Tocopheryl acetate	0.2
C	Isododecane	9
	Alcohol	8
	Phenoxyethanol	0.36
	Ethylhexylglycerin	0.04

2.3. Physical Characterization

2.3.1. Micromorphological Characterization of the Co-Assembled Sunscreen Film

Scanning electron microscopy (SEM, Apreo S HiVac, Thermo Scientific, Waltham, MA, USA) was used to analyze the film-forming toughness with different combinations of components in the co-assembled sunscreen film. The raw material trimethylsiloxysilicate was recorded as sample A; polysiloxane-15 as sample B; polysiloxane-15 and trimethylsiloxysilicate mixed in a 2:3 mass ratio as sample C; and polysiloxane-15, trimethylsiloxysilicate, and VDSC in a 2:3:2 mix as sample D. The samples were prepared by dispersing A, B, C, and D into isododecane (volatile solvent), taking 50 μL of the samples and adding it dropwise on the conductive adhesive substrates. The conductive adhesive substrates loaded with samples were then transferred to a vacuum-drying oven at 40 °C for 24 h to form a membrane. Finally, all samples were stretched at the same strength to observe the microscopic film structure after stretching. Moreover, SEM images of VDSC and talc were also observed.

2.3.2. Fourier-Transform Infrared Spectroscopy and X-Ray Photoelectron Spectroscopy Measurements

Fourier-transform infrared (FTIR, IRPrestige-21, Shimadzu, Kyoto, Japan) spectroscopy measurements were performed on VDSC, talc, and the film-forming agents (polysiloxane-15 and trimethylsiloxysilicate in a 2:3 mass ratio). To investigate the interaction between the powder and the film-forming agents, the powders after mixing and treatment with the film-forming agents were also tested. The treatment process of VDSC and talc was similar to the preparation process of the sunscreen samples, specifically, the powders and film-forming agent solutions were mixed according to a mass ratio of 2:5 and heated to 78~80 °C under stirring to create homogeneous dispersions. Then, the powders were homogenized at 6000 rpm for 3 min and centrifuged when the temperature was cooled to below 40 °C. Subsequently, the powders obtained by centrifugation were washed with deionized water and ethanol and dried at 60 °C in an oven. Furthermore, X-ray photoelectron spectroscopy (XPS, Thermo Scientific K-Alpha, Waltham, MA, USA) tests were performed on VDSC and film-forming agent-treated VDSC. All XPS spectra were calibrated using the C 1s signal at 284.8 eV.

2.3.3. Optical Profilometer and Contact Angle Measurements

The film-forming uniformity of each sunscreen sample after film formation was assessed using an optical profilometer (NT9100, Veeco, Plainview, NY, USA). Firstly, the profilometer test was carried out by selecting a 1 mm × 1 mm area on the blank pigskin surface, thus serving as the profile mapping prior to the application of sunscreen samples. Then, 5 μL of the treated sunscreen samples was added dropwise to the selected area using a pipette. After the samples were completely dry, the profilometer test was again performed on the area. The wettability of VDSC and talc to the film-forming agent solutions was determined using an optical contact angle meter (OCA15EC, Dataphysics, Stuttgart, Germany), and the powders were pressed at 8 MPa for 5 min.

2.4. *Performance Measurement*

2.4.1. Anti-Stretching Test

The sunscreen samples of 0.25 g were uniformly coated on the elastic silicone rubber belt with an area of 2 cm × 4 cm, and then left for 2.5 h to allow the sunscreens to form films. The initial condition of the film layer before stretching was first recorded using an optical microscope (MXB-2016Z, Hirox, Tokyo, Japan). After that, a unidirectional axial force was applied to all samples, which were stretched laterally to a length of 7.4~7.5 cm (deformation rate of about 85%) and fixed on the carrier stage with a G-clamp, observing whether holes and cracks appeared in the sunscreen film during stretching, thus evaluating the film-forming toughness of each sunscreen sample.

2.4.2. Anti-Migrating Test

A 0.1 g aliquot of the sunscreen formulation samples BC, CA, and NP was added dropwise onto a polymethylmethacrylate (PMMA) plate and then spread evenly with a spatula. The distribution of the coatings during the evaporation of samples (film formation) was observed and recorded every 15 min, thus evaluating the anti-migrating property of the co-assembled sunscreen film.

2.4.3. Anti-Rubbing Test

The formulation samples were evenly applied to the inner side of the arm with a loading of about 2 mg/cm^2 and an application area of 2.5 cm × 2.5 cm. Subsequently, the area of the applied sample was rubbed with a paper towel loaded with an iron block (gravity of about 1.4 N). UV LOOK (the sunscreen application area behaves as black) was used to observe the scratches on the edge of the application area after rubbing and the residue on the surface of the paper towel, thereby evaluating the rubbing resistance of the samples.

2.4.4. In Vitro Sun Protection Value Test

The method for in vitro testing of the sun protection value was based on the ISO 24443:2021 standard [32]. Specifically, samples CA, BC, NS, NH, and NP were added dropwise to a HD6-type PMMA plate (5 cm × 5 cm) with a loading of 1.3 mg/cm^2, after which the samples were pushed away with a finger wearing a finger cuff in a circular motion to uniformly spread the samples on the PMMA plate. Finally, the PMMA panels coated with samples were dried at room temperature without direct sunlight for 30 min, and the SPF and protection factor of UVA (PFA) were determined using a UV transmittance analyzer (UV2000, Labsphere, North Charleston, SC, USA). Four PMMA plates were prepared for each sample and the average value was recorded as the initial sun protection value of the sample. In addition, the PMMA panels after testing were immersed in a water bath at 35 °C under low-speed agitation for 40 min, and then the sun protection value of the dried samples was tested and recorded as the after-bath sun protection value. The differences in the formulations of sunscreens CA, BC, NS, NH, and NP were variations in the three components, polysiloxane-15, trimethylsiloxysilicate, and VDSC (as shown in Table 2), while the other components were consistent with Table 1.

Table 2. Differences in the composition of co-assembled sunscreen film in sunscreen formulations.

Component	Content (%)				
	CA	BC	NS	NH	NP
Polysiloxane-15	2	/	/	2	2
Trimethylsiloxysilicate	3	/	3	/	3
VDSC	2	/	2	2	/

3. Results and Discussion

3.1. Micromorphological Analysis of the Co-Assembled Sunscreen Film

Firstly, the micromorphological variation of the co-assembled sunscreen film was observed by SEM, thus qualitatively evaluating the role of each component in enhancing the toughness of the sunscreen film from the raw material level. Figure 1A shows the longitudinal and transverse cracks in the membrane layer, indicating that trimethylsiloxysilicate used as the hard film was a brittle material, which was prone to reaching failure strain upon stretching, leading to the membrane layer cracking and forming fragments. Figure 1B demonstrates that the membrane cracking of polysiloxane-15 was different from that of trimethylsiloxysilicate, showing a localized hole-type breakage (the exposed black area is the conductive adhesive substrate). The reason was that although polysiloxane-15 possessed the toughness of a flexible material, its inherent lack of strength led to holes during the stretching process. In addition, large areas of the membrane layer in Figure 1C remained basically intact with only a few smaller pores, indicating that the combination of polysiloxane-15 and trimethylsiloxysilicate significantly improved the film-forming toughness of the samples. In this case, the flexible chain segments of polysiloxane-15 could reduce the crosslinking density of the film groups, thereby increasing the free space of the siloxane chains to absorb impact energy and disperse stress [33,34]. Strikingly, after the addition of VDSC, based on the soft and hard film-compounded sample, the integrity of the film layer in Figure 1D was further enhanced compared to that in Figure 1C. No pores and cracks were found, and VDSC was observed to be uniformly dispersed and embedded in the film layer, indicating that the introduction of the functional powder further enhanced the film-forming toughness of the co-assembled sunscreen film through generating interfacial effects [35].

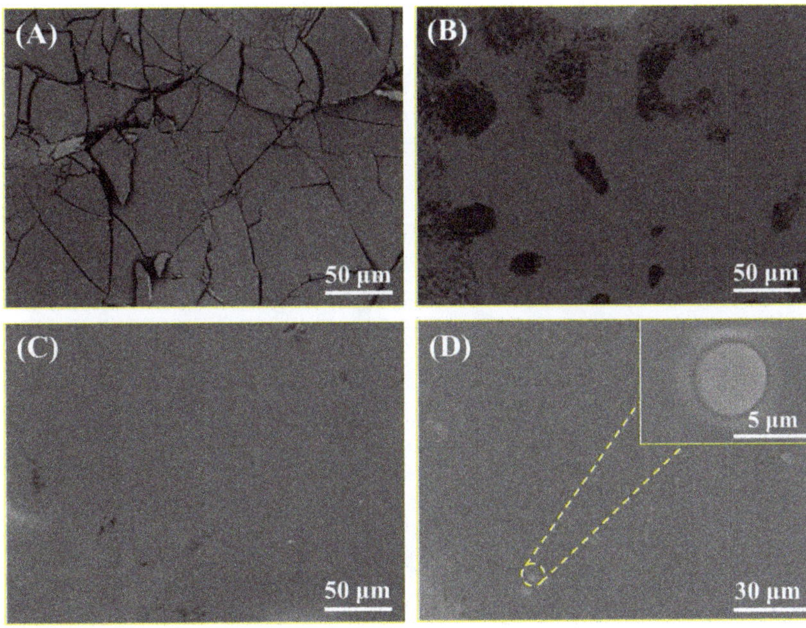

Figure 1. SEM images of (**A**) sample A, (**B**) sample B, (**C**) sample C, and (**D**) sample D after stretching.

3.2. Effect of the Co-Assembled Sunscreen Film on Anti-Stretching Property

The film-forming toughness of each sunscreen formulation was reflected by tensile tests. As shown in Figure 2, different combinations of film-forming agents and powders resulted in significant differences in the film-forming toughness of the samples. Figure 2A shows that the coating of sample CA containing the co-assembled sunscreen film remained uniform and intact during the stretching process, with no holes or cracks, indicating that the co-assembled sunscreen film constructed by the specific film-forming agents and the functional powder could effectively enhance the toughness of the sunscreen product against damage induced by movement of the facial skin. Figure 2B demonstrates that the film layer of blank sample BC was already broken before stretching, and the sunscreen coating further cracked with visible fractures during stretching, which undoubtedly worsened the UV absorption effect. When the specific combination of film-forming agents and powder in the co-assembled sunscreen film was varied, the coatings of sunscreen samples were found to be broken to varying degrees, which was consistent with the results in Figure 1. Specifically, when the soft (polysiloxane-15) and hard (trimethylsiloxysilicate) films were removed, the film layers of the corresponding samples NS (Figure 2C) and NH (Figure 2D) showed holes, cracks, and other damage after stretching. Surprisingly, when the functional powder (VDSC) was removed in the co-assembled sunscreen film, the sample NP (Figure 2E) was found to exhibit small cracks during stretching as well, further confirming that the functional powder played an important role in enhancing the toughness of the sunscreen film. As the powder had a similar silicone-oxygen bonded structure as the soft and hard films, its excellent compatibility allowed it to work synergistically with the film-forming agents to jointly resist the damage to the sunscreen caused by facial skin movement.

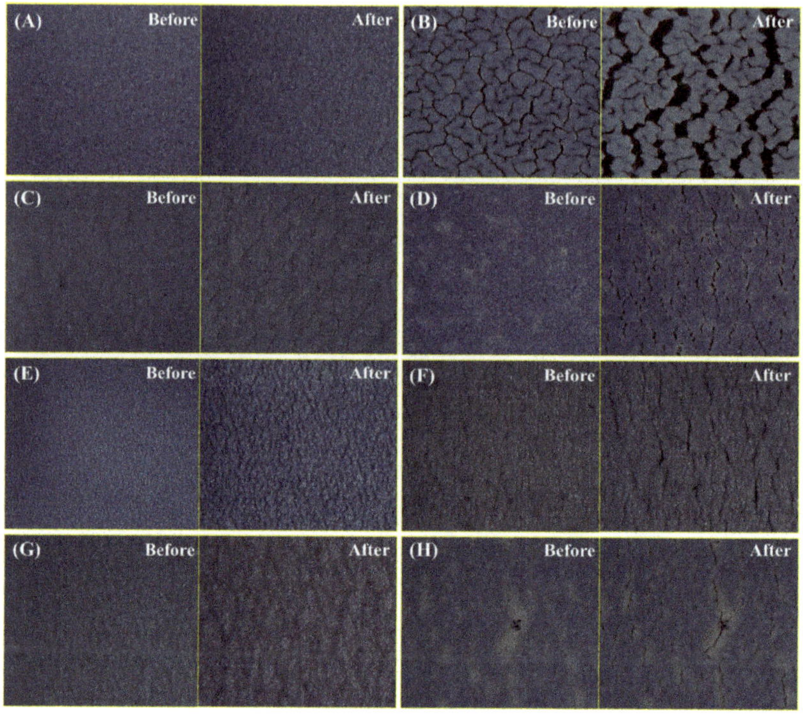

Figure 2. Film-forming conditions of sunscreen samples (**A**) CA, (**B**) BC, (**C**) NS, (**D**) NH, (**E**) NP, (**F**) RS, (**G**) RH, and (**H**) RP before and after stretching.

To investigate the reasons why the components of the co-assembled sunscreen film enhanced the toughness, each component was replaced separately. Figure 2F shows the film-forming conditions of sample RS with the soft film replaced by dextrin palmitate before and after stretching, revealing that the film layer of sample RS showed obvious cracking after stretching, which was probably caused by the poor compatibility of dextrin palmitate and trimethylsiloxysilicate, resulting in insufficient binding of the film-forming agents in the formulation and a decrease in the anti-stretching property. Sample RH in Figure 2G reconfirmed that the structural similarity between the sunscreen films played a key role in enhancing the mechanical properties. When the less compatible polysiloxane-15 and VP/eicosene copolymer were combined as soft and hard films, the sunscreen film was found to have difficulty resisting external stretching, and the same cracking condition occurred. Notably, when the functional powder in the co-assembled sunscreen film was replaced with talc, which was found to exhibit poor dispersion in the system according to Figure 2H, sample RP before stretching exhibited obvious particles and powder agglomerates on the surface of the coating, and clear cracks appeared in the coating during stretching. The above results indicated that the primary reason for the excellent toughness of the co-assembled sunscreen film was the high structural similarity of the components, and the excellent compatibility allowed them to fully combine and undergo inter-component interactions.

In addition, to reveal the strengthening mechanism of the mechanical properties of the sunscreen film by the functional powders, the microscopic morphology of different powders was analyzed by SEM and compared. The SEM images of VDSC are shown in Figures 3A and S4, exhibiting ~70 nm nanoparticles attached to the surface of ~5 μm microspheres, which possessed a large specific surface area. The rough nanoparticle edges of VDSC could act as "tentacles" to grasp and embed the film layer formed by the film-forming agents, which could enhance its interfacial combination with film-forming agents through chemical bonding to realize film-forming reinforcement [35]. By contrast, the microscopic morphology of talc presented a large nano-sheet structure (Figure 3B), the highly stacked structure of which led to a substantial decrease in specific surface area as well as poor compatibility and bonding with the soft and hard films, thus worsening the film-forming toughness of the sunscreen film after replacing the VDSC with talc. Furthermore, the wettability of pure VDSC and talc to the film-forming agents was compared through contact angle tests (Figure 3C), which were 59.3° and 107.4° for VDSC and talc, respectively. The smaller contact angle was attributed to the fact that VDSC possessed a similar silicone-oxygen bonded structure as silicone-based film-forming agents, predicting high compatibility and strong interactions with the film-forming agents, which further confirmed the dominant role of structural similarity in enhancing the binding of film-forming agents and powders.

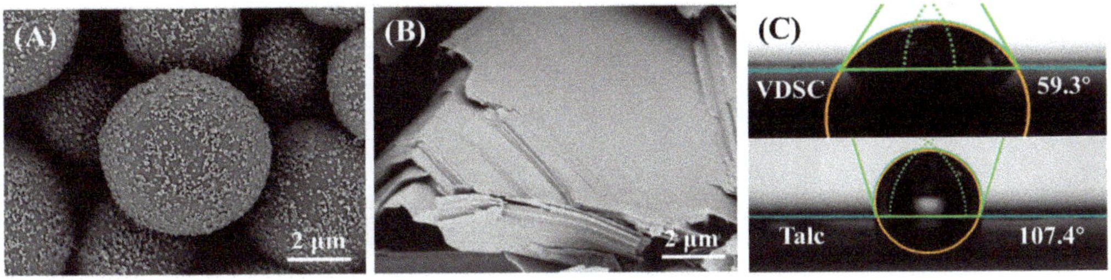

Figure 3. SEM images of (**A**) VDSC and (**B**) talc. (**C**) Contact angles of pure VDSC and talc to film-forming agent solution.

3.3. Effect of the Co-Assembled Sunscreen Film on Anti-Migrating Property

The anti-migrating property of the co-assembled sunscreen film was next analyzed by tracking the coating distribution during the evaporation of the sunscreen formulations. As shown in Figure 4A, the blank sample BC showed agglomeration of the coating visible to the naked eye after 15 min, with non-uniform distribution of the coating and significant cracking after 45 min, which was attributed to the severe aggregation of the paste (including powders, sunscreen agents, etc.) during the evaporation process [18]. After introducing the co-assembled sunscreen film, sample CA (Figure 4B) had a uniform coating distribution throughout the volatilization process without any cracking, indicating that the presence of the co-assembled sunscreen film could effectively anchor the paste to avoid its migration and aggregation during the volatilization process, which contributed to a uniform distribution of sunscreen agents in the film layer to maximize the sunscreen effect. The anti-migrating effect of the functional powder in the co-assembled sunscreen film was also investigated, and it was found that the paste at the edge of sample NP (Figure 4C) showed obvious migration to the interior after the removal of VDSC. Therefore, the introduction of the powder could be inferred to inhibit the dehumidification of the sunscreen cream and make the sunscreen agents stable [18], facilitating the formation of a uniform sunscreen film, which proved that the film-forming agents and functional powder of the co-assembled sunscreen film exerted a full synergistic effect in terms of anti-migration.

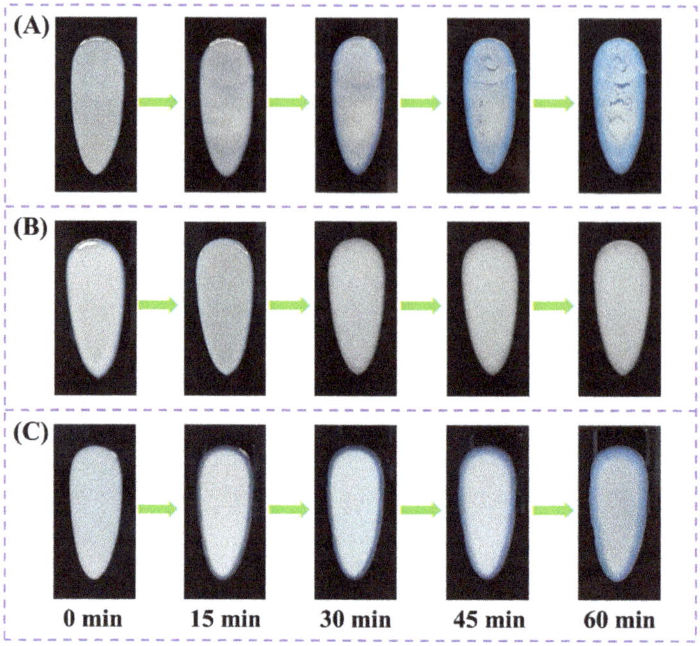

Figure 4. The migration conditions of coatings during drying of sunscreen samples (**A**) BC, (**B**) CA, and (**C**) NP. The time interval for each graph from left to right is 15 min.

3.4. Effect of the Co-Assembled Sunscreen Film on Anti-Rubbing Property

The effectiveness of the different sunscreen formulations against external friction was evaluated by means of anti-friction tests. As shown in Figure 5, different combinations of film-forming agents and powders had a significant effect on the anti-rubbing properties of the sunscreen samples. Figure 5A presents the film-forming conditions before and after rubbing of the sample containing the co-assembled sunscreen film. Sample CA was able to greatly maintain the initial state of the film layer after the rubbing treatment, with no obvious scratches on the edges and only trace residues on the tissue paper. However, a large amount of sunscreen product was carried away when blank sample BC (Figure 5B) was subjected to friction, the coating was severely damaged, and obvious scratches appeared on the edges. The above results indicated that the interaction of film-forming agents and functional powder with similar silicone-oxygen bonded structures significantly enhanced the friction-resistant property of the sunscreens. To verify the necessity of each component in the co-assembled sunscreen film, the anti-rubbing properties of the samples after removing the soft film, hard film, and functional powder were compared, respectively. Samples NS (Figure 5C) and NH (Figure 5D) were found to have different degrees of scratches on the coating edges after rubbing, and the film layer of sample NH was more obviously broken after rubbing, which indicated that trimethylsiloxysilicate provided a better anti-rubbing effect than polysiloxane-15. When the functional powder was removed, Figure 5E shows that the membrane edges of sample NP were scratched after rubbing, and part of the sunscreen agents were removed by the paper towel, indicating that the presence of VDSC had a further reinforcing effect on the soft and hard film groups, which was consistent with the role of the functional powder in the anti-stretching and anti-migrating tests, as described previously.

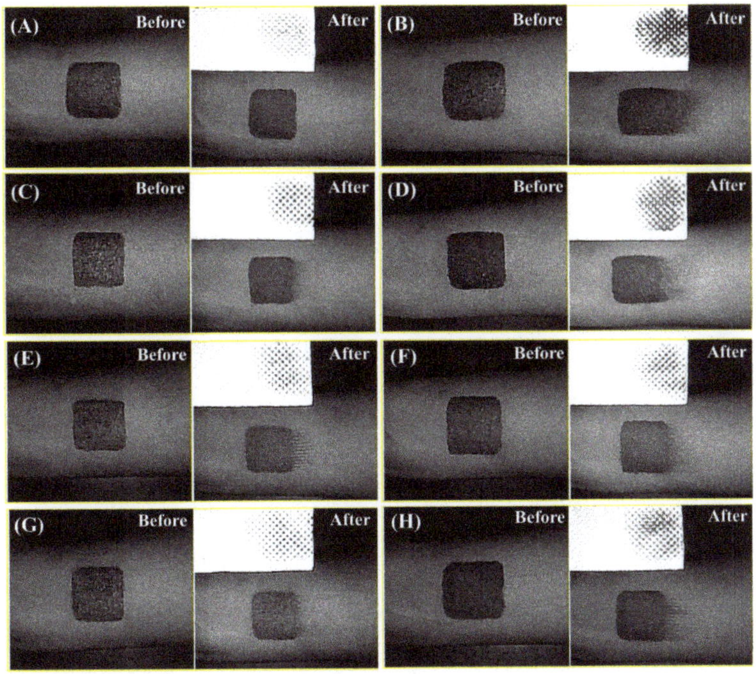

Figure 5. Film-forming conditions of sunscreen samples (**A**) CA, (**B**) BC, (**C**) NS, (**D**) NH, (**E**) NP, (**F**) RS, (**G**) RH, and (**H**) RP before and after rubbing.

To further confirm the effectiveness of structurally similar-oriented construction of sunscreen membranes, the components in the co-assembled sunscreen film were replaced by film-forming agents and powder with non-siloxane structures. Figure 5F shows the film-forming conditions of sample RS with the soft film replaced by dextrin palmitate before and after rubbing treatment, in which scratches were found to appear on the film edges after rubbing. Residual amounts of the sunscreen paste on the tissue paper were observed for samples CA and NS, indicating that dextrin palmitate could provide some anti-friction ability but was not as effective as polysiloxane-15, which was probably due to the poor compatibility between dextrin palmitate and trimethylsiloxysilicate leading to worse bonding between the soft and hard films, thus deteriorating the anti-rubbing performance of the sunscreen film. Similar results are shown in Figure 5G, where sample RH also showed scratches on the film edges after rubbing, which was attributed to the poor compatibility between VP/eicosene copolymer and polysiloxane-15, again confirming the importance of structural similarity in enhancing the mechanical properties of sunscreen films. Notably, Figure 5H demonstrates that the amount of paste removed by the tissue paper for sample RP after rubbing was significantly higher than that for sample NP, which was attributed to the poor compatibility of talc with the hard and soft film groups, which made it difficult to achieve sufficient embedding and exert a reinforcing effect on the film groups. Meanwhile, the poor dispersion of talc led to a rough and grainy film, so that the sunscreen agents were more likely to be carried away with the paste when the sample was subjected to friction, and scratches appeared on the edges. The above results indicated that the anti-friction property of the co-assembled sunscreen film mainly originated from the structural similarity of the components, which strengthened the film-forming agent–powder bonding to resist external friction.

3.5. FTIR and XPS Analyses

The excellent performance of the co-assembled sunscreen film was mainly attributed to the film-forming agent–powder interaction, which was investigated by a combination of FTIR and XPS tests. As seen in Figure 6A, the absorption peaks (positioned by the pink dashed lines) of VDSC changed after mixed treatment with film-forming agents, the relative intensities of the Si–O stretching vibrational absorption peaks at 1018 and 1095 cm^{-1} varied [36,37], and a new vibrational absorption peak appeared at 840 cm^{-1} [38], which proved that strong interactions between VDSC and the film-forming agents occurred in the sunscreen film. By contrast, no change in absorption peaks was observed for talc with a non-silicone-oxygen bonded structure (Figure 6B), implying no intrinsic interactions between talc and film-forming agents, further confirming the specificity among the components of the co-assembled sunscreen film. In addition, the interaction between VDSC and film-forming agents was further revealed by XPS tests. Figure 6C shows the Si 2p XPS spectra of VDSC before and after treatment. The signal at 102.6 eV was assigned to the Si–O bond [39,40]. After mixed treatment with film-forming agents, the Si–O bond in VDSC shifted toward higher binding energy (positioned by the black dashed lines). The signals located at 532.4 and 533.9 eV in the O 1s XPS plot were attributed to the Si–O bond and absorbed water, respectively (Figure 6D) [41,42], demonstrating that the same binding energy enhancement occurred in the Si–O bond of VDSC after treatment, which again confirmed the existence of strong interactions between VDSC and the film-forming agents [43,44]. These results indicated that structural similarity between film-forming agents and powder was essential to generating the interaction that reinforced the sunscreen film, and that such interactions were primarily reflected in the enhancement of the Si–O bond binding energy. The research showed that when the sunscreen film generated cracks under external force, the VDSC powder could absorb a large amount of destructive energy and slow down or terminate the spread of cracks at the interface, thereby realizing a strengthening and toughening effect on the sunscreen film [45,46].

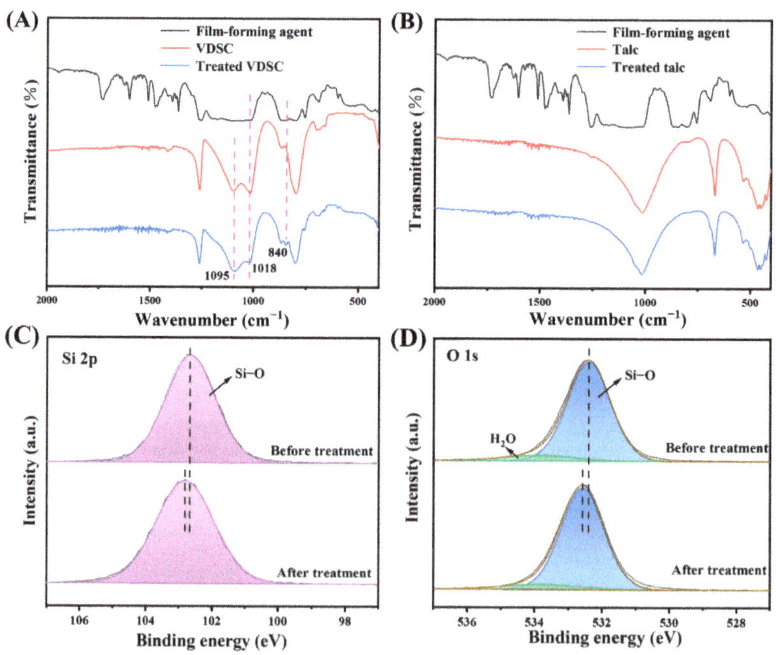

Figure 6. FTIR spectra of film-forming agent, (**A**) VDSC, and (**B**) talc before and after treatment. (**C**) Si 2p and (**D**) O 1s XPS spectra of VDSC before and after treatment.

3.6. Effect of Co-Assembled Sunscreen Film on In Vitro Sun Protection Values

To investigate the effect of the co-assembled sunscreen film on the sun protection value, UV absorption tests were performed on samples CA, BC, NS, NH, and NP, respectively. As shown in Figure 7A, sample CA exhibited the best UV absorption capacity, while the absorbance of blank sample BC was obviously weakened and showed the worst performance, which implied that the introduction of the co-assembled sunscreen film could significantly enhance the UV-shielding effect of the sunscreen formulations. Compared to the blank sample, the SPF and PFA values of sunscreen CA were increased by 61.58% and 43.84%, respectively, which demonstrated superior UV protection enhancement over other works [23,47,48]. The specific SPF and PFA values of each sample are given in Figure 7B, which showed that after removing the soft film, hard film, and functional powder in the co-assembled sunscreen film, respectively, the sun protection values of the corresponding samples NS, NH, and NP were reduced to a certain extent, suggesting that the enhanced UV absorptive capacity of the co-assembled sunscreen film originated from the synergistic interaction between film-forming agents and the powder.

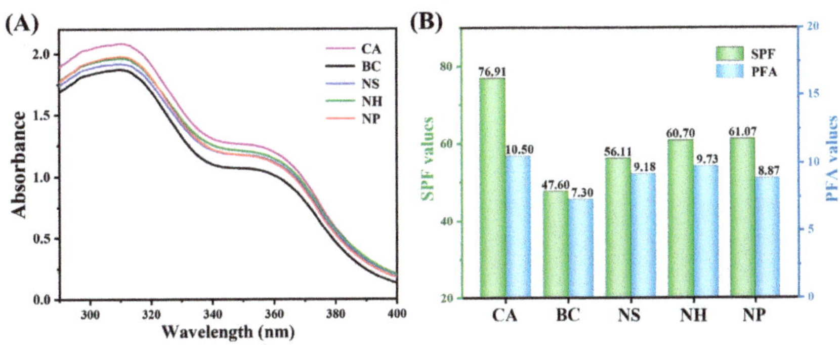

Figure 7. (**A**) UV absorption curves and (**B**) sun protection value test results of each sample.

To reveal the mechanism of the co-assembled sunscreen film to enhance the sun protection value, the film-forming conditions after the application of sunscreen samples were compared by optical profilometry. As seen in Figure 8A,B, the surface contour of the pigskin was highly consistent before and after the application of sample CA containing the co-assembled sunscreen film, and the high overlap of the X (Figure 8C) and Y (Figure 8D) profile curves reconfirmed the film-forming uniformity of sample CA. Uniform sunscreen films had been reported to maximize the effectiveness of sunscreen [25]. By contrast, the yellow dashed areas in Figure 8E,F as well as the green dashed areas in Figure 8G,H indicated that the depressions in the skin were filled with sunscreen after the application of sample BC, meaning that the sunscreen gathered in the skin grooves and formed a non-uniform film, resulting in a decrease in sun protection value. Moreover, the role of film-forming agents in the co-assembled sunscreen film was revealed by observing the film-forming conditions of the sunscreen after removing the film-forming agents. Figures S5 and S6 show that the skin-fit property of sunscreen without film-forming agents was reduced compared to sample CA, suggesting that the anti-migration effect of the film-forming agents mitigated the agglomeration and inhomogeneous distribution of the sunscreens to some extent. After removing the functional powder VDSC from the co-assembled sunscreen film, a similar decrease in film-forming uniformity was observed in Figures S7 and S8, indicating that the introduction of VDSC powder could mediate the volatilization process of the sunscreen film and allowed for uniform film formation. On the other hand, the introduction of the functional powder could enhance the solid content of the sunscreen cream, so that the film thickness after film formation increased, while relying on its diffuse reflective property through extending the optical distance to promote the value of sun protection [49].

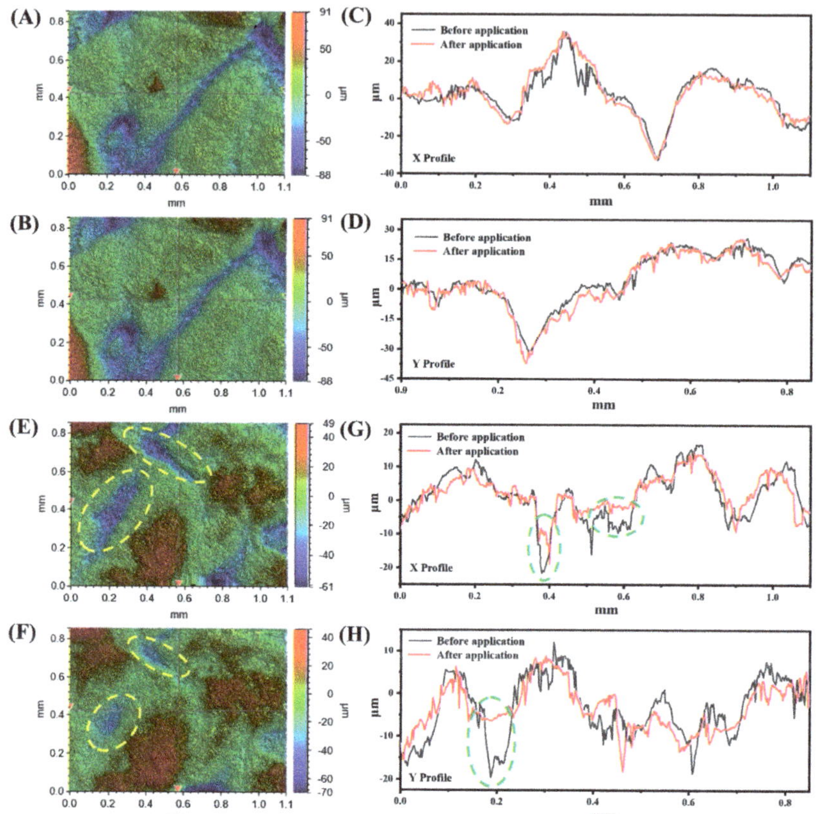

Figure 8. Three-dimensional contour images of pig skin (**A**) before and (**B**) after application of sunscreen CA, and contour curves of (**C**) X and (**D**) Y profiles. Three-dimensional contour images of pig skin (**E**) before and (**F**) after application of sunscreen BC, and contour curves of (**G**) X and (**H**) Y profiles.

In addition, the ability of the co-assembled sunscreen film to enhance the water resistance of sunscreens was assessed by testing the retention rate of sunscreen values after bathing treatment of each sample. According to Figure 9A, the after-bath UV absorption curve of sample CA was highly consistent with the initial curve, with only a slight attenuation of UVB absorption, and the retention rates of the SPF and PFA values after bathing were 93.95 and 99.62% (Table S1), respectively. Figure 9B demonstrates the UV absorption test results of the blank sample BC, which showed that the overall absorbance after bathing exhibited obvious attenuation. The retention rates of the SPF and PFA values at this time were 68.74 and 76.99% respectively, which were significantly lower compared to sample CA, indicating that the introduction of the co-assembled sunscreen film could confer excellent water resistance to the sunscreen formulation. By analyzing the UV absorption test results of Figure 9C–E, it could be seen that the water resistance of the samples corresponding to removal of the soft film, hard film, and functional powder were all reduced compared with sample CA. The specific changes in the sun protection value are shown in Figure 9F, indicating that the enhancement of the water-resistant performance originated from the combined action of the components in the co-assembled sunscreen film, for which the hydrophobic silica-oxygen bonded structure endowed the sunscreen with excellent water resistance.

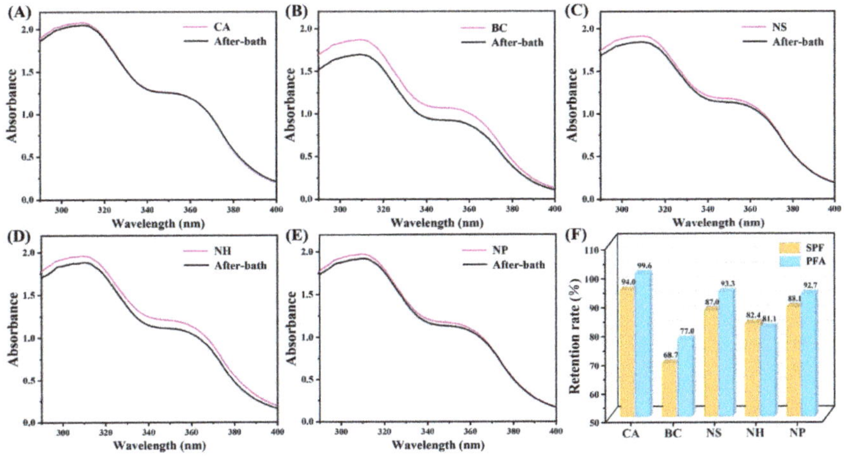

Figure 9. Comparison of initial and post-bath UV absorption curves of (**A**) CA, (**B**) BC, (**C**) NS, (**D**) NH, and (**E**) NP, and (**F**) sun protection value test results.

4. Conclusions

In summary, we constructed a co-assembled sunscreen film consisting of silicone polymer-based film-forming agents and powder with structural similarity as the guideline, which achieved the effective integration of different functional components and thus endowed the sunscreen formulation with excellent performance. The performance enhancement was mainly due to the structural similarity between the film-forming agents and functional powder, specifically, the combination of soft and hard films modulated the crosslinking density of the silicone-oxygen chains through the formation of a good interpenetrating network, thus optimizing the film strength and toughness simultaneously. The introduction of the functional powder strengthened the interfacial bonding with the film-forming agents through silicone-oxygen bonding, which further reinforced the anti-stretching, anti-migration, and anti-friction properties of the sunscreen. In addition, the film-forming agents and functional powder jointly mediated the film-forming process of the sunscreen to avoid agglomeration, resulting in a homogeneous film. Meanwhile, the introduction of the functional powder enhanced film thickness after film formation and increased contact with sunscreen agents, reflecting UV light at multiple levels, which in turn improved the sun protection property. The co-assembled sunscreen film showed an outstanding potentiation effect on sun protection values, respectively enhancing the SPF and PFA values by 61.58 and 43.84% compared with the blank sample. Benefiting from the hydrophobicity of the silicone-oxygen bonded structure, the co-assembled sunscreen film effectively enhanced the water resistance of the sunscreen formulation, with a retention rate of 93.95% for SPF and 99.62% for PFA after bathing. This work will broaden the application of polymers in cosmetics and provides a new strategy for the development of durable and efficient sunscreens.

Supplementary Materials: The following supporting information can be downloaded at: https://www.mdpi.com/article/10.3390/polym16233317/s1, Figure S1: Structural formula of polysiloxane-15; Figure S2: Structural formula of trimethylsiloxysilicate; Figure S3: Structural formula of VDSC; Figure S4: SEM image of VDSC surface-attached nanoparticles; Figure S5: Three-dimensional contour images of pig skin (A) before and (B) after application of sunscreen without film-forming agents; Figure S6: Contour curves of (A) X and (B) Y profiles before and after application of sunscreen without film-forming agents; Figure S7: Three-dimensional contour images of pig skin (A) before and (B) after application of sunscreen without VDSC; Figure S8: Contour curves of (C) X and (D)

Y profiles before and after application of sunscreen without VDSC; Table S1: Retention rate of sun protection values for each sample after water bath treatment.

Author Contributions: Conceptualization, Y.C., H.X. and X.Y.; methodology, Y.C. and Y.L.; software, Q.F. and P.Z.; validation, H.X. and X.Y.; formal analysis, J.Z.; investigation, H.D.; resources, X.Y.; data curation, Y.C.; writing—original draft preparation, Y.C.; writing—review and editing, H.X. and X.Y.; visualization, Y.L.; supervision, X.Y.; project administration, Q.F. and P.Z. All authors have read and agreed to the published version of the manuscript.

Funding: This research received no external funding.

Institutional Review Board Statement: Not applicable.

Data Availability Statement: Data are contained within the article and the Supplementary Information.

Acknowledgments: Y.C. thanks the School of Materials & Environmental Engineering, Hangzhou Dianzi University, for providing test support in the study.

Conflicts of Interest: Author Yuyan Chen, Hanwen Xu, Yuhang Liu, Qiuting Fu, Pingling Zhang, Jie Zhou, Hongyu Dong were employed by the company Proya Cosmetics Co., Ltd. The remaining authors declare that the research was conducted in the absence of any commercial or financial relationships that could be construed as a potential conflict of interest.

References

1. Verma, A.; Zanoletti, A.; Kareem, K.Y.; Adelodun, B.; Kumar, P.; Ajibade, F.O.; Silva, L.F.O.; Phillips, A.J.; Kartheeswaran, T.; Bontempi, E.; et al. Skin protection from solar ultraviolet radiation using natural compounds: A review. *Environ. Chem. Lett.* **2024**, *22*, 273–295. [CrossRef]
2. Byun, K.-A.; Lee, S.Y.; Oh, S.; Batsukh, S.; Jang, J.-W.; Lee, B.-J.; Rheu, K.-M.; Li, S.; Jeong, M.-S.; Son, K.H.; et al. Fermented Fish Collagen Attenuates Melanogenesis via Decreasing UV-Induced Oxidative Stress. *Mar. Drugs* **2024**, *22*, 421. [CrossRef] [PubMed]
3. Liu, Y.; Qin, D.; Wang, H.; Zhu, Y.; Bi, S.; Liu, Y.; Cheng, X.; Chen, X. Effect and mechanism of fish scale extract natural hydrogel on skin protection and cell damage repair after UV irradiation. *Colloids Surf. B Biointerfaces* **2023**, *225*, 113281. [CrossRef] [PubMed]
4. Griffin, G.K.; Booth, C.A.G.; Togami, K.; Chung, S.S.; Ssozi, D.; Verga, J.A.; Bouyssou, J.M.; Lee, Y.S.; Shanmugam, V.; Hornick, J.L.; et al. Ultraviolet radiation shapes dendritic cell leukaemia transformation in the skin. *Nature* **2023**, *618*, 834–841. [CrossRef]
5. Salminen, A.; Kaarniranta, K.; Kauppinen, A. Photoaging: UV radiation-induced inflammation and immunosuppression accelerate the aging process in the skin. *Inflamm. Res.* **2022**, *71*, 817–831. [CrossRef]
6. Lee, S.C.; Yoo, E.; Lee, S.H.; Won, K. Preparation and application of light-colored lignin nanoparticles for broad-spectrum sunscreens. *Polymers* **2020**, *12*, 699. [CrossRef]
7. Wang, C.; Wang, D.; Dai, T.; Xu, P.; Wu, P.; Zou, Y.; Yang, P.; Hu, J.; Li, Y.; Cheng, Y. Skin pigmentation-inspired polydopamine sunscreens. *Adv. Funct. Mater.* **2018**, *28*, 1802127. [CrossRef]
8. Girard, V.; Fragnières, L.; Chapuis, H.; Brosse, N.; Marchal-Heussler, L.; Canilho, N.; Parant, S.; Ziegler-Devin, I. The Impact of Lignin Biopolymer Sources, Isolation, and Size Reduction from the Macro-to Nanoscale on the Performances of Next-Generation Sunscreen. *Polymers* **2024**, *16*, 1901. [CrossRef]
9. Cardillo, D.; Sencadas, V.; Devers, T.; Islam, M.; Tehei, M.; Rosenfeld, A.; Boutard, T.; Rocher, E.; Barker, P.J.; Konstantinov, K. Attenuation of UV absorption by poly (lactic acid)-iron oxide nanocomposite particles and their potential application in sunscreens. *Chem. Eng. J.* **2021**, *405*, 126843. [CrossRef]
10. Lee, S.J.; Lee, D.; Park, S.A.; Park, J.J.; Park, W.H. Hyaluronic acid/polyphenol sunscreens with broad-spectrum UV protection properties from tannic acid and quercetin. *Int. J. Biol. Macromol.* **2024**, *257*, 128585. [CrossRef]
11. Zhang, Z.; Wang, Y.; Li, T.; Wu, J.; Huang, J.; Jiang, J.; Chen, M.; Dong, W. Mussel-inspired anti-permeation hybrid sunscreen with reinforced UV-blocking and safety performance. *Colloids Surf. A Physicochem. Eng. Aspects* **2023**, *676*, 132140. [CrossRef]
12. van Bodegraven, M.; Kröger, M.; Zamudio Díaz, D.F.; Lohan, S.B.; Moritz, R.K.; Möller, N.; Knoblich, C.; Vogelsang, A.; Milinic, Z.; Hallhuber, M.; et al. Redefine photoprotection: Sun protection beyond sunburn. *Exp. Dermatol.* **2024**, *33*, e15002. [CrossRef] [PubMed]
13. Hayag, M.V.; Chartier, T.; DeVoursney, J.; Tie, C.; Machler, B.; Taylor, J. A high SPF sunscreen's effects on UVB-induced immunosuppression of DNCB contact hypersensitivity. *J. Dermatol. Sci.* **1997**, *16*, 31–37. [CrossRef] [PubMed]
14. Williams, J.D.; Maitra, P.; Atillasoy, E.; Wu, M.-M.; Farberg, A.S.; Rigel, D.S. SPF 100+ sunscreen is more protective against sunburn than SPF 50+ in actual use: Results of a randomized, double-blind, split-face, natural sunlight exposure clinical trial. *J. Am. Acad. Dermatol.* **2018**, *78*, 902–910. [CrossRef] [PubMed]
15. Li, H.; Colantonio, S.; Dawson, A.; Lin, X.; Beecker, J. Sunscreen application, safety, and sun protection: The evidence. *J. Cutan. Med. Surg.* **2019**, *23*, 357–369. [CrossRef]
16. Diffey, B.L. When should sunscreen be reapplied? *J. Am. Acad. Dermatol.* **2001**, *45*, 882–885. [CrossRef]
17. Keshavarzi, F.; Knudsen, N.; Brewer, J.R.; Ebbesen, M.F.; Komjani, N.M.; Moghaddam, S.Z.; Jafarzadeh, S.; Thormann, E. In vitro skin model for characterization of sunscreen substantivity upon perspiration. *Int. J. Cosmet. Sci.* **2021**, *43*, 359–371. [CrossRef]

18. Ou-Yang, H.; Jiang, L.I.; Meyer, K.; Wang, S.Q.; Farberg, A.S.; Rigel, D.S. Sun protection by beach umbrella vs sunscreen with a high sun protection factor: A randomized clinical trial. *JAMA Dermatol.* **2017**, *153*, 304–308. [CrossRef]
19. Ruvolo, E.; Aeschliman, L.; Cole, C. Evaluation of sunscreen efficacy over time and re-application using hybrid diffuse reflectance spectroscopy. *Photodermatol. Photoimmunol. Photomed.* **2020**, *36*, 192–199. [CrossRef]
20. Binks, B.P.; Fletcher, P.D.; Johnson, A.J.; Marinopoulos, I.; Crowther, J.M.; Thompson, M.A. Evaporation of particle-stabilized emulsion sunscreen films. *ACS Appl. Mater. Interfaces* **2016**, *8*, 21201–21213. [CrossRef]
21. Giacomoni, P.U.; Teta, L.; Najdek, L. Sunscreens: The impervious path from theory to practice. *Photochem. Photobiol. Sci.* **2010**, *9*, 524–529. [CrossRef] [PubMed]
22. Ngoc, L.T.N.; Van Tran, V.; Moon, J.-Y.; Chae, M.; Park, D.; Lee, Y.-C. Recent trends of sunscreen cosmetic: An update review. *Cosmetics* **2019**, *6*, 64. [CrossRef]
23. Infante, V.H.P.; Campos, P.M.; Calixto, L.; Darvin, M.; Kröger, M.; Schanzer, S.; Lohan, S.; Lademann, J.; Meinke, M. Influence of physical–mechanical properties on SPF in sunscreen formulations on ex vivo and in vivo skin. *Int. J. Pharm.* **2021**, *598*, 120262. [CrossRef] [PubMed]
24. Kakuda, L.; Campos, P.M.B.G.M.; Zanin, R.B.; Favaro, L.N. Development of multifunctional sunscreens: Evaluation of physico-mechanical and film-forming properties. *Int. J. Pharm.* **2023**, *635*, 122705. [CrossRef] [PubMed]
25. Binks, B.P.; Brown, J.; Fletcher, P.D.; Johnson, A.J.; Marinopoulos, I.; Crowther, J.M.; Thompson, M.A. Evaporation of sunscreen films: How the UV protection properties change. *ACS Appl. Mater. Interfaces* **2016**, *8*, 13270–13281. [CrossRef]
26. Keshavarzi, F.; Knudsen, N.; Komjani, N.M.; Ebbesen, M.F.; Brewer, J.R.; Jafarzadeh, S.; Thormann, E. Enhancing the sweat resistance of sunscreens. *Skin Res. Technol.* **2022**, *28*, 225–235. [CrossRef]
27. Sohn, M.; Herzog, B.; Osterwalder, U.; Imanidis, G. Calculation of the sun protection factor of sunscreens with different vehicles using measured film thickness distribution—Comparison with the SPF in vitro. *J. Photochem. Photobiol. B Biol.* **2016**, *159*, 74–81. [CrossRef]
28. Tan, N.C.; Djordjevic, I.; Malley, J.A.; Kwang, A.L.; Ikhwan, S.; Šolić, I.; Singh, J.; Wicaksono, G.; Lim, S.; Steele, T.W. Sunlight activated film forming adhesive polymers. *Biomater. Adv.* **2021**, *127*, 112240. [CrossRef]
29. Yu, S.; Lu, Y.; Guo, S.; Guo, T.; Takagi, A.; Kamkar, M.; Rojas, O.J. Lignin-Polylactide Reverse Emulsions for Water and UV-Resistant Composite Films. *ACS Sustain. Chem. Eng.* **2023**, *11*, 12503–12513. [CrossRef]
30. Sohn, M.; Buehler, T.; Imanidis, G. Repartition of oil miscible and water soluble UV filters in an applied sunscreen film determined by confocal Raman microspectroscopy. *Photochem. Photobiol. Sci.* **2016**, *15*, 861–871. [CrossRef]
31. Li, P.; Wang, S.; Zhou, S. Comfortable skin sunscreens based on waterborne cross-linkable polydimethylsiloxane coatings. *J. Mater. Chem. C* **2020**, *8*, 17383–17394. [CrossRef]
32. *ISO 24443:2021*; Cosmetics—Determination of sunscreen UVA photoprotection in vitro. ISO: Geneva, Switzerland, 2021.
33. Zhang, Y.; Yang, K.; Liu, R.; Yao, J.; Yan, H. Superior tough, highly wear durable and self-lubricating epoxy composite co-enhanced by soft and hard nanomaterials. *Chem. Eng. J.* **2023**, *460*, 141773. [CrossRef]
34. Yamawake, K.; Hayashi, M.; Nobukawa, S. Preparation of All Amorphous PMMA Resins Based on the Graft Architecture with a Flexible Main Chain for Simultaneous Enhancement of Thermal and Mechanical Toughness. *Macromol. Chem. Phys.* **2022**, *223*, 2200255. [CrossRef]
35. Gilbert, E.N.; Hayes, B.S.; Seferis, J.C. Interlayer toughened unidirectional carbon prepreg systems: Effect of preformed particle morphology. *Compos. Part A Appl. Sci. Manuf.* **2003**, *34*, 245–252. [CrossRef]
36. Yin, Y.; Yin, J.; Zhang, W.; Tian, H.; Hu, Z.; Ruan, M.; Xu, H.; Liu, L.; Yan, X.; Chen, D. FT-IR and micro-Raman spectroscopic characterization of minerals in high-calcium coal ashes. *J. Energy Inst.* **2018**, *91*, 389–396. [CrossRef]
37. Jakobsson, S. Determination of Si/Al ratios in semicrystalline aluminosilicates by FT-IR spectroscopy. *Appl. Spectrosc.* **2002**, *56*, 797–799. [CrossRef]
38. Yusuf, M.O. Bond characterization in cementitious material binders using Fourier-transform infrared spectroscopy. *Appl. Sci.* **2023**, *13*, 3353. [CrossRef]
39. Coxon, P.R.; Coto, M.; Juzeliunas, E.; Fray, D.J. The use of electro-deoxidation in molten salts to reduce the energy consumption of solar grade silicon and increase the output of PV solar cells. *Prog. Nat. Sci. Mater. Int.* **2015**, *25*, 583–590. [CrossRef]
40. Kaur, A.; Chahal, P.; Hogan, T. Selective fabrication of SiC/Si diodes by excimer laser under ambient conditions. *IEEE Electron Device Lett.* **2015**, *37*, 142–145. [CrossRef]
41. Xiao, Z.; Yu, C.; Lin, X.; Chen, X.; Zhang, C.; Jiang, H.; Zhang, R.; Wei, F. TiO_2 as a multifunction coating layer to enhance the electrochemical performance of SiO_x@ TiO_2@C composite as anode material. *Nano Energy* **2020**, *77*, 105082. [CrossRef]
42. Krishnan, P.; Liu, M.; Itty, P.A.; Liu, Z.; Rheinheimer, V.; Zhang, M.-H.; Monteiro, P.J.M.; Yu, L.E. Characterization of photocatalytic TiO_2 powder under varied environments using near ambient pressure X-ray photoelectron spectroscopy. *Sci. Rep.* **2017**, *7*, 43298. [CrossRef] [PubMed]
43. Chan, M.-L.; Lau, K.-T.; Wong, T.; Cardona, F. Interfacial bonding characteristic of nanoclay/polymer composites. *Appl. Surf. Sci.* **2011**, *258*, 860–864. [CrossRef]
44. Bagus, P.S.; Nelin, C.J.; Brundle, C.R. Chemical significance of x-ray photoelectron spectroscopy binding energy shifts: A Perspective. *J. Vac. Sci. Technol. A* **2023**, *41*, 068501. [CrossRef]
45. Shin, H.; Kim, B.; Han, J.-G.; Lee, M.Y.; Park, J.K.; Cho, M. Fracture toughness enhancement of thermoplastic/epoxy blends by the plastic yield of toughening agents: A multiscale analysis. *Compos. Sci. Technol.* **2017**, *145*, 173–180. [CrossRef]

46. Picu, C.R.; Krawczyk, K.K.; Wang, Z.; Pishvazadeh-Moghaddam, H.; Sieberer, M.; Lassnig, A.; Kern, W.; Hadar, A.; Constantinescu, D.M. Toughening in nanosilica-reinforced epoxy with tunable filler-matrix interface properties. *Compos. Sci. Technol.* **2019**, *183*, 107799. [CrossRef]
47. Peres, D.D.; Sarruf, F.D.; de Oliveira, C.A.; Velasco, M.V.R.; Baby, A.R. Ferulic acid photoprotective properties in association with UV filters: Multifunctional sunscreen with improved SPF and UVA-PF. *J. Photochem. Photobiol. B Biol.* **2018**, *185*, 46–49. [CrossRef]
48. Meaudre, H.; Aubrun, O.; Boitte, J.; Douezan, S.; Josso, M.; Le Verge, D.; Renoux, P.; Rondepierre, G. New formulation technology to boost sun protection. *Int. J. Cosmet. Sci.* **2023**, *45*, 802–814. [CrossRef]
49. Oshina, I.; Spigulis, J. Beer–Lambert law for optical tissue diagnostics: Current state of the art and the main limitations. *J. Biomed. Opt.* **2021**, *26*, 100901. [CrossRef]

polymers

MDPI

Article

Flexural Behavior of Innovative Glass Fiber-Reinforced Composite Beams Reinforced with Gypsum-Based Composites

Yiwen Liu, Bo Su * and Tianyu Zhang

Faculty of Civil Engineering and Mechanics, Jiangsu University, Zhenjiang 212000, China; liuyiwen0426@163.com (Y.L.); zty9152024@163.com (T.Z.)
* Correspondence: subo@ujs.edu.cn

Abstract: Glass Fiber-Reinforced Composite (GFRP) has found widespread use in engineering structures due to its lightweight construction, high strength, and design flexibility. However, pure GFRP beams exhibit weaknesses in terms of stiffness, stability, and local compressive strength, which compromise their bending properties. In addressing these limitations, this study introduces innovative square GFRP beams infused with gypsum-based composites (GBIGCs). Comprehensive experiments and theoretical analyses have been conducted to explore their manufacturing process and bending characteristics. Initially, four types of GBIGC—namely, hollow GFRP beams, pure gypsum, steel-reinforced gypsum, and fiber-mixed gypsum-infused beams—were designed and fabricated for comparative analysis. Material tests were conducted to assess the coagulation characteristics of gypsum and its mechanical performance influenced by polyvinyl acetate fibers (PVAs). Subsequently, eight GFRP square beams (length: 1.5 m, section size: 150 mm × 150 mm) infused with different gypsum-based composites underwent four-point bending tests to determine their ultimate bending capacity and deflection patterns. The findings revealed that a 0.12% dosage of protein retarder effectively extends the coagulation time of gypsum, making it suitable for specimen preparation, with initial and final setting times of 113 min and 135 min, respectively. The ultimate bending load of PVA-mixed gypsum-infused GFRP beams is 203.84% higher than that of hollow beams, followed by pure gypsum and steel-reinforced gypsum, with increased values of 136.97% and 186.91%, respectively. The ultimate load values from the theoretical and experimental results showed good agreement, with an error within 7.68%. These three types of GBIGCs with significantly enhanced flexural performance can be filled with different materials to meet specific load-bearing requirements for various scenarios. Their improved flexural strength and lightweight characteristics make GBIGCs well suited for applications such as repairing roof beams, light prefabricated frames, coastal and offshore buildings.

Keywords: gypsum; polyvinyl acetate fibers; GFRP; composite beams; coagulation characteristics

1. Introduction

Glass Fiber-Reinforced Composite (GFRP) is a material composed of resin and glass fiber, which is valued in the construction industry for its high strength and cost-effectiveness. The decreasing cost of raw materials and advancements in composite manufacturing technologies, such as pultrusion or vacuum introduction, have contributed to the growing popularity of GFRP in civil engineering [1–3]. Noteworthy applications of GFRP in civil engineering include the use of GFRP sandwich panels for bridge decks [4], lightweight slabs for the restoration of wooden or masonry buildings [5,6], composite crashworthy devices [7], ship deck panels [8], emergency housing, and various new construction projects [9]. However, it is important to note that GFRP, being inherently brittle, exhibits weaknesses in terms of stiffness, stability, and durability [10–12]. Many researchers are actively exploring ways to enhance its structural and mechanical performance [13,14].

An effective strategy for enhancing the mechanical performance of GFRP involves combining it with concrete to create composite structures. Robinson et al. [15] conducted a study on the bonding effects of short-span rectangular GFRP beams filled with concrete. The results demonstrated that the strength and stiffness of the concrete-filled GFRP beams were twice as high as those of unbonded GFRP beams. In a similar vein, Ahmed Abouzied et al. [16] investigated the flexural behavior of rectangular FRP beams filled with reinforced concrete. Their findings revealed that the strength of concrete and the thickness of GFRP beams were the primary factors influencing the flexural performance of infilled FRP beams. Additionally, Ahmed et al. [17] and Majid Muttashar et al. [18] highlighted improvements in the ultimate bearing capacity, ductility, and energy absorption of concrete-infilled GFRP beams with an increase in FRP beam thickness and concrete compressive strength. It is worth noting, however, that the incorporation of concrete significantly increases the weight of the structure, which may have detrimental effects on its overall performance.

In recent research efforts, there has been a focus on using lighter infilling materials instead of concrete to reduce structural weight. For instance, Huang Li et al. [19] introduced three types of wood–GFRP beams (plate form, slot form, and inverted T-shaped form). Their study revealed that slot GFRP–wood beams exhibited the best mechanical properties, showing a 26.43% increase in the ultimate load-carrying capability compared to that of the reference GFRP beams. C. Yoganantham et al. [20] conducted research on the flexural performance of pultruded glass fiber-reinforced polymer (GFRP) beams filled with a large volume of fly ash engineering cementitious composite (HVFAECC). The results indicated that the bearing capacity of tension GFRP beams filled with HVFAECC was significantly higher than that of hollow GFRP beams. Fubin Zhang et al. [21] proposed a novel composite sandwich structure that incorporates GFRP skins, cold-formed profiled steel plate, and lightweight balsa wood. The test findings demonstrated that the ultimate bearing capacity of the composite structure increased by 68–194% compared to that of hollow structures. These studies highlight the potential of using alternative, lighter materials to enhance the mechanical properties of GFRP structures while addressing concerns related to structural weight.

Gypsum, which is recognized as a cementing material, finds widespread use in the construction industry due to its excellent features such as its light weight, fire resistance, and fast setting time. Gypsum can be sourced from natural reserves or industrial by-products, and thus presents an effective means of waste utilization. The properties of fiber-reinforced materials have become increasingly common in recent years. Palomo IRI et al. [22] highlight the significant potential of utilizing ultra-high-performance fibers to enhance the mechanical properties of concrete, thereby improving the structural performance of reinforced concrete (RC) beam–column joints. In efforts to enhance gypsum's physical and mechanical properties, researchers have conducted experiments in which they have added various fibers to gypsum [23,24]. These fibers include paper, polystyrene, polyester, seagrass [25,26], and recycled waste gypsum [27–31]. Cong Zhu et al. [32] investigated the physical and mechanical properties of polyvinyl alcohol and polypropylene fiber-reinforced gypsum-based composites. Their results demonstrated a significant improvement in the flexural strength of hardened gypsum-based composites with the addition of polyvinyl alcohol fiber. Aldi Kuqo et al. [33] studied the mechanical properties of two natural fibers: Mediterranean seaweed and pine gypsum. They discovered that seaweed fiber and wood fiber could enhance crucial properties of gypsum, including its compressive strength, impact resistance, and hardness. Zhenxing Li et al. [34] explored the effects of basalt fiber, glass fiber, and polyvinyl alcohol fiber on the setting time, extension diameter, water absorption, and bending strength of fiber-reinforced gypsum-based composites. Their contributions have established the positive impact of fibers on the physical and mechanical properties of gypsum composites. However, there remains a gap in research concerning the flexural behavior of gypsum-based composite structures.

In summary, while load-bearing GFRP pipes can meet the necessary weight requirements, they still exhibit certain limitations in terms of their flexural and compressive performance compared to traditional reinforced concrete beams. Focusing on both lightweight characteristics and sufficient strength, this study selected gypsum as the matrix material and incorporated reinforcement and PVA to address the brittleness of the structure. This enhanced the ultimate load-bearing capacity while introducing certain plasticity characteristics, thereby meeting the requirements of practical engineering applications. For example, this composite material system could be widely applied in structural elements such as beams and columns in factory buildings, emergency structures, prefabricated buildings, and other construction scenarios that demand both light weight and a high load-bearing capacity.

This research focuses on three types of gypsum-based composite GFRP beams: pure gypsum, steel-reinforced gypsum, and fiber-reinforced gypsum composites. The primary focus of the investigation lies in the bending properties of these composite beams. Section 2 provides a detailed description of the material preparation process, including tests to evaluate the setting characteristics of gypsum and the influence of polyvinyl acetate (PVA) on the mechanical properties of gypsum blocks. Additionally, comprehensive details of the four-point bending test procedure are provided. Section 3 presents the theoretical calculation formulas for the load-bearing capacity of four types of components. Subsequently, Section 4 analyzes the results of the four-point bending experiments, offering a detailed examination of the failure modes observed during the experiments, and compares the theoretical and experimental results. Section 5 outlines the innovations of this study and summarizes the bending behavior exhibited by square GFRP beams filled with gypsum-based composites.

2. Materials and Methodology

The main materials used for the fabrication of GBIGCs include GFRP, gypsum, PVA, and steel bars.

2.1. Pultruded GFRP Beams' Properties

As shown in Figure 1a, the hollow pultruded GFRP square beams (length: 1.5 m, section area: 150 mm × 150 mm) used in this study were manufactured by Si Tong FRP Co., Ltd. (Soochow, China). In the pultrusion process, the material matrix is a thermosetting resin, and the fibers used are glass fibers. Through traction pultrusion and heating, the material is cured and formed into a single unit without differences in interlayer properties in the thickness direction. However, fibers are typically aligned along the axial direction of the profile. This fiber arrangement results in high strength and stiffness in the axial direction, allowing the profile to effectively withstand tensile, compressive, and bending mechanical stresses [35–37]. In this study, the GFRP material adopted this pultrusion process, which demonstrated the differences in mechanical properties in different directions. Therefore, the mechanical properties test was carried out in two directions of 0° and 90°, and the tensile, compression, and shear properties were tested, respectively. The direction is shown in Figure 1b. The direction of the extended square tube was set to 0°, and the direction perpendicular to 0° was 90°. The test samples are shown in Figure 2. The testing standards utilized were as follows: the in-plane shear performance testing standard was ASTM D5379-19; the in-plane tensile performance testing standard was ASTM D3039-07, and the in-plane compression performance testing standard was ASTM D3410-16 [38–40]. The three performance tests easily demonstrated the characteristics of orthotropic stratification, and the elastic modulus in the 0° direction was higher than that in the 90° direction, as shown in Table 1.

Table 1. Tensile properties of GFRP.

Direction	Tensile Strength/MPa	Modulus of Elasticity/GPa	Compressive Strength/MPa	Modulus of Elasticity/GPa	Shear Strength/MPa	Modulus of Elasticity/GPa
0°	321.7	22.00	266.7	21.28	71.7	9.87
90°	28.8	2.65	57.8	4.45	52.4	9.76

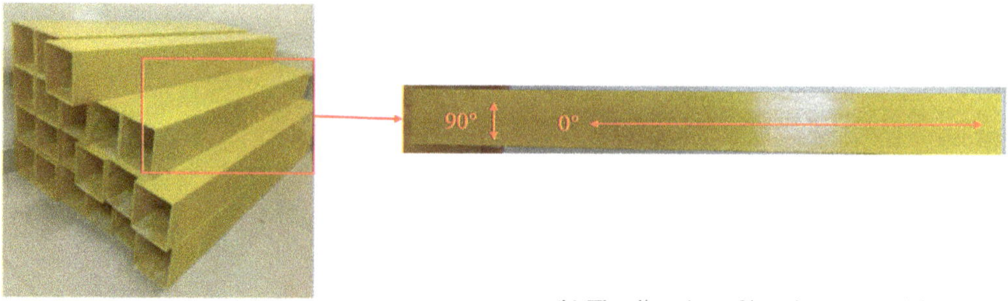

(a) GFRP beams

(b) The direction of interlayer material distribution of GFRP beam

Figure 1. GFRP beams and directions.

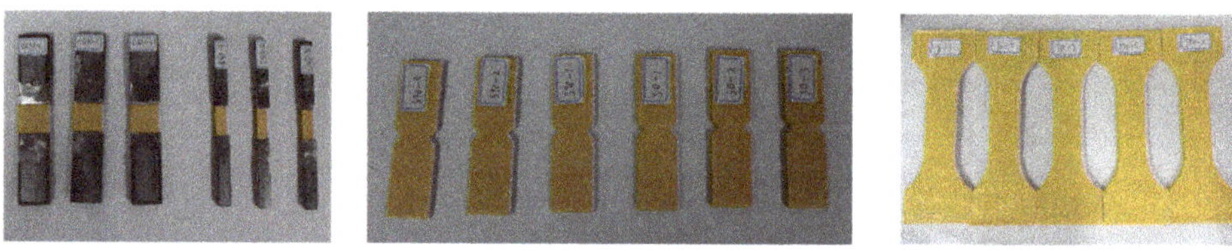

(a) Compression samples (b) Shear samples (c) Tensile samples

Figure 2. Samples in two directions of GFRP.

2.2. Desulfurization Gypsum

The gypsum was desulfurized and provided by Shandong Dong Taiyuan Gypsum Technology Co. Ltd. (Zaozhuang, China), and its physical properties are shown in Table 2.

Table 2. Physical parameters of gypsum.

Material	Water Requirement for Normal Consistency	Setting Times		Powdery Degree/200 Mesh	Whiteness/%
		Initial	Final		
Desulfurized gypsum	70%	5 min	10 min	0.01	50

2.3. PVA

The PVA was provided by Shanghai Chenqi Chemical Technology Co., Ltd. (Shanghai, China) and was used to improve gypsum's performance in light of its high bonding strength, corrosion resistance, and hydrophilicity. The physical parameters and morphology of PVA are shown in Table 3.

Table 3. Physical parameters and morphology of PVA.

Name	Diameter/μm	Length/mm	Density/g/cm^3	Tensile Strength/MPa	Modulus of Elasticity/GPa	Elongation at Break
PVA	15.09	12	1.29	1830	40	6.9%

2.4. Steel Bars

Hot rolled steel bars of grade HRB400 [41] were used for the fabrication of steel-reinforced gypsum beams; these were provided by Kunshan Guibang Steel Co., Ltd. (Kunshan, China). The diameters of the longitudinal reinforcement and stirrup were

8 mm and 6 mm, respectively. The basic mechanical performance of the steel bars is listed in Table 4.

Table 4. Mechanical parameters of Rebars.

Grade	Modulus of Elasticity/MPa	Poisson's Ratio	Yield Stress/MPa
HRB400 steel bars	200,000	0.3	400

2.5. Coagulation Test

The initial and final setting times of pure gypsum are 5 min and 10 min, respectively; these times were insufficient for the pouring process. Therefore, a protein retarder (manufactured by Beijing Long Tengda Chemical Company) was introduced to extend the setting time, and coagulation tests were conducted to assess the retarder's impact on the condensation time. The key steps of the main test, as illustrated in Figure 3, are as follows:

(a) Weigh 300 g of gypsum, 210 g of water, and the specified ratio of retarder. Thoroughly mix the gypsum and retarder, and within 5 s, pour the sample into water to achieve a uniform slurry, as shown in Figure 3a.
(b) Pour the evenly stirred slurry into the ring mold, ensuring the slurry flushes to the upper end of the ring mold, as depicted in Figure 3b.
(c) Record the initial setting time, defined as the period from the contact of the sample with water to the first instance when the steel needle cannot touch the glass-bottom plate, as demonstrated in Figure 3c.
(d) Record the final setting time, defined as the period from the contact of the sample with water to the time when the depth of the steel needle inserted into the slurry is below 1 mm for the first time, as shown in Figure 3d.

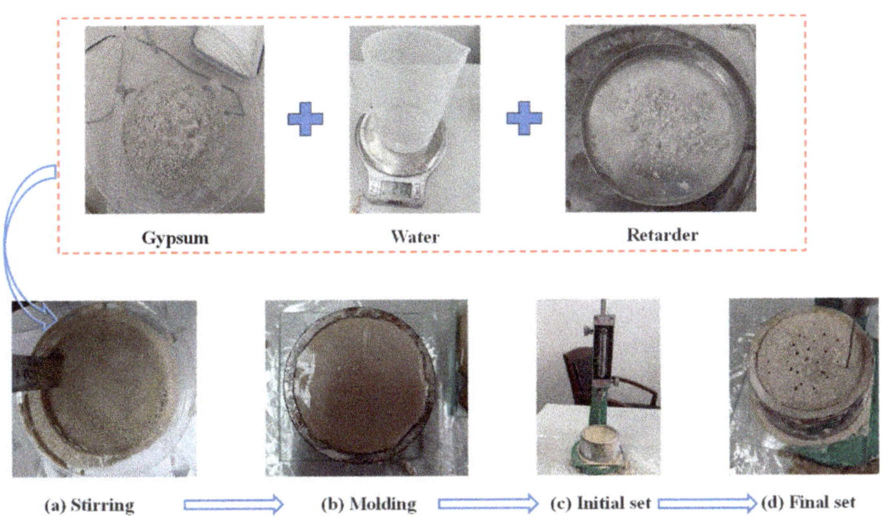

Figure 3. Gypsum coagulation tests.

2.6. Fabrication of Gypsum Blocks for Mechanical Test

As depicted in Figure 4, PVA gypsum test blocks were fabricated in the laboratory, incorporating varying dosages (0~1.5%) of PVAs. The primary steps involved in this process are as follows:

(a) Weighing: Measure 864 g of gypsum, 605 g of water, 1.04 g of retarder, and PVAs with dosages ranging from 0% to 1.5%.

(b) Mixing and stirring: Pour the aforementioned materials into the blender and stir until a homogeneous mixture is achieved.
(c) Brush oil onto the inner surfaces of the molds.
(d) Pour the mixture into the molds, allowing the blocks to take shape over the course of a day.
(e) Demold the blocks and cure them in a curing box for a duration of 3 days.

Figure 4. Manufacturing process of PVA gypsum blocks.

The gypsum blocks were subjected to flexural and compressive tests according to the standard GB/T50081-2019 [42], and we evaluated how the dosage of PVAs affected its mechanical performance.

2.7. Four-Point Bending Tests

As outlined in Table 5, four types of GFRP square beams were designed, comprising the hollow GFRP beam (H-G beam), pure gypsum GFRP beam (G-G beam), reinforced gypsum GFRP beam (R-G beam), and PVA gypsum GFRP beam (P-G beam). Two identical test specimens were created for each type, resulting in a total of eight GFRP square beams that underwent four-point bending tests.

The material properties of the gypsum, retarder, and PVAs are as described in Section 2.1.

The four-point bending tests were conducted using the 50 T self-balancing device at Jiangsu University (ZF-F500). In preparation, seven strain gauges and three dial indicators were strategically installed, as shown in Figure 5. Among the seven strain gauges, five (No. 2~No. 6) are evenly distributed along the beam web, while two (No. 1 and No. 7) are positioned at the midpoint of the upper and lower panels. The three dial indicators were employed to record deflection, with two placed at the bottom ends and one situated at the bottom midpoint. The test standard is GB/T 1449 [43].

Table 5. Four kinds of GFRP square beams.

Types	Length/mm	Width/mm	Height/mm	End Face Style
K-G beam	1500	150	150	
S-G beam	1500	150	150	
G-G beam	1500	150	150	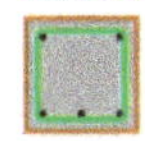
P-G beam	1500	150	150	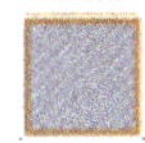

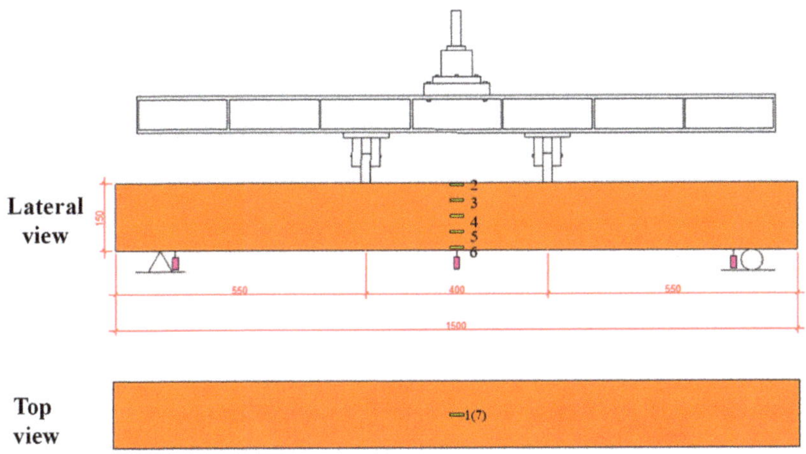

Figure 5. The 50 T self-balancing device and layout of sensors.

The primary test procedures are as follows:

(a) Prepare square GFRP beams filled with gypsum-based composites and affix the strain gauges using silicone rubber.
(b) Install the GFRP beams in the vertical loading reaction system device, adjusting the position of supports and their height.
(c) Connect the static resistance strain gauges with wires to form a quarter bridge and install displacement indicators.
(d) Preload the specimen with 2.5 kN to ensure full contact between the loading device and the test beam, then unload it back to zero.
(e) Apply a continuous load on the specimen with loading steps of 2.5 kN until failure occurs. Record load levels, strain, and displacement values throughout the entire loading process.

3. Theoretical Study

The bending and shearing deformation value of GBIGCs can be calculated out according to Equations (1) and (2), respectively [2,15].

$$\delta_b = \frac{Pa}{48EI}\left(3l^2 - 4a^2\right) \quad (1)$$

$$\delta_s = \frac{Pa}{GA} f_s \tag{2}$$

By adding Equations (1) and (2), the deflection of GBIGCs can be written as shown in Equation (3) [2,15].

$$\delta = \delta_b + \delta_s = \frac{Pa}{48EI}\left(3l^2 - 4a^2\right) + \frac{Pa}{GA} f_s \tag{3}$$

where δ is the deflection of the composite beam; δ_b is the bending deflection; δ_s is the shear deflection; P is the maximum load; E is the elastic modulus; I is the moment of inertia of the section; a is the shear span; l is the calculation span of the simply supported beam; G is the shear modulus; A is the section area; and f_s is the shape coefficient of cross-section shear stress.

The values of EI, GA and f_s can be derived by Equations (4)–(6) [14,44].

$$EI = \sum E_i I_i \tag{4}$$

$$GA = \sum G_i A_i \tag{5}$$

$$f_s = \frac{A}{I^2} \int_A \frac{S_i^2}{b_i^2} d_A \tag{6}$$

where E_i is the elastic modulus of each component; I_i is the moment of inertia of each component; G_i is the shear modulus of each component; A_i is the area of each component; S_i is the first moment of the area of each component; and b_i is the width of each component of the composite beam.

According to the above Equations (3)–(6), the ultimate load value of GFRP square beams can be deduced by Equation (7).

$$P = \frac{\delta}{\frac{a(3l^2 - 4a^2)}{48EI} + \frac{a}{GA} f_s} \tag{7}$$

4. Results and Discussion

4.1. Coagulation Test Results

The relationship between the protein retarder dosage and the gypsum setting time is illustrated in Figure 6. The graph indicates that as the retarder dosage increases, both the initial and final setting times of gypsum extend. In the subsequent experiments, a protein retarder dosage of 0.12% is employed for gypsum or gypsum-based composites. The initial and final setting times are recorded as 113 min and 135 min, respectively, and these times are deemed suitable for specimen preparation.

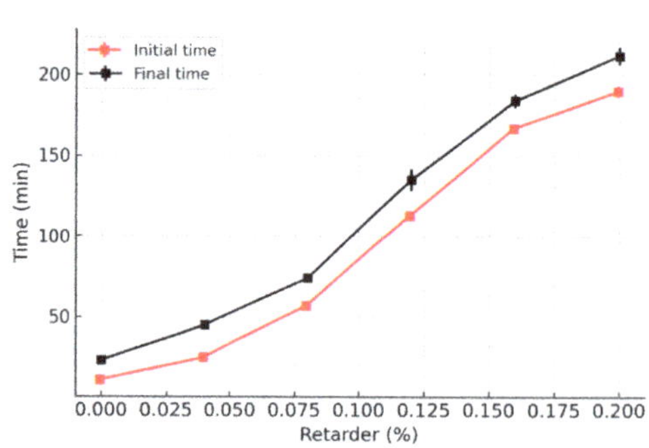

Figure 6. Relation between the retarder and the setting time of gypsum.

4.2. Results of the Flexural and Compressive Strength Tests on Gypsum Blocks

The test results are presented in Figure 7. As depicted in the figure, with the increase in PVA fiber content (0% to 1.2%), the flexural and compressive strength of PVA gypsum material increase, reaching maximum values of 2.97 MPa and 3.4 MPa, respectively, at a PVA dosage of 1.2%. However, the strength exhibits a downward trend when the dosage of PVA increases to 1.5%. Therefore, a PVA dosage of 1.2% is selected for the subsequent tests. When the PVA content increases from 0% to 1.2%, both flexural strength and compressive strength significantly improve. This is primarily due to the fiber's crack-bridging effect, which enhances the material's toughness and slows the propagation of microcracks, thus improving its load-bearing capacity. However, when the fiber content increases to 1.5%, the strength begins to decrease. This may be attributed to fiber agglomeration leading to uneven distribution and causing stress concentration. Moreover, an excess of fibers could disrupt the continuity of the gypsum matrix, reducing the effective bonding area of the matrix. Additionally, the bond strength at the fiber matrix interface may weaken, leading to increased interface debonding. These results suggest that the optimal fiber content is 1.2%; at this percentage, a balance between enhanced performance and material uniformity is achieved, maximizing the material's overall properties.

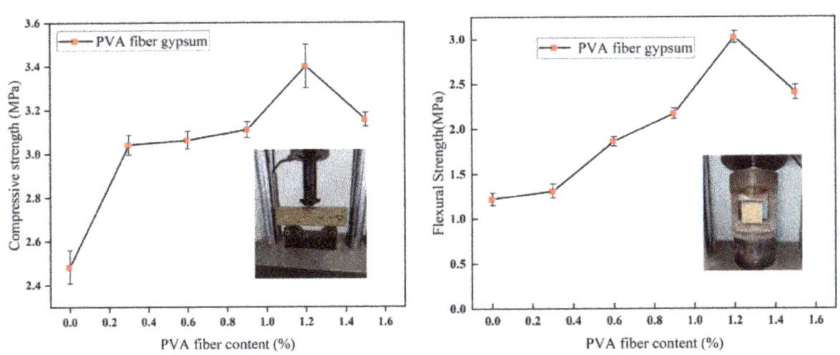

(**a**) Flexural strength (**b**) Compressive strength

Figure 7. Mechanical performance of PVA gypsum blocks.

4.3. Results of the Four-Point Bending Tests

As illustrated in Figure 8, the hollow GFRP specimens undergo loading from two symmetric concentrated loads. As the loading level increases, the specimens generate a more pronounced sound, indicating deeper fiber fractures. Ultimately, all the specimens lose their strength due to the crushing failure of the GFRP panel at the loading points. Moreover, as shown in Figure 8, localized fiber damage leads to fracture failure, with a simultaneous loss of local load-bearing capacity, accompanied by noticeable buckling on the sides. This results in overall instability of the component. Therefore, to prevent buckling, the study introduced a series of filler materials, such as the gypsum matrix, to mitigate this issue.

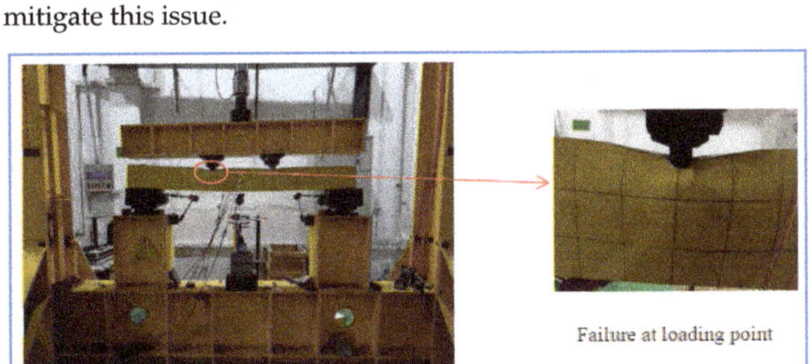

Figure 8. Loading of GFRP beams.

The hollow GFRP or GBIGC specimens exhibit different failure patterns around the loading points. In the case of the H-G beam shown in Figure 9a, the GFPR panel experiences intensive crushing at the loading points, especially at the junction between the upper flange and the web. This induces oblique and horizontal tearing failure, accompanied by an obvious upward bulge near the loading points. For the G-G beam in Figure 9b, it also displays intensive crushing at the loading points, but without oblique or horizontal tearing failure. This may be due to the good compatibility between the two materials in the G-G beam. The gypsum matrix provides supporting strength, which helps to avoid an excessive concentration of tensile stress to some extent. When the internal gypsum undergoes damage, it leads to localized crushing, while the tearing failure is mitigated. In the case of the R-G beam in Figure 9c, it not only shows intensive crushing at the loading points but also exhibits obvious oblique tearing failure below the crushing zone. When initial cracks occur locally, both the gypsum and the reinforcement can still function effectively. As the load increases, the gypsum gradually crushes and loses its effectiveness, but since the reinforcement remains intact, the material, primarily composed of GFRP, still has residual stress. The overall structure retains load-bearing capacity, which leads to further damage after GFRP failure. Compared to other structures, localized fractures in this case are more pronounced. However, for the P-G beam in Figure 9d, it displays relatively light local compression failure at the loading points, indicating better local compressive mechanical performance. Compared to ordinary gypsum, PVA-reinforced gypsum exhibits a significant improvement in both compressive and flexural strength. Additionally, the gypsum matrix shows good compatibility with GFRP, which effectively enhances the stiffness and strength at the loading points, reducing the occurrence of concentrated stress.

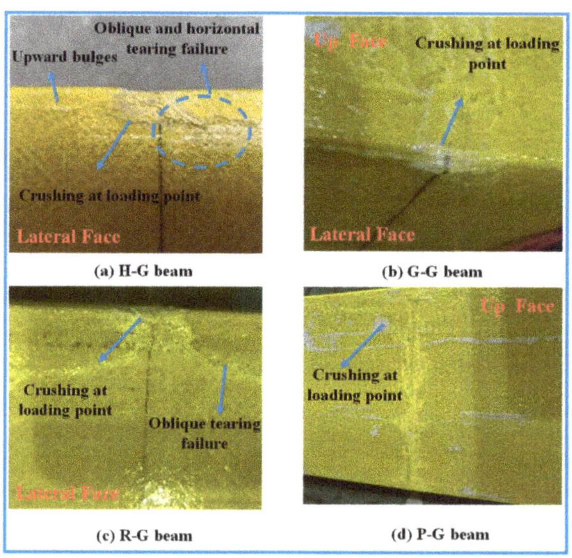

Figure 9. Local failure pattern at loading points of hollow GFRP or GBIGC specimens.

4.3.1. Ultimate Limit State Analysis

The load–deflection curves of the GFRP square beams are presented in Figure 10. The curves illustrate that the gradient of the curve, which represents rigidity, for G-G beam, R-G beam, and P-G beam is greater than that of H-G beam. R-G beam exhibits the greatest stiffness, followed by P-G beam and G-G beam. The ultimate load values of the infilled beams are significantly improved compared to that of H-G beam. It was also found that the stiffness of the R-G beam is the highest, but its ultimate strength is 7% lower than that of the P-G beam. This is because the strength of the reinforcement itself is higher than that of the

PVAs, which results in higher stiffness at the initial stage of the experiment. However, as the load increases, the bond between the gypsum and reinforcement in the R-G beam is not optimal, leading to a failure similar to over-reinforcement, where the bending resistance of the materials prevents them from working together effectively. In contrast, the fibers and gypsum in the P-G beam show good compatibility, with the fibers having strong ductility. This results in higher load-bearing capacity and greater displacement at the monitoring points, making the P-G beam more suitable for practical weight-reduction applications.

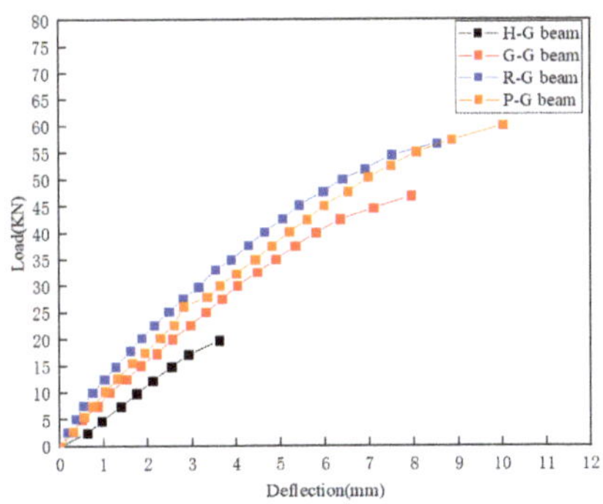

Figure 10. Mid-span deflection of test beams.

The test results for the ultimate load values and their corresponding deflections are summarized in Table 6. The ultimate loads for H-G, G-G, R-G, and P-G beams are 19.759 kN, 46.822 kN, 56.690 kN, and 60.035 kN, respectively. The maximum displacements occur at the mid-span, measuring 3.610 mm, 7.928 mm, 8.513 mm, and 10.013 mm, respectively. Consequently, the ultimate loads of G-G, R-G, and P-G beams are increased by 136.97%, 186.91%, and 203.84%, respectively, compared with that of the H-G beam. The increase in ultimate load is attributed to the synergistic effect between the gypsum matrix and the GFRP interface. Additionally, after gypsum filling, its compressive strength provides strong support for the GFRP, leading to stress redistribution on the GFRP's surface. This enhances the local compressive capacity and increases the overall load-bearing capacity. The reinforcement and PVAs, with their high tensile strength, further enhance the overall strength by combining with the compressive performance of the gypsum.

Table 6. Test results of H-G and GBIGC specimens.

Sample	Ultimate Load/kN	Average Value/kN	Mid-Span Deflection/mm	Average Value/mm
H-G-1	19.537	19.759	3.51	3.610
H-G-2	19.980		3.71	
G-G-1	45.130	46.822	7.62	7.928
G-G-2	48.514		8.235	
R-G-1	53.880	56.690	8.635	8.513
R-G-2	59.500		8.39	
P-G-1	60.070	60.035	9.495	10.013
P-G-2	60.000		10.53	

Note: Each type of GFRP square beam has two specimens, represented by the number 1 or 2.

4.3.2. Strain Distribution Analysis

The load–strain curves of the four types of specimens are drawn in Figure 11. The strain values of No.1~No.3 gauges are negative values, as they are installed in the compression zone, and the strain values of No.5~No.7 gauges are positive values because they are installed in the tensile zone. The strain value of No.4 gauge is near zero, as it is located at a neutral axis. Generally, the absolute strain values at all gauges of G-G, R-G and P-G beams increase with the loading level, which demonstrates the linear load–strain relationship.

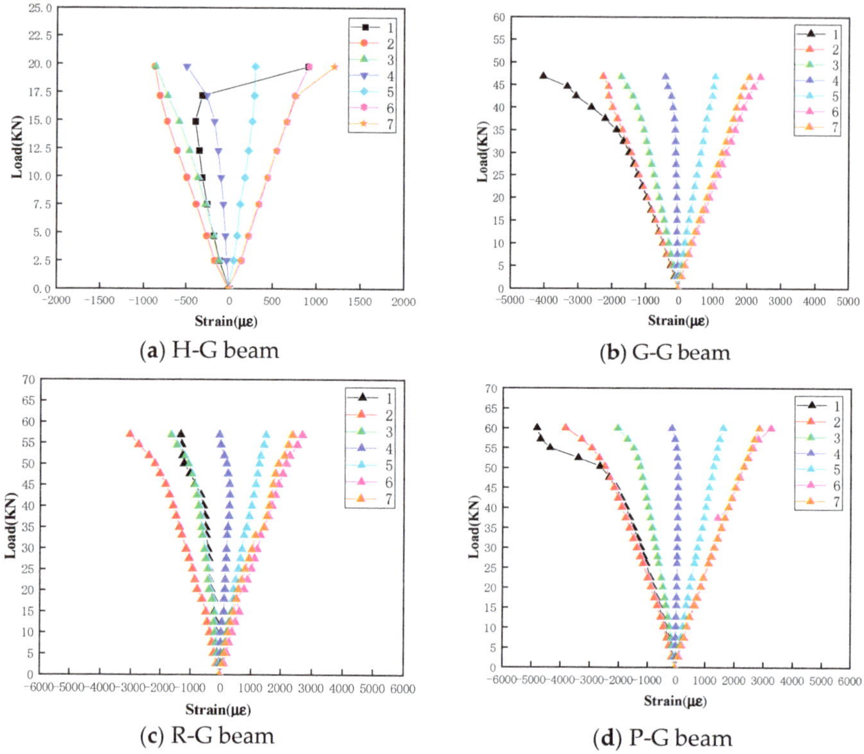

Figure 11. Load–strain curves of different gauges.

As shown in Figure 11a, for the No.1 gauge of the H-G beam, the strain increases with higher loading values and reaches its minimum negative value at a loading value of 14.4 kN. However, as the load level continues to increase, the absolute value of the strain decreases and suddenly jumps to a larger positive value (tensile strain) when the loading value reaches 19.759 kN. This behavior is attributed to local buckling and the upward bulge of the GFRP upper surface around the No.1 gauge near the failure stage, which induces tensile strain. For the G-G, R-G, and P-G beams, there is no such strain jump phenomenon, as shown in Figure 11c,d.

4.4. Discussion

4.4.1. Height-Wise Strain Distribution Analysis

Based on the strain data from gauges (No.2~No.5) installed on the webs of the specimens, strain distribution curves along the height can be derived, as depicted in Figure 12. Generally, the strain along the web height conforms to a linear distribution rule, aligning with the plane section assumption of beam deformation in material mechanics [40]. In Figure 12a, it can be observed that the curves at 4.638 kN and 9.856 kN are linear, indicating that the material is still in the elastic phase. As the applied load increases, the curves at 12.276 kN, 17.154 kN, and 19.759 kN exhibit nonlinear characteristics, with some regions entering the plastic phase. At higher load levels, strain changes rapidly at the top, indicating that yielding has occurred.

In Figure 12b, only the 46.822 kN curve gradually shows nonlinear characteristics, but compared to the H-G beam, the strain range is larger. This suggests that using gypsum as the matrix material effectively improves overall integrity, reducing the impact of localized damage on the overall strength.

Figure 12c,d show that the ultimate strain can reach 3000 and 3400, respectively, which is approximately 200% higher than the strain of 1000 observed in the hollow pipe, demonstrating a significant improvement. Additionally, it can be seen that the PVA gypsum matrix beams have a larger strain range, indicating that the tensile performance provided by the PVAs in the gypsum matrix has been fully utilized.

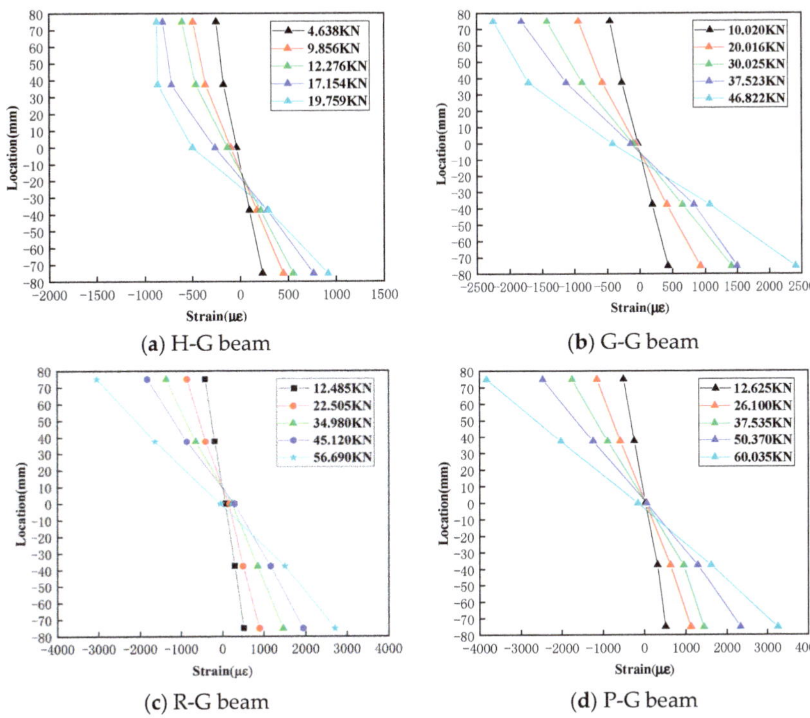

(**a**) H-G beam (**b**) G-G beam (**c**) R-G beam (**d**) P-G beam

Figure 12. Strain distribution along web height.

4.4.2. Theoretical Analysis

The theoretical calculation results of the ultimate load values are provided in Table 7, and experimental results are also included for comparison. As shown in the table, the theoretical ultimate load values for H-G, G-G, R-G, and P-G beams are 19.983 kN, 50.428 kN, 53.935 kN, and 65.027 kN, respectively. The relative errors between theoretical and experimental values are 1.12%, 7.15%, 5.11%, and 7.68%, respectively, indicating a good match between theoretical predictions and experimental results. The results presented by the four stress–strain curves are generally consistent with the assumption of plane sections, so the errors in the calculated results are all within 8%. The relatively low error for the H-G beam indicates that the theoretical model can effectively reflect the beam's load-bearing behavior under the given conditions. However, due to the nonlinear characteristics of the gypsum matrix material, there is some error in the other three materials. The error in the P-G beam is also closely related to the distribution pattern of the PVAs within the gypsum. Nevertheless, the overall error for all three materials is within an acceptable range, and this calculation theory generally meets engineering calculation requirements.

Table 7. Theoretical calculation.

Type	f/mm	Experimental Ultimate Load P_u/kN	Theoretical Ultimate Load P_{ul}/kN	Relative Error
H-G beam	3.610	19.759	19.983	1.12%
G-G beam	7.928	46.822	50.428	7.15%
R-G beam	8.513	56.690	53.935	5.11%
P-G beam	10.013	60.035	65.027	7.68%

5. Conclusions

The bending properties of novel GFRP square beams infilled with gypsum-based composites (GBIGCs) were comprehensively studied through experimental and theoretical analyses. The main conclusions are summarized as follows:

- Protein retarder sensitivity: The setting time of gypsum was found to be sensitive to protein retarder. An optimal dosage of 0.12% protein retarder effectively prolongs the initial and final setting time from 5 min and 10 min to 113 min and 135 min, respectively, making it suitable for specimen preparation.
- PVA composite gypsum strength: The flexural and compressive test results of gypsum blocks indicate that as the PVA content increases from 0% to 1.2%, the strength of PVA gypsum continuously increases. However, a downward trend is observed as the PVA dosage changes from 1.2% to 1.5%. The optimal dosage of PVA is determined to be 1.2%, enhancing the flexural and compressive strength of PVA composite gypsum from 1.17 MPa and 2.50 MPa to 2.97 MPa and 3.4 MPa, respectively.
- Ultimate load values of GRFP square beams: In the four-point bending tests, the ultimate load values of pure gypsum, reinforced gypsum, and PVA gypsum infilled GFRP square beams are 46.822 kN, 56.690 kN, and 60.035 kN, respectively. Compared with the ultimate load value of hollow GFRP square beams (19.759 kN), the ultimate bending capacity of GFRP square beams infilled with gypsum, reinforced gypsum, and PVA gypsum increases by 136.97%, 186.91%, and 203.84%, respectively. Among them, PVA gypsum GFRP square beams exhibit the best mechanical performance.
- Strain distribution and theoretical equations: The strain distribution along the height at the middle span section generally meets the plane section assumption under different load levels. Theoretical equations are established to calculate the ultimate load values, whose results align well with experimental values.

The beam dimensions studied in this paper have certain limitations, as the beam length is 1500 mm, while in actual building structures, beam lengths typically exceed 2 m. The mechanical performance of such beam structures, particularly when they have varying cross-sections and span large distances, requires further investigation. Secondly, the study of the microscopic mechanical properties is somewhat limited. Specifically, how the internal reinforcement behaves under mechanical action and failure during four-point bending with gypsum-based filling materials, as well as the distribution characteristics of PVAs, are key areas for future research. Finally, the response of GBIGCs under different load conditions, such as uniformly distributed loads, dynamic impacts, and seismic performance, needs to be further studied and addressed for practical applications.

Author Contributions: Conceptualization, Y.L.; methodology, Y.L.; data curation, Y.L., B.S. and T.Z.; writing—original draft preparation, Y.L. and T.Z.; writing—review and editing, Y.L., B.S. and T.Z.; visualization, B.S. and T.Z.; supervision, B.S.; project administration, B.S.; funding acquisition, B.S. All authors have read and agreed to the published version of the manuscript.

Funding: The research in this paper was supported by three foundations: (*i*) the National Natural Science Foundation of China (grant number 51108210); (*ii*) the special guiding fund for building energy conservation and green building in Zhenjiang City (grant number zk20220001; JSU10SK2024ZX003); (*iii*) the youth support foundation of Jiangsu University (grant number 5621480001).

Institutional Review Board Statement: Not applicable

Data Availability Statement: The raw data supporting the conclusions of this article will be made available by the authors on request.

Conflicts of Interest: The authors declare that they have no conflicts of interest.

References

1. Alnahhal, W.; Aref, A. Structural performance of hybrid fiber reinforced polymer–concrete bridge superstructure systems. *Compos. Struct.* **2008**, *84*, 319–336. [CrossRef]
2. Chakrabortty, A.; Khennane, A.; Kayali, O.; Morozov, E. Performance of outside filament-wound hybrid FRP-concrete beams. *Compos. Part B Eng.* **2011**, *42*, 907–915. [CrossRef]
3. Hadi, M.N.; Wang, W.; Sheikh, M.N. Axial compressive behaviour of GFRP beams reinforced concrete columns. *Constr. Build. Mater.* **2015**, *81*, 198–207. [CrossRef]
4. Keller, T.; Schollmayer, M. Plate bending behavior of a pultruded GFRP bridge deck system. *Compos. Struct.* **2004**, *64*, 285–295. [CrossRef]
5. Evernden, M.C.; Mottram, J.T. A case for houses to be constructed of fibre reinforced polymer components. *Proc. Inst. Civ. Eng.-Constr. Mater.* **2012**, *165*, 3–13. [CrossRef]
6. Yanes-Armas, S.; Keller, T. Structural concept and design of a GFRP-polyurethane sandwich roof structure. In Proceedings of the Eighth International Conference on Fibre-Reinforced Polymer (FRP) Composites in Civil Engineering, Hong Kong, China, 14–16 December 2016.
7. Wang, J.; Song, Y.; Wang, W.; Tu, L. Evaluation of composite crashworthy device for pier protection against barge impact. *Ocean Eng.* **2018**, *169*, 144–158. [CrossRef]
8. Alagusundaramoorthy, P.; Reddy, R.V.S. Testing and evaluation of GFRP composite deck panels. *Ocean. Eng.* **2008**, *35*, 287–293. [CrossRef]
9. Abdolpour, H.; Garzón-Roca, J.; Escusa, G.; Sena-Cruz, J.M.; Barros, J.A.; Valente, I.B. Development of a composite prototype with GFRP profiles and sandwich panels used as a floor module of an emergency house. *Compos. Struct.* **2016**, *153*, 81–95. [CrossRef]
10. Ndukwe, C.O.; Ezurike, B.O.; Okpala, P.C. Comparative studies of experimental and numerical evaluation of tensile properties of Glass Fibre Reinforced Polyester (GFRP) matrix. *Heliyon* **2021**, *7*, e06887. [CrossRef]
11. Otoom, O.F.; Lokuge, W.; Karunasena, W.; Manalo, A.C.; Ozbakkaloglu, T.; Thambiratnam, D. Experimental and numerical evaluation of the compression behaviour of GFRP-wrapped infill materials. *Case Stud. Constr. Mater.* **2021**, *15*, e00654. [CrossRef]
12. Shekarchi, M.; Yekrangnia, M.; Biniaz, A.; Raftery, G.M. Effect of elevated temperatures on the compressive behavior of timber filled steel and pultruded GFRP Beams. *Compos. Struct.* **2021**, *271*, 114135. [CrossRef]
13. Al-Saadi, A.U.; Aravinthan, T.; Lokuge, W. Structural applications of fibre reinforced polymer (FRP) composite tubes: A review of columns members. *Compos. Struct.* **2018**, *204*, 513–524. [CrossRef]
14. Zhang, F.; Liu, W.; Fang, H.; Xie, G. Experimental study on flexural behavior of GFRP panel-cold-formed thin-walled steel composite beams. *J. Build. Struct.* **2018**, *39*, 104–111.
15. Robinson, M.J.; Melby, I.H. Effects of bonding in short-span rectangular concrete filled GFRP Beams. *Compos. Struct.* **2015**, *133*, 131–139. [CrossRef]
16. Abouzied, A.; Masmoudi, R. Flexural behavior of rectangular FRP-Beams filled with reinforced concrete: Experimental and theoretical studies. *Eng. Struct.* **2017**, *133*, 59–73. [CrossRef]
17. Ahmed, A.A.; Hassan, M.; Masmoudi, R. Effect of concrete strength and beams thickness on the flexural behavior of prestressed rectangular concrete-filled FRP Beams beams. *Eng. Struct.* **2020**, *205*, 110112. [CrossRef]
18. Muttashar, M.; Manalo, A.; Karunasena, W.; Lokuge, W. Influence of infill concrete strength on the flexural behaviour of pultruded GFRP square beams. *Compos. Struct.* **2016**, *145*, 58–67. [CrossRef]
19. Huang, P.; Yu, H.; Feng, J.; Hao, G.; Li, Z. Experimental study on flexural behavior of GFRP profile-wood composite beam. *J. Hebei Univ. Sci. Technol.* **2022**, *43*, 560–570.
20. Yoganantham, C.; Joanna, P. Flexural behaviour of pultruded GFRP beams infilled with HVFA ECC. *Mater. Today Proc.* **2021**, *45*, 5978–5981. [CrossRef]
21. Zhang, F.; Lu, Z.; Wang, D.; Fang, H. Mechanical properties of the composite sandwich structures with cold formed profiled steel plate and balsa wood core. *Eng. Struct.* **2024**, *300*, 117256. [CrossRef]
22. Palomo, I.R.I.; Frappa, G.; de Almeida, L.C.; Trautwein, L.M.; Pauletta, M. Analytical and numerical models to determine the strength of RC exterior beam–column joints retrofitted with UHPFRC. *Eng. Struct.* **2024**, *312*, 118244. [CrossRef]
23. Alyousef, R.; Abbass, W.; Aslam, F.; Shah, M.I. Potential of waste woven polypropylene fiber and textile mesh for production of gypsum-based composite. *Case Stud. Constr. Mater.* **2023**, *18*, e02099. [CrossRef]
24. Balti, S.; Boudenne, A.; Dammak, L.; Hamdi, N. Mechanical and thermophysical characterization of gypsum composites reinforced by different wastes for green building applications. *Constr. Build. Mater.* **2023**, *372*, 130840. [CrossRef]
25. Chen, C.; Fang, L.; Wang, Y.; Jiu, S.; Chen, Y. Mechanical, thermal and microscopic properties of gypsum matrix composites containing capric acid-palmitic acid/urea-formaldehyde microcapsules. *Case Stud. Constr. Mater.* **2023**, *18*, e02084. [CrossRef]
26. Ding, X.; Huang, W.; Li, Y.; Hu, Z.; Shan, Z. Study on retarding feature and retardation mechanism of various retarding materials on gypsum as a construction material: A review. *J. Build. Eng.* **2023**, *72*, 106569. [CrossRef]

27. Wu, C.; He, J.; Wang, K.; Yang, L.; Wang, F. Enhance the mechanical and water resistance performance of flue gas desulfurization gypsum by quaternary phase. *Constr. Build. Mater.* **2023**, *387*, 131565. [CrossRef]
28. Babu, K.S.; Ratnam, C. Mechanical and thermophysical behavior of hemp fiber reinforced gypsum composites. *Mater. Today Proc.* **2021**, *44*, 2245–2249. [CrossRef]
29. Alcaraz, J.S.; Belda, I.M.; Sanchis, E.J.; Borrell, J.M.G. Mechanical properties of plaster reinforced with yute fabrics. *Compos. Part B Eng.* **2019**, *178*, 107390. [CrossRef]
30. Gou, M.; Zhao, M.; Zhou, L.; Zhao, J.; Hou, W.; Ma, W.; Hou, Z. Hydration and mechanical properties of FGD gypsum-cement-mineral powder composites. *J. Build. Eng.* **2023**, *69*, 106288. [CrossRef]
31. Ma, X.; Tan, L.; Lu, Y.; Yao, W.; Wei, Y. Upcycling of waste plasterboard for the synthesis of high-quality gypsum-based 3D printing powder. *Constr. Build. Mater.* **2023**, *373*, 130846. [CrossRef]
32. Zhu, C.; Zhang, J.; Peng, J.; Cao, W.; Liu, J. Physical and mechanical properties of gypsum-based composites reinforced with PVA and PP fibers. *Constr. Build. Mater.* **2018**, *163*, 695–705. [CrossRef]
33. Kuqo, A.; Mai, C. Mechanical properties of lightweight gypsum composites comprised of seagrass *Posidonia oceanica* and pine (*Pinus sylvestris*) wood fibers. *Constr. Build. Mater.* **2021**, *282*, 122714. [CrossRef]
34. Li, Z.; Wang, X.; Yan, W.; Ding, L.; Liu, J.; Wu, Z.; Huang, H. Physical and mechanical properties of gypsum-based composites reinforced with basalt, glass, and PVA fibers. *J. Build. Eng.* **2023**, *64*, 105640. [CrossRef]
35. Elkin, A.; Konev, S.; Safonov, A.; Gusev, S.; Sergeichev, I. Compressive residual strength of the pultruded glass-fiber composite after tension-compression fatigue. *Compos. Part C Open Access* **2024**, *14*, 100456. [CrossRef]
36. Guo, R.; Xian, G.; Li, F.; Li, C.; Hong, B. Hygrothermal resistance of pultruded carbon, glass and carbon/glass hybrid fiber reinforced epoxy composites. *Constr. Build. Mater.* **2021**, *315*, 125710. [CrossRef]
37. Shaowei, L.U.; Xie, H.; Chen, P. Simulation and optimization of GFRP pultrusion process. *J. Compos. Mater.* **2008**, *1*, 46–51.
38. *ASTM D5379-19*; Standard Test Method for Shear Properties of Composite Materials by the V-Notched Beam Method. ASTM International: West Conshohocken, PA, USA, 2019.
39. *ASTM D3039-07*; Standard Test Method for Tensile Properties of Polymer Matrix Composite Materials. ASTM International: West Conshohocken, PA, USA, 2007.
40. *ASTM D3410-16*; Standard Test Method for Compressive Properties of Polymer Matrix Composite Materials with Unsupported Gage Section by Shear Loading. ASTM International: West Conshohocken, PA, USA, 2016.
41. *GB/T 1499.2-2018*; Steel for Reinforced Concrete Part 2, Hot Rolled Ribbed Bars. China Iron and Steel Industry Association: Beijing, China, 2018.
42. *GB 50081*; Standard for Test Methods of Physical and Mechanical Properties of Concrete. China Academy of Building Research Co., Ltd.: Beijing, China, 2019.
43. *GB/T 1449-2005*; Standard for Test Methods of Flexural Properties of Fiber-Reinforced Plastics. China Building Materials Industry Association: Beijing, China, 2005.
44. Lejia, L. Theoretical Bifurcation and Numerical Simulation of Flexural Behavior of FRP-Concrete Composite Beams. Master's Thesis, Hunan University, Changsha, China, 2014.

polymers

MDPI

Article

Investigation of the Flexural and Charpy Impact Properties of Polymer Composites Reinforced with Tururi (*Manicaria saccifera*) Fibrous Fabric

Avener Gleidson Andrade Santos [1], Damares da Cruz Barbosa Nascimento [1], Felipe Perissé Duarte Lopes [2], Noan Tonini Simonassi [2], Sérgio Neves Monteiro [3], Alisson Clay Rios da Silva [1] and Verônica Scarpini Candido [1,*]

1. Materials Science and Engineering Program, Federal University of Pará—UFPA, Ananindeua 67130-660, Brazil; eng.avenersantos@gmail.com (A.G.A.S.); damares.barbosa62@gmail.com (D.d.C.B.N.); alissonrios@ufpa.br (A.C.R.d.S.)
2. Materials Science Program, North Fluminense State University—UENF, Campos dos Goytacazes 28013-602, Brazil; perisse@uenf.br (F.P.D.L.); noansimonassi@uenf.br (N.T.S.)
3. Materials Science Program, Military Institute of Engineering—IME, Rio de Janeiro 22290-270, Brazil; snevesmonteiro@gmail.com

* Correspondence: scarpini@ufpa.br

Abstract: The search for new natural, sustainable, economical and biodegradable reinforcements for composite materials has increased in recent years, highlighting the importance of fibers from the natural environment. This work evaluates the use of tururi fibrous fabric as a reinforcement in a polymer matrix, using Fourier transform infrared spectroscopy, X-ray diffraction, thermogravimetry and scanning electron microscopy. The mechanical and fractographic performance of composites reinforced with 2.5, 5.0 and 7.5% mass fraction of tururi in a polyester matrix is also investigated. The FTIR and XRD results identified groups characteristic of natural fibers and the presence of elemental constituents such as cellulose, hemicellulose and lignin. Thermogravimetry indicated good thermal stability near 246 °C. The morphology of the fibrous fabric is irregular and formed by tangles of threads. The mechanical behavior of the composites in bending revealed a variation in stress with the increase in the percentage of fabric in the matrix, explained by defects and failures due to low interfacial adhesion between the phases. Impact tests indicated that increasing the percentage of fabric in the matrix improves impact energy absorption, reflecting better adhesion and load distribution. Thus, the development of this natural composite is promising for applications in green and sustainable products.

Keywords: tururi fabric; characterization; mechanical properties

1. Introduction

Sustainable development is a determining environmental issue in the design of new materials. In this context, composite materials have been attracting attention on a global scale, as they have interesting environmental and mechanical requirements for various applications [1–3]. These materials are used on a large industrial scale due to their high resistance to stress and low specific mass. Industries such as aerospace, automotive, shipbuilding, packaging, and medicine and biomedicine use composite materials as a basis for the development of high- and low-intensity activities [4–8].

Increasing concerns about climate change and excessive carbon dioxide emissions have driven the development of sustainable, biodegradable and less environmentally aggressive materials [9]. As a result, the use of natural inputs from environmental reutilization, most notably natural fibers as reinforcements in polymer matrix composites, has leveraged the technical–scientific field in the manufacture of new products. The use of natural reinforcements has a number of advantages, such as low density, low purchase price, biodegradability and non-toxicity, as well as great environmental availability [7,10,11].

Lignocellulosic fibers from large and small plants have cellulose, hemicellulose, pectin and lignin as elementary components, in varying proportions [12–15]. For this reason, the use of environmentally recyclable natural inputs, particularly natural fibers as reinforcements in polymer matrix composites, is driving innovation in the technical–scientific field, leading to the manufacture of new products. The use of natural reinforcements has a number of advantages, such as low density, low purchase value, biodegradability and non-toxicity, as well as great environmental availability [7,10,11,16–18].

The main properties of composite materials reinforced with natural fibers are high mechanical strength due to the presence of cellulosic material, low density, high rigidity, low abrasiveness, wear resistance and a lower specific weight [19–24]. In addition, the arrangement and characteristics of the fibers, whether they are continuous or in the form of yarns or mats, make the composite material even more attractive for various industrial applications, with a view to replacing petroleum-derived fibers. Studies have been carried out to understand how the insertion of fabrics made from natural and synthetic fibers affects the properties of polymer matrix composite materials [25,26].

Oliveira et al. (2020) [27] developed polybenzozaxine composites reinforced with tururi fabric and observed a bending elastic modulus of 2 GPa and elevated glass transition temperatures. Ng et al. (2022) [28] observed that the manufacture of fabrics using natural and synthetic pineapple and glass fibers, respectively, results in lightweight composites with good mechanical properties that are environmentally interesting for application in the transport sector. In another study, Ng et al. (2024) [29] observed that the constitution of a hybrid fabric of ramie and pineapple fibers oriented in different directions shows high resistance to impact, traction and bending, due to the easy distribution of the load along the layers. In addition, they explain that hybrid fabrics have great potential for use, due to the environmentally friendly aspect of the material. Monteiro et al. [30] characterized tururi fabric and found tensile strength values of 17.6 MPa, as well as functional groups corresponding to cellulose, hemicellulose and lignin. Porras et al. [31] also characterized tururi fabric for use as a reinforcement in composites and concluded that cellulose and lignin were the main constituents in higher amounts, and that the fabric surface was coated with waxes and fats, suggesting that a chemical treatment could promote the removal of these compounds and improve interfacial adhesion. Midani et al. [32] studied the effects of the structural parameters of tururi fabric on the impact properties of multilayer composites and observed that increasing the number of layers led to a reduction in impact absorption capacity, with failure occurring due to matrix delamination, indicating lower fabric wettability by the resin.

On the scene of discoveries of new reinforcements for composite materials is tururi fabric. This fabric is extracted from the Ubuçu palm (*Manicaria saccifera*), a tree common in Latin America and used by native communities for a variety of applications. Straw and candles are made from the leaves, while drinks and other medicinal remedies are produced from the fruit, as well as hats and fishing accessories. The bracts form sacs that give rise to a fibrous fabric, which can be obtained in the form of an involucre [30,31]. The use of a natural fabric covers three dimensions (economic, social and sustainable development), requirements that make it interesting for use as a reinforcement. Although it is a fabric that has already been used as a reinforcement in composite materials [26,27,30,32] for the first time, the mechanical and dynamic mechanical behavior, as well as the Charpy impact energy, of polyester matrix composites reinforced with tururi fabric at mass percentages of 2.5, 5.0 and 7.5% are investigated, making it a relevant study in the field of composites reinforced with natural fiber fabrics, thus making it more attractive for the manufacture of new materials. Additionally, although the literature [26,27,30,32] provides relevant information on the fabric and composites, none of them investigated the reinforcement potential of tururi fabric in a polyester matrix, making the present study important for elucidating issues such as mechanical behavior as a function of temperature, the influence of fabric addition on the glass transition temperature of the polymer, and the absorbed impact energy. In this context, the objective of this work is to evaluate, through a series of analyses and tests, the potential use of a tururi fibrous fabric reinforcement for polyester polymer matrix composites.

2. Materials and Methods

The tururi fibrous fabric was obtained commercially at the Ver-o-Peso market in Belém-PA, Brazil. The material was naturally obtained, without any previous chemical treatment. Figure 1 shows the ubuçu plant, the region where it was extracted and the fabric used.

(c)

Figure 1. Ubuçu plant (**a**) [33], fabric extraction site [33] (**b**) and tururi fabric (**c**).

The matrix used was a medium-viscosity isophthalic polyester resin and methyl ethyl ketone peroxide (MEK) hardening agent, supplied by Center Glass (Belém, PA, Brazil). The polyester resin was mixed with the hardening agent in a proportion of 1.5% by weight.

Table 1 shows the chemical composition of tururi fabric according to all TAPPI T203 and T222.

Table 1. Chemical analysis of tururi fabric.

	Cellulose (%)	Hemicellulose (%)	Lignin (%)	Reference
Tururi Fabric	74.1	12.0	31.1	[31]

2.1. Composites Preparation

The composites were produced manually using the hand lay-up process, which consists of the dimensions of the ASTM D 790 [34] and ASTM D6110 [35] standards. The specimens were made using mass proportions of 2.5, 5.0 and 7.5% by weight of tururi fabric reinforcement, manually cut according to the specifications of each standard. The composites were produced manually using a silicone mold, where the tururi fabric was aligned longitudinally. After curing for 24 h, the specimens were demolded and sanded with 80 and 600 mesh grain sizes, followed by flexural and impact tests.

2.2. X-Ray Diffraction (XRD)

The qualitative crystalline phases were obtained by XRD in a Proto Manufacturing, XRD Powder Diffraction System: 30 kV and 2 mA generator, Cu-Kα1 radiation, angular step of 0.0149°, 0.5 s time interval, 47 min scan and 2θ ranging from 5° to 60°. The crystallinity index was calculated using Equation (1) [36].

$$CI\% = \left(1 - \frac{I_1}{I_2}\right) \times 100 \quad (1)$$

where the following are defined:

CI: crystallinity index; I_1: maximum intensity associated with the crystalline plane of cellulose; I_2: maximum intensity associated with the amorphous plane of cellulose.

Microfibril Angle (MFA)

The microfibrillar angle for tururi fabric was calculated using the Origin® software program, based on the method described by Cave [37], considering the cellulose peak (0 0 2). Based on the relationship between the Gauss curve and the first- and second-order curves, the "T" parameter is obtained and then applied to the following expression:

$$MFA = -12.19 \times T^3 + 113.67 \times T^2 - 348.40 \times T + 358.09 \quad (2)$$

2.3. Thermogravimetric Analysis (TGA)

Thermogravimetric analysis was carried out in a NETZSCH (Berlin, Germany) model STA 449 F3 analyzer. The atmosphere used was nitrogen a flow rate of 50 mL/min, a heating rate of 10 °C/min, and a temperature range from 25 to 500 °C.

2.4. Fourier Transform Infrared Spectroscopy (FTIR)

The FTIR of the fabric was carried out using BRUKER equipment, model VERTEX 70V, using an infrared range of 4000–400 cm^{-1}. The FTIR analysis for the composites was carried out on equipment produced by Thermo Scientific, model NICOLET 6700 using an infrared range of 4000–500 cm^{-1}.

2.5. SEM Analysis

A morphological analysis of the tururi fabric and fractographic analysis of the composites were carried out using a model Mira3 FEG 250 TESCAN microscope (Brno, Czech republic) operating with secondary electrons at 5 KV and a working distance ranging from 25 to 10 mm.

2.6. Mechanical Properties Tururi Fabric

The tururi fabric samples were tested according to the ASTM D5035, using an iM50 Universal Electromechanical Machine (Mogi das Cruzes, SP, Brazil), using a 5 kN load cell at a tensile and bending speed of 10 mm/min. Twenty specimens cut by hand according to the dimensions of the standard used were tested.

2.7. DMA

The DMA test was carried out using the single-cantilever configuration. The test was carried out on a TA instrument Q800 V21.3, at a temperature rate of 20 to 200 °C with heating of 3 °C per minute and frequency 1 Hz, in a nitrogen atmosphere. The dimensions used were 12 × 46 × 3 mm.

2.8. Flexure Test

The flexure test was carried out on EMIC electromechanical universal equipment model (DL500), with a 5 KN load cell and a loading rate of 2 mm/min. The test was carried out in compliance with ASTM D790–17 [34].

2.9. Charpy Impact Test

The Charpy impact test was carried out on instrumented pendulum equipment from PANTEC, model XC-50, 1× 200 V × 60 Hz with a pendulum of 0.7 J. The specimens were produced with prismatic dimensions and a 2.54 mm deep notch. The test was carried out in accordance with ASTM D6110 [35].

2.10. Statistical Analysis

A statistical validation of the data was carried out using the analysis of variance (ANOVA) tool, with a 95% confidence interval ($p < 0.05$). Mean values were compared using the Tukey test.

3. Results

3.1. Characterization of Tururi Fabric

3.1.1. XRD Analysis

Figure 2 shows the x-ray diffractogram obtained for tururi fabric.

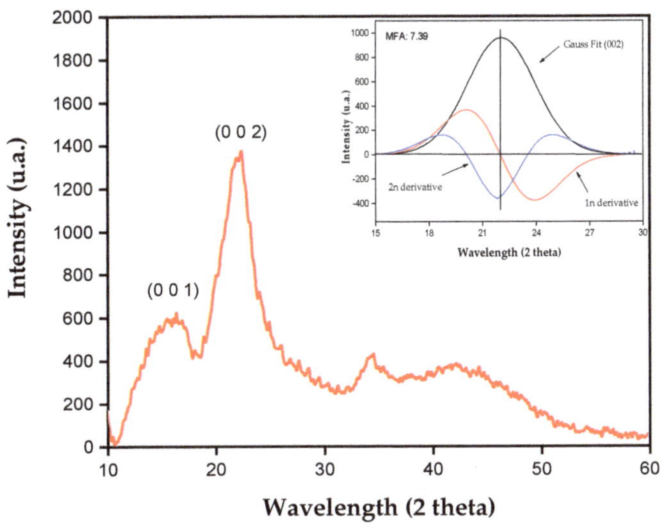

Figure 2. XRD pattern of the tururi fabric.

The XRD analysis shows halos at angles of 16.23° and 22.32°, characteristic of the 1 0 1 and 0 0 2 planes, respectively. The peaks at 16.23° and 22.32° indicate the presence of amorphous constituents such as hemicellulose, lignin and pectins, in addition to crystalline cellulose. These results are similar to those found for natural fibers such as guaruman [38,39], ubim [40] and periquiteira [41].

The crystallinity index of the tururi fabric was 46.67%.

Table 2 compares the crystallinity index results of different natural fibers.

Table 2. Comparative results of crystallinity indices (CIs) of different natural fibers and tururi fabric.

Fiber	CI (%)	References
Tururi Fabric	46.67	Present Work
Carnauba	86.90	[42]
Hemp	82.10	[43–45]
Jack Tree Fiber	86.00	[46]
Curauá	75.60	[47]
Pineapple	38.00	[48]
Ubim	83.00	[40]
Periquiteira	70.49	[41]
Buriti	63.00	[49]
Cereus hildmannianus	40.19	[50]
Calamus manan	48.28	[51]
Citrullus lanatus	33.33	[52]

Although the tururi fabric has a lower crystallinity index, it still outperforms fibers such as pineapple, *Citrullus lanatus* and *Cereus hildmannianus*, which have exhibited even lower indices [48,50,52].

Natural fibers have compatible microfibril structural arrangements, which determine the microfibril angle (MFA). Fibers with a low MFA, above 0°, tend to be stronger and stiffer, while fibers with a high MFA, above 90°, are more flexible and susceptible to fracture [45].

The microfibrillar angle of the tururi fabric, calculated from the peak intensity of 22.32° [48], was 7.39°. This value is close to the value obtained for other lignocellulosic fibers, such coconut palm fiber, *Eurcraea foetida*, Ubim and periquiteira [40,41,53,54]. According to JUNIO et al. [41], low MFA values indicate good mechanical properties, making these fibers suitable for reinforcements.

The results show that tururi fabric has structural characteristics typical of natural fibers. The presence of these constituents reinforces the potential use of the fabric in polymer matrix composite materials.

3.1.2. TGA

The thermogravimetric curve (TG) and the thermogravimetric derivative (DTG) for tururi fibrous fabric are shown in Figure 3.

The thermogravimetric (TG) curve shows three distinct mass loss events. The first occurs at around 102 °C, indicating the loss of residual moisture, as reported by Dalmis et al. [55] and Khan et al. [56]. The second event occurs at approximately 247 °C, associated with the initial thermal degradation of hemicellulose and pectins present in lignocellulosic fibers, in addition to the release of all structural water [57,58]. The third and final thermal degradation event occurs at around 334 °C, corresponding to the beginning of the degradation of cellulose and other organic components [59]. This process involves the depolymerization, decarboxylation and decomposition of the glucose units present in the structure of lignocellulosic fibers [60,61]. The thermogram shows that tururi fibrous fabric maintains good thermal stability up to approximately 246 °C. Above this temperature, the thermal degradation of the fabric begins. The results suggest that tururi fabric has adequate thermal stability, which makes it promising for applications in polymeric matrices.

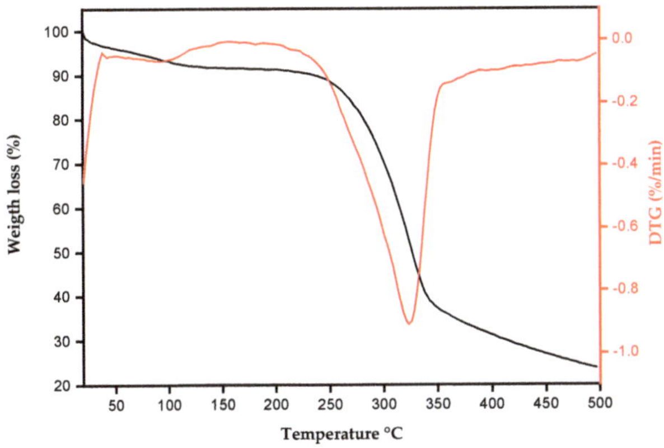

Figure 3. TG and DTG for tururi fabric.

3.1.3. FTIR Analysis

Figure 4 shows the spectrum for tururi fibrous fabric.

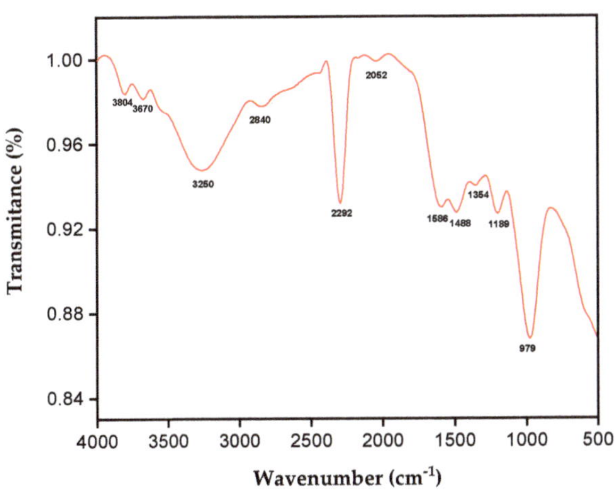

Figure 4. FTIR of tururi fabric.

FTIR is an important characterization technique for identifying the functional groups or absorption bands of constituents. Between 3600 and 3100 cm^{-1}, two vibrational bands are observed that can be attributed to the stretching of the hydroxyl group (OH-) [62,63]. At 2840 cm^{-1}, the vibrational stretching of the CH and CH_2 bonds, constituents of cellulose and hemicellulose, is observed [64]. At 2355 cm^{-1}, there is stretching of the C=C bonds and the presence of waxes [41,65]. The absorption band at 2052 cm^{-1} corresponds to the stretching of the C≡N and C≡C bonds [66]. At 1586 cm^{-1}, the stretching of the C-C bonds of the aromatic lignin ring is observed [31]. The peak observed at 1488 cm^{-1} corresponds to the C-H bond bending frequency of cellulose [62]. At 1354 cm^{-1}, there is stretching of the acetyl group of lignin [31]. The band at 1189 cm^{-1} corresponds to the stretching of the C-O-C group of the lignin and hemicellulose constituents [51]. At 979 cm^{-1}, it is associated with the crystalline cellulose constituent [60]. Tururi fabric shows absorption bands characteristic of natural fibers.

3.1.4. SEM Analysis of Tururi Fabric

The morphology of tururi fabric was analyzed using scanning electron microscopy, as shown in Figure 5.

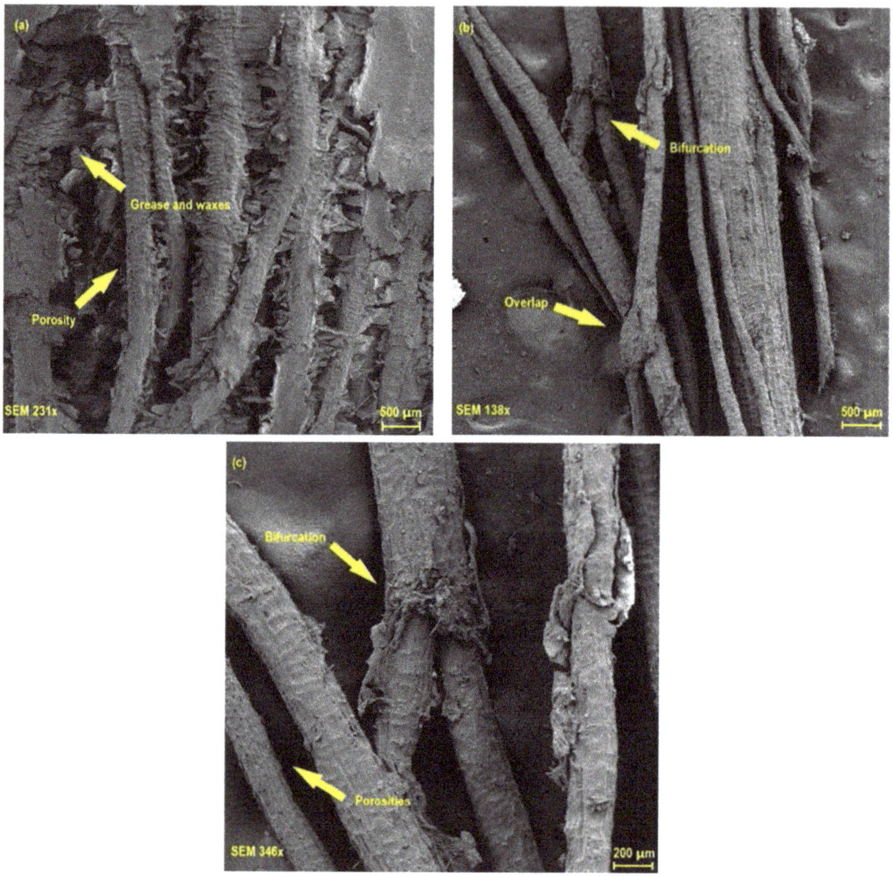

Figure 5. Scanning electron microscopy obtained for the tururi fabric. (**a**) tururi fabric yarns oriented on different surfaces, 231×, (**b**) tuturi fabric with magnification of 138× and (**c**) tuturi fabric with magnification of 348×.

Figure 5a shows that the tururi fibrous fabric is structured by yarns oriented on different surfaces, as well as having heterogeneous and irregular surfaces formed by roughness, pores and organic matter such as grease and wax [64,67]. Although the presence of pores and imperfections limits the mechanical performance of the fabric, these aspects can contribute positively to better anchoring and interfacial adhesion between the fabric and the polymer matrix, favoring the formation of more resistant and rigid green materials [27,49]. Oliveira et al. [68] and Porras et al. [31] also observed similar behavior when investigating the properties of tururi fabric. In Figure 5b,c, areas can be observed composed of interlacing, bifurcations and overlapping of threads that present characteristics of the fabric. The bifurcations have a fibrous structure and help with load distribution, the breaking mode and the interaction between the matrix and the fabric. Karthik et al. [69] and Gao et al. [70] explained that the structural organization of yarns or fibers directly determines the properties of composite materials.

Tururi fabric has morphological characteristics similar to those of other natural fibers already investigated in the literature, such as *Vachellia farnesiana, Citrullus lanatus* and *Cortaderia selloana* [52,56,71].

3.1.5. Mechanical Properties Tururi Fabric

Figure 6 illustrates the tensile strength of the tururi fabric as a function of strain.

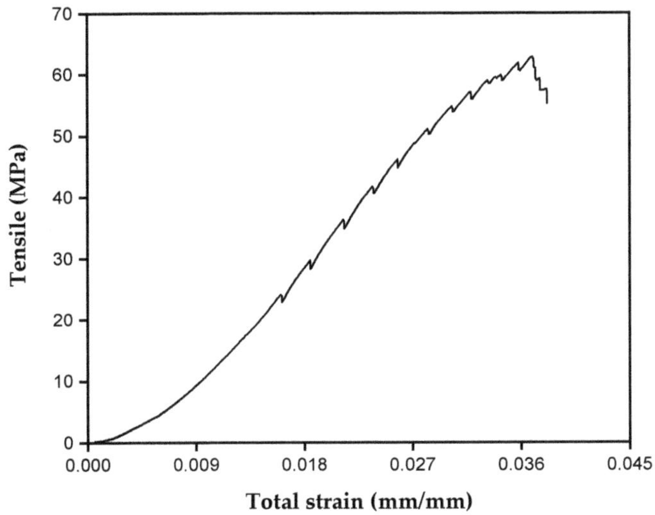

Figure 6. Tension versus strain curve of natural tururi fabric.

The tururi fabric had a maximum tensile strength peak of 60 MPa and an average Young's modulus of 1.1 GPa. After reaching the point of maximum tension, the filaments of the fabric tend to fail in different directions, due to the redistribution of tensions and the breaking of the fibers. This characteristic is associated with the structural design of tururi fabric, which comprises natural fibers arranged in multiple directions or interlaced, as described by Porras et al. [31].

The results obtained for tururi fabric are superior to those of other natural fibers, such as *Corypha taliera* [66] and coconut [72]. However, studies carried out by Akter et al. (2025) [73] indicated that fabrics made from jute fibers have superior mechanical performance compared to tururi fabric. However, the properties of tururi fabric, combined with its environmental characteristics, reinforce its potential as a reinforcement material in sustainable composite applications.

3.2. Characterization of Composite Materials

3.2.1. FTIR Composite Analysis

Figure 7 shows the spectrum obtained for the composites reinforced with tururi fabric with different percentages of incorporation.

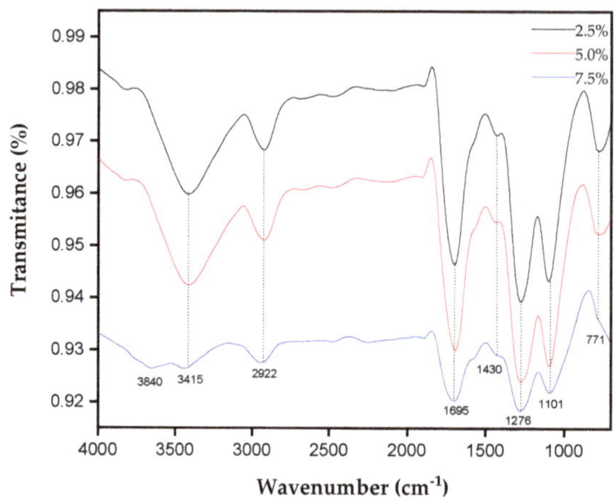

Figure 7. FTIR of composite materials reinforced with tururi fabric.

The band observed at 3415 cm^{-1} is characteristic of the stretching of the hydroxyl cluster [2]. The band observed at 2922 cm^{-1} corresponds to the asymmetric stretching of the CH bond. At 1695 cm^{-1}, an expressive peak is presented which is related to the grouping between the C=C bond of the carbonyl [2]. The band observed at 1430 cm^{-1} corresponds to the stretching of the bonds in the aromatic ring of lignin [74]. At 1276 cm^{-1}, it refers to the stretching of the CO bonds of the acetyl ketone group [55]. At 771 cm^{-1}, it corresponds to the stretching of the C=H bond [2,75]. As the percentage of the fabric increases, the transmission bands decrease, confirming that the percentage variation had an influence on the composition of the material.

3.2.2. DMA of Composites

Figure 8 shows the dynamic storage modulus (E′), tan delta (δ) and loss storage modulus (E″) for the composites with 2.5, 5.0 and 7.5% mass fraction of fabric in relation to temperature.

The DMA curves show that the composites reinforced with 2.5% fabric had an E′ equal to 1195 MPa, while the composites reinforced with 5.0 and 7.5% exhibit an E′ of 2077 and 2107 MPa, respectively. The increase in the percentage of fabric, as well as the increase in temperature, influenced the decrease in the storage modulus of the composites. These results suggest that increasing the percentage of reinforcement can provide greater rigidity and thermal and mechanical stability, as well as influencing movement in the vitreous transition region. In addition, the formation of bonds prevents the movement of the polymer chains, increasing the storage modulus. In this case, high stress is required to break the bonds between the interfaces and cause the material to rupture [76].

The loss of storage behavior shows that the 2.5% curve has a lower peak and a lower vitreous transition temperature. However, the composites with 5 and 7.5% reinforcement showed higher E″ peaks and a higher vitreous transition temperature compared to the composites with 2.5%. This indicates that the increase in reinforcement restricts the movement of the polymer chains due to the density of the cross-links, suggesting that there is an interaction between the fabric and the matrix [15].

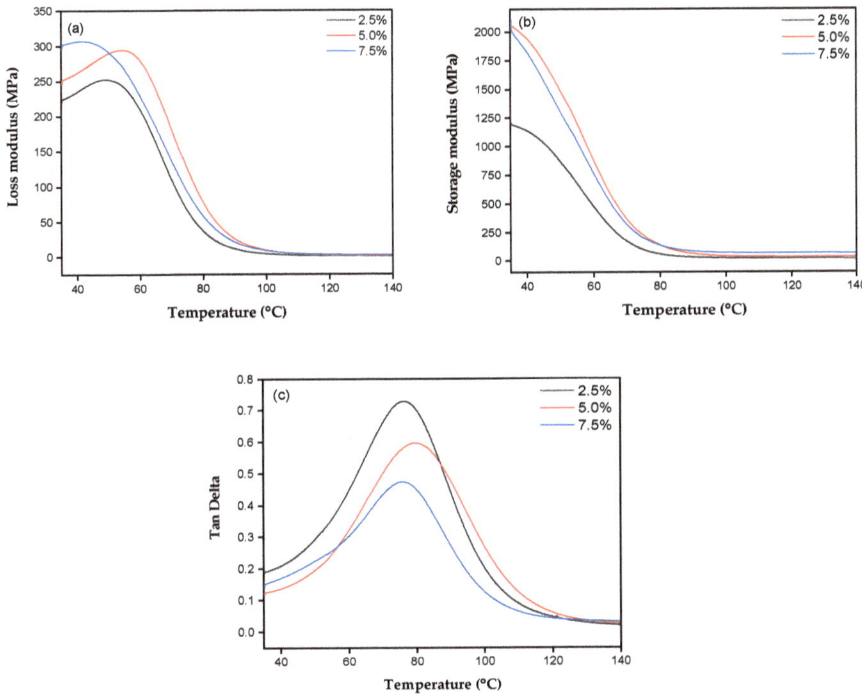

Figure 8. DMA curves of the composites of (**a**) loss modulus, (**b**) storage modulus and (**c**) tan delta.

The tan delta, which evaluates the material's viscoelastic energy dissipation, showed that higher reinforcement percentages of 5 and 7.5% tururi fabric dissipated less energy during strain, resulting in a rigid and mechanically efficient material. This behavior suggests that increasing the reinforcement content contributes positively to the structure of the material, reducing the exclusive dependence on the polymer matrix for mechanical performance.

3.2.3. Flexure Test

The flexural strength of the polyester matrix composites reinforced with tururi fibrous fabric is shown in Figure 9.

Table 3 shows the flexural strength, modulus of elasticity and strain results for the composites reinforced with natural tururi fabric.

Table 3. Flexural strength results for tururi/polyester composites.

Sample	Flexural Strength (MPa)	Modulus Elastic (GPa)	Strain (mm/mm)
P0	49.47 ± 4.99	2.49 ± 0.28	0.065 ± 0.005
P2.5	65.97 ± 7.41	3.64 ± 0.26	0.058 ± 0.008
P5.0	47.55 ± 2.58	2.08 ± 0.13	0.080 ± 0.003
P7.5	44.57 ± 9.20	3.95 ± 0.18	0.031 ± 0.011

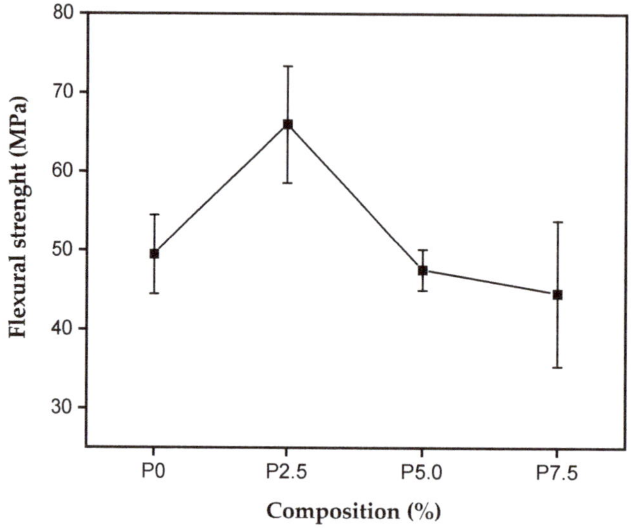

Figure 9. Flexural strength of composites reinforced with tururi fabrics.

The composition with 2.5% reinforcement showed the best flexural strength results, with an average of 65 MPa. The compositions with 5.0% and 7.5% reinforcement showed lower results than the polyester matrix and the composition with 2.5%. The relative decrease in flexural strength may be associated with the presence of organic compounds in the fibrous fabric, which make it difficult to anchor the reinforcement to the matrix [77,78]. In addition, it is suggested that the presence of pores, microstructural damage and bubbles act as stress concentrators, reducing the mechanical strength of the composite material [79]. Karthik et al. [69] explain that conventional interfacial bonding between the fiber or fabric and the matrix results in low flexural strength.

Figure 10 shows the results of the modulus of elasticity and strain of the composites reinforced with tururi fabric.

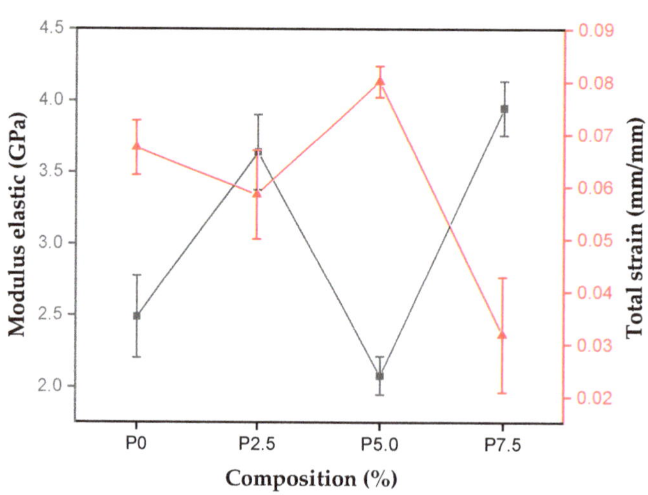

Figure 10. Modulus of elasticity and strain results for composites with tururi fabric.

The composites produced with 7.5% reinforcement had a higher modulus of elasticity, with an average of 3.95 GPa. The increase in modulus of elasticity is associated with the uniformity and homogeneity of the microfibrils of the tururi fabric, helping to distribute the loads throughout the matrix [80].

It should be noted that as the percentage of reinforcement increases, there is also an increase in the deformation of the material, resulting in an increase in flexural strength and modulus of elasticity. This behavior can be explained by the increase in the percentage of fabric and the increase in interfaces, which reduces the deformation capacity. The random arrangement of the yarns in the fabric reduces delaminations and increases the flexural strength of the material [81,82]. NG et al. [29] explain that, during bending tests, the top layer of the fabric tends to support compression loads, while the lower layers support the tension offered. In this way, the fabrics in the outer layers tend to support more bending loads than the other layers of the material.

The morphology of the flexural fractures for the composites reinforced with 2.5% and 5.0% tururi fabric is shown in Figures 11a and 11b, respectively.

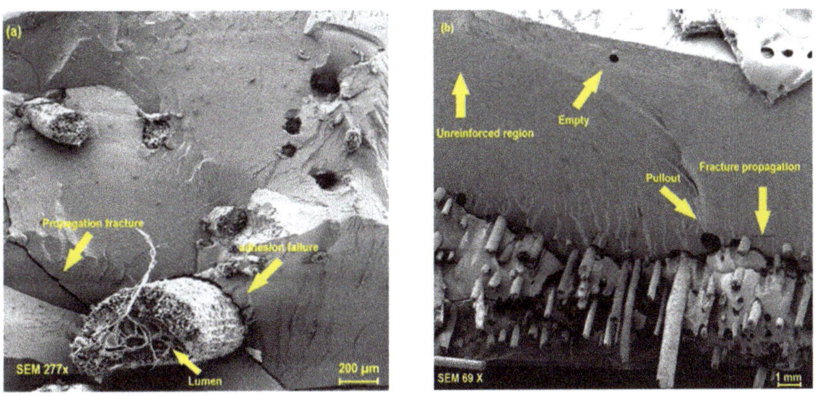

Figure 11. Scanning electron microscopy obtained for the composites in percentages of 2.5 (**a**) and 7.5% (**b**) in mass fraction of tururi fabric.

It is possible to observe the presence of river marks which indicate that the fracture is of the brittle type for the composites with 2.5 and 5.0% fabric percentage. In addition, regions characteristic of the pull-out effect can be seen, as well as voids and pores, indicating that there was low adhesion between the fabric and the polymer matrix. This interaction results in a reduction in the flexural strength of the 2.5% and 5.0% compositions. Simonassi et al. [83] explains that the pull-out effect and the presence of voids are attributed to interfacial phenomena between the fiber and the polymer matrix, which tend to reduce the mechanical properties of composite materials [81]. Other characteristics associated with the hydrophobic characteristics between the matrix and the reinforcement can promote low interaction between the phases, causing flaws and defects along the length of the material [84]. In addition, when subjected to static loads or dynamic events, they lead to early rupture. These effects reduce the mechanical properties of composite materials reinforced with natural tururi fabric.

3.2.4. Charpy Impact Test

The results of the impact resistance test of the polymer matrix composites with different percentages of tururi fabric are shown in Figure 12.

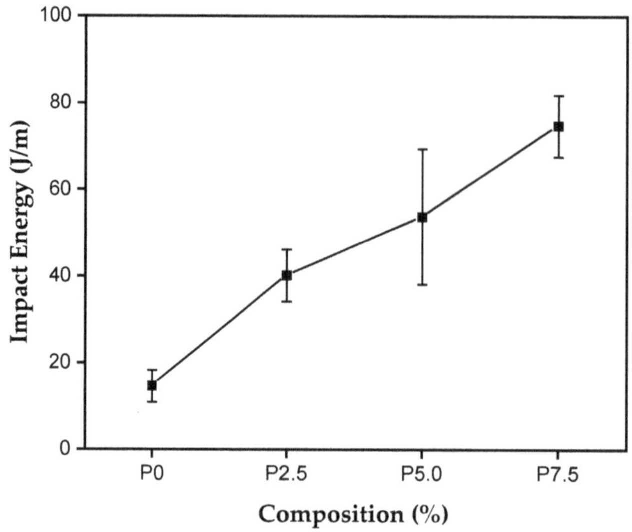

Figure 12. Impact energy for polyester matrix and composites reinforced with tururi fabric.

Table 4 shows the Charpy impact energy results for the composites reinforced with natural tururi fabric.

Table 4. Charpy impact results for Tururi/polyester composites.

Sample	Charpy Impact Resistance (J/m)
P0	14.61 ± 3.69
P2.5	40.13 ± 5.60
P5.0	53.73 ± 15.63
P7.5	74.72 ± 7.13

The mean results show that the matrix and the 7.5% composition had around 14.61 J/m and 74.72 J/m, respectively. The composites produced with 7.5% showed an increase of 471.2% compared to the matrix. Therefore, the increase in impact resistance can be attributed to the increase in the percentage of reinforcement in the polymer matrix, resulting in an improvement in mechanical properties [85]. Pereira et al. [86], Pereira et al. [87] and Candido et al. [88] observed that the increase in impact energy absorption capacity is high when the percentage of reinforcement in the matrix is increased. From this, it can be said that the toughness resistance is increased by incorporating tururi fabric into the polyester matrix.

Fabrics made with natural or synthetic fibers oriented in the longitudinal and transverse directions have greater impact and load resistance due to the alignment of the fibers. In addition, fabrics, whether natural or synthetic in structure, tend to absorb more energy when compared to composites made with unidirectionally aligned fibers [89]. Chaves et al. [90] explain that since fibers are stiffer and stronger than the matrix, these properties result in stiffer and stronger composites, proportionally to the increase in the percentage of fibers.

The significant increase in impact energy is intrinsically linked to the characteristics and particularities of natural tururi fabric.

Micrographs of the fractures of composites subjected to the Charpy impact test are shown in Figure 13.

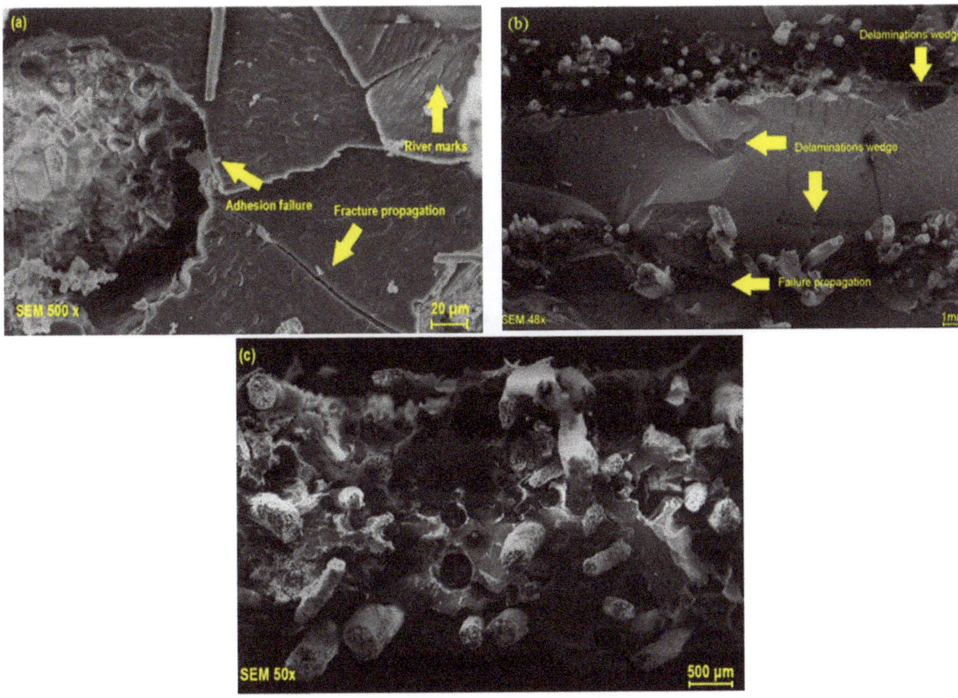

Figure 13. Micrographs of composites reinforced with 2.5% (**a**), 5.0% (**b**) and 7.5% (**c**) mass fraction of tururi fabric.

Figure 13a shows marks that appear as rivers, as well as the propagation of cracks, which indicate brittle fracture. Figure 13b shows the presence of voids and propagated cracks, which result from low energy dissipation along the matrix and reinforcement. Figure 13c reveals the presence of voids or areas marked by the pull-out effect, which manifests itself when the material is subjected to the transfer of impact energy [91]. Ng et al. [29] explains that the stacking and microstructural characteristics of the fabric can lead to separations, delaminations and the pull-out effect. These factors are related to the interfacial bonding between the phases, which may be insufficient to ensure that the material absorbs even more energy, resulting in the pull-out and rupture of the material when subjected to high loads [90,92].

In addition to the failure mechanisms, the structural organization of tururi fabric facilitates better adhesion with the polymer matrix and, consequently, a better distribution of stresses throughout the material. According to Gao et al. [70] and Karthik et al. [69], fabrics comprising multilayers of synthetic or natural fibers with randomly arranged yarns favor an increase in the energy absorption and impact resistance of the composite material. Furthermore, as the relative percentage of tururi reinforcement increases, there is a reduction in apparent defects.

The use of a natural fabric reinforces the development of lightweight and sustainable composite materials with exceptional properties for various applications.

3.2.5. Statistical Analysis

Table 5 presents the analysis of variance for the mechanical characterization of composites reinforced with tururi fabric.

Table 5. Analysis of variance for results.

Source	Sum squares	Degree of freedom	Mean of squares	F (calculated)	p-value	F-critical
Flexural Strengh (MPa)						
Between the groups	110,683.7	3	368.945	8.629	0.002	3.490
Inside the groups	513.064	12	42.755			
Total	161,990.2	15				
Elastic Modulus (GPa)						
Source	Sum squares	Degree of freedom	Mean of squares	F (calculated)	p-value	F-critical
Between the groups	9.652	3	3.217	28.951	8.88×10^{-6}	3.490
Inside the groups	1.333	12	0.111			
Total	10.986	15				
Total Strain (mm/mm)						
Source	Sum squares	Degree of freedom	Mean of squares	F (calculated)	p-value	F-critical
Between the groups	0.005	3	0.001	17.288	1.181×10^{-4}	3.490
Inside the groups	0.001	12	9.916×10^{-5}			
Total	0.006	15				
Charpy Impact Resistance (J/m)						
Source	Sum squares	Degree of freedom	Mean of squares	F (calculated)	p-value	F-critical
Between the groups	7618.892	3	2539.631	22.099	3.54×10^{-5}	3.490
Inside the groups	1379.001	12	114.916			
Total	8997.893	15				

The ANOVA results for flexural strength indicate that the calculated F (8.629) is higher than the critical F (3.490). For deformation, the calculated F was (17.288) and the critical F was equal to (3.490). For the modulus of elasticity, the calculated F (28.951) was higher than the critical F (3.490). The impact resistance result showed a calculated F (22.099) greater than the critical F (3.490). Based on these results, the null hypothesis of equality between the means of the properties evaluated was rejected, with a 95% confidence level. After rejecting equality on the basis of the ANOVA result, Tukey's test was applied to identify which means differed significantly in terms of the variation in the percentage of reinforcement in the polymer matrix.

Table 6 shows the results of Tukey's test.

The minimum significant difference (msd) is a parameter that differentiates the averages obtained in the tests. For the properties of flexural strength, modulus of elasticity, deformation and impact resistance, the msd values were 14.810, 0.755, 0.022 and 24.280, respectively.

The results of Tukey's test for flexural strength indicated statistical differences between the composites produced with 2.5% fabric and the other compositions. These data suggest that the composites with 2.5% fabric are less ductile but have greater strength. The differences observed in the modulus of elasticity averages confirm that adding more than 2.5% fabric significantly increases the rigidity of the material.

The Charpy impact resistance test results showed statistical differences. Composites with 5.0% and 7.5% tururi showed significant differences compared to composites with the lower percentages of 0% and 2.5%. These results indicate that increasing the percentage of fabric is associated with better impact properties.

Table 6. Presents the results of the Tukey Test.

Flexural Strength (m.s.d: 14.810)				
Sample	**P0**	**P2.5**	**P5.0**	**P7.5**
P0	0	16.018 *	1.925	4.905
P2.5	16.501 *	0	18.427 *	21.407 *
P5.0	1.925	18.427 *	0	2.980
P7.5	4.905	21.407 *	2.980	0
Elastic modulus (m.s.d: 0.755)				
Sample	**P0**	**P2.5**	**P5.0**	**P7.5**
P0	0	1.155 *	0.405	1.462 *
P2.5	1.155 *	0	1.56 *	0.307
P5.0	0.405 *	1.56 *	0	1.867 *
P7.5	1.462 *	0.307	1.867 *	0
Total Strain (m.s.d: 0.022)				
Sample	**P0**	**P2.5**	**P5.0**	**P7.5**
P0	0	0.007	0.015	0.034 *
P2.5	0.088 *	0	0.022	0.027 *
P5.0	0.015 *	0.022	0	0.049 *
P7.5	0.034 *	0.027 *	0.049 *	0
Impact Energy (m.s.d: 24.280)				
Sample	**P0**	**P2.5**	**P5.0**	**P7.5**
P0	0	25.526 *	39.125 *	60.118 *
P2.5	25.526 *	0	13.598	34.591 *
P5.0	39.125	13.598	0	20.993
P7.5	60.118 *	34.591 *	20.993	0

* Statistically different values.

Thus, incorporating tururi fabric into the polymer matrix tends to improve the properties of flexural strength, deformation, modulus of elasticity and impact resistance.

The characterization of tururi revealed that the fabric exhibited thermal stability up to approximately 250 °C, indicating that it can be used as a reinforcement agent for more thermally stable matrices. Furthermore, its use as a reinforcement agent in polymeric matrices showed low flexural strength values, as observed by [69,78], although these are still considered relatively high values for this property. Due to its energy absorption capacity upon impact and its visual appearance, it is suggested that the produced composite could be employed in the construction sector for the manufacturing of cladding panels and exposed bricks wall.

4. Conclusions

The characterizations of the tururi fabric and the composites revealed the following observations:

FT-IR indicated the presence of functional groups characteristic of the constituents present in tururi fabric, such as cellulose, hemicellulose and lignin. The thermogram showed the water loss from and degradation of the elemental constituents present in the fabric, as well as thermal stability at temperatures of up to 220 °C. From the XRD analysis, it was possible to identify the microfibrillar angle (7.39°) and the crystallinity index (46.67%). A morphological analysis of the tururi fabric revealed the presence of imperfections, flaws, tangles and irregular microstructural features in the fabric, as well as bifurcation regions. In the mechanical characterization by flexural test, the composites with 2.5% reinforcement showed a flexural strength and modulus of elasticity of 65 MPa and 3.76 GPa, respectively. A DMA indicated that increasing the percentage of fabric in the polymer matrix influenced the changes in vitreous temperature and the final properties of the composite material. The Charpy impact test indicated that the composition with 7.5% reinforcement showed an increase in energy absorption capacity of approximately 471% compared to the matrix. This result suggests that the energy absorption capacity increases with the incorporation of reinforcement. A fracture analysis of the composites indicated the presence of microstructural imperfections, such as porosity, voids and poor distribution of the reinforcement, attributed to the processing phase. The decrease in mechanical strength is intrinsically related to the morphological characteristics of the fabric. Compared to other studies using natural fibers as reinforcement, the new composite reinforced with fibrous fabric in a polyester polymer matrix shows great potential for applications subject to dynamic loads. Moreover, due to its good mechanical properties, tururi is an attractive alternative for use in composite materials because it is a completely natural fabric derived from a renewable natural resource.

Author Contributions: Investigation, methodology, resource, conceptualization, A.G.A.S.; investigation, conceptualization, methodology, D.d.C.B.N. and N.T.S.; visualization, methodology, F.P.D.L. and A.C.R.d.S.; Formal analysis, validation, and funding acquisition S.N.M.; visualization, curation, conceptualization, investigation, writing—review and editing, and validation, V.S.C. All authors have read and agreed to the published version of the manuscript.

Funding: The Amazon Foundation for Studies and Research (Fapespa), call for proposals N°. 15/2021, Brazil, and Dean of Research and Graduate Studies (Propesp), Federal University of Para, financed this research.

Institutional Review Board Statement: Not applicable.

Data Availability Statement: The original contributions presented in this study are included in the article. Further inquiries can be directed to the corresponding author.

Acknowledgments: Amazon Foundation for Studies and Research (Fapespa), Coordination for the Improvement of Higher Education Personnel (Capes), Dean of Research and Graduate Studies (Propesp), Materials Science and Engineering Program (PPGCEM), Federal University of Para.

Conflicts of Interest: The authors declare no conflicts of interest.

References

1. Balla, V.K.; Kate, K.H.; Satyavolu, J. Additive Manufacturing of Natural Fiber Reinforced Polymer Composites: Processing and Prospects. *Compos. Part B* **2019**, *174*, 106956. [CrossRef]
2. Hossain, M.D.S.; Mobarak, M.B.; Rony, F.K.; Sultana, S.; Mahmud, M.; Ahmed, S. Fabrication and Characterization of Banana Fiber Reinforced Unsaturated Polyester Resin Based Composites. *Nano Hybrids Compos.* **2020**, *29*, 84–92. [CrossRef]
3. Mekonnen, B.Y.; Mano, Y.J. Tensile and Flexural Analysis of a Hybrid Bamboo/Jute Fiber-Reinforced Composite with Polyester Matrix as a Sustainable Green Material for Wind Turbine Blades. *Int. J. Eng.* **2020**, *33*, 314–319. [CrossRef]
4. Ilyas, R.A.; Khalina, A.; Nurazzi, N.M.; Lee, S.H.; Lee, C.H. Uma Revisão Abrangente em Tecido Sustentável Avançado Compósitos de polímeros de fibra natural. *Polímeros* **2021**, *13*, 471. [CrossRef]
5. Cavalcanti, D.K.K.; Banea, M.D.; Neto, J.S.S. Comparative Analysis of the Mechanical and Thermal Properties of Polyester and Epoxy Natural Fiber-Reinforced Hybrid Composites. *J. Compos. Mater.* **2021**, *55*, 1683–1692. [CrossRef]
6. Dipen, K.R.; Durgesh, D.P.; Ravinder, K. Recent Progress of Reinforcement Materials: A Comprehensive Overview of Composite Materials. *J. Mater. Res. Technol.* **2019**, *8*, 6354–6374. [CrossRef]
7. Das Subrata, C.; Debasree, P.; Sotirios, A.I.; Siddiquee, M.A.B.; Papatzani, S.; Koralli, P.; Islam, J.M.M.; Khan, M.A.; Shauddin, S.M.; Khan, R.A.; et al. Effect of Stacking Sequence on the Performance of Hybrid Natural/Synthetic Fiber Reinforced Polymer Composite Laminates. *Compos. Struct.* **2021**, *276*, 114525. [CrossRef]
8. Mindermann, P.; Gil Pérez, M.; Knippers, J.; Gresser, G.T. Investigation of the Fabrication Suitability, Structural Performance, and Sustainability of Natural Fibers in Coreless Filament Winding. *Materials* **2022**, *15*, 3260. [CrossRef]
9. Kumar, S.; Prasad, L.; Bijlwan, P.P.; Yadav, A. Thermogravimetric Analysis of Lignocellulosic Leaf-Based Fiber-Reinforced Thermosets Polymer Composites: An Overview. *Biomass Convers. Biorefinery* **2024**, *14*, 12673–12698. [CrossRef]

10. Zhao, X.; Li, K.; Wang, Y.; Tekinalp, H.; Richard, A.; Webb, E.; Ozcan, S. Bio-Treatment of Poplar via Amino Acid for Interface Control in Biocomposites. *Compos. Part B Eng.* **2020**, *199*, 108276. [CrossRef]
11. Zhao, X.; Copenhaver, K.; Wang, L.; Korey, M.; Gardner, D.J.; Li, K.; Lamm, M.E.; Kishore, V.; Bhagia, S.; Tajvidi, M.; et al. Recycling of Natural Fiber Composites: Challenges and Opportunities. *Resour. Conserv. Recycl.* **2022**, *177*, 105962. [CrossRef]
12. Asrofi, M.; Sapuan, S.M.; Ilyas, R.A.; Ramesh, M. Characteristic of Composite Bioplastics from Tapioca Starch and Sugarcane Bagasse Fiber: Effect of Time Duration of Ultrasonication (Bath-Type). *Mater. Today Proc.* **2020**, *46*, 1626–1630. [CrossRef]
13. Sabaruddin, F.A.; Paridah, M.T.; Sapuan, S.M.; Ilyas, R.A.; Lee, S.H.; Abdan, K.; Mazlan, N.; Roseley, A.S.M.; Abdul Khalil, H.P.S. The Effects of Unbleached and Bleached Nanocellulose on the Thermal and Flammability of Polypropylene-Reinforced Kenaf Core Hybrid Polymer Bionanocomposites. *Polymers* **2020**, *13*, 116. [CrossRef]
14. Gohal, H.; Kumar, V.; Jena, H. Study of Natural Fiber Composite Material and Its Hybridization Techniques. *Mater. Today Proc.* **2020**, *26*, 1368–1372. [CrossRef]
15. Haris, N.I.N.; Hassan, M.Z.; Ilyas, R.A.; Suhot, M.A.; Sapuan, S.M.; Dolah, R.; Mohammad, R.L.; Asyraf, M.R.M. Dynamic Mechanical Properties of Natural Fiber Reinforced Hybrid Polymer Composites: A Review. *J. Mater. Res. Technol.* **2022**, *19*, 167–182. [CrossRef]
16. Karimah, A.; Ridho, M.R.; Munawar, S.S.; Adi, D.S.; Damayanti, R.; Subiyanto, B.; Fatriasari, W.; Fudholi, A. A review on natural fibers for development of eco-friendly bio-composite: Characteristics, and utilizations. *J. Mater. Res. Technol.* **2021**, *13*, 2442–2458. [CrossRef]
17. Alhijazi, M.; Safaei, B.; Zeeshan, Q.; Asmael, M.; Eyvazian, A.; Qin, Z. Recent Developments in Luffa Natural Fiber Composites: Review. *Sustainability* **2020**, *12*, 7683. [CrossRef]
18. Ribeiro, M.M.; Pinheiro, M.A.; Rodrigues, J.d.S.; Ramos, R.P.B.; Corrêa, A.d.C.; Monteiro, S.N.; da Silva, A.C.R.; Candido, V.S. Comparison of Young's Modulus of Continuous and Aligned Lignocellulosic Jute and Mallow Fibers Reinforced Polyester Composites Determined Both Experimentally and from Theoretical Prediction Models. *Polymers* **2022**, *14*, 401. [CrossRef]
19. Kumar, S.S. Effect of Natural Fiber Loading on Mechanical Properties and Thermal Characteristics of Hybrid Polyester Composites for Industrial and Construction Fields. *Fibers Polym.* **2020**, *21*, 1508–1514. [CrossRef]
20. Sarikaya, E.; Çallioğlu, H.; Demirel, H. Production of Epoxy Composites Reinforced by Different Natural Fibers and Their Mechanical Properties. *Compos. Part B Eng.* **2019**, *167*, 461–466. [CrossRef]
21. Bachchan, A.A.; Das, P.P.; Chaudhary, V. Effect of Moisture Absorption on the Properties of Natural Fiber Reinforced Polymer Composites: A Review. *Mater. Today Proc.* **2021**, *49*, 3403–3408. [CrossRef]
22. Gazem de Carvalho, J.P.R.; Duarte Lopes, F.P.; Simonassi, N.T.; Neves, A.C.C.; Atem de Carvalho, E.; Monteiro, S.N.; Vieira, C.M.F. Methodological Analysis of Composites Green Polyurethane Resin Reinforced with Jute Fabric. *Case Stud. Constr. Mater.* **2022**, *17*, e01512. [CrossRef]
23. Mishra, T.; Mandal, P.; Rout, A.K.; Sahoo, D. A State-of-the-Art Review on Potential Applications of Natural Fiber-Reinforced Polymer Composite Filled with Inorganic Nanoparticle. *Compos. Part C Open Access* **2022**, *9*, 100298. [CrossRef]
24. Akhil, U.V.; Radhika, N.; Saleh, B.; Krishna, S.A.; Nobre, N.; Rajeshkumar, L. Uma Revisão Abrangente Sobre Compósitos Poliméricos Reforçados A comprehensive review on plant-based natural fiber reinforced polymer composites: Fabrication, properties, and applications. *Polím. Compós.* **2023**, *44*, 2598–2633. [CrossRef]
25. Liu, S.; Dong, C.; Yuan, C.; Bai, X.; Tian, Y.; Zhang, G. A new polyimide matrix composite to improve friction-induced chatter performance through reducing fluctuation in friction force. *Compos. Part B Eng.* **2021**, *217*, 108887. [CrossRef]
26. Monteiro, A.S.; Dantas, D.; Yojo, T. Preparation of Amazonian Palm Tree Fiber (*Manicaria saccifera Gaertn.*) for Composite Materials. *U.Porto J. Eng.* **2021**, *7*, 31–36. [CrossRef]
27. Oliveira, J.R.; Kotzebue, L.R.V.; Freitas, D.B.; Mattos, A.L.A.; da Costa Júnior, A.E.; Mazzetto, S.E.; Lomonaco, D. Towards Novel High-Performance Bio-Composites: Polybenzoxazine-Based Matrix Reinforced with *Manicaria saccifera* Fabrics. *Compos. Part B Eng.* **2020**, *194*, 108060. [CrossRef]
28. Ng, L.F.; Yahya, M.Y.; Muthukumar, C. Mechanical Characterization and Water Absorption Behaviors of Pineapple Leaf/Glass Fiber-Reinforced Polypropylene Hybrid Composites. *Polym. Compos.* **2022**, *43*, 203–214. [CrossRef]
29. Ng, L.F.; Yahya, M.Y.; Muthukumar, C.; Parameswaranpillai, J.; Ma, Q.; Muhammad Asyraf, M.R.; Abdul Majid, R. Mechanical Characterization, Water Absorption, and Thickness Swelling of Lightweight Pineapple Leaf/Ramie Fabric-Reinforced Polypropylene Hybrid Composites. *Polymers* **2024**, *16*, 1847. [CrossRef]
30. Monteiro, A.S.; Leonardi, B.; Savastano, H., Jr.; Baruque-Ramos, J. Tururi Palm Fibrous Material (*Manicaria saccifera* Gaertn.) Characterization. *Green Mater.* **2015**, *3*, 120–131. [CrossRef]
31. Porras, A.; Maranon, A.; Ashcroft, I.A. Characterization of a Novel Natural Cellulose Fabric from *Manicaria saccifera* Palm as Possible Reinforcement of Composite Materials. *Compos. Part B Eng.* **2015**, *74*, 66–73. [CrossRef]
32. Midani, M.; Seyam, A.F.M.; Monteiro, A.S.; Baruque-Ramos, J. Effect of Structural Parameters on the Impact Properties of Multilayer Composites from Tururi Palm (*Manicaria saccifera* Gaertn.) Fibrous Material. *J. Nat. Fibers* **2018**, *17*, 284–297. [CrossRef]

33. Botanical Garden Lancetilla. Manicaria Saccifera Gaertn. 2014. Available online: http://jblancetilla.org/manicaria_saccifera_gaertn.htm (accessed on 27 October 2024).
34. *ASTM D790-17*; Standard Test Method for Flexural Properties of Unreinforced and Reinforced Plastics and Electrical Insulating Materials. American Society for Testing and Materials (ASTM): West Conshohocken, PA, USA, 2017.
35. *ASTM D6110-18*; Standard Test Method for Determining the Charpy Impact Resistance of Notched Specimens of Plastics. American Society for Testing and Materials (ASTM): West Conshohocken, PA, USA, 2010.
36. Segal, L.A.; Creely, J.J.; Martin, A.E.; Conrad, C.M. An Empirical Method for Estimating the Degree of Crystallinity of Native Cellulose Using the X-ray Diffractometer. *Text. Res. J.* **1959**, *29*, 786–794. [CrossRef]
37. Cave, I.D. X-ray Microfibril Angle Measurement. *For. Prod. J.* **1966**, *16*, 37–42.
38. Pinheiro, M.A.; Gomes, L.G.; da Silva, A.C.R.; Candido, V.S.; Reis, R.H.M.; Monteiro, S.N. Guaruman: A Natural Amazonian Fiber with Potential for Polymer Composite Reinforcement. *Mater. Res.* **2019**, *22*, e20190092. [CrossRef]
39. Reis, R.H.M.; Nunes, L.F.; Oliveira, M.S.; Veiga Júnior, V.F.; Garcia Filho, F.C.; Pinheiro, M.A.; Cândido, V.S.; Monteiro, S.N. Fibra Guaruman: Another possible reinforcement in composites. *J. Mater. Res. Technol.* **2020**, *9*, 6. [CrossRef]
40. Marchi, B.Z.; Oliveira, M.S.; Bezerra, W.B.A.; de Sousa, T.G.; Candido, V.S.; da Silva, A.C.R.; Monteiro, S.N. Ubim Fiber (*Geonoma baculífera*): A Less Known Brazilian Amazon Natural Fiber for Engineering Applications. *Sustainability* **2022**, *14*, 421. [CrossRef]
41. Pinheiro, M.A.; Ribeiro, M.M.; Rosa, D.L.S.; Nascimento, D.d.C.B.; da Silva, A.C.R.; dos Reis, M.A.L.; Monteiro, S.N.; Candido, V.S. Periquiteira (*Cochlospermum orinocense*): A Promising Amazon Fiber for Application in Composite Materials. *Polymers* **2023**, *15*, 2120. [CrossRef] [PubMed]
42. Junio, R.F.P.; de Mendonça Neuba, L.; Souza, A.T.; Pereira, A.C.; Nascimento, L.F.C.; Monteiro, S.N. Thermochemical and Structural Characterization of Promising Carnauba Novel Leaf Fiber (*Copernicia prunifera*). *J. Mater. Res. Technol.* **2022**, *18*, 4714–4723. [CrossRef]
43. Stevulova, N.; Cigasova, J.; Estokova, A.; Terpakova, E.; Geffert, A.; Kacik, F.; Singovszka, E.; Holub, M. Properties Characterization of Chemically Modified Hemp Hurds. *Materials* **2014**, *7*, 8131–8150. [CrossRef] [PubMed]
44. Viscusi, G.; Barra, G.; Gorrasi, G. Modification of Hemp Fibers Through Alkaline Attack Assisted by Mechanical Milling: Effect of Processing Time on the Morphology of the System. *Cellulose* **2020**, *27*, 8653–8665. [CrossRef]
45. da Silveira, P.H.P.M.; Santos, M.C.C.d.; Chaves, Y.S.; Ribeiro, M.P.; Marchi, B.Z.; Monteiro, S.N.; Gomes, A.V.; Tapanes, N.d.L.C.O.; Pereira, P.S.d.C.; Bastos, D.C. Characterization of Thermo-Mechanical and Chemical Properties of Polypropylene/Hemp Fiber Biocomposites: Impact of Maleic Anhydride Compatibilizer and Fiber Content. *Polymers* **2023**, *15*, 3271. [CrossRef]
46. Hossain, S.; Jalil, M.A.; Islam, T.; Rahman, M.M. A Low-Density Cellulose Rich New Natural Fiber Extracted from the Bark of Jack Tree Branches and Its Characterizations. *Heliyon* **2022**, *8*, e11667. [CrossRef] [PubMed]
47. Costa, U.O.; Nascimento, L.F.C.; Garcia, J.M.; Bezerra, W.B.A.; da Luz, F.S.; Pinheiro, W.A.; Monteiro, S.N. Mechanical Properties of Composites with Graphene Oxide Functionalization of Either Epoxy Matrix or Curaua Fiber Reinforcement. *J. Mater. Res. Technol.* **2020**, *9*, 13390–13401. [CrossRef]
48. Rekha, B.; NagarajaGanesh, B. X-ray Diffraction: An Efficient Method to Determine Microfibrillar Angle of Dry and Matured Cellulosic Fibers. *J. Nat. Fibers* **2022**, *19*, 3689–3696. [CrossRef]
49. Demóstenes, L.C.C.; Nascimento, L.F.C.; Monteiro, S.N.; Costa, U.O.; da Costa Garcia Filho, F.; da Luz, F.S.; Oliveira, M.S.; Ramos, F.J.H.T.V.; Pereira, A.C.; Braga, F.O. Thermal and Structural Characterization of Buriti Fibers and Their Relevance in Fabric-Reinforced Composites. *J. Mater. Res. Technol.* **2020**, *9*, 115–120. [CrossRef]
50. Subramanian, S.G.; Rajkumar, R.; Ramkumar, T. Characterization of Natural Cellulosic Fiber from *Cereus hildmannianus*. *J. Nat. Fibers* **2021**, *18*, 343–354. [CrossRef]
51. Ding, L.; Han, X.; Li, H.; Han, J.; Cao, L.; Chen, Y.; Ling, Z.; He, S.; Jiang, S. Characterization of Novel Natural Fiber from Manau Rattan (*Calamus manan*) as a Potential Reinforcement for Polymer-Based Composites. *J. Bioresour. Bioprod.* **2022**, *7*(3), 190–200. [CrossRef]
52. Khan, A.; Vijay, R.; Singaravelu, D.L.; Sanjay, M.R.; Siengchin, S.; Jawaid, M.; Asiri, A.M. Extraction and Characterization of Natural Fibers from *Citrullus lanatus* Climber. *J. Nat. Fibers* **2020**, *19*, 621–629. [CrossRef]
53. Selvan, M.T.G.A.; Binoj, J.S.; Moses, J.T.E.J.; Sai, N.P.; Siengchin, S.; Sanjay, M.R.; Liu, Y. Extraction and Characterization of Natural Cellulosic Fiber from Fragrant Screw Pine Prop Roots as Potential Reinforcement for Polymer Composites. *Polym. Compos.* **2022**, *43*, 320–329. [CrossRef]
54. Manimaran, P.; Senthamaraikannan, P.; Sanjay, M.R.; Marichelvam, M.K.; Jawaid, M. Study on Characterization of *Furcraea foetida* New Natural Fiber as Composite Reinforcement for Lightweight Applications. *Carbohydr. Polym.* **2018**, *181*, 650–658. [CrossRef] [PubMed]
55. Dalmis, R.; Kilic, G.B.; Seki, Y.; Koktas, S.; Keskin, O.Y. Characterization of a Novel Natural Cellulosic Fiber Extracted from the Stem of *Chrysanthemum morifolium*. *Cellulose* **2020**, *27*, 8621–8634. [CrossRef]

56. Khan, A.; Vijay, R.; Singaravelu, D.L.; Sanjay, M.R.; Siengchin, S.; Verpoort, F.; Alamry, K.A.; Asiri, A.M. Characterization of Natural Fibers from *Cortaderia selloana* Grass (Pampas) as Reinforcement Material for the Production of Composites. *J. Nat. Fibers* **2021**, *18*, 1893–1901. [CrossRef]
57. Mannai, F.; Elhleli, H.; Dufresne, A.; Elaloui, E.; Moussaoui, Y. *Opuntia* (Cactaceae) Fibrous Network-Reinforced Composites: Thermal, Viscoelastic, Interfacial Adhesion, and Biodegradation Behavior. *Fibers Polym.* **2020**, *21*, 2353–2363. [CrossRef]
58. Lai, T.S.M.; Jayamani, E.; Soon, K.H. Comparative Study on Thermogravimetric Analysis of Banana Fibers Treated with Chemicals. *Mater. Today Proc.* **2023**, *78*, 458–461. [CrossRef]
59. Heckadka, S.S.; Ballambat, R.P.; Bhagavath, P.; Kini, M.V.; Sinha, R.K.; Sonali, M.K.; Sen, D. Thermogravimetric Analysis of Flax, Jute, and UHMWPE Fibers and Their Composites with Melamine and Phenol Formaldehyde Resins. *Cogent Eng.* **2023**, *10*, 2209990. [CrossRef]
60. Jaruwan, M.; Klinkosum, P.; Chaichanasongkram, T.; Sarak, S.; Kaewtatip, K. Caracterização de uma nova fibra de celulose natural de *Enhalus acoroides* e sua potencial aplicação. *Ind. Crops Prod.* **2022**, *186*, 115285. [CrossRef]
61. Asim, M.; Paridah, M.T.; Chandrasekar, M.; Shahroze, R.M.; Jawaid, M.; Nasir, M.; Siakeng, R. Thermal Stability of Natural Fibers and Their Polymer Composites. *Iran. Polym. J.* **2020**, *29*, 625–648. [CrossRef]
62. Raja, A.K.A.; Geethan, K.A.V.; Kumar, S.S.; Kumar, P.S. Influência de atributos mecânicos, absorção de água, características de deflexão de calor e caracterização de compósitos híbridos epóxi reforçados com fibras naturais para uma aplicação de engenharia. *Fibers Polym.* **2021**, *22*, 3444–3455. [CrossRef]
63. Sathishkumar, T.P.; Shah, M.A.; Panchal, H.; Sharma, K.; Gopinath, R.; Sanjay, M.R.; Siengchin, S.; Rajesh Kumar, L.; Rampradheep, G.S. Characterization of New Cellulose Fiber Extracted from Second Generation Bitter Albizia Tree. *Sci. Rep.* **2024**, *14*, 1693. [CrossRef]
64. Madhu, P.; Sanjay, M.R.; Senthamaraikannan, P.; Pradeep, S.; Siengchin, S.; Jawaid, M.; Kathiresan, M. Effect of Various Chemical Treatments of *Prosopis juliflora* Fibers as Composite Reinforcement: Physicochemical, Thermal, Mechanical, and Morphological Properties. *J. Nat. Fibers* **2020**, *17*, 833–844. [CrossRef]
65. Maheshwaran, M.V.; Hyness, N.R.J.; Senthamaraikannan, P.; Saravanakumar, S.S.; Sanjay, M.R. Characterization of Natural Cellulosic Fiber from *Epipremnum aureum* Stem. *J. Nat. Fibers* **2017**, *15*, 789–798. [CrossRef]
66. Tamanna, T.A.; Belal, S.A.; Shibly, M.A.H.; Khan, A.N. Characterization of a New Natural Fiber Extracted from *Corypha taliera* Fruit. *Sci. Rep.* **2021**, *11*, 7622. [CrossRef]
67. Rajeshkumar, G.; Arvindh Seshadri, S.; Devnani, G.L.; Sanjay, M.R.; Siengchin, S.; Prakash Maran, J.; Al-Dhabi, N.A.; Karuppiah, P.; Mariadhas, V.A.; Sivarajasekar, N.; et al. Environment friendly, renewable and sustainable poly lactic acid (PLA) based natural fiber reinforced composites—A comprehensive review. *J. Clean. Prod.* **2021**, *310*, 127483. [CrossRef]
68. Oliveira, A.K.F.; d'Almeida, J.R.M. Characterization of Ubuçu (*Manicaria saccifera*) Natural Fiber Mat. *Polym. Renew. Resour.* **2014**, *5*, 13–28. [CrossRef]
69. Karthik, A.; James, D.J.; Vijayan, V.; Ahmad, Z.; Rajkumar, S.; Sharma, S.; Sharma, K.P.; Singh, R.; Li, C.; Eldin, S.M. Study on the Physicomechanical, Fracture-Strain, Interface-Adhesion, and Water-Absorption Properties of Twill Fabric Cotton-Bamboo/Epoxy Composites. *J. Mater. Res. Technol.* **2023**, *24*, 8429–8442. [CrossRef]
70. Gao, I.; Fezzaa, K.; Chen, W. Multiscale dynamic experiments on fiber-reinforced composites with damage assessment using high-speed synchrotron X-ray phase-contrast imaging. *NDT E Int.* **2022**, *129*, 102636. [CrossRef]
71. Vijay, R.; James Dhilip, J.D.; Gowtham, S.; Harikrishnan, S.; Chandru, B.; Amarnath, M.; Khan, A. Characterization of Natural Cellulose Fiber from the Barks of *Vachellia farnesiana*. *J. Nat. Fibers* **2020**, *19*, 1343–1352. [CrossRef]
72. Alves Fidelis, M.E.; Pereira, T.V.C.; Gomes, O.D.F.M.; De Andrade, S.F.; Toledo Filho, R.D. The effect of fiber morphology on the tensile strength of natural fibers. *J. Mater. Res. Technol.* **2013**, *2*, 149–157. [CrossRef]
73. Akter, N.; Das, S.C.; Fahad, M.M.; Islam, D.; Khan, M.A.; Shamsuddin, S.M. Tailoring the performance of cellulosic textiles by chemical treatment and ionizing radiation: Assessment of physical, mechanical, thermal, crystal and morphological properties. *Next Mater.* **2025**, *7*, 100372. [CrossRef]
74. Triki, A.; Dittmer, J.; Hassen, M.B.; Arous, M.; Bulou, A.; Gargouri, M. Spectroscopy analyses of hybrid unsaturated polyester composite reinforced by Alfa, wool, and thermo-binder fibers. *Polym. Sci. Ser. A* **2016**, *58*, 255–264. [CrossRef]
75. Ali, M.F.; Hossain, M.S.; Moin, T.S.; Ahmed, S.; Chowdhury, A.M.S. Utilization of waste chicken feather for the preparation of eco-friendly and sustainable composite. *Clean. Eng. Technol.* **2021**, *4*, 100190. [CrossRef]
76. Alshahrani, H.; Arun Prakash, V.R. Mechanical, thermal, viscoelastic, and hydrophobicity behavior of complex grape stalk lignin and bamboo fiber reinforced polyester composite. *Int. J. Biol. Macromol.* **2022**, *223*, 851–859. [CrossRef]
77. Prasad, L.; Singh, V.; Patel, R.V.; Yadav, A.; Kumar, V.; Winczek, J. Physical and Mechanical Properties of *Rambans* (Agave) Fiber Reinforced with Polyester. *J. Nat. Fibers* **2021**, *19*, 6104–6118. [CrossRef]
78. Dhibar, B.; Singh, S.V.; Anwar, S.; Singh, A. Sugarcane Bagasse Reinforced Polyester Composites. *Int. Res. J. Eng. Technol. (IRJET)* **2018**, *5*, 4204–4211. [CrossRef]

79. Ganesan, K.; Kailasanathan, C.; Rajini, N.; Ismail, S.O.; Ayrilmis, N.; Mohammad, F.; Al-Lohedan, H.A.; Tawfeek, A.M.; Issa, Z.A.; Aldhayan, D.M. Assessment on hybrid jute/coir fibers reinforced polyester composite with hybrid fillers under different environmental conditions. *Constr. Build. Mater.* **2021**, *301*, 124117. [CrossRef]
80. Patel, U.; Ray, R.; Mohapatra, A.; Das, S.N.; Das, H.C. Effect of Different Chemical Treatments on Surface Morphology, Thermal, and Traction Strength of *Bauhinia Vahlii* (BV) Stem Fibers. *J. Nat. Fibers* **2020**, *19*, 280–291. [CrossRef]
81. Neto, J.; Queiroz, H.; Aguiar, R.; Lima, R.; Cavalcanti, D.; Banea, M.D. A Review of Recent Advances in Hybrid Natural Fiber Reinforced Polymer Composites. *J. Renew. Mater.* **2022**, *10*, 561. [CrossRef]
82. Meliande, N.M.; Silveira, P.H.P.M.d.; Monteiro, S.N.; Nascimento, L.F.C. Tensile Properties of Curaua–Aramid Hybrid Laminated Composites for Ballistic Helmet. *Polymers* **2022**, *14*, 2588. [CrossRef]
83. Simonassi, N.T.; Pereira, A.C.; Monteiro, S.N.; Margem, F.M.; Rodríguez, R.J.S.; Deus, J.F.D.; Vieira, C.M.F.; Drelich, J. Reinforcement of Polyester with Renewable Ramie Fibers. *Mater. Res.* **2017**, *20* (Suppl. S2), 51–59. [CrossRef]
84. Vignesh, V.; Balaji, A.N.; Nagaprasad, N.; Sanjay, M.R.; Khan, A.; Asiri, A.M.; Ashraf, G.M.; Siengchin, S. Indian mallow fiber reinforced polyester composites: Mechanical and thermal properties. *J. Mater. Res. Technol.* **2021**, *11*, 274–284. [CrossRef]
85. Hossain, M.T.; Hossain, M.S.; Uddin, M.B.; Khan, R.A.; Chowdhury, A.S. Preparation and characterization of sodium silicate–treated jute-cotton blended polymer–reinforced UPR-based composite: Effect of γ-radiation. *Adv. Compos. Hybrid Mater.* **2021**, *4*, 257–264. [CrossRef]
86. Pereira, A.C.; Monteiro, S.N.; Assis, F.S. Charpy Toughness Behavior of Fique Fabric Reinforced Polyester Matrix Composites. In *Characterization of Minerals, Metals, and Materials 2017*; Ikhmayies, S., Li, B., Carpenter, J.S., Li, J., Hwang, J.-Y., Monteiro, S.N., Firrao, D., Zhang, M., Peng, Z., Escobedo-Diaz, J.P., et al., Eds.; The Minerals, Metals & Materials Series; Springer: Cham, Switzerland, 2017. [CrossRef]
87. Pereira, A.C.; Assis, F.S.; Filho, F.C.G. Evaluation of the Projectile's Loss of Energy in Polyester Composite Reinforced with Fique Fiber and Fabric. *Mater. Res.* **2019**, *22* (Suppl. S1), e20190146. [CrossRef]
88. Candido, V.S.; Silva, A.C.R.; Simonassi, N.T.; da Luz, F.S.; Monteiro, S.N. Toughness of Polyester Matrix Composites Reinforced with Sugarcane Bagasse Fibers by Charpy Impact Tests. *J. Mater. Res. Technol.* **2017**, *6*, 334–338. [CrossRef]
89. Kayaaslan, M.; Coskun, T.; Unlu, U.M.; Sahin, O.S. Effects of Thickness, Fiber Orientation and Fabric Textile on the Low-Velocity Impact Performances of Thermoset and Thermoplastic Composites. *J. Thermoplast. Compos. Mater.* **2023**, *36*, 4408–4429. [CrossRef]
90. Chaves, Y.S.; Monteiro, S.N.; Nascimento, L.F.C.; Rio, T.G.-d. Mechanical and Ballistic Properties of Epoxy Composites Reinforced with Babassu Fibers (*Attalea speciosa*). *Polymers* **2024**, *16*, 913. [CrossRef]
91. Sheng, C.; He, G.; Hu, Z.; Chou, C.; Shi, J.; Li, J.; Meng, Q.; Ning, X.; Wang, L.; Ning, F. Yarn on Yarn Abrasion Failure Mechanism of Ultrahigh Molecular Weight Polyethylene Fiber. *J. Eng. Fibers Fabr.* **2021**, *16*, 1925832385. [CrossRef]
92. Vallejos, M.E.; Vilaseca, F.; Méndez, J.A.; Espinach, F.X.; Aguado, R.J.; Delgado-Aguilar, M.; Mutjé, P. Response of Polypropylene Composites Reinforced with Natural Fibers: Impact Strength and Water-Uptake Behaviors. *Polymers* **2023**, *15*, 900. [CrossRef] [PubMed]

polymers

MDPI

Article

Mercury Adsorption by Ca-Based Shell-Type Polymers Synthesized by Self-Assembly Mineralization

Yang Peng [1,*], Chuxuan Zhang [2], Xiaomin Li [3], Tianyi Feng [3] and Xun Gong [3]

1 School of Low-Carbon Energy and Power Engineering, China University of Mining and Technology, No. 1, Daxue Road, Xuzhou 221116, China
2 Department of Electrical Engineering, Xi'an University of Technology, Xi'an 710054, China; zhangchuxuan@xaut.edu.cn
3 State Key Laboratory of Coal Combustion, School of Energy and Power Engineering, Huazhong University of Science and Technology, Wuhan 430074, China; lixmin2024@163.com (X.L.); fty204971161@163.com (T.F.); gx@hust.edu.cn (X.G.)
* Correspondence: ypeng@cumt.edu.cn

Abstract: Adsorption is one of the most promising strategies for heavy metal removal. For Hg(II) removal, mineralized Ca-based shell-type self-assembly beads (MCABs) using alginate as organic polymer template were synthesized in this work. The adsorbent preparation consists of gelation of a Ca-based spherical polymer template (CAB) and rate-controlled self-assembly mineralization in bicarbonate solution with various concentrations. The comparative study demonstrates that 1% (MCAB-1) is the optimal concentration of bicarbonate. Based on this condition, the maximum adsorption capacity (48 ± 4 mg/g) of MCAB-1 was observed at pH = 5 in a batch test, which was 2.67 times more than that of the unmodified one, CAB, at 18 ± 1 mg/g. Long-duration (10 h) adsorption tests showed that MCAB-1 exhibited remarkable performance stability and anti-wear ability (43.2% removal efficiency and 74.3% mass retention, compared to 2.7% and 38.6% for CAB at pH = 3, respectively). The morphology determination showed that a shell-type porous amorphous carbonate layer was formed at the surface of the organic polymer template by rate-controlled self-assembly mineralization. This transition not only promotes the pore structure and activated cation binding functional sites, but also improves the anti-wear ability of materials effectively.

Keywords: surface modification; alginate polymer template; calcium-based adsorbents; heavy metal removal

1. Introduction

Nowadays, human health is seriously menaced by heavy metal emissions [1–5]. Mercury contamination of wastewater, especially from coal combustion [6,7] and gold mining, including divalent mercury (Hg(II)) and methyl mercury (Me-Hg), is an extreme threat to human health because of bioaccumulation in the food chain [8–10]. Therefore, the treatments for mercury-contaminated wastewater have attracted wide attentions in recent decades [11–13]. The maximum contaminant level (MCL) for mercury in potable water proposed by the World Health Organization (WHO) and U.S. Environmental Protection Agency (EPA) is set under 0.001 mg/L [14]. Several physical and chemical methods, such as precipitation, coagulation, ion exchange, electrochemical methods, adsorption, membrane processes, and ultra-filtration, have been developed to overcome this problem [11,14–16]. Among them, adsorption is the most widely used because of its remarkably high efficiency, easy access and operation, and relatively low cost. The development of adsorbents with high efficiency and environmental protection for specific pollutants is still one of the dominant directions of wastewater treatment technology.

Traditional adsorbents are mainly powder particles. For example, nowadays, the dominant commercial adsorbents used for heavy metal treatment in sewage are activated carbon powders and their derived products. Hadi et al. reviewed the research status of aqueous mercury adsorption by activated carbons [17,18]. Several aspects including the preparation of activated carbon, the effect of treatment techniques on mercury removal (physical and chemical activation, sulfurization), the effect of adsorption parameters (equilibrium contact time, initial concentration, pH value, temperature, adsorption dosage, and particle size), functional groups, and equilibrium adsorption isotherms were exhaustively reviewed. Moreover, activated carbon derived from various types of biomass has been reported, such as rice hulls [19], coconut shells [20], apricot stones [21], apple peels [22], pecan shells [23], and so on. In addition to activated carbon, various types of biomass and microorganisms are also potential feedstock for heavy metal adsorbents, including chitosan [24–30], bacteria [31–38], fungus [39,40], algae [41–45], and so on.

Adsorption materials with excellent properties can be obtained by selecting different raw materials and corresponding modification methods [46]. However, there are some fatal problems in the large-scale commercial utilization of powder adsorbents. Firstly, in the application process of powder adsorbents, rapid and efficient separation is often required, which inevitably leads to increases in energy consumption and operation costs [47]. Meanwhile, the recovery efficiency of the adsorbent is relatively low. In view of this, the immobilization of adsorbents is quite essential for heavy metal treatment in sewage [48–50]. One of the most widely employed materials is Ca-alginate, which is an ideal immobilized matrix in the classical "egg-box" model [51]. However, the adsorption capacity of calcium alginate itself is extremely low. To obtain the satisfied adsorption capacity, it is necessary to load additional materials with excellent adsorption performance, which will inevitably lead to a significant increase in the preparation cost of the adsorbent. It is a challenge to slash the costs while retaining the formation and adsorption properties.

At the same time, calcium alginate is a crosslinked organic polymer formed by the chelation of calcium ions. The structure and stability of the materials are seriously affected by the decomposition of the long-chain polymer group under strong acid conditions because metal ions are easily replaced by hydrogen ions. Damage to this structure will seriously reduce the stability and repeatability of the adsorbent in industrial utilization. Therefore, it is necessary to evaluate the long-term adsorption stability of the forming adsorbent in the solid–liquid environment. Unfortunately, research in both areas has not been reported.

Given this situation, a self-assembly mineralization method to modify calcium alginate beads is proposed in this work, aiming to synthesize a shell-type Ca-based adsorbent for mercury adsorption. It involved a sol–gel process for Ca-alginate preparation followed by impregnation by an NH_4HCO_3 solution. The characterization of the modified beads was conducted by BET, XRD, FTIR, and stereoscopic microscope analysis. The performance of the Hg(II) adsorption process was assessed based on the Hg(II) isotherms and kinetic adsorption analysis. Furthermore, the material structural stability was assessed by long-duration adsorption tests. This study presents a novel approach to enhance the adsorption performance and durability of calcium alginate-based adsorbents through controlled self-assembly mineralization, addressing critical challenges such as low adsorption efficiency, poor stability under acidic conditions, and high preparation costs. The innovative shell-type porous structure formed on the beads not only significantly improves Hg(II) uptake but also ensures superior anti-wear and long-term operational stability, making it a potential breakthrough for sustainable and cost-effective wastewater treatment technologies.

2. Materials and Methods

2.1. Chemicals and Material Preparation

A total of 2 g of sodium alginate (Aladdin Ltd., AR, Shanghai, China) powder was dissolved in 100 mL of deionized water followed by being stirred for 40 min at 85 °C. After that, the sodium alginate sol was dropped into 2% wt. $CaCl_2$ (Aladdin Ltd.) solution by a syringe pump to form a kind of spherical Ca-based polymer, Ca-alginate beads (raw sample, named as CABs). These hydrogel beads were then impregnated and mineralized by 100 mL NH_4HCO_3 solutions (Aladdin Ltd.) with certain concentrations (0.5%, 1%, and 1.5% wt., which were named as MCAB-0.5, MCAB-1, and MCAB-1.5, respectively) in a stable-temperature oscillation shaker at 150 rpm for a number of hours. After that, these samples were washed until Ca^{2+} and Cl^- ions were undetected in the leachates, followed by drying in a freezer dryer for testing (80 Pa, −50 °C).

2.2. Morphological Characterization

A high-megapixel camera (D5200, Nikon, Shinagawa-ku, Japan) was adopted to observe the changes in the surface morphology before and after modification. The sectional views and micro-topography were obtained separately by a stereo microscope (Stemi 508, Carl Zeiss, Jena, Germany). The surface area and pore size distribution were tested by the BET method, which was conducted at 77 K in a N_2 atmosphere using an ASAP 2460 instrument (Micromeritics, Norcross, GA, USA) after drying at 120 °C for 12 h. In addition, the changes in functional groups during adsorption were analyzed using a VERTEX 70 FTIR system (Bruker, Bunde, Germany) with the ATR method. The surface composition and morphology of the adsorbents were characterized by X-ray diffraction (XRD) using Cu Kα radiation (Empyrean, PANalytical B.V., Almelo, The Netherlands), with a 2θ scan range from 10° to 80°.

2.3. The Verification of Adsorbent Uniformity

Given the larger particle sizes of raw and modified samples than those of commercial adsorbent powders in this work, the average mass of the materials should be measured to ensure the uniformity of the materials selected in the subsequent parallel experiments. The random sampling method was adopted for uniformity measurement: 10 groups, with 5 spherical particles in each group, were selected randomly from each sample. The single bead mass m_{ij} of each group of samples was measured (i is the serial number of sampling group; j is the serial number of single bead in each group; $1 \leq i \leq 10$, $1 \leq j \leq 5$). The average mass M_i and error level e_i of each group of samples can be obtained as

$$M_i = \frac{\sum_{j=1}^{5} m_{ij}}{5} \tag{1}$$

$$e_i = \frac{\sum_{j=1}^{5} \left(m_{ij} - M_i\right)}{5} \tag{2}$$

2.4. Batch Adsorption Testing

A standard solution of Hg(II) (1000 mg/L, GR, Aladdin Ltd.) was used as the metal stock solution. Serial dilutions were carried out to obtain the desired concentration. Although it was proven that the temperature of the aqueous system affected the binding ability of heavy metals for some adsorbents [52–54], the pH value, metal ion concentration, and adsorbent dosage were still the dominate influencing factors for the adsorption process [55]. The temperature in this study was fixed at 30 °C, which was consistent with actual industrial operations [56].

Batch adsorption testing was conducted in conical flasks, which contained a certain amount of adsorbent and a corresponding volume of Hg(II) ion solution for various solid-to-liquid ratios (0.1, 0.3, 0.5 1, 3, 5, 10 g/L) and involved stirring at 150 rpm. The effect of different pH values on the adsorption was evaluated by controlling the initial pH values to be 3.07, 4.07, 5.01, and 6.06, which was achieved by mixing 0.1 mol/L HNO_3 and NaOH solutions. The study of the adsorption isotherms was carried out with a variety of initial ion concentrations (1, 2, 5, 10, 15, 20, 25 mg/L) under the optimal conditions. For the adsorption kinetics study, two groups with 1 mL Hg(II) solution were prepared for kinetics testing. Sampling was arranged at 2, 5, 10, 30, 60, 120, and 180 min, respectively. The initial pH and dosage were fixed at the optimal conditions according to the tests above. The mixed solutions after adsorption were centrifuged and filtered by a 0.45 microporous membrane filter and then tested by an inductively coupled plasma optical emission spectrometry (ICP-OES) system (SPECTRO ARCOS, Kleve, Germany). Each case was tested 3 times to ensure accuracy.

The equilibrium adsorption capacity of the adsorbent was calculated by the following formula [57]:

$$q_e = \frac{(C_0 - C_e) \cdot V}{m} \quad (3)$$

where q_e is the equilibrium adsorption capacity in mg/g, C_0 and C_e are the initial and equilibrium Hg(II) solution concentrations in mg/L, respectively, V is the volume of the Hg(II) solution in L, and m is the mass weight of adsorbent in g.

To evaluate the Hg(II) removal efficiency of the adsorbent, another formula was applied as follows:

$$\eta = \frac{(C_0 - C_t)}{C_0} \times 100\% \quad (4)$$

where η is the removal efficiency of the adsorbent in % at time t, C_t is the concentration of Hg(II) in mg/L at time t, and C_0 is the same parameter as mentioned in Equation (3).

2.5. Adsorption Stability Evaluation of Adsorbents

Long-duration (10 h) adsorption tests were conducted in one 500 mL conical flask containing 300 adsorbent beads and certain volume of the 1 mg/L Hg(II) ion solution (optimal solid–liquid ratio maintained) at 30 °C with various pH values, to simulate harsh corrosion conditions. After each hour, 3 mL of solution and 3 adsorbent beads were sampled for analysis until the test finished. Two parameters, the variation in the adsorption capacity and the mass weight, were employed to assess the stability of the MCABs for Hg(II) adsorption in the liquid system.

3. Results

3.1. Sample Characterization

Sectional views of four samples were observed using the stereo microscope, as shown in Figure 1. The images show that the unmodified beads show a loose network structure. This is due to the crosslinking reaction between calcium ions and alginate polymers. However, this structure could be easily decomposed by ion exchange, ligand exchange, or hydrogen-bond interactions in aqueous systems [58]. As a comparison, a dense homogeneous shell formed on the surface of the modified pellet. Combined with the XRD spectra shown in Figure 2, there is a mineralized calcium carbonate layer on the surface of the shell-type calcium alginate sphere. This is due to the confined crystallization between calcium and carbonate ions from bicarbonate secondary ionization [59]. As for the bicarbonate, the secondary ionization constant is much more faint than the first ionization constant; this highlighted that the surface mineralization process of calcium alginate would be milder due to the relatively slow release of carbonate [60]. The concentration of bicarbonate solution was determined to be a significant factor in the formation of the mineralized shell. From the cross section of modified samples in Figure 1, lower concentration (0.5% wt.) could lead an irregular mineralized layer to form, while higher concentration (1.5% wt.) could result in reuniting and oversaturated crystallization [61].

Table 1 exhibits the porous characteristics of CAB, MCAB-0.5, MCAB-1, and MCAB-1.5 by BET analysis. The specific surface area of MCAB-1, 15.319 m^2/g, is proven to be the highest among these four samples compared with those of CAB, 8.201 m^2/g; MCAB-0.5, 12.972 m^2/g; and MCAB-1.5, 14.772 m^2/g, respectively. Obviously, however, both the pore volume and average pore size of CAB are slightly larger than that of the modified samples, which might be due to blocking of the network structure on Ca-alginate polymers by the formation and growth of crystal layers [62]. In this work, 1% of NH_4HCO_3 was proven to be the optimal concentration for mineralization.

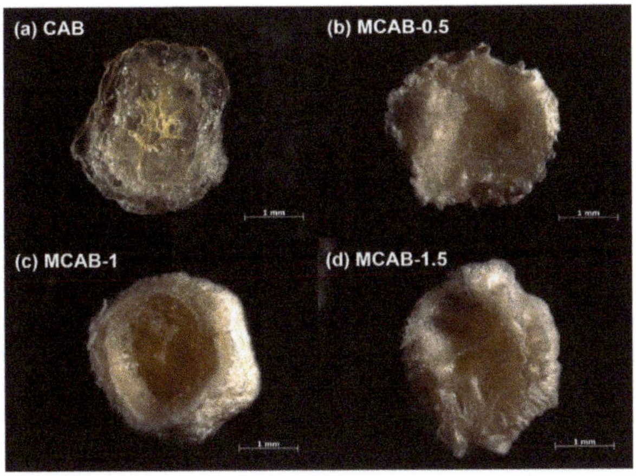

Figure 1. Sectional views of CAB and the three MCAB samples.

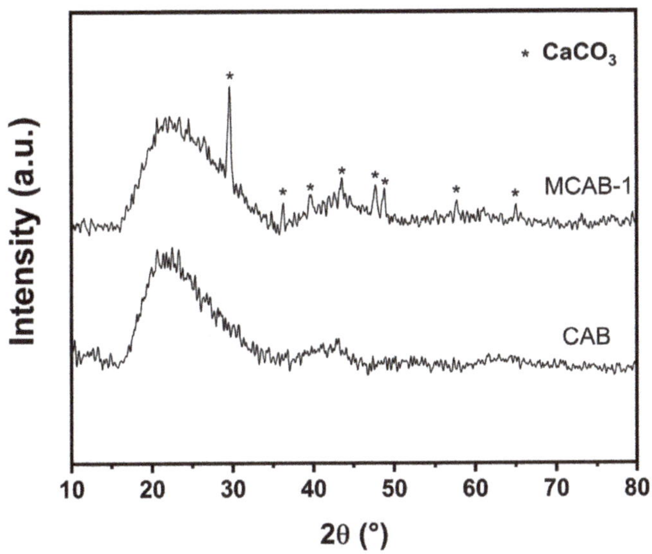

Figure 2. XRD spectra of CAB and MCAB-1.

Table 1. BET analysis for CAB, MCAB-0.5, MCAB-1, and MCAB-1.5.

Sample	BET Surface Area (m^2/g)	Pore Volume (cm^3/g)	Average Pore Width (nm)
CAB	8.201	0.013	6.361
MCAB-0.5	12.972	0.009	5.992
MCAB-1	15.319	0.007	5.672
MCAB-1.5	14.772	0.006	5.773

The statistical mean mass weights of the four samples are shown in Figure 3. The average mass weight of single CAB, MCAB-0.5, MCAB-1, and MCAB-1.5 beads is 1.157 ± 0.038 mg, 1.250 ± 0.043 mg, 1.256 ± 0.048 mg, and 1.340 ± 0.073 mg, respectively. It is obvious that the mass weight of the single modified bead increases with the increasing impregnation concentration of NH_4HCO_3. Moreover, the mass of each sample is statistically constant within approximately 5% error, which implies that the total sampling quality is in line with the quantity of sampling beads. The quality difference between the adsorbent pellets in the same sample is negligible.

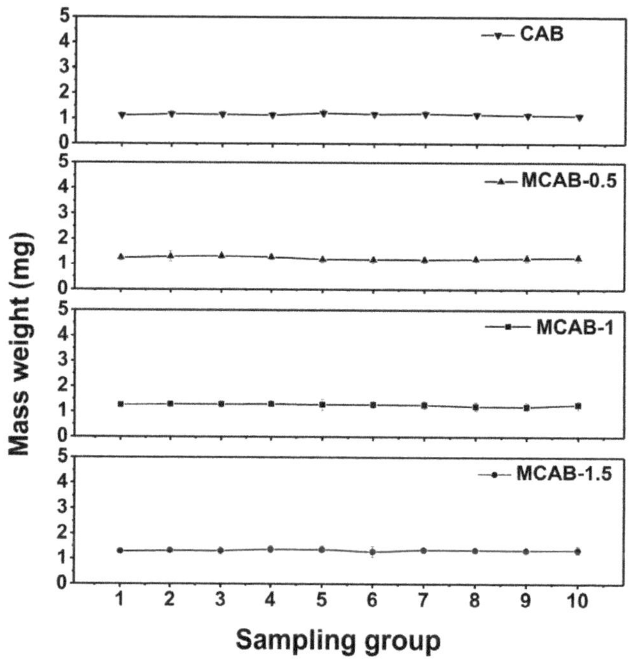

Figure 3. Statistical analysis of the average mass of a single adsorbent particle.

3.2. *Effect of Single Factors on the Hg(II) Adsorption Efficiency*

3.2.1. Dosage of Adsorbent

It is widely reported that the dosage of the adsorbent has significant impacts on the capability and economic efficiency of metal adsorption. Given that the dispersibility of large-particle-forming adsorbents is not as good as that of powder materials, it is necessary to adjust the solid–liquid ratio to meet the requirement of higher adsorption efficiency [8,63,64]. Figure 4 gives the effect of dosage on the Hg(II) adsorption efficiency of CAB and the three MCAB samples. With the increasing dosage from 0.1 g/L to 1 g/L, the efficiency of each sample increased remarkably. Among them, the efficiency of MCAB-0.5, MCAB-1, and MCAB-1.5 increased from 36.3.4 $\pm$ 3.2%, 43.2 $\pm$ 2.5%, and 39.3 $\pm$ 2.4% to 78.4 $\pm$ 2.1%, 84.3 $\pm$ 3.7%, and 82.2 $\pm$ 2.5%, respectively, compared to that of CAB, from 12.4 $\pm$ 1.2% to 21.6 $\pm$ 3.5%. This indicates that the modified samples can provide more adsorption sites under the same dosage. In further increasing the dosage from 1 g/L, the tendency of efficiency to improve slowed down significantly, which implied that the adsorption sites tend to be saturated and the adsorption system gradually reaches dynamic equilibrium [54,65].

3.2.2. Effect of pH

pH (especially acidic pH conditions) is a very important influencing variable in the adsorption process because it not only affects the adsorption efficiency, but also has a great influence on the stability of the material. As shown in Figure 5, the adsorption efficiency increased from pH = 3 to pH = 5 followed by a slight decline from pH = 5 to pH = 6. To be specific, the efficiency of MCAB-0.5, MCAB-1, and MCAB-1.5 increased from 38.9 $\pm$ 2.3%, 58.1 $\pm$ 0.9%, and 53.2 $\pm$ 1.2% at pH = 3 to 84.2 $\pm$ 3.1%, 89.7 $\pm$ 3.3%, and

83.3 ± 2.3% at pH = 5, and then decreased to 81.7 ± 4.2%, 84.1 ± 3.2%, and 82.1 ± 3.1% at pH = 6, respectively. Comparing the modified samples, the efficiency of CAB increased from 11.2 ± 0.7% at pH = 3 to 19.2 ± 0.7% at pH = 5, then decreased to 17.1 ± 0.8% at pH = 6. This trend is mainly due to the competitive adsorption of H^+, H_3O^+, and heavy metal ions under different pH conditions. Under low-pH conditions, H^+ and H_3O^+ occupy the adsorption sites on the adsorbent surface, resulting in the inhibition of Hg(II) during the adsorption process. With the increase in pH, the competitive effect is gradually weakened due to the decrease in H^+ and the increase in OH^- in the system. However, under the condition of relatively high pH, Hg(II) is partially converted into neutral or electronegative hydroxides due to hydrolysis, which is also unfavorable for the binding of electronegative functional groups such as hydroxyl groups and carboxyl groups on the adsorbent surface [66].

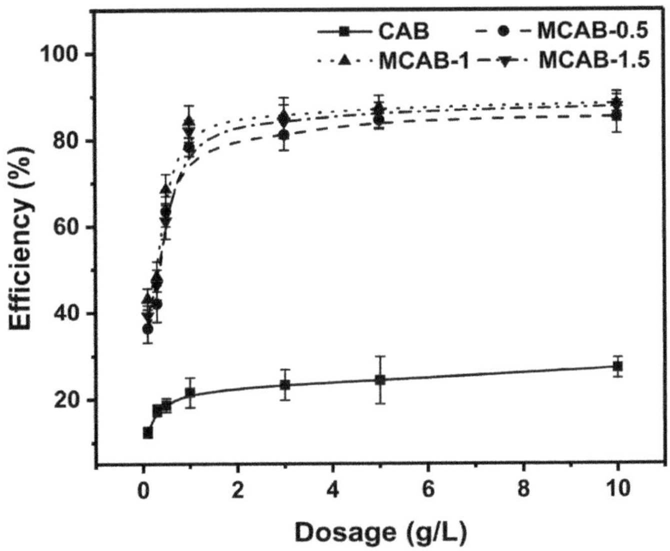

Figure 4. Effect of the dosage on the Hg(II) adsorption efficiency (initial Hg(II) concentration = 1 mg/L, oscillation time = 180 min, pH = 5, temperature = 30 °C).

According to the single-factor variable control experiment, it can be seen that under the same conditions, both the adsorption ability and environmental tolerance of the mineralized modified samples are significantly better than the original samples. To maximize performance, the combination of 1 g/L and pH = 5 is demonstrated as the optimal operation condition.

3.3. *The Assessment of Adsorbent Performance in Batch Adsorption*

3.3.1. Kinetics Analysis

Pseudo-first-order, pseudo-second-order, and Weber–Morris (W-M) [67–69] models were used to describe the kinetic process. The pseudo-first-order model can be expressed as follows:

$$q_t = q_e(1 - e^{-k_1 t}) \tag{5}$$

where t in min is the contact time between the solid and aqueous phase, q_e in mg/g is the adsorption capacity of the adsorbent at equilibrium, q_t in mg/g is the adsorption capacity of the adsorbent at time t, and k_1 in min^{-1} is the pseudo-first-order rate constant.

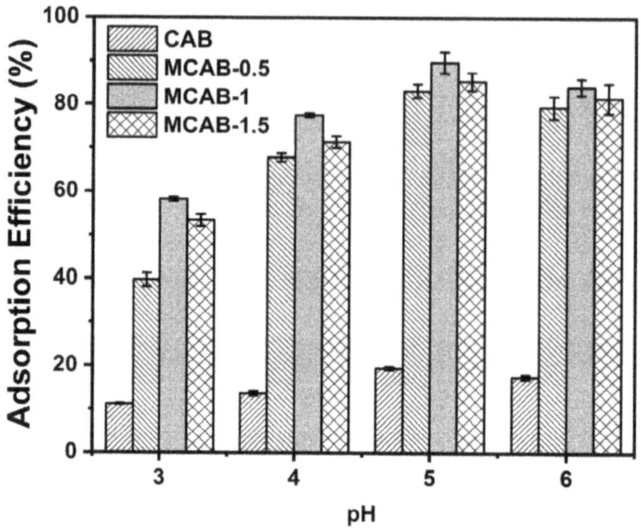

Figure 5. Effect of pH on the Hg(II) adsorption efficiency (initial Hg(II) concentration = 1 mg/L, oscillation time = 180 min, dosage = 1 g/L, temperature = 30 °C).

The pseudo-second-order model can be presented as

$$q_t = q_e \frac{q_e k_2 t}{1 + q_e k_2 t} \tag{6}$$

where q_e, q_t and t are the same parameters as in Equation (5) and k_2 in g/mg·min is the pseudo-second-order rate constant.

The W-M model can ben presented as

$$q_t = k_{id} t^{0.5} + C \tag{7}$$

where k_{id} is the intraparticle diffusion rate constant (mg/g·min$^{0.5}$), and C is a constant related to the speed of the intraparticle diffusion process.

The experimental data fitting with two kinetic models is shown in Figure 6. The relevant parameters were calculated (shown in Table 2) using Equations (5)–(7). The correlation coefficients (r^2) indicate that in addition to the W-M model, both the first-order and second-order kinetic models are well fitted to the experimental data. The results suggest that the controlling step of the adsorption process is not intraparticle diffusion but rather surface reactions or intermolecular interactions. This implies that the adsorption process is likely governed by surface chemical processes, the number of available adsorption sites on the adsorbent, and the interactions between the adsorbate and adsorbent, with diffusion playing a less significant role in limiting the overall rate. Nonetheless, compared with the pseudo-second-order model (R^2 = 0.994 ± 0.002, 0.991 ± 0.004, 0.993 ± 0.001, and 0.998 ± 0.006), the pseudo-first-order model (R^2 = 0.994 ± 0.002, 0.991 ± 0.004, 0.993 ± 0.001, and 0.998 ± 0.006 for CAB, MCAB-0.5, MCAB-1, and MCAB-1.5, respectively) slightly underestimates the metal uptake of Hg(II) by both adsorbents and is inaccurate in characterizing the kinetic mechanism. Furtherly, the equilibrium capacity is 0.18 ± 0.01, 0.57 ± 0.01, 0.79 ± 0.01, and 0.78 ± 0.02 for CAB, MCAB-0.5, MCAB-1, and MCAB-1.5, respectively, calculated by the pseudo-second-order fitting model. This indicates that the intrinsic adsorption capacity was improved notably by self-assembly mineralization based on the comparison of adsorption rate constants of the pseudo-second-order model, i.e., k_2 (MCAB-0.5) < k_2 (MCAB-1.5) ~ k_2 (MCAB-1) << k_2 (CAB), implying that the adsorption rate was significantly improved by modification. As for the relatively higher k_2 of MCAB-0.5 compared to the other two modified samples, this is probably due to the lower adsorption capacity requiring a shorter time to reach equilibrium.

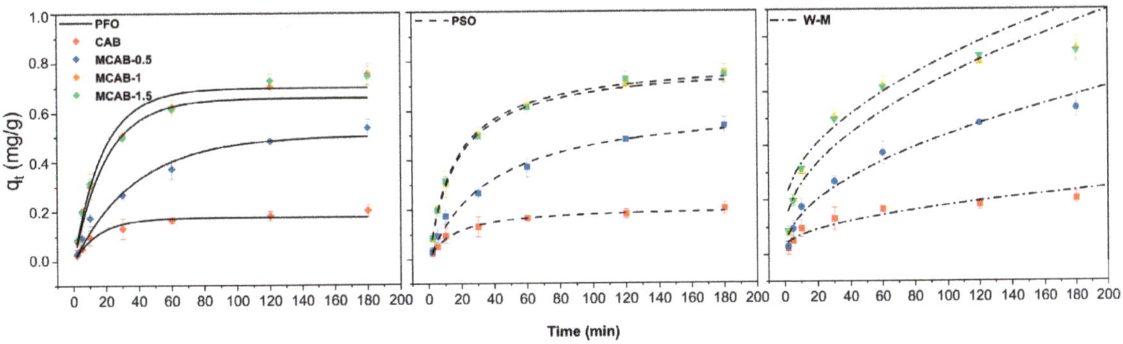

Figure 6. Adsorption kinetics models (pH = 5, initial Hg(II) concentration = 1 mg/L, adsorbent dosage = 1 g/L, temperature = 30 °C).

Table 2. Calculated parameters with models from kinetic data.

		CAB	MCAB-0.5	MCAB-1	MCAB-1.5
Pseudo-first-order model	q_e (mg/g)	0.19 ± 0.01	0.54 ± 0.03	0.77 ± 0.02	0.75 ± 0.03
	k_1 (/min)	0.065 ± 0.004	0.026 ± 0.001	0.055 ± 0.003	0.050 ± 0.003
	r^2	0.972 ± 0.004	0.989 ± 0.001	0.992 ± 0.002	0.963 ± 0.007
Pseudo-second-order model	q_e (mg/g)	0.18 ± 0.01	0.57 ± 0.01	0.79 ± 0.01	0.78 ± 0.02
	k_2 (gmg^{-1} min^{-1})	3.151 ± 0.003	0.419 ± 0.009	0.750 ± 0.001	0.741 ± 0.005
	r^2	0.994 ± 0.002	0.991 ± 0.004	0.993 ± 0.001	0.998 ± 0.006
W-M model	K_{id} (mg/g·min$^{0.5}$)	1.561 ± 0.207	4.266 ± 0.248	5.794 ± 0.352	5.712 ± 1.083
	C	2.444 ± 1.498	1.344 ± 0.248	7.911 ± 2.485	15.571 ± 7.268
	r^2	0.963 ± 0.017	0.951 ± 0.003	0.967 ± 0.015	0.977 ± 0.024

3.3.2. Adsorption Isotherms

Langmuir, Freundlich, and Temkin models were applied to calculate the isotherm parameters. The Langmuir isotherm is described by the following equation [57,70]:

$$q_e = \frac{q_m k_L C_e}{1 + k_L C_e} \tag{8}$$

where q_e is the equilibrium capacity of the adsorbent in mg/g, C_e is the concentration of Hg(II) ions in the liquid at equilibrium in mg/L, and q_m in mg/g is the maximum adsorption capacity of the adsorbent by calculation. k_L, the so-called Langmuir isotherm adsorption constant in L/mg, refers to the bonding energy of sorption.

The Freundlich isotherm model is expressed by the following equation:

$$q_e = K_F {C_e}^{1/n} \tag{9}$$

where K_F in (mg/g)(1/mg)$^{1/n}$ is the Freundlich constant and q_e and C_e are the same variables as in Equation (8); n is an empirical exponent.

The Temkin model is expressed by

$$q_e = \frac{RT}{b}\ln(K_T) + \frac{RT}{b}\ln(C_e) \tag{10}$$

where K_T is the Temkin adsorption constant, indicating the affinity of the adsorbent for the adsorbate (L/g), R is the universal gas constant (8.314 J/(mol·K)), T is the temperature (K), and b is a constant related to the adsorption heat, which indicates the strength of the adsorbate–adsorbent interactions.

The surface properties and affinities of the adsorbent can be expressed by certain constants that characterize the adsorption isotherms, presented in Figure 7. The adsorption isotherm constants are presented in Table 3. Compared to the Langmuir and Freundlich isotherm models, the Temkin model generally exhibits lower fitting accuracy, primarily due to its overly simplified assumptions. Specifically, the Temkin model assumes that the adsorption heat decreases linearly with increasing adsorption, which may not accurately reflect the actual variation in adsorption heat, especially in cases where the energy distribution of adsorption sites is heterogeneous. Additionally, the Temkin model assumes that the adsorption sites on the adsorbent surface have similar energy levels, failing to fully account for surface heterogeneity, which limits its applicability to systems with significantly heterogeneous surfaces. In contrast, the Langmuir model assumes uniform adsorption sites and is suitable for describing monolayer adsorption, while the Freundlich model introduces a logarithmic function to better accommodate heterogeneous surfaces, leading to more accurate fits in many adsorption processes. Therefore, the Temkin model may not capture the non-ideal behavior of the adsorption process in complex systems, resulting in lower fitting accuracy. The maximum capacity q_m determined from the Langmuir isotherm indicates the limiting capacity of the absorbent for Hg(II) ions. It was found that the maximum capacity q_m for the MCAB-1 sample of 48 ± 1 mg/g was much higher than that for CAB of 18 ± 1 mg/g. The correlation coefficients (R^2) for MCABs and CABs show that the Langmuir isotherm model is properly used to describe both adsorbents, which indicates that monolayer adsorption of heavy metal ions is formed on the outer surface of adsorbents [52]. The Freundlich model is used to characterize the adsorption on a heterogeneous surface and is not limited to mono-shell formation. However, the values of the correlation coefficients (R^2) calculated by the Freundlich isotherm model (0.970 ± 0.006, 0.968 ± 0.003, 0.965 ± 0.003, and 0.960 ± 0.009) are lower than those from Langmuir model (0.996 ± 0.002, 0.995 ± 0.004, 0.998 ± 0.003, and 0.999 ± 0.006 for CAB, MCAB-0.5, MCAB-1, and MCAB-1.5, respectively), which indicates that the Freundlich model is unacceptable for describing the adsorption process.

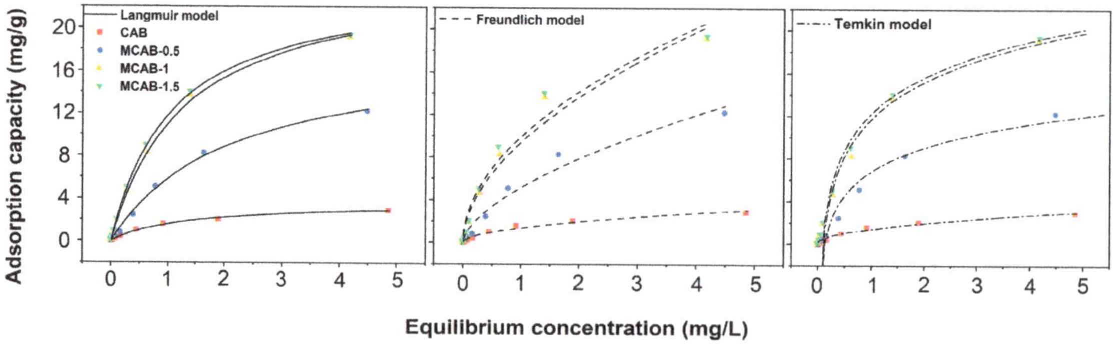

Figure 7. Adsorption isotherm fitting curves (pH = 5; oscillation time = 180 min; adsorbent dosage = 1 g/L; test temperature = 30 °C).

Table 3. Langmuir and Freundlich parameters calculated from experimental data.

		CAB	MCAB-0.5	MCAB-1	MCAB-1.5
Langmuir	q_m (mgg^{-1})	18 ± 1	32 ± 2	48 ± 1	46 ± 4
	K_L (Lmg^{-1})	0.76 ± 0.01	0.47 ± 0.04	0.76 ± 0.01	0.92 ± 0.01
	R^2	0.996 ± 0.002	0.995 ± 0.004	0.998 ± 0.003	0.999 ± 0.006
Freundlich	K_F (L^{n}g^{-1}mg^{1-n})	1.6 ± 0.2	4.0 ± 0.1	10.0 ± 0.5	11.8 ± 0.3
	$1/n$	0.52 ± 0.07	0.61 ± 0.02	0.54 ± 0.06	0.52 ± 0.01
	R^2	0.970 ± 0.006	0.968 ± 0.003	0.965 ± 0.003	0.960 ± 0.009
Temkin model	K_T (Lmg^{-1})	0.181 ± 0.051	0.082 ± 0.022	0.103 ± 0.021	0.121 ± 0.022
	b (Jmg^{-1})	331.819 ± 34.816	61.494 ± 6.500	39.462 ± 2.855	40.429 ± 2.390
	R^2	0.964 ± 0.007	0.981 ± 0.009	0.987 ± 0.003	0.989 ± 0.004

3.4. Long-Term Anti-Wear Test

Long-duration (10 h) adsorption was conducted to investigate the effect of mineralization on the stability of the adsorption efficiency and retardation of material wear. Figure 8 shows the changes in mass weight and adsorption efficiency during long-duration adsorption at pH = 3. Both indices of all samples decreased as the adsorption progressed. Specifically, after 10 h of adsorption, the mass weights of CAB, MCAB-0.5, MCAB-1, and MCAB-1.5 were 38.6%, 61.2%, 74.3%, and 74.3% of each initial mass, respectively. Meanwhile, the efficiency of these four samples decreased from 10.1%, 38.2%, 57.4%, and 56.4% to 2.7%, 21.4%, 43.2%, and 39.7%, respectively. In addition, the dynamic changes in quality and efficiency indicate that there is a positive correlation between them. This improvement indicated that mineralization not only improved the adsorption efficiency but also enhanced the stability of adsorbent performance.

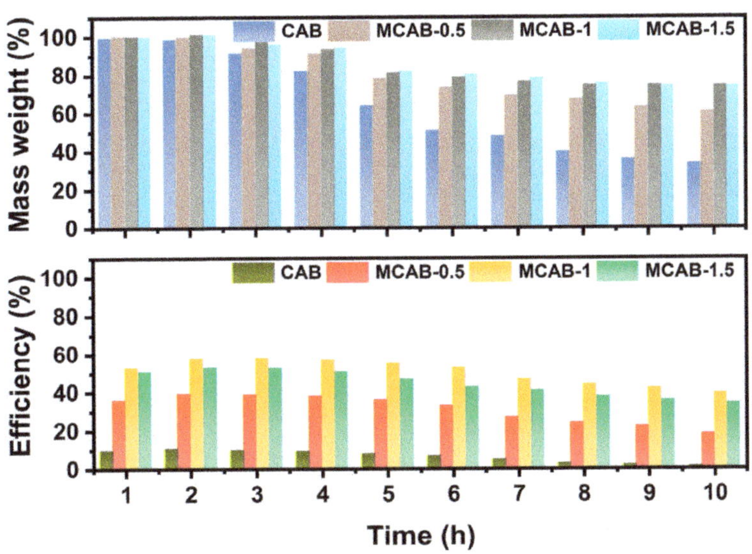

Figure 8. Mass weight and adsorption efficiency during adsorption process (pH = 3, initial Hg(II) concentration = 1 mg/L, dosage = 1 g/L, temperature = 30 °C).

4. Discussion

4.1. The Evolution of Functional Groups During Batch Adsorption

FTIR measurements on MCABs and CABs before and after Hg(II) adsorption were investigated as well, as shown in Figure 9. The bands at 3266 cm^{-1} in sample (a) were attributed to the hydroxyl stretching vibration mode. The same results were reflected for (b) at 3193 cm^{-1}, (c) at 3206 cm^{-1} and (d) at 3185 cm^{-1}, which were observed to move to lower frequencies because of the hydrogen-bond interaction during the impregnation and adsorption. The weak bands observed at 2925 and 2852 cm^{-1} for sample (a) were due to the symmetric and asymmetric stretching bands of -CH_2, which moved to 2988 and 2910 cm^{-1} along with 2995 and 2921 cm^{-1} for samples (b) and (d), respectively. The bands at 1598 and 1415 cm^{-1} for sample (a) were attributed to the symmetric and asymmetric stretching bands of C=O, which shifted to 1590 and 1410 cm^{-1} for both (b) and (d), while they disappeared for sample (c). The wide peak at 1414 cm^{-1} for sample (c) was found to be N-H bending vibration due to the NH_4HCO_3 coating. The C-O-C stretching vibration peak was observed at 1028 cm^{-1} for all four of these samples, showing that the condensation between monomer G and M blocks cannot be interrupted by either NH_4HCO_3 modification or Hg(II) adsorption. Additionally, the outer bending vibration peak of $CaCO_3$ was observed at 882 cm^{-1} for sample (c), which indicated that Ca^{2+} ions trapped by -COO^- in polymers before modification might be grabbed by hydrolyzed CO_3^{2-} ions from NH_4HCO_3 to further form the $CaCO_3$ microcrystalline structure [59], which is consistent with the study mentioned above.

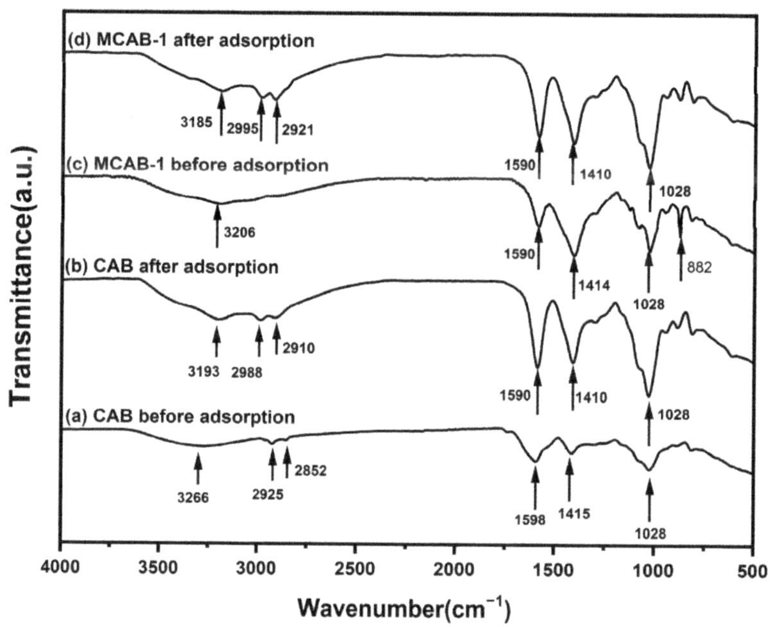

Figure 9. FTIR spectra of CAB and MCAB-1 before and after Hg(II) adsorption (pH = 5, oscillation time = 180 min, adsorbent dosage = 1 g/L, test temperature = 30 °C).

4.2. *Effect of Mineralization on the Stability of the Adsorption Efficiency and Material Structure*

Macroscopic morphological changes during long-duration (10 h) adsorption for CAB and MCABs are represented in Figure 10. For CAB, the raw bead is a ball with good sphericity. After 10 h of adsorption, different degrees of disintegration are observed in various pH conditions. At a pH = 3, the bead is totally collapsed, which is in accordance with the weight ratio tests. After mineralization, the surfaces of raw modified beads, MCAB-0.5, MCAB-1, and MCAB-1.5, are whiter and more substantial compared with CAB. Additionally, the completeness of the materials is much better than that of CAB. The concentration of NH_4HCO_3 shows a dominant effect on the morphological changes. This is mainly because different carbonate concentrations lead to different calcium carbonate crystallization rates in the growth process of microcrystals, so the microcrystalline structure bears different degrees of local stress in the confined crystallization process. As the mineralization unit grows, the stress changes the local structure more obviously and finally leads to different macroscopic mineralization layers [59].

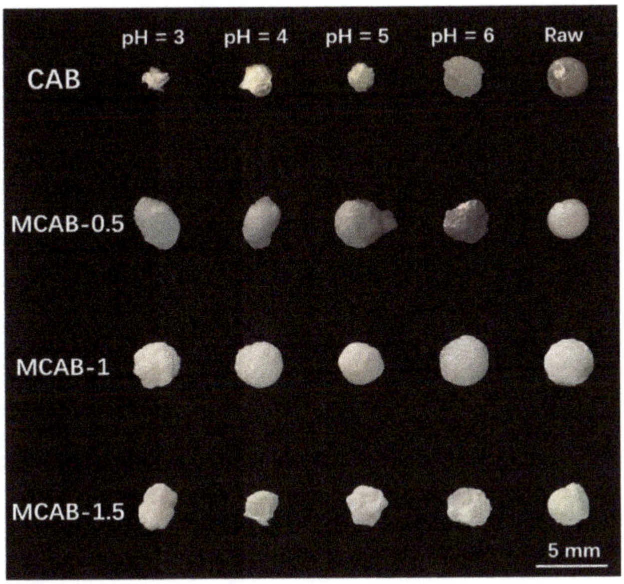

Figure 10. Effect of pH on the morphology changes of CAB and the three MCAB samples (initial Hg(II) concentration = 1 mg/L, dosage = 1 g/L, temperature = 30 °C, oscillation time = 10 h).

4.3. Mechanism of Mercury Adsorption by Self-Assembly Mineralized Ca-Based Shell-Type Beads

Based on the above discussion, the potential mercury adsorption mechanism of the polymeric materials in this study can be inferred, as shown in Figure 11. The initial structure of calcium alginate beads is a macromolecular reticular polymer which is crosslinked by calcium ions. With the impregnation of NH_4HCO_3, the calcium ions and carbonate ions produced by ammonium bicarbonate secondary hydrolysis were gradually combined and crystallized by self-assembly mineralization. Due to the existence and hindrance of the uronic acid polymer chain, the mineralized crystallization is gradually transformed to the confined crystallization, leading to the formation of a large number of amorphous calcium carbonate mineralized layers. Moreover, with the calcium ions in the outer layer of the microsphere mineralized firstly, the difference between the internal and external concentrations will lead to diffusional pressure difference. This difference results in the transfer of inner ions to the external environment, leading to mineralization and the formation of a shell-type spherical structure. To some extent, the formation of the mineralized layer promotes the transformation of the structure from dense polymer crosslinking to an organic–inorganic doped skeleton with a large specific surface area and high porosity. On the other hand, a large number of chelated carboxylic acids on the polymer chain are released and activated because of the trapping of calcium ions by carbonates, which also provides abundant binding sites for mercury ions.

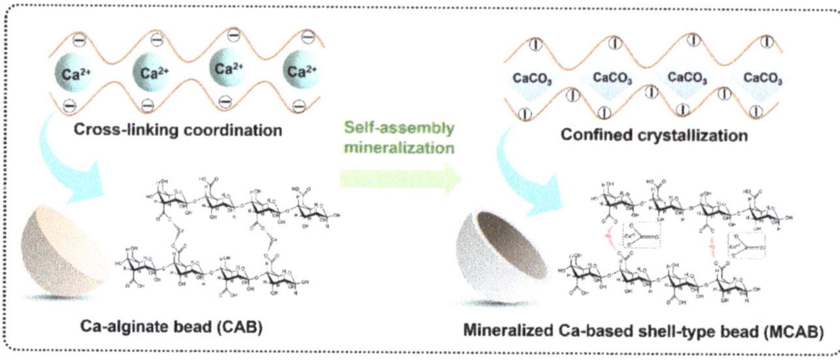

Figure 11. Potential adsorption mechanism.

5. Conclusions

In the process of sewage treatment by adsorption, there are a number of drawbacks such as high energy consumption in solid–liquid separation, serious wear of materials, and recovery of adsorbent particles. Given this, it is imperative to develop high-performance anti-wear-forming adsorbents. In this work, a series of self-assembly mineralized Ca-based shell-type adsorbents (MCABs) were synthesized and employed to remove Hg(II) ions from an aqueous solution. It was found that the optimal sample set, MCAB-1, showed significant enhancement for Hg(II) adsorption with a maximum adsorption of 48 ± 1 mg/g in 3 h at pH = 5, which is 2.67 times more than that of unmodified one. Long-duration adsorption tests showed that MCABs exhibited remarkable stability regarding their capacity and anti-fraying ability. By analyzing the structure of the material, it is demonstrated that the material is transformed from a dense crosslinking polymer to an organic–inorganic mineralization framework with a large specific surface area and high porosity by self-assembly mineralization. Meanwhile, with the confined crystallization of calcium ions, the carboxylic groups in the structure can be activated, making more binding sites for mercury ions.

Author Contributions: Conceptualization, Y.P.; methodology, X.L.; software, X.L. and C.Z.; validation, T.F.; formal analysis, C.Z.; investigation, Y.P.; resources, T.F.; data curation, C.Z.; writing—original draft preparation, Y.P.; writing—review and editing, X.G.; visualization, C.Z.; supervision, X.G.; project administration, X.G.; funding acquisition, Y.P. and X.G. All authors have read and agreed to the published version of the manuscript.

Funding: This research was funded by the National Natural Science Foundation of China, grant numbers 52106242 and 52176188.

Institutional Review Board Statement: Not applicable.

Data Availability Statement: The original contributions presented in this study are included in the article. Further inquiries can be directed to the corresponding author.

Acknowledgments: The authors appreciate Gong for the support that made this project possible.

Conflicts of Interest: The authors declare no conflicts of interest.

References

1. Alexandre, J.; Poulain, T.B. Cracking the Mercury Methylation Code. *Science* **2013**, *339*, 1280–1281.
2. Menger-Krug, E.; Niederste-Hollenberg, J.; Hillenbrand, T.; Hiessl, H. Integration of microalgae systems at municipal wastewater treatment plants: Implications for energy and emission balances. *Environ. Sci. Technol.* **2012**, *46*, 11505–11514. [CrossRef] [PubMed]
3. Zhang, B.; Xu, P.; Qiu, Y.; Yu, Q.; Ma, J.; Wu, H.; Luo, G.; Xu, M.; Yao, H. Increasing oxygen functional groups of activated carbon with non-thermal plasma to enhance mercury removal efficiency for flue gases. *Chem. Eng. J.* **2015**, *263*, 1–8. [CrossRef]

4. Xu, M.; Qiao, Y.; Zheng, C.; Li, L.; Liu, J. Modeling of homogeneous mercury speciation using detailed chemical kinetics. *Combust. Flame* **2003**, *132*, 208–218. [CrossRef]
5. Luo, G.; Yao, H.; Xu, M.; Gupta, R.; Xu, Z. Identifying modes of occurrence of mercury in coal by temperature programmed pyrolysis. *Proc. Combust. Inst.* **2011**, *33*, 2763–2769. [CrossRef]
6. Xu, M.; Yu, D.; Yao, H.; Liu, X.; Qiao, Y. Coal combustion-generated aerosols: Formation and properties. *Proc. Combust. Inst.* **2011**, *33*, 1681–1697. [CrossRef]
7. Yu, D.; Xu, M.; Yao, H.; Sui, J.; Liu, X.; Yu, Y.; Cao, Q. Use of elemental size distributions in identifying particle formation modes. *Proc. Combust. Inst.* **2007**, *31*, 1921–1928. [CrossRef]
8. Zhang, Y.; Kogelnig, D.; Morgenbesser, C.; Stojanovic, A.; Jirsa, F.; Lichtscheidl-Schultz, I.; Krachler, R.; Li, Y.; Keppler, B.K. Preparation and characterization of immobilized [A336][MTBA] in PVA-alginate gel beads as novel solid-phase extractants for an efficient recovery of Hg (II) from aqueous solutions. *J. Hazard. Mater.* **2011**, *196*, 201–209. [CrossRef]
9. Li, F.; Zhang, J.; Jiang, W.; Liu, C.; Zhang, Z.; Zhang, C.; Zeng, G. Spatial health risk assessment and hierarchical risk management for mercury in soils from a typical contaminated site, China. *Environ. Geochem. Health* **2017**, *39*, 923. [CrossRef]
10. Huang, J.; Li, F.; Zeng, G.; Liu, W.; Huang, X.; Xiao, Z.; Wu, H.; Gu, Y.; Li, X.; He, X.; et al. Integrating hierarchical bioavailability and population distribution into potential eco-risk assessment of heavy metals in road dust: A case study in Xiandao District, Changsha city, China. *Sci. Total. Environ.* **2015**, *541*, 969. [CrossRef]
11. Farooq, U.; Kozinski, J.A.; Khan, M.A.; Athar, M. Biosorption of heavy metal ions using wheat based biosorbents--a review of the recent literature. *Bioresour. Technol.* **2010**, *101*, 5043–5053. [CrossRef] [PubMed]
12. Gong, X.; Wu, T.; Qiao, Y.; Xu, M. In Situ Leaching of Trace Elements in a Coal Ash Dump and Time Dependence Laboratory Evaluation†. *Energy Fuels* **2010**, *24*, 84–90. [CrossRef]
13. Lin, G.; Zeng, B.; Li, J.; Wang, Z.; Wang, S.; Hu, T.; Zhang, L. A systematic review of metal organic frameworks materials for heavy metal removal: Synthesis, applications and mechanism. *Chem. Eng. J.* **2023**, *460*, 141710. [CrossRef]
14. He, J.; Chen, J.P. A comprehensive review on biosorption of heavy metals by algal biomass: Materials, performances, chemistry, and modeling simulation tools. *Bioresour. Technol.* **2014**, *160*, 67–78. [CrossRef] [PubMed]
15. Moreno-Garrido, I. Microalgae immobilization: Current techniques and uses. *Bioresour. Technol.* **2008**, *99*, 3949–3964. [CrossRef]
16. He, T.; Li, Q.; Lin, T.; Li, J.; Bai, S.; An, S.; Kong, X.; Song, Y.-F. Recent progress on highly efficient removal of heavy metals by layered double hydroxides. *Chem. Eng. J.* **2023**, *462*, 142041. [CrossRef]
17. Kyzas, G.Z.; Bomis, G.; Kosheleva, R.I.; Efthimiadou, E.K.; Favvas, E.P.; Kostoglou, M.; Mitropoulos, A.C. Nanobubbles effect on heavy metal ions adsorption by activated carbon. *Chem. Eng. J.* **2019**, *356*, 91–97. [CrossRef]
18. Pejman Hadi, M.-H.T.; Hui, C.-W.; Lin, C.S.K.; McKay, G. Aqueous mercury adsorption by activated carbons. *Water Res.* **2015**, *73*, 37–55. [CrossRef]
19. Teker, M.; Glu, I.; Saltabas, O. Adsorption of Copper and Cadmium Ions by Activated Carbon From Rice Hulls. *Turk. J. Chem.* **1999**, *23*, 185–192.
20. Khosravi, R.; Azizi, A.; Ghaedrahmati, R.; Gupta, V.K.; Agarwal, S. Adsorption of gold from cyanide leaching solution onto activated carbon originating from coconut shell—Optimization, kinetics and equilibrium studies. *J. Ind. Eng. Chem.* **2017**, *54*, 464–471. [CrossRef]
21. Kobya, M.; Demirbas, E.; Senturk, E.; Ince, M. Adsorption of heavy metal ions from aqueous solutions by activated carbon prepared from apricot stone. *Bioresour. Technol.* **2005**, *96*, 1518–1521. [CrossRef] [PubMed]
22. Enniya, I.; Rghioui, L.; Jourani, A. Adsorption of hexavalent chromium in aqueous solution on activated carbon prepared from apple peels. *Sustain. Chem. Pharm.* **2018**, *7*, 9–16. [CrossRef]
23. Bansode, R.R.; Losso, J.N.; Marshall, W.E.; Rao, R.M.; Portier, R.J. Adsorption of metal ions by pecan shell-based granular activated carbons. *Bioresour. Technol.* **2003**, *89*, 115–119. [CrossRef] [PubMed]
24. Wang, J.; Chen, C. Chitosan-based biosorbents: Modification and application for biosorption of heavy metals and radionuclides. *Bioresour. Technol.* **2014**, *160*, 129–141. [CrossRef] [PubMed]
25. Inoue, K.; Ohto, K.; Yoshizuka, K.; Shinbaru, R.; Kina, K. Adsorption behaviors of some metal ions on chitosan modified with EDTA-type ligand. *Bunseki Kagaku* **1995**, *44*, 283–287. [CrossRef]
26. Ngah, W.S.W.; Fatinathan, S. Adsorption of Cu(II) ions in aqueous solution using chitosan beads, chitosan–GLA beads and chitosan–alginate beads. *Chem. Eng. J.* **2008**, *143*, 62–72. [CrossRef]
27. Liu, B.; Wang, D.; Yu, G.; Meng, X. Adsorption of heavy metal ions, dyes and proteins by chitosan composites and derivatives—A review. *J. Ocean Univ. China* **2013**, *12*, 500–508. [CrossRef]
28. Crini, G.; Badot, P.-M. Application of chitosan, a natural aminopolysaccharide, for dye removal from aqueous solutions by adsorption processes using batch studies: A review of recent literature. *Prog. Polym. Sci.* **2008**, *33*, 399–447. [CrossRef]
29. Zhu, Y.; Hu, J.; Wang, J. Competitive adsorption of Pb(II), Cu(II) and Zn(II) onto xanthate-modified magnetic chitosan. *J. Hazard. Mater.* **2012**, *221–222*, 155–161. [CrossRef]
30. Nan, L.; Bai, R. Copper adsorption on chitosan–cellulose hydrogel beads: Behaviors and mechanisms. *Sep. Purif. Technol.* **2005**, *42*, 237–247.
31. Alvarez, A.; Saez, J.M.; Davila Costa, J.S.; Colin, V.L.; Fuentes, M.S.; Cuozzo, S.A.; Benimeli, C.S.; Polti, M.A.; Amoroso, M.J. Actinobacteria: Current research and perspectives for bioremediation of pesticides and heavy metals. *Chemosphere* **2017**, *166*, 41–62. [CrossRef]

32. Talaiekhozani, A.; Rezania, S. Application of photosynthetic bacteria for removal of heavy metals, macro-pollutants and dye from wastewater: A review. *J. Water Process. Eng.* **2017**, *19*, 312–321. [CrossRef]
33. Vijayaraghavan, K.; Yun, Y.S. Bacterial biosorbents and biosorption. *Biotechnol. Adv.* **2008**, *26*, 266–291. [CrossRef]
34. Gupta, P.; Diwan, B. Bacterial Exopolysaccharide mediated heavy metal removal: A Review on biosynthesis, mechanism and remediation strategies. *Biotechnol. Rep.* **2017**, *13*, 58–71. [CrossRef] [PubMed]
35. Etesami, H. Bacterial mediated alleviation of heavy metal stress and decreased accumulation of metals in plant tissues: Mechanisms and future prospects. *Ecotoxicol. Environ. Saf.* **2018**, *147*, 175–191. [CrossRef] [PubMed]
36. Liu, S.H.; Zeng, G.M.; Niu, Q.Y.; Liu, Y.; Zhou, L.; Jiang, L.H.; Tan, X.F.; Xu, P.; Zhang, C.; Cheng, M. Bioremediation mechanisms of combined pollution of PAHs and heavy metals by bacteria and fungi: A mini review. *Bioresour. Technol.* **2017**, *224*, 25–33. [CrossRef] [PubMed]
37. Kang, C.-H.; Kwon, Y.-J.; So, J.-S. Bioremediation of heavy metals by using bacterial mixtures. *Ecol. Eng.* **2016**, *89*, 64–69. [CrossRef]
38. Feng, Y.; Yu, Y.; Wang, Y.; Lin, X. Biosorption and bioreduction of trivalent aurum by photosynthetic bacteria Rhodobacter capsulatus. *Curr. Microbiol.* **2007**, *55*, 402. [CrossRef]
39. Martinez-Juarez, V.M.; Cardenas-Gonzalez, J.F.; Torre-Bouscoulet, M.E.; Acosta-Rodriguez, I. Biosorption of Mercury (II) from Aqueous Solutions onto Fungal Biomass. *Bioinorg. Chem. Appl.* **2012**, *2012*, 156190. [CrossRef]
40. Xiao, H.; Zhong, J.-J. Production of Useful Terpenoids by Higher-Fungus Cell Factory and Synthetic Biology Approaches. *Trends Biotechnol.* **2016**, *34*, 242–255. [CrossRef]
41. Bayramoğlu, G.; Tuzun, I.; Celik, G.; Yilmaz, M.; Arica, M.Y. Biosorption of mercury(II), cadmium(II) and lead(II) ions from aqueous system by microalgae Chlamydomonas reinhardtii immobilized in alginate beads. *Int. J. Miner. Process.* **2006**, *81*, 35–43. [CrossRef]
42. Chang, H.X.; Fu, Q.; Huang, Y.; Xia, A.; Liao, Q.; Zhu, X.; Zheng, Y.P.; Sun, C.H. An annular photobioreactor with ion-exchange-membrane for non-touch microalgae cultivation with wastewater. *Bioresour. Technol.* **2016**, *219*, 668–676. [CrossRef] [PubMed]
43. Ibrahim, W.M. Biosorption of heavy metal ions from aqueous solution by red macroalgae. *J. Hazard. Mater.* **2011**, *192*, 1827–1835. [CrossRef] [PubMed]
44. Liu, Y.; Cao, Q.; Luo, F.; Chen, J. Biosorption of Cd^{2+}, Cu^{2+}, Ni^{2+} and Zn^{2+} ions from aqueous solutions by pretreated biomass of brown algae. *J. Hazard. Mater.* **2009**, *163*, 931–938. [CrossRef]
45. Vilar, V.J.; Botelho, C.M.; Loureiro, J.M.; Boaventura, R.A. Biosorption of copper by marine algae Gelidium and algal composite material in a packed bed column. *Bioresour. Technol.* **2008**, *99*, 5830–5838. [CrossRef]
46. Jabar, J.M.; Adebayo, M.A.; Taleat, T.A.A.; Yılmaz, M.; Rangabhashiyam, S. Ipoma batatas (*Sweet potato*) leaf and leaf-based biochar as potential adsorbents for procion orange MX-2R removal from aqueous solution. *J. Anal. Appl. Pyrolysis* **2025**, *185*, 106876. [CrossRef]
47. Chen, S. Collision-induced breakage of agglomerates in homogenous isotropic turbulence laden with adhesive particles. *J. Fluid Mech.* **2020**, *902*, A28. [CrossRef]
48. Akhtar, N.; Iqbal, M.; Zafar, S.I.; Iqbal, J. Biosorption characteristics of unicellular green alga Chlorella sorokiniana immobilized in loofa sponge for removal of Cr(III). *J. Environ. Sci.* **2008**, *20*, 231–239. [CrossRef]
49. Sheng, P.X.; Wee, K.H.; Ting, Y.P.; Chen, J.P. Biosorption of copper by immobilized marine algal biomass. *Chem. Eng. J.* **2008**, *136*, 156–163. [CrossRef]
50. Arıca, M.Y.; Arpa, Ç.; Kaya, B.; Bektaş, S.; Denizli, A.; Genç, Ö. Comparative biosorption of mercuric ions from aquatic systems by immobilized live and heat-inactivated Trametes versicolor and Pleurotus sajur-caju. *Bioresour. Technol.* **2003**, *89*, 145–154. [CrossRef]
51. Grant, G.T.; Morris, E.R.; Rees, D.A.; Smith, P.J.C.; Thom, D. Biological interactions between polysaccharides and divalent cations: The egg-box model. *FEBS Lett.* **1973**, *32*, 195–198. [CrossRef]
52. Bulgariu, D.; Bulgariu, L. Equilibrium and kinetics studies of heavy metal ions biosorption on green algae waste biomass. *Bioresour. Technol.* **2012**, *103*, 489–493. [CrossRef] [PubMed]
53. Sarı, A.; Uluozlü, Ö.D.; Tüzen, M. Equilibrium, thermodynamic and kinetic investigations on biosorption of arsenic from aqueous solution by algae (*Maugeotia genuflexa*) biomass. *Chem. Eng. J.* **2011**, *167*, 155–161. [CrossRef]
54. Khataee, A.R.; Vafaei, F.; Jannatkhah, M. Biosorption of three textile dyes from contaminated water by filamentous green algal *Spirogyra* sp.: Kinetic, isotherm and thermodynamic studies. *Int. Biodeterior. Biodegrad.* **2013**, *83*, 33–40. [CrossRef]
55. Plazinski, W. Equilibrium and kinetic modeling of metal ion biosorption: On the ways of model generalization for the case of multicomponent systems. *Adsorption* **2013**, *19*, 659–666. [CrossRef]
56. Huang, Y.H.; Peddi, P.K.; Tang, C.; Zeng, H.; Teng, X. Hybrid zero-valent iron process for removing heavy metals and nitrate from flue-gas-desulfurization wastewater. *Sep. Purif. Technol.* **2013**, *118*, 690–698. [CrossRef]
57. Hadavifar, M.; Bahramifar, N.; Younesi, H.; Li, Q. Adsorption of mercury ions from synthetic and real wastewater aqueous solution by functionalized multi-walled carbon nanotube with both amino and thiolated groups. *Chem. Eng. J.* **2014**, *237*, 217–228. [CrossRef]
58. Zia, K.M.; Zia, F.; Zuber, M.; Rehman, S.; Ahmad, M.N. Alginate based polyurethanes: A review of recent advances and perspective. *Int. J. Biol. Macromol.* **2015**, *79*, 377–387. [CrossRef]

59. Li, X.; Shen, Q.; Su, Y.; Tian, F.; Zhao, Y.; Wang, D. Structure−Function Relationship of Calcium Alginate Hydrogels: A Novel Crystal-Forming Engineering. *Cryst. Growth Des.* **2009**, *9*, 3470–3476. [CrossRef]
60. Diaz-Rodriguez, P.; Garcia-Triñanes, P.; Echezarreta López, M.M.; Santoveña, A.; Landin, M. Mineralized alginate hydrogels using marine carbonates for bone tissue engineering applications. *Carbohydr. Polym.* **2018**, *195*, 235–242. [CrossRef]
61. Ma, Y.; Zhang, J.; Guo, S.; Shi, J.; Du, W.; Wang, Z.; Ye, L.; Gu, W. Biomimetic mineralization of nano-sized, needle-like hydroxyapatite with ultrahigh capacity for lysozyme adsorption. *Mater. Sci. Eng. C* **2016**, *68*, 551–556. [CrossRef] [PubMed]
62. Zhao, J.; He, X.; Wang, L.; Tian, J.; Wan, C.; Jiang, C. Addition of NH4HCO3 as pore-former in membrane electrode assembly for PEMFC. *Int. J. Hydrogen Energy* **2007**, *32*, 380–384. [CrossRef]
63. Vilar, V.J.P.; Botelho, C.M.S.; Boaventura, R.A.R. Influence of pH, ionic strength and temperature on lead biosorption by Gelidium and agar extraction algal waste. *Process Biochem.* **2005**, *40*, 3267–3275. [CrossRef]
64. Ponce, S.C.; Prado, C.; Pagano, E.; Prado, F.E.; Rosa, M. Effect of solution pH on the dynamic of biosorption of Cr(VI) by living plants of Salvinia minima. *Ecol. Eng.* **2015**, *74*, 33–41. [CrossRef]
65. Bakatula, E.N.; Cukrowska, E.M.; Weiersbye, I.M.; Mihaly-Cozmuta, L.; Peter, A.; Tutu, H. Biosorption of trace elements from aqueous systems in gold mining sites by the filamentous green algae (*Oedogonium* sp.). *J. Geochem. Explor.* **2014**, *144*, 492–503. [CrossRef]
66. David Kratochvil, B.V. Advances in the biosorption of heavy metals. *Trends Biotechnol.* **1998**, *16*, 291–300. [CrossRef]
67. Qiu, H.; Lv, L.; Pan, B.c.; Zhang, Q.j.; Zhang, W.m.; Zhang, Q.x. Critical review in adsorption kinetic models. *J. Zhejiang Univ. Sci. A* **2009**, *10*, 716–724. [CrossRef]
68. Vilar, V.J.; Botelho, C.M.; Boaventura, R.A. Kinetics and equilibrium modelling of lead uptake by algae Gelidium and algal waste from agar extraction industry. *J. Hazard. Mater.* **2007**, *143*, 396–408. [CrossRef]
69. Weber Walter, J.; Morris, J.C. Kinetics of Adsorption on Carbon from Solution. *J. Sanit. Eng. Div.* **1963**, *89*, 31–59. [CrossRef]
70. Foo, K.Y.; Hameed, B.H. Insights into the modeling of adsorption isotherm systems. *Chem. Eng. J.* **2010**, *156*, 2–10. [CrossRef]

polymers

Article

Residual Stress Analysis at the Conductor–Insulator Interface During the Curing Process of Hair-Pin Motors

Mingze Ma [1], Hongyi Gan [2], Xiao Shang [2], Linsen Song [1,*], Yiwen Zhang [1,3,*], Jingru Liu [2], Chunbai Liu [4], Yanzhong Hao [4] and Xinming Zhang [1,5]

1 School of Mechanical and Electrical Engineering, Changchun University of Science and Technology, Changchun 130022, China; mmz195797@163.com (M.M.); zxm@fosu.edu.cn (X.Z.)
2 FAW Tooling Die Manufacturing Co., Ltd., Changchun 130000, China; ganhongyi@faw.com.cn (H.G.); shangxiao_td@faw.com.cn (X.S.); liuqr_td@faw.com.cn (J.L.)
3 University-Enterprise Joint Innovation Laboratory of Intelligent Manufacturing, Assembly and Inspection Technology and Equipment for Automotive Parts and Components, Changchun University of Science and Technology, Changchun 130022, China
4 FAW Casting Co., Ltd., Changchun 130013, China; liuchunbai@faw.com.cn (C.L.); haoyanzhong@faw.com.cn (Y.H.)
5 School of Mechatronic Engineering and Automation, Foshan University, Foshan 528225, China
* Correspondence: songlinsen@cust.edu.cn (L.S.); zhyw@cust.edu.cn (Y.Z.)

Abstract: The curing process of hair-pin motor stator insulation is critical, as residual stress increases the risk of partial discharge and shortens a motor's lifespan. However, studies on the stress-induced defects during insulation varnish curing remain limited. This research integrates three-dimensional numerical simulations and experimental analysis to develop a curing model based on unsaturated polyester imide resin, aiming to explore the mechanisms of residual stress formation and optimization strategies. A dual fiber Bragg grating (FBG) sensor system is employed for simultaneous temperature and strain monitoring, while curing kinetics tests confirm the self-catalytic nature of the process and yield the corresponding kinetic equations. The multi-physics simulation model demonstrates strong agreement with the experimental data. The results show that optimizing the curing process reduces the maximum stress from 45.1 MPa to 38.6 MPa, effectively alleviating the stress concentration. These findings highlight the significant influence of the post-curing temperature phase on residual stress. The proposed model offers a reliable tool for stress prediction and process optimization in various insulating materials, providing valuable insights for motor insulation system design.

Keywords: curing residual stress; unsaturated polyester imide resin (UPIR); curing kinetics; finite element analysis; hairpin motors

1. Introduction

With the rapid development of the automobile industry, the market share of new energy vehicles has increased markedly [1]. As the core component of new energy vehicles, the performance of the motor directly determines the overall vehicle performance [2,3]. Due to their high slot fullness and power density, hair-pin motors are widely used in the manufacturing of new energy vehicles. In the manufacturing process of hair-pin motor stators, stator drip coating is a key step; its quality directly affects the insulation performance of the stator. Common motor insulation materials are mostly polymeric resin lacquers, such as polyester lacquers, epoxy lacquers, and polyurethane lacquers [4,5]. These resins are thermosetting polymers that, upon curing, form cross-linked structures which are infusible and insoluble [6]. When selecting insulating varnish materials, both the insulating properties and the environmental impact, specifically the emission of volatile organic compounds (VOCs), must be considered [7,8]. VOCs in polyester resins are typically released during the polymerization or curing process by toxic crosslinking monomers. These monomers act as solvents for the base polyester to reduce its viscosity and are reactive during the curing process [9]. This issue can be avoided by using solvent-free insulating varnishes. Another critical factor in selecting insulation materials is thermal performance. According to international standards [7], thermal classification testing specifies the maximum temperature that the motor can withstand during operation. Unsaturated polyester imide resins, due to the introduction of imide functional groups, offer higher thermal ratings and improved insulating properties [10]. Among these, unsaturated polyester imide resin (UPIR) is often used as the main material for motor stator insulation because of its superior insulating properties, mechanical strength, and thermal stability. However, during the curing process of resin composite insulating varnish, due to the dissimilar material properties between the insulating varnish and the stator copper conductor, residual stresses are generated at the interface [11–14], resulting in cracks or gaps in the insulating varnish and partial discharges, which can seriously affect the motor's lifespan.

Numerous studies have focused on analyzing the curing residual stresses in resin composites, particularly regarding their composition and other characteristics. However, relatively few studies have examined the interfacial residual stresses during the curing process of stator insulating varnish in hair-pin motors. Dewangan et al. employed X-ray diffraction (XRD) techniques to study the curing of polyester resins under high pressure [15]. The results were validated through three-dimensional numerical simulations, and a method to increase porosity was proposed. Hao et al. used fiber Bragg grating (FBG) technology to detect cure residual strains in the E51/W93 epoxy resin system with varying Al_2O_3 content. Their primary focus was on examining the effect of Al_2O_3 content on gel temperature and gel time, and they proposed a method to simultaneously detect temperature and curing residual strain using thermocouple temperature compensation combined with an FBG sensor [16]. Yang et al. designed a device to measure the elastic modulus of epoxy resin potting materials based on a viscoelastic model; they derived an expression for the viscoelastic intrinsic properties of epoxy resins [17]. Wang et al. identified the cracking issues in insulating surfaces caused by curing stresses during the secondary curing of epoxy resins, and investigated the impact of curing agent content on curing stresses [18]. Qu et al. investigated the curing stresses of modified epoxy resins with varying toughener contents. They analyzed the material properties of the epoxy resin curing process and curing residual stresses using differential scanning calorimetry (DSC), Dynamic Mechanical Analysis (DMA), Fourier-Transform Infrared Spectroscopy (FTIR), and Thermogravimetric Analysis (TGA). The curing thermal stresses were calculated using the Stoney equation [19].

The above studies primarily focus on the analysis of the residual stresses during the curing process of resin systems, with a particular emphasis on the effects of the material composition and curing parameters. These studies utilize techniques such as DSC and FBG to measure the curing-induced residual strain and stress under various conditions, highlighting the importance of precise stress measurement and control. By investigating the curing characteristics of resin materials, they offer valuable insights into the stress mitigation and crack prevention in insulating materials. This body of work underscores the critical role of the material properties and curing conditions in influencing the residual stresses and mechanical integrity, and it has inspired the dual-FBG curing strain experiment in this study. However, these studies predominantly focus on single-resin systems or simple geometries, leaving a gap in understanding the residual stresses at complex interfaces, such as those in motor stators. This gap highlights the necessity of the present research, which aimed to investigate the interfacial residual stresses during the curing process of insulating varnish in hair-pin motor stators.

Based on previous studies, the numerical simulation method can be regarded as a universal approach for stress analysis during the curing process of resin materials. Ye et al. conducted a multi-physics numerical simulation of L-shaped composite laminates, developing a model that incorporates the resin's transient flow compaction to predict the curing behavior of resin composites [20]. Balaji et al. combined the data from DSC experiments with neural networks to predict the deformation of resin materials during curing. They verified their model's accuracy through finite element simulations, which demonstrated greater realism than the conventional curing kinetics models at temperature rise rates below 3 K/min [21]. Wu et al. studied the deformation of polyurethane composites during the curing when used as battery bezels, focusing on the factors affecting curing deformation; withholding pressure was identified as the most significant factor [22].

The above studies demonstrate the value of numerical simulation in analyzing the curing behavior of resin systems. They show that numerical methods are highly effective for predicting the residual stresses and deformation during curing, improving the accuracy and efficiency of stress analysis. Additionally, combining simulations with experimental validation has proven to be a reliable way to understand the curing process under different conditions and have provided important guidance for this research. Building on these insights, this study applies numerical simulation to investigate the curing behavior of insulating varnish in hair-pin motor stators, with a focus on addressing the residual stress issues at complex geometric interfaces.

The simulation and analysis of the curing residual stresses in resin-insulating varnish, one of the most commonly used motor-insulating materials, are crucial. Numerical simulation not only reduces the experimental costs but also predicts the curing defects caused by various factors and assesses their impact on the electrical performance of motors. In this paper, we focus on the residual stresses, temperature, and strain during the curing process of stator-insulating varnish for hair-pin motors and explore the characteristics and mechanisms of the curing residual stresses through a combination of simulation and experimental verification.

This paper will follow the process outlined in Figure 1.

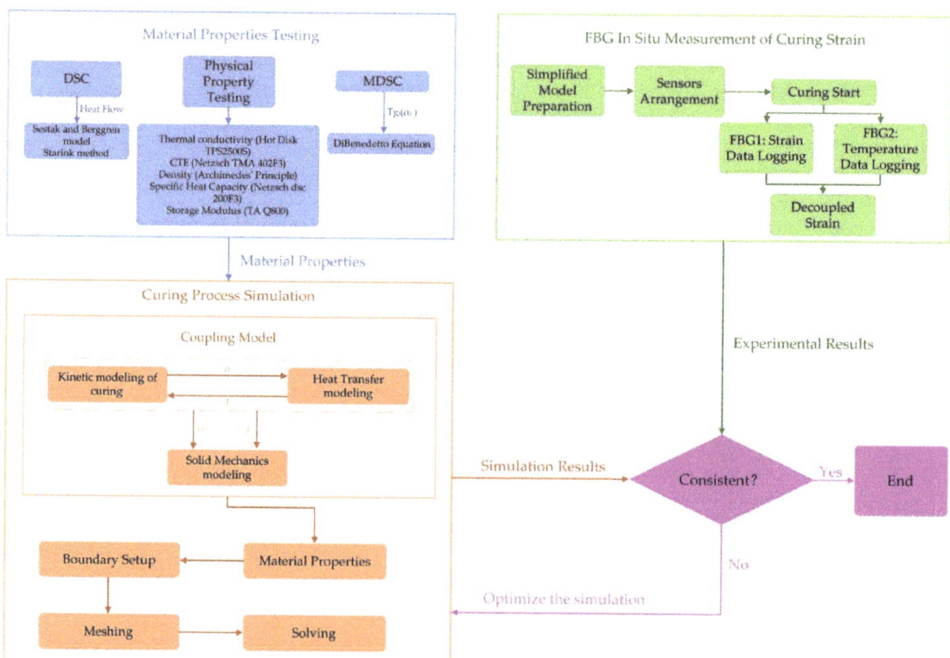

Figure 1. Flowchart of the experimental and simulation process.

This study presents a novel approach to analyze the curing residual stresses at the conductor–insulator interface of hairpin motors. Research on the curing residual stresses of unsaturated polyester imide resin insulating varnishes is limited. Additionally, unlike the traditional methods using thermocouple temperature compensation, this study uniquely employs encapsulated FBG for temperature monitoring, which leads to more accurate experimental results. Furthermore, by considering the temperature or curing degree dependence of the material properties, a three-dimensional simulation model is developed. The model established in this study can be applied to the residual stress analysis during the curing process of other thermoset resin materials, thus contributing to a better understanding of the residual stress generation behavior of resin-based insulating materials in various fields.

2. FBG In Situ Measurement of Curing Strain

2.1. Materials

The insulating varnish used in this study is a one-component, low-volatility impregnating varnish, Voltatex® 4200 (Axalta Coating Systems Ltd., Philadelphia, PA, USA). The main component is an unsaturated polyester imide resin, which offers excellent insulating properties for the motor stator. According to the official performance specifications, Voltatex® 4200 demonstrates a dielectric strength of 85 kV/mm at 155 °C, significantly higher than that of comparable epoxy resin insulating varnishes, which typically range from 20 to 30 kV/mm [23,24]. Furthermore, the thermal class of unsaturated polyester imide resin materials is generally higher than that of epoxy resin-based insulating varnishes [4]. Additionally, UPIR has a lower initial viscosity compared to typical epoxy resins, which is more favorable for the contact between the insulation varnish and the insulated surface during the motor insulation process.

2.2. Principle of the FBG Sensor

The FBG sensor is a single-mode fiber that detects reflections at specific wavelengths as light passes through the FBG fiber [25]. The Bragg wavelength expression is presented in Equation (1):

$$\lambda_B = 2n_{eff}\Lambda \tag{1}$$

where λ_B is the wavelength reflected by the Bragg grating, n_{eff} is the effective reflection coefficient of the fiber, and Λ is the grating period.

FBG is sensitive to external changes, and for FBG sensors affected by both temperature and strain, λ_B is linearly correlated with changes in temperature and strain. The Bragg wavelength shift can be expressed as:

$$\Delta\lambda_B = k_\varepsilon \varepsilon + k_T \Delta T \tag{2}$$

where k_ε is the coefficient related to strain and the photo-elastic coefficient, and k_T is the coefficient related to the thermal-optic coefficient and thermal expansion coefficient of the optical fiber.

Due to the characteristics of unsaturated polyester imide resin, in the early stage of insulating varnish curing, the varnish exists as a viscous liquid. Upon reaching the gel point, the molecules gradually crosslink, and the resin hardens into a solid. At this stage, the shrinkage strain caused by curing and the thermal expansion and contraction due to temperature changes can be transferred to the FBG sensor. The curing strain of the resin can then be determined by monitoring the change in the wavelength of the FBG.

2.3. Experimental Design

Equation (2) clearly shows that when using FBG to measure the strain influenced by both temperature and strain, temperature compensation must be applied. In this experiment, the temperature compensation is provided by a packaged FBG temperature sensor, which offers more accurate real-time temperature readings. The wavelength changes due to strain are obtained through the fiber Bragg grating demodulator. The test equipment is illustrated in Figure 2a.

The FBG sensor used is the OSC-1100 (Xi'an Huance Automation Co., Ltd., Xi'an, China). The FBG demodulator is the HOSI-1000E (Huaying Instrument Equipment Co., Ltd.,

Xi'an, China). The oven model is DZF-6020 (Shanghai Yiheng Scientific Instrument Co., Ltd., Shanghai, China).

The high slot fullness of the hair-pin motor's stator is attributed to the geometry of the flat copper wire and the winding method, which allows the stator slots to be filled. The stator conductor is made of copper, while the stator core (including the stator slots) is composed of silicon steel, which offers superior magnetic properties and lower energy loss. The motor consists of 8 strands of conductors, which can provide higher power.

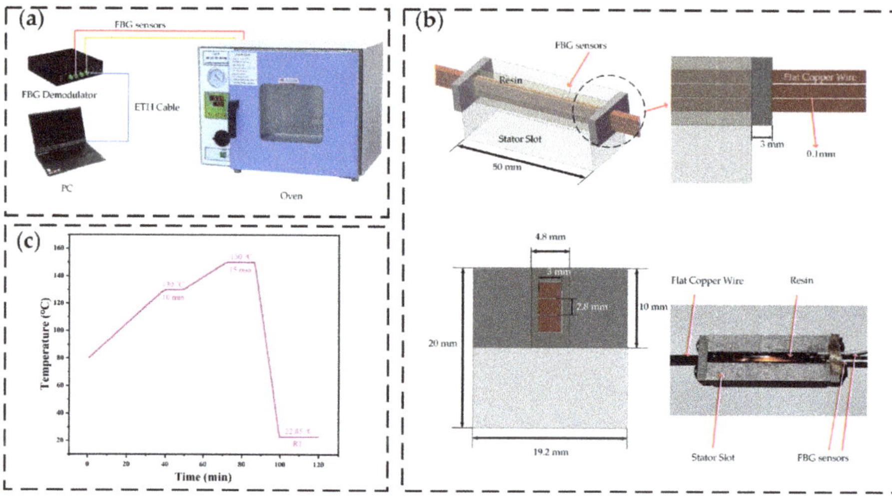

Figure 2. FBG in situ measurement. (**a**) Test equipment; (**b**) curing test model; (**c**) curing procedure.

In this paper, to simplify the model, only the stresses generated at the interface of the flat copper wires by the insulating varnish between the three conductors are investigated. The simplified model is illustrated in Figure 2b.

The slot structure surrounding the copper wire is made of silicon steel and simulates the stator slots. During the experiment, insulating varnish is filled into the groove. After ensuring complete immersion, two FBG sensors are inserted on either side of the copper wire to measure temperature and strain data during the curing process of the insulating varnish at the same location.

Figure 2c illustrates the curing process of the drop-dip test. First, the stator must be heated to 80 °C, after which the insulating varnish is drop-dipped onto it. The temperature is then increased from 80 °C to 130 °C and held for 10 min, during which the curing reaction takes place. Subsequently, the temperature is raised to 150 °C and held for 15 min to cure the insulating varnish fully and eliminate thermal stress. Finally, the stator is cooled to room temperature to complete the curing process.

It should be noted that some sources of error may arise during the experiment. First, since a packaged FBG is used for temperature compensation, it is essential to ensure that the insulating varnish does not seep into the encapsulation tube. Second, environmental vibrations or fiber bending may cause signal fluctuations or distortion [26]. Third, misalignment of the optical fiber could result in minor measurement deviations. These factors were carefully considered during the experimental setup, and measures such as sensor calibration and environmental stabilization were implemented to minimize their impact.

2.4. Experimental Results and Analysis

As described in Section 2.2, the FBG sensor is affected by both temperature and strain. When using the FBG sensor to measure the temperature, it is essential to measure without applying force. Teflon capillary-encapsulated FBGs are used to compensate for the temperature effects on strain. Two FBGs are placed symmetrically on the left and right sides of the flat copper wire for the test. From Equation (2), by knowing the sensitivity coefficients k_ε and k_T, and the temperature change ΔT of the optical fiber, the true cure strain ε can be determined according to Equation (3):

$$\varepsilon = \frac{\Delta\lambda_B - k_T \Delta T}{k_\varepsilon} \tag{3}$$

where the sensitivity coefficients k_ε and k_T of the optical fiber are provided by the manufacturer, and the temperature change ΔT is obtained from the FBG sensor used for temperature compensation.

The stress and temperature test results for the insulating varnish curing process are illustrated in Figure 3. The initial strain value was set in the FBG demodulator's PC software Hippo 3.1.4 when the specimen reached 80 °C, and the curing degree curve was later obtained from the curing kinetics model analysis. The figure shows that there is no significant change in the strain during the early stage of curing, as the resin remains in a flow state at this point. Due to preheating to 80 °C, the resin's viscosity decreases significantly, but cross-linking has not yet occurred. The molecular chains mainly rely on physical interactions, allowing the internal molecules to move freely.

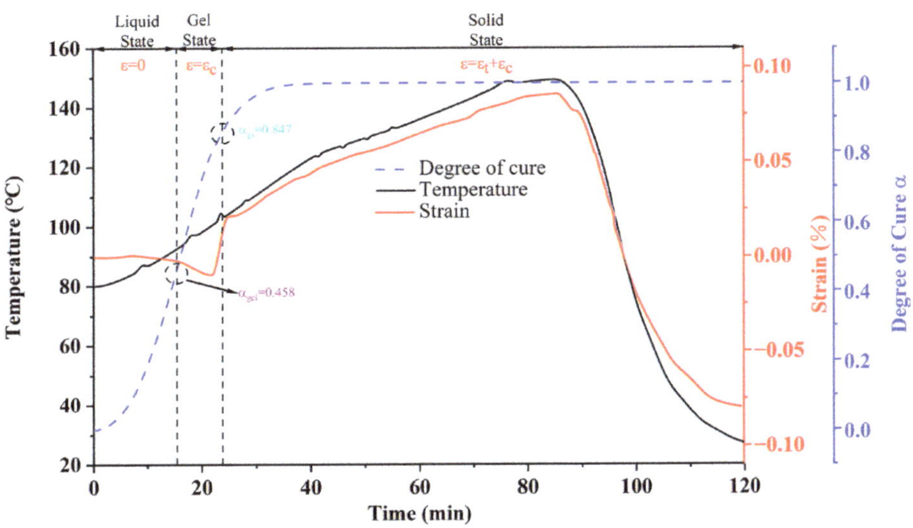

Figure 3. Curing strain test results of insulating varnish.

When the strain curve decreases significantly, it indicates that the resin has undergone considerable volume contraction during curing. Since the primary component of the material is unsaturated polyester imide resin, the curing process is accompanied by significant contraction. Even though both thermal strain and chemical contraction occur simultaneously, chemical contraction predominates, resulting in the observed decrease in the strain curve. When the strain curve begins to rise, it indicates that the material has gradually transitioned from a viscous-fluid state to a glassy state. Most cross-linking reactions within the material have been completed, and the mobility of the molecular chains is significantly reduced. Due to the further temperature increase, the material undergoes thermal expansion, causing the strain curve to rise.

During the heat preservation stage and the subsequent warming stage, the temperature profile does not fully align with the preset temperature profile. One reason for this is that the resin's curing reaction is exothermic, and the heat released during curing leads to a local temperature increase. Another reason is that the stator slot has a low thermal conductivity, causing slow heat transfer from the outside, which results in noticeable thermal hysteresis.

In the post-temperature rise stage, the strain and temperature curves are roughly linear because the curing reaction of the insulating varnish is nearly complete, and the strain is primarily affected by thermal strain. However, the strain and temperature curves are not completely linear because the coefficient of thermal expansion (CTE) of the cured resin exhibits a noticeable nonlinear characteristic with changing temperature. The coefficient of thermal expansion of the cured resin was measured to be 45.93×10^{-6}/K at 25 °C, increasing to 152.55×10^{-6}/K at 150 °C, as measured by the Netzsch TMA 402F3 (Netzsch-Gerätebau GmbH, Selb, Germany). Another reason is that the higher crosslink density of the material hinders molecular chain movement, slowing the release of internal stresses [27,28].

During cooling, when the strain curve of the resin decreases more slowly than the temperature curve, it indicates that the material is passing through the glass transition temperature (T_g). At this point, the material transforms from a rubbery state to a glassy state, further restricting molecular chain movement and hampering the release of internal stresses. Thus, a partially nonlinear relationship between the strain curve and the temperature curve is observed during the cooling phase. As discussed in Section 3.1, the glass transition temperature of this insulating varnish after complete curing is 97 °C, which aligns with the figure.

3. Three-Dimensional Numerical Simulation of the Curing Process

Due to the compact structure of the stator flat-wire winding, measuring the data related to the curing process of the insulating varnish is challenging, especially as the distance between the flat wires is only 0.1 mm. This small gap limits the ability of sensors to detect the curing residual stress directly. Numerical simulation provides an effective solution to this problem. The simulations in this study were performed using COMSOL Multiphysics. The simulation of the insulating varnish curing process involves multi-physical field coupling, including the temperature field, curing degree field, and stress–strain field. The relevant mathematical model includes a heat conduction model, a curing kinetics model, and a mechanical model of the material.

The temperature field and the curing field are bi-directionally coupled. By solving the relationship between the degree of curing and temperature, the calculated temperature and degree of curing are coupled with the stress–strain field to predict the curing deformation of the insulating varnish and the interfacial residual stresses. The coupling relationship is illustrated in Figure 4.

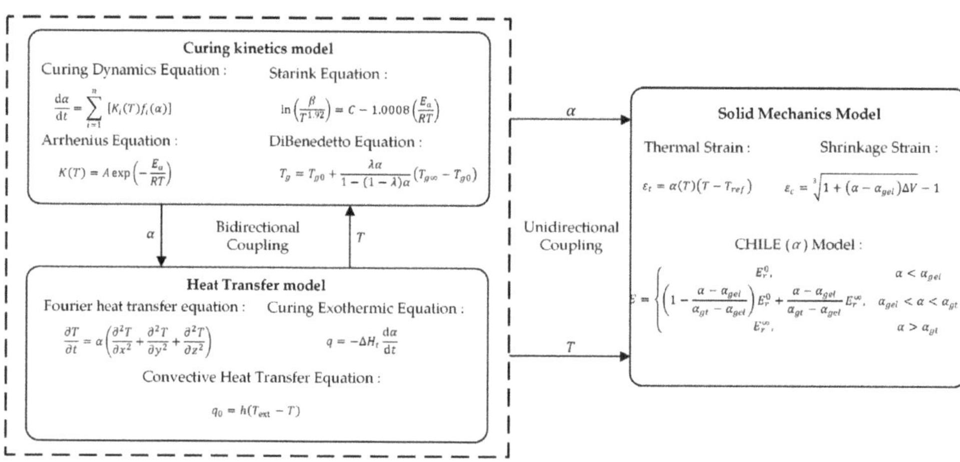

Figure 4. Coupling relationship.

It should be noted that the distribution of insulating varnish is assumed to be uniform during the curing of drip varnish on flat-wire windings. Since the insulating varnish coating is thin and the curing time is short, the resin exhibits almost no mobility during curing. Therefore, the effect of the flow compaction model can be neglected.

3.1. Curing Kinetics Model

3.1.1. Data Analysis and Model Derivation

The cure rate of insulating varnish reflects the dynamic behavior of the curing process. There are two approaches to studying this process: a macroscale phenomenological model and a microscale mechanistic model [29]. The phenomenological model considers only the overall process of the reaction, and a single rate equation, $d\alpha/dt$, is used to express this model. In contrast, the mechanistic model can better predict the curing process and explain the reaction mechanism. However, due to the complexity of the insulating varnish curing reaction, the mechanistic model requires more parameters to accurately describe the curing process, leading to a significant increase in the computational demands and making it difficult to obtain a highly accurate model. Therefore, in this paper, we adopt the phenomenological model, expressed as Equation (4) [30–33]:

$$\frac{d\alpha}{dt} = \sum_{i=1}^{n}[K_i(T)f_i(\alpha)] \tag{4}$$

where α is the degree of cure, $d\alpha/dt$ is the curing rate, T is the temperature, and $K_i(T)$ and $f_i(\alpha)$ represent the reaction rate equation and the curing reaction equation, respectively. The reaction rate equation $K(T)$ is parameterized by the Arrhenius formula shown in Equation (5) [34].

$$K(T) = A\,exp\left(-\frac{E_a}{RT}\right) \tag{5}$$

where T, A, E_a, and R represent the temperature, pre-exponential factor, activation energy, and the molar gas constant, respectively.

The primary component of the insulating varnish is unsaturated polyester imide resin, and the degree of curing is commonly used to describe the extent of the resin's curing reaction. The degree of curing is a key parameter in simulating the curing process, as it significantly affects the final performance of the resin material (e.g., mechanical strength, chemical resistance, and thermal stability). To develop a kinetic model for the resin's curing process, the curing behavior was analyzed using differential scanning calorimetry (DSC) in this study.

Four samples were heated from 25 °C to 180 °C at heating rates of 5 K/min, 10 K/min, 15 K/min, and 20 K/min using a Netzsch DSC 200F3. The sample mass ranged from 7 to 10 mg, and the nitrogen flow rate was set at 50 mL/min. The heat flow-temperature curves obtained by DSC are illustrated in Figure 5a. (Note: the heat flow curve obtained by the device is exothermic downward; it was multiplied by −1 to display the exothermic peak upward).

The curing onset temperature T_i, peak temperature T_p, termination temperature T_f, and total heat of reaction ΔH_t obtained for the curing process of the insulating varnish samples at different heating rates are presented in Table 1.

It is observed that the exothermic peak of the curing reaction shifts significantly to the right as the heating rate of the insulating varnish increases. The curing onset temperature T_i, peak temperature T_p, and termination temperature T_f also increase significantly with the higher heating rate. Equation (6) can be used to calculate the exothermic quantity $\Delta H(t)$ of the curing reaction. This integral represents the area under the heat flow–time curve and the baseline from the onset to the end of the curing reaction.

$$\Delta H(t) = \int_0^t \phi(t)\cdot \mathrm{d}t \tag{6}$$

where $\phi(t)$ is the heat flow function.

The total exothermic heat of curing for the insulating varnish remained relatively consistent across different heating rates, with an average total heat of reaction ΔH_t = 192.625 J·g^{-1}.

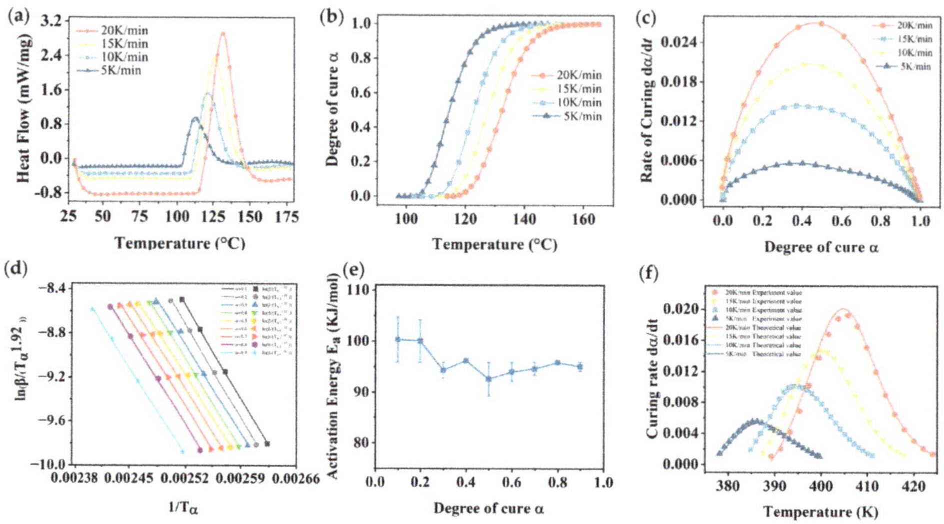

Figure 5. DSC data analysis. (**a**) Heat flow–temperature curves; (**b**) degree of cure–temperature curves; (**c**) rate of curing at different degrees of cure; (**d**) fitting curves of the Starink equation; (**e**) reaction activation energy E_a at different degrees of cure; (**f**) model validation curves.

Table 1. Characteristic temperature and total reaction heat at different heating rates.

Heating Rate (K/min)	T_i (°C)	T_p (°C)	T_f (°C)	ΔH_t(J·g^{-1})
5	103.6	112.8	127.8	198.2
10	110.2	121.0	135.8	187.2
15	114.5	127.4	141.3	192.5
20	118.4	132.1	145.8	192.6

The complete curing of the insulating varnish is achieved when the reaction exotherm $\Delta H(t)$ equals the total reaction heat ΔH_t. Therefore, the degree of cure α at any given time can be defined as the ratio between the exothermic heat $\Delta H(t)$ and the total reaction heat ΔH_t from the beginning of the curing reaction to a specific moment, as shown in Equation (7).

$$\alpha(t) = \frac{\Delta H(t)}{\Delta H_t} \tag{7}$$

Figure 5b presents the relationship between the degree of cure and temperature. It exhibits a distinct S-shaped curve, where the cure rate increases initially and then decreases, consistent with the autocatalytic model [35–37].

To further validate the results, the time derivative of the degree of cure was calculated to obtain the curing rate. Figure 5c illustrates the relationship between the curing rate $d\alpha/dt$ and the degree of cure $\alpha(t)$.

It is evident that, at various heating rates, the curing rate first increases and subsequently decreases as the degree of cure increases. The peak of the curing rate occurs at approximately a degree of cure of 0.4, further indicating that the curing of this insulating varnish conforms to the self-catalytic model [38].

The curing process of insulating varnishes is usually influenced by two factors: curing kinetics control and diffusion control. According to the studies in the literature [39], at the beginning of the isothermal cure, the curing rate of the resin is controlled solely by the curing kinetics, and with an increasing degree of cure, the material becomes partially glassy, hindering the cross-linking reaction within the resin; thus, the process becomes diffusion-controlled. According to previous thermal analysis methods [39,40], the first step involves determining the activation energy of the cure reaction and assessing whether a single model can describe the curing kinetics by analyzing the relationship between the activation energy and the degree of cure. Commonly used methods for calculating the activation energy include the Starink model [41,42], the Kissinger model [42], the Ozawa model [43,44], with the Starink model considered more accurate compared to other model-free isotropic methods [45].

By fitting the experimental data with the Starink equation shown in Equation (8), the activation energy (E_a) of the insulating varnish reaction can be determined.

$$\ln\left(\frac{\beta}{T^{1.92}}\right) = C - 1.0008\left(\frac{E_a}{RT}\right) \quad (8)$$

where β represents the rate of ramping, T denotes the reaction temperature, C is a constant, E_a stands for the reaction activation energy, and R refers to the molar gas constant.

According to the Starink equation, $\ln(\beta/T^{1.92})$ was linearly fitted against $1/T$ for different heating rates, with the results presented in Figure 5d.

The obtained slopes of the curves for various degrees of cure and the corresponding reaction activation energies are shown in Table 2. The R-squared values for different degrees of cure are all greater than 0.99, indicating a strong fit. The average reaction activation energy (E_a) is calculated to be 95.96 kJ/mol.

Table 2. Data fitted by the Starink method.

Degree of Cure α	Slope	E_a (kJ/mol)	R-Squared
0.1	−12,080.9	100.3607	0.99956
0.2	−12,050.2	100.1052	0.99912
0.3	−11,364.3	94.40691	0.99532
0.4	−11,590.8	96.28893	0.99952
0.5	−11,152.3	92.64607	0.99761
0.6	−11,324	94.07278	0.99747
0.7	−11,398.3	94.68944	0.99751
0.8	−11,545.9	95.91561	0.99851
0.9	−11,449.3	95.11346	0.99884

The curing process of insulating varnish resin is controlled by two mechanisms: reaction control and diffusion control. According to the study in [39], when the activation energy required for the reaction changes significantly with the degree of cure, it indicates that the reaction is controlled by diffusion. In this case, using a single model to describe the reaction would lead to errors, and the reaction can only be described using a model that includes a diffusion term [46–48] (such as the Kamal model [49]). According to previous thermal analysis methods [40], the activation energy of the curing reaction must first be determined, and the relationship between the activation energy and degree of cure should be examined to determine whether a single model can describe the curing kinetics.

The reaction activation energies at different curing degrees are presented in Figure 5e, and the activation energy (E_a) remains relatively constant once the degree of cure exceeds 0.5. Therefore, it can be concluded that the curing behavior of this insulating varnish is controlled solely by the reaction kinetics.

There are multiple models for describing the curing kinetics of autocatalytic composites. Notably, the Prout–Tompkins (PT) model [50–53] incorporates diffusion terms, thereby providing a more accurate depiction of the cure front propagation under conditions of high fiber content or complex thermal conduction. This makes the PT model particularly effective in capturing the autocatalytic characteristics of the reaction. However, the PT model is associated with high computational complexity. In contrast, the Sestak and Berggren (SB) model [54] offers a simplified description of the reaction mechanism, resulting in enhanced computational efficiency. Consequently, the SB model is well suited for large-scale or complex structural numerical simulations within specific resin systems and processing conditions. Therefore, in this paper, we chose to use the SB model to determine the curing kinetics of insulating varnishes. By combining the temperature-rate-dependent equation of Sestak and Berggren with the classical autocatalytic reaction model, the fitted curing kinetics model, represented by Equation (9), was obtained.

$$\frac{d\alpha}{dt} = Aexp\left(-\frac{E_a}{RT}\right)\alpha^m(1-\alpha)^n \tag{9}$$

where A represents the pre-exponential factor, and $m,\ n$ are the reaction orders.

The average reaction activation energy (E_a) obtained by the Starink method was substituted into Equation (5) and fitted using the least squares method to derive the kinetic parameters presented in Table 3.

Table 3. Kinetic parameters at different heating rates.

Heating Rate β (K/min)	ln$A$	m	n	R-Squared
5	25.6134	0.35678	1.22978	0.99772
10	25.72109	0.44207	1.26424	0.99917
15	25.72341	0.47535	1.24681	0.99984
20	25.70551	0.50076	1.26355	0.99992
Average	25.69085	0.44374	1.251095	0.99916

Ultimately, the kinetic model equation for the curing of this insulating paint can be expressed as Equation (10):

$$\frac{d\alpha}{dt} = e^{25.69085}exp\left(-\frac{95955.45}{RT}\right)\alpha^{0.44}(1-\alpha)^{1.25} \tag{10}$$

To verify the model's accuracy, the curing rate–temperature curves obtained from the model at different heating rates were compared with the experimentally obtained curves, as shown in Figure 5f. The model curves showed a higher degree of consistency with the experimental curves.

The glass transition temperature (T_g) is one of the most important thermodynamic parameters in the study of thermosetting resins [31]. When a material exceeds its glass transition temperature, its physical and chemical properties change significantly. Previous studies have demonstrated that the glass transition temperature increases as the degree of cure increases. Therefore, the glass transition temperatures of insulating varnishes at various degrees of cure were tested using modulated DSC, and DiBenedetto's equation was applied to identify these temperatures.

$$T_g = T_{g0} + \frac{\lambda\alpha}{1-(1-\lambda)\alpha}\left(T_{g\infty} - T_{g0}\right) \tag{11}$$

where T_{g0} represents the glass transition temperature of the uncured state, $T_{g\infty}$ denotes the glass transition temperature of the fully cured state, and λ is the fitting parameter.

First, five samples were thermostated at 90 °C for different durations and then rapidly cooled to −40 °C. The degree of cure was determined from the exothermic peak of the irreversible heat flow curve using MDSC. The samples were then heated from −40 °C to 200 °C at a rate of 3 K/min with a temperature amplitude of 1 K and a period of 60 s. The glass transition temperature (T_g) of each sample was determined from the post-elevated reversible heat flow curve, following ASTM E1356 (3 K/min, inflection point method). The relationship between the glass transition temperature and curing of insulating varnish determined by Equation (11) is shown in Figure 6.

The equation fitting yields an R^2 value of 0.99735, indicating a high degree of fit. The fitted parameter λ is 0.48, suggesting that the molecular chain segments of the insulating varnish possess a certain degree of mobility, resulting in a relatively low sensitivity of the glass transition temperature to the degree of cure.

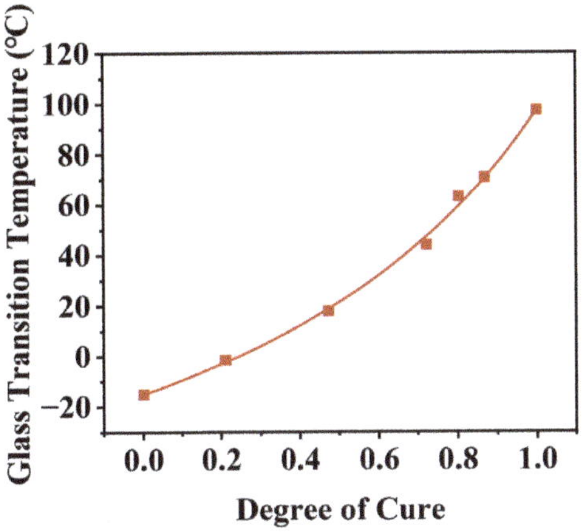

Figure 6. Fitting curve of the DiBenedetto equation.

3.1.2. Boundary Setup

In the simulation software, the ODE and Differential-Algebraic Equation (DAE) modules, in conjunction with the Heat Transfer module, can be used to solve the temperature-dependent curing kinetics equations by inputting the expression of Equation (1) directly into the ODE module.

The default equation form of the ODE module is Equation (12):

$$e_a \frac{\partial^2 alpha}{\partial t^2} + d_a \frac{\partial alpha}{\partial t} = f \quad (12)$$

where e_a is the mass coefficient, d_a is the damping coefficient, and f is the source term equation. In comparison to Equation (1), e_a is set to 0 and d_a is set to 1, and f is set to $e^{25.69085} exp\left(-\frac{95955.45}{RT}\right)\alpha^{0.44}(1-\alpha)^{1.25}$.

The initial value of the degree of cure is set to 1×10^{-20}, which helps prevent errors in the solver caused by computing the negative exponent of zero.

3.2. Heat Transfer Model

During the insulation curing process of the hair-pin motor stator, the heating device provides the thermal energy required for curing the insulating resin. Owing to the differing thermal conductivities of the flat copper wire and insulating varnish, the varnish is not heated uniformly. Since the curing process is exothermic, it further exacerbates the temperature non-uniformity. In addition, the material properties of the insulating varnish evolve with the degree of cure, and these factors collectively contribute to the generation of residual stresses and strains during the curing process.

3.2.1. Model Derivation and Parameter Determination

The thermal field during the insulating varnish curing process is essentially a nonlinear heat transfer problem with an internal heat source. This problem arises from the exothermic reaction of the unsaturated polyester imide resin curing in the insulating varnish [55]. This nonlinear heat transfer problem can be characterized by the Fourier heat transfer equation coupled with the curing exothermic Equation (13).

$$\rho C_p \frac{dT}{dt} = \frac{d}{dx}\left(K_{xx}\frac{dT}{dx}\right) + \frac{d}{dy}\left(K_{yy}\frac{dT}{dy}\right) + \frac{d}{dz}\left(K_{zz}\frac{dT}{dz}\right) + \rho \Delta H_t \frac{d\alpha}{dt} \quad (13)$$

where ρ represents the density, C_p is the specific heat, T is the temperature, ΔH_t is the total reaction heat, α is the degree of cure, and K_{xx}, K_{yy} and K_{zz} represent the thermal conductivities in the three directions.

Assuming that the insulating varnish material is isotropic, the thermal conductivities in all three directions are equal. This equation can be solved through the coupling of the Heat Transfer module with the Curing Kinetics module.

Thermal conductivity, specific heat capacity, and density are important material properties of insulating varnish materials. The properties of insulating varnish obtained from experimental measurements are presented in Table 4.

Table 4. Thermodynamic parameters of insulating varnish.

Material Properties	Unit	Before Curing	Cured
Thermal Conductivity $k(\alpha)$	W/(m·K)	0.21	0.23
Specific Heat Capacity $C_p(\alpha)$	J/(kg·K)	1908	1273
Density $\rho(\alpha)$	kg/m^3	1.12	1.169

The material properties of the insulating varnish are linearly related to the degree of cure, and the relationship between these parameters and the degree of cure can be expressed by Equation (14) [56].

$$\begin{cases} k(\alpha) = k_a(1-\alpha) + k_b\alpha \\ C_P(\alpha) = C_{Pa}(1-\alpha) + C_{Pb}\ \alpha \\ \rho(\alpha) = \rho_a(1-\alpha) + \rho_b\alpha \end{cases} \tag{14}$$

where subscript a represents the parameters before curing, and subscript b represents the parameters after curing.

3.2.2. Boundary Setup

During the curing process, an external oven heats the insulating varnish, and the heat transfer between the insulating varnish and the hot air occurs through natural convection, with the convective heat transfer described by Equation (15).

$$q_0 = h(T_{\text{ext}} - T) \tag{15}$$

where q_0 is the heat flux, h is the convective heat transfer coefficient, T_{ext} is the external temperature, and T is the temperature of the insulating varnish; T_{ext} can be defined as the heating temperature via a custom function.

During the drop-dip curing process, the insulating varnish directly comes into contact with the flat copper wires and heats them together, so the heat flux boundaries are all the boundaries in contact with the outside during the numerical simulation. The convective heat transfer coefficient h is set to 25 W/(m^2·K) based on an assumption. The heat source is insulating paint, and the heat source expression is the rightmost term $\rho\Delta H_t d\alpha/dt$ in Equation (11). And the reference temperature is set to a preset value of 80 °C.

3.3. Solid Mechanics Model

During the curing process, the material properties of the insulating varnish change with the degree of cure and temperature, leading to the generation of internal stresses at the interface between the flat copper wire and the insulating varnish. The curing process of the insulating varnish material undergoes a viscous flow state, a highly elastic state, and a glassy state. Cure shrinkage occurs during the highly elastic stage. During the cooling stage, the varnish shrinks unevenly due to structural differences, cooling rates, and the mismatch in the coefficients of thermal expansion of varnish and the stator, resulting in the formation of internal stresses, thereby affecting the insulation performance.

3.3.1. Model Derivation and Parameter Determination

The strain during the curing process of insulating varnish is composed of two parts: one part is the thermal strain induced by heat, and the other part is the shrinkage strain caused by the curing of the resin material, as represented in Equation (16):

$$\varepsilon = \varepsilon_t + \varepsilon_c \tag{16}$$

where ε represents the total strain, ε_t denotes the thermal strain, and ε_c refers to the cure shrinkage strain.

Studies have shown that the shrinkage of resin materials during curing occurs during the glassy state [18,29].

The thermal strain ε_t is a function of the temperature and the coefficient of thermal expansion, as expressed in Equation (17):

$$\varepsilon_t = \alpha(T)\left(T - T_{ref}\right) \tag{17}$$

where $\alpha(T)$ represents the coefficient of thermal expansion of the material at temperature T and T_{ref} denotes the reference temperature.

The coefficient of thermal expansion of the insulating varnish after curing can be measured using a linear thermal expansion meter, with the results presented in Figure 7.

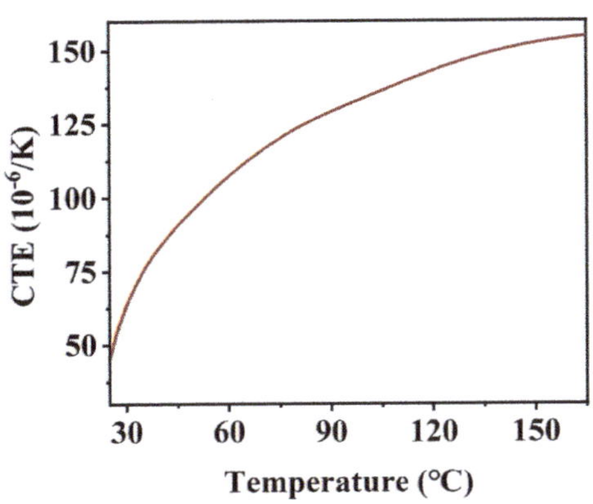

Figure 7. Thermal expansion coefficient curve of cured insulating varnish.

The coefficient of thermal expansion throughout the curing process is given by Equation (18) [57]:

$$\alpha_{CTE}(T,\alpha) = \begin{cases} \alpha^{v}_{\mathrm{CTE}}, & \alpha < \alpha_{\mathrm{gel}} \\ \alpha^{r}_{\mathrm{CTE}}, & \alpha \geq \alpha_{\mathrm{gel}}, \quad T \geq T_{\mathrm{g}}(\alpha) \\ \alpha^{g}_{\mathrm{CTE}}, & \alpha \geq \alpha_{T_g}, \quad T < T_{\mathrm{g}}(\alpha) \end{cases} \tag{18}$$

where $\alpha^{v}_{\mathrm{CTE}}$, $\alpha^{r}_{\mathrm{CTE}}$, and $\alpha^{g}_{\mathrm{CTE}}$ represent the thermal expansion coefficients for the viscous state, rubbery state, and glassy state, respectively, α_{gel} denotes the degree of cure at the gel point, and α_{T_g} represents the degree of cure at the glass transition point.

Assuming that the insulating varnish is isotropic, the shrinkage is uniform in all directions during the curing process. The following equation can describe the chemical shrinkage strain of the resin:

$$\varepsilon_c = \sqrt[3]{1 + \Delta v} - 1 \tag{19}$$

where ε_c represents the curing strain, and Δv denotes the volume change rate.

As shrinkage strain occurs mainly after the gel state; it can be expressed as a function of the degree of cure and total volume change ΔV, as shown in Equation (20):

$$\Delta v = \left(\alpha - \alpha_{gel}\right)\Delta V \tag{20}$$

where α represents the current degree of cure and α_{gel} denotes the degree of cure at the gel point.

Thus, the expression for the chemical shrinkage strain of the insulating varnish can be derived, as shown in Equation (21) [58].

$$\varepsilon_c = \sqrt[3]{1 + \left(\alpha - \alpha_{gel}\right)\Delta V} - 1 \tag{21}$$

The linear elastic constitutive model is commonly used to predict the curing deformation of composite materials. Its general form is given by:

$$\{\sigma\} = [E](\{\varepsilon\} - \{\varepsilon_0\}) + \{\sigma_0\} \tag{22}$$

where $\{\sigma\}$ and $\{\sigma_0\}$ represent the stress and initial stress to be solved, respectively; $[E]$ is the stiffness matrix; $\{\varepsilon\}$ and $\{\varepsilon_0\}$ are the total strain and initial strain in tensor form.

During the curing process of insulating resin materials, the material properties evolve, depending on both temperature and curing degree. Consequently, the formula in Equation (22) cannot fully capture the material behavior during curing. To address this limitation, the CHILE (Cure Hardening Instantaneous Linear Elastic) model is introduced, which incorporates the time-dependent characteristics of the elastic modulus to account for the dynamic changes in the material properties throughout the curing process.

Bogetti et al. proposed the CHILE (α) model, based on the curing degree, which assumes that the modulus of the resin varies linearly with the curing degree [58]. This model describes the evolution of the elastic modulus during the curing of composite materials by considering the material properties at different curing stages. The fundamental stress–strain relationship in this model is given by Equation (23):

$$\{\sigma\} = [Q(\alpha, T)]\{\varepsilon\} \tag{23}$$

where $[Q(\alpha, T)]$ is the time-dependent stiffness matrix, which is a function of both temperature and curing degree.

The Poisson's ratio ν of the insulating varnish is an important parameter in the stiffness matrix. To this end, uniaxial tensile tests were conducted on fully cured insulating varnish samples according to the ASTM D638 standard, and the deformation was measured using strain gauges. The measured Poisson's ratio of the fully cured insulating varnish was 0.37. It has been found that the variation in Poisson's ratio during curing has a negligible impact on the residual stress accumulation [59–61]. Therefore, in this study, for the sake of model simplification, the Poisson's ratio of the insulating varnish is assumed to be 0.37.

The change in the modulus during the curing process of insulating varnish can be divided into three stages: In the first stage, the epoxy resin behaves as a fluid with a low storage modulus. In the second stage, as the curing reaction progresses, the insulating varnish transitions from a viscoelastic state to a gel state, and then to a glassy state. The storage modulus in this stage is a function of the curing degree. In the third stage, after the curing reaction is complete, when the temperature is below the glass transition temperature, the resin undergoes minimal deformation under external forces. The storage modulus becomes large and remains almost constant. In this study, we used the CHILE (α) model to describe the modulus of elasticity of the resin during curing, E, as shown in Equation (24) [58,62]:

$$E = \begin{cases} E_r^0, & \alpha < \alpha_{gel} \\ \left(1 - \frac{\alpha - \alpha_{gel}}{\alpha_{gt} - \alpha_{gel}}\right) E_r^0 + \frac{\alpha - \alpha_{gel}}{\alpha_{gt} - \alpha_{gel}} E_r^\infty, & \alpha_{gel} < \alpha < \alpha_{gt} \\ E_r^\infty, & \alpha > \alpha_{gt} \end{cases} \tag{24}$$

E_r^0 and E_r^∞ represent the modulus of elasticity of the resin before and after curing, α_{gel} and α_{gt} denote the gel point cure and glass point cure of the resin.

Since the viscosity of the insulating varnish is not considered, the storage modulus is used as a substitute for the elastic modulus and it was determined through DMA using a TA Q800 instrument. The test was conducted at a frequency of 1 Hz with a three-point bending setup, and with a heating rate of 3 °C/min from 0 °C to 200 °C, and the resulting storage modulus and loss factor are shown in Figure 8. The insulating varnish was determined to have E_r^0 = 2.7 MPa and E_r^∞ = 2729 MPa.

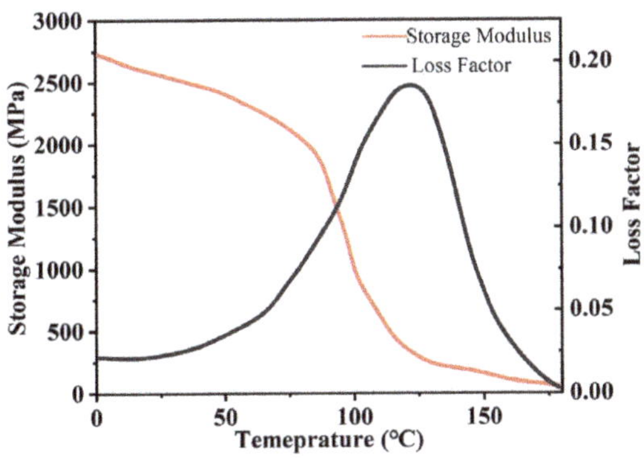

Figure 8. DMA test results.

3.3.2. Boundary Setup

Considering only the insulating varnish curing process on a single copper wire during the stator dripping process, the surface between the copper wire and the insulating varnish is set as a continuous boundary, with a fixed constraint applied on the other side of the copper wire. In the simulation software, the solid heat transfer and solid mechanics nodes can be used to simulate the thermal expansion of the cured insulating varnish. However, since the total strain ε during curing is the sum of the thermal strain ε_t due to thermal expansion and the contraction strain ε_c due to the curing of the insulating varnish, it is necessary to add an external strain sub-node under the solid mechanics–linear elastic material node. Equation (21) is incorporated into this node to simulate the cure shrinkage strain. In addition, to improve the convergence of the model, the "free rigid body inhibition" node can be added. This node functions to prevent the solid mechanics module from generating multiple non-converging solutions, and the deformation results from the simulation with this node better align with the real deformation results.

3.4. Simulation Setup

The simulation was conducted using COMSOL's automatic tetrahedral meshing. Differentiated mesh sizes were applied to the copper conductor, insulating varnish, and external stator slot regions. To ensure higher resolution at the interface between the insulating varnish and copper conductor, this region was assigned a maximum element size of 0.1 mm and a minimum of 0.02 mm, while the external stator slot used a maximum size of 0.4 mm and a minimum of 0.2 mm. The final mesh contained 2,112,621 elements. The mesh result is shown in Figure 9.

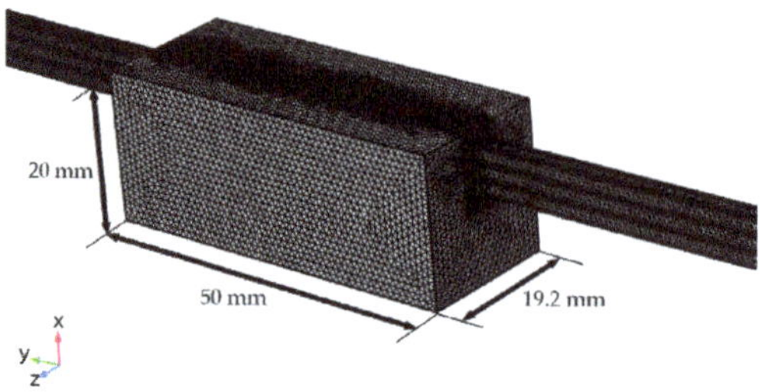

Figure 9. Finite element mesh for the simplified curing module.

The glass transition temperature of the material was determined using the DiBenedetto equation. The parameters for the equation, including λ, were fitted using glass transition temperatures obtained at various curing degrees through MDSC experiments. In the simulation, the glass transition temperature before curing (T_{g0}), after full curing (T_{gu}), and the parameter λ were defined.

To model the material behavior during curing, the elastic modulus and the coefficient of thermal expansion (CTE) were defined based on the curing state of the insulating paint. The CTE after curing was modeled as a temperature-dependent function, fitted to the experimental thermal expansion data using an analytical function in the software. Additionally, the heat capacity, thermal conductivity, and density were defined as functions of the degree of cure, consistent with the assumptions outlined in the manuscript.

For the curing shrinkage, the curing strain was defined as a function of the degree of cure based on Equation (21). This strain was implemented as an external strain in the solid mechanics module to simulate the effects of curing shrinkage. The curing volumetric shrinkage was calculated based on the density change before and after curing.

The simulation in this study uses a step-by-step solution approach. The transient solver in COMSOL is applied to solve the bidirectional coupling of the curing kinetics model and the heat transfer model. Then, the steady-state solver is used to inherit the solution from the transient solver to calculate the solid mechanics model.

Given the spatial effects involved in the curing process, a 3D model was employed to ensure accurate simulation. A 2D model would not capture the asymmetry of the stator slots or the heat transfer characteristics effectively. The use of a 3D model provides a more realistic representation of the curing process. While the results are presented in 2D plots, they clearly show the temperature and stress distribution within the stator slots.

4. Results Discussion

Using the aforementioned physical field setup and the material properties obtained from experimental data, a Multiphysics-coupled simulation under the same conditions as the experiment is conducted. The temperature and strain results from both the experiment and simulation are compared to analyze the stress–strain behavior during the insulating varnish curing process.

4.1. Simulation Results Validation

The temperature–time curve and strain–time curve of the simulation results were compared with the experimental values, as shown in Figure 10. The comparison indicates that the simulation model can generally reflect the strain and temperature conditions of the actual curing process of the insulating paint. Some data points do not fully match the experimental data because the strain test curve was obtained through decoupled calculations, and approximations in the calculation process may have affected some values. Additionally, some material properties were derived from empirical formulas, which may not correspond to the actual values during the curing process. These discrepancies can be addressed through subsequent optimization of the model.

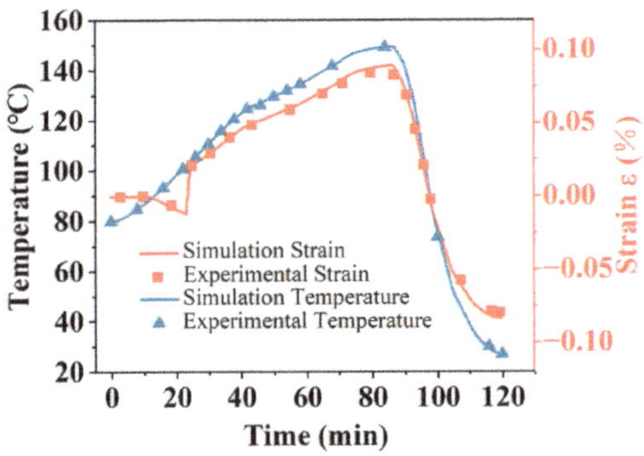

Figure 10. Comparison of strain and temperature simulation curves with experimental data.

4.2. Results Analysis

Figure 11 shows the stress distribution of the copper wire cross-section at the end of curing and after cooling to room temperature. The figure illustrates that the cured insulating varnish experiences the highest stress at the interface with the copper wire, reaching 45.1 MPa, with a difference of nearly 40 MPa compared to the minimum stress within the insulating varnish, indicating a stress concentration phenomenon. This is a key factor contributing to insulation defects.

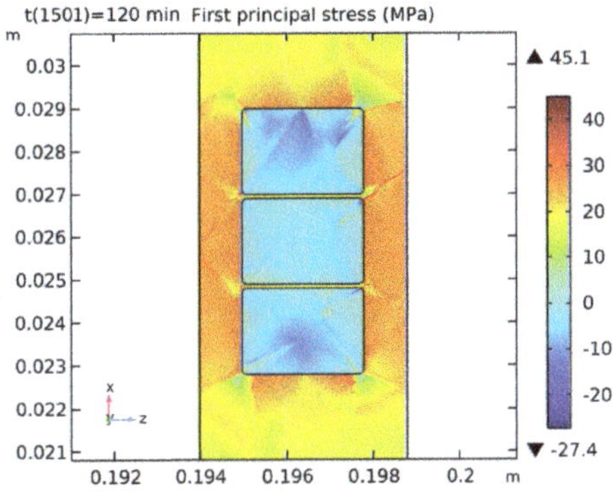

Figure 11. Stress distribution in the cross-section of copper wire and insulating varnish.

Figures 12 and 13 show the distribution of the curing degree and temperature in the initial stage, mid-stage, complete curing, and final stage of the insulating varnish curing process.

In the initial stage of curing, the temperature of the insulating varnish near the copper wires rises first due to the high thermal conductivity of the copper, allowing the varnish to reach the external preset temperature more quickly. During the 15–31 min, the varnish begins curing, with a temperature distribution that is lower at the edges and higher in the center. By 83 min, when the external preset temperature reaches 150 °C and has been maintained for almost 10 min, the temperature distribution becomes more uniform. At this point, the varnish is nearly fully cured, and the temperature is close to the preset value. In the cooling phase after 83 min, the varnish near the copper wire cools faster, similar to the heating phase.

The curing and crosslinking of the varnish are temperature driven, so the distribution of the curing degree is similar to that of the temperature. As the temperature increases, the varnish near the copper wire begins to cure first, with the curing degree following a stepwise pattern. By 31 min, the curing reaction weakens or disappears, and the curing degree is nearly complete. Although the temperature starts to drop after 83 min, the low thermal conductivity of the varnish causes a noticeable thermal lag during both the heating and cooling phases. As a result, the temperature and curing degree distributions align more closely, reaching a steady state and indicating that the curing process is nearing completion.

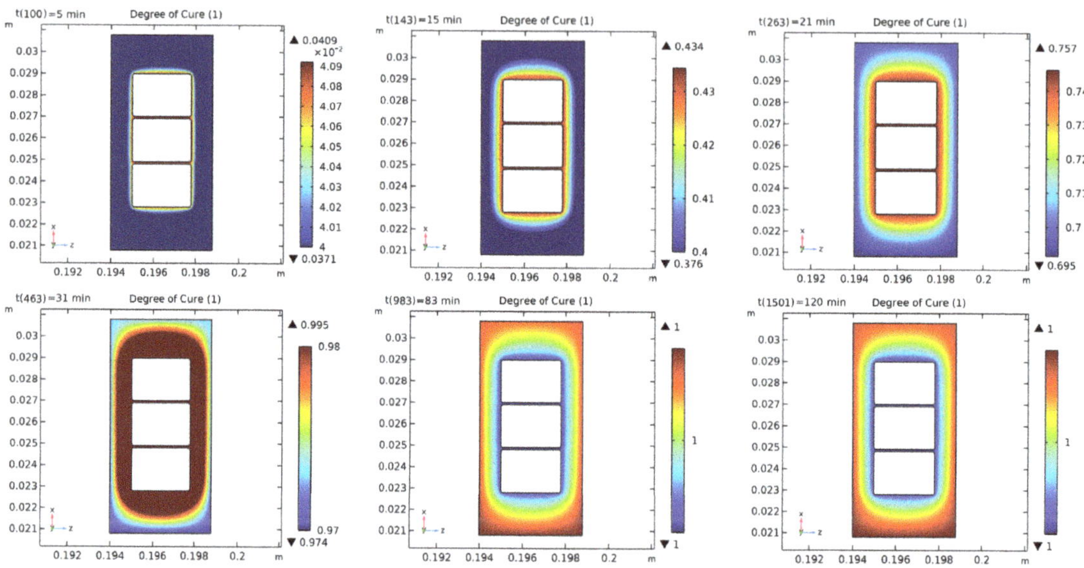

Figure 12. Degree of cure distribution in the cross-section of copper wire and insulating varnish.

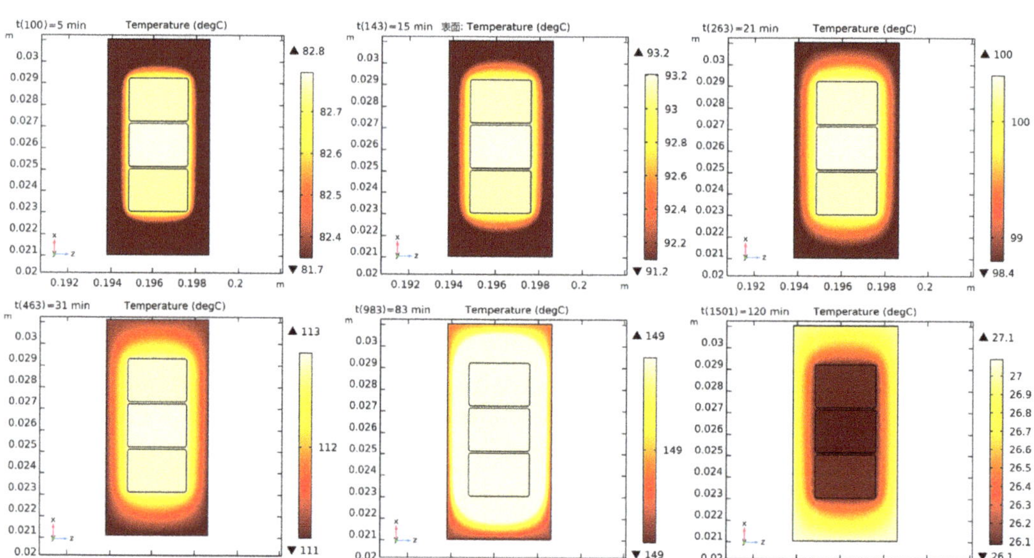

Figure 13. Temperature distribution in the cross-section of copper wire and insulating varnish.

The strains in the curing process of insulating varnish primarily consist of the shrinkage strain generated during curing and the thermal strain caused by temperature changes. While the material composition influences the shrinkage strain, the thermal strain is affected by the material's thermal expansion coefficient and the temperature difference. This paper presents a preliminary study of the effect of the cure temperature on the cure stress.

The curing process was modified by adjusting the maximum temperature of the post-rise phase to 130 °C. The resulting stress distribution is shown in Figure 14a. The results show that when the maximum cure temperature is reduced to 130 °C, the maximum stress value decreases to 38.6 MPa, representing a 14.4% reduction.

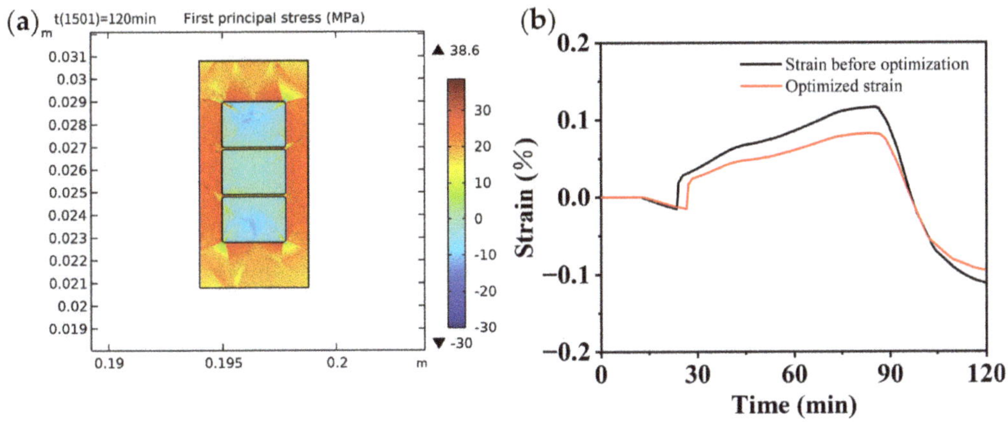

Figure 14. (**a**) Stress distribution; (**b**) strain curve of measurement points.

Figure 14b shows the strain curves of the measured points after optimization. Following the reduction in the post-heating stage temperature, the glass transition of the insulating varnish exhibits hysteresis, and the strains observed during and at the end of curing are smaller than those before optimization. This indicates that the post-heating stage temperature has a significant effect on the residual stress.

5. Conclusions

In this study, the curing process of insulating varnish for a hair-pin motor stator was monitored with respect to temperature and strain using FBG sensors. The curing kinetics of the insulating varnish material was investigated, and a simulation model of the process was developed using the finite element method, leading to the following conclusions:

(1) Both strain and temperature influence FBG sensors, and the temperature can be monitored using packaged FBG sensors. By analyzing the wavelength shift due to temperature changes, it is possible to decouple the strain variations. Through non-isothermal differential scanning calorimetry (DSC) experiments, the curing kinetics model of Voltatex® 4200 insulating varnish, primarily composed of unsaturated polyimide resin, was investigated. The curing reaction of this insulating varnish is characterized as a self-catalyzed reaction with a relatively large pre-exponential factor, allowing for rapid curing. Additionally, a comparison was conducted between the curing kinetics model obtained using the Starink method and the experimental data, which demonstrated a high degree of consistency.

(2) Based on the material properties of the insulating varnish after curing and the DSC test data, a three-dimensional simulation model of the curing process was established to analyze the temperature, degree of cure, and stress distribution. Given that the thermal conductivity of the flat copper wire is significantly higher than that of the resin and the silicon steel stator slot, there is a pronounced hysteresis phenomenon in heat transfer within the stator slot during curing. The temperature and degree of cure of the insulating varnish near the flat copper wire are greater than those near the stator slot.

In terms of the strain field, thermal strain is the primary influencing factor as curing nears completion, while shrinkage strain predominates during the initial stages of curing. The temperature and degree of cure of the insulating varnish near the flat copper wire are higher than those near the stator slot. Regarding the strain field, the strain in the insulating varnish during the initial curing phase is primarily shrinkage strain. In contrast, thermal strain becomes the primary influencing factor as curing approaches completion.

(3) After curing, the maximum stress occurs at the interface between the flat copper wire and the insulating varnish, with a maximum stress value of 45.1 MPa. There is a significant phenomenon of stress concentration, which is a major factor contributing to insulation defects. By modifying the temperature during the post-heating stage and using the established simulation model for stress prediction, the results indicate that the temperature of the post-heating stage also affects the residual stress. Specifically, when the maximum temperature during the post-heating stage is reduced to 130 °C, the maximum residual stress decreases by 14.4%.

However, there are some limitations to the present study that should be considered for future research:

(1) The relationships between density, constant pressure specific heat, and thermal conductivity with the degree of cure are based on assumptions, which may limit the accuracy of the predictions.

(2) The influence of flow consolidation, which could affect the material's behavior during curing, was not incorporated into the simulation model.

(3) The simulation model employs a modified linear elastic constitutive model to represent the material's behavior, which replaces the viscoelastic model. This simplification might not fully capture the time-dependent behavior of the material during the curing process.

(4) The long-term effects, such as thermal fatigue, were not specifically investigated. Thermal fatigue can be a critical factor in materials subjected to cyclic thermal loading, as the mismatched thermal expansion between the copper wire and the insulating varnish can lead to further stress accumulation and degradation over time. This could affect the material's long-term performance and reliability, which warrants further investigation in future research.

Author Contributions: Conceptualization, L.S. and Y.Z.; methodology, H.G. and M.M.; software, M.M. and Y.Z.; validation, M.M., X.S., and H.G.; formal analysis, M.M. and J.L.; investigation, J.L. and C.L.; resources, L.S. and Y.Z.; data curation, Y.H.; writing—original draft preparation, M.M.; writing—review and editing, L.S. and Y.Z.; visualization, M.M. and X.Z.; supervision, L.S., Y.Z., and H.G.; project administration, L.S. and Y.Z.; funding acquisition, L.S. and Y.Z. All authors have read and agreed to the published version of the manuscript.

Funding: The research in this paper was supported by two foundations: (i) the Jilin Province and Changchun City Major Science and Technology Special, Project: 20220301029GX; (ii) the Jilin Province and Changchun City Major Science and Technology Special, Project: 20230301005ZD; (iii) Scientific Research Project of Education Department of Guangdong Province, Project: 2022KCXTD029.

Institutional Review Board Statement: Not applicable.

Data Availability Statement: The original contributions presented in this study are included in the article. Further inquiries can be directed to the corresponding authors.

Acknowledgments: The authors extend their gratitude to Shiyanjia Lab (www.shiyanjia.com) for providing invaluable assistance with the DSC analysis. Jiabin Li from the College of Chemistry, Jilin University, is acknowledged for his assistance with material analysis.

Conflicts of Interest: Authors Hongyi Gan, Xiao Shang and Jingru Liu were employed by the company: "FAW Tooling Die Manufacturing Co., Ltd., Changchun 130000, China". Authors Chunbai Liu and Yanzhong Hao were employed by the company: "FAW Casting Co., Ltd., Changchun 130013, China". The remaining authors declare that the research was conducted in the absence of any commercial or financial relationships that could be construed as a potential conflict of interest.

References

1. Li, Q.; Wang, S.; Yu, C.; Cang, Y.; Yuan, C. Study on Paint Removal Welding Process of Flat Wire Motors for New Energy Vehicles. *Automob. Technol. Mater.* **2023**, 1–7. [CrossRef]
2. Juergens, J.; Fricasse, A.; Marengo, L.; Gragger, J.; De Gennaro, M.; Ponick, B.; IEEE. Innovative Design of an Air Cooled Ferrite Permanent Magnet Assisted Synchronous Reluctance Machine for Automotive Traction Application. In Proceedings of the 22nd IEEE International Conference on Electrical Machines (ICEM), Lausanne, Switzerland, 4–7 September 2016; pp. 803–810.
3. Wang, X.; Wand, L. Review of Performance Test Research on Enamelled Copper Flat Wire for New Energy Vehicle Drive Motors. *Wire Cable* **2024**, *67*, 8–14. [CrossRef]
4. Calabrese, E.; Raimondo, M.; Catauro, M.; Vertuccio, L.; Lamberti, P.; Raimo, R.; Tucci, V.; Guadagno, L. Thermal and Electrical Characterization of Polyester Resins Suitable for Electric Motor Insulation. *Polymers* **2023**, *15*, 1374. [CrossRef] [PubMed]
5. Budrugeac, P.; Cucos, A.; Dascalu, R.; Paraschiv, C.; Mitrea, S.; Sbarcea, B.G. The Use of Thermal Analysis Methods for Predicting the Thermal Endurance of an Epoxy Resin Used as Electrical Insulator. *J. Therm. Anal. Calorim.* **2021**, *146*, 1791–1801. [CrossRef]
6. Fetouhi, L.; Martinez-Vega, J.; Petitgas, B.; IEEE. Dielectric and Mechanical Properties Correlated to Physico-Chemical Characteristics of a Polyester-Imide Resin Used in Rotating Machines Insulations. In Proceedings of the IEEE International Conference on Dielectrics (ICD), Montpellier, France, 3–7 July 2016.
7. Frost, N.; IEEE. IEEE Standards Related to Materials Used in Motors and Generators. In Proceedings of the IEEE Electrical Insulation Conference (EIC), Knoxville, TN, USA, 19–23 June 2022.
8. Jiang, Q.B.; Deng, Q.S.; Gao, J.; Wang, Y.Q. Study on the HEA-blocking Unsaturated Polyester Solventless Impregnating Varnish. In Proceedings of the China Academic Conference on Green Printing and Packaging Materials, Harbin, China, 20–23 August 2011.
9. Goetter, R.; Winkeler, M. New developments in unsaturated polyester resins used for electrical insulation. In Proceedings of the Proceedings: Electrical Insulation Conference and Electrical Manufacturing and Coil Winding Conference (Cat. No.01CH37264), Cincinnati, OH, USA, 18 October 2001.
10. Fetouhi, L.; Martinez-Vega, J.; Petitgas, B. Electric Conductivity, Aging and Chemical Degradation of Polyesterimide Resins Used in the Impregnation of Rotating Machines. *IEEE Trans. Dielectr. Electr. Insul.* **2018**, *25*, 294–305. [CrossRef]
11. Sun, J.H.; Qian, L.J.; Li, J. Toughening and Strengthening Epoxy Resin with Flame Retardant Molecular Structure Based on Tyrosine. *Polymer* **2021**, *230*, 124045. [CrossRef]
12. Sun, W.; Wang, L.Z.; Zhu, N.; Xin, J.Y.; Luo, Y.C.; Jiang, X.R.; Fan, G.H.; Chen, M. Characterization of Packaging Warpage, Residual Stress and Their Effects on the Mechanical Reliability of IGBT Power Modules. *Eng. Fail. Anal.* **2023**, *152*, 107517. [CrossRef]
13. Xing, Y.; Wang, Z.; Liu, L.; Xu, Y.; Yang, Y.; Liu, S.; Zhou, F.; He, S.; Li, C. Defects and Failure Types of Solid Insulation in Gas-insulated Switchgear: In Situ Study and Case Analysis. *High Volt.* **2022**, *7*, 158–164. [CrossRef]
14. Li, J.; Liang, H.; Chen, Y.; Du, B. Promising Functional Graded Materials for Compact Gaseous Insulated Switchgears/Pipelines. *High Volt.* **2020**, *5*, 231–240. [CrossRef]
15. Dewangan, B.; Chakladar, N.D. Modelling of Residual Stress During Curing of A Polymer Under Autoclave Conditions and Experimental Validation. *Comput. Mater. Sci.* **2024**, *241*, 113038. [CrossRef]
16. Hao, Y.P.; Zhang, Y.Y.; Shen, Z.K.; Du, D.Y.; Huang, S.L.; Liu, L.; Huang, L. Fiber Bragg Grating Detection of Gel Time and Residual Strain of Curing in Epoxy Resins. *Polym. Eng. Sci.* **2024**, *64*, 3101–3108. [CrossRef]
17. Yang, L.; Ding, A.X.; Xu, M.; Li, Y.F.; Zhao, X.H.; Peng, J.X.; Li, X. Characterization of Potting Epoxy Resins Performance Parameters Based on a Viscoelastic Constitutive Model. *Polymers* **2024**, *16*, 930. [CrossRef] [PubMed]
18. Wang, C.; Zhu, W.Y.; Zhou, G.; Wang, Y.; Chen, C.; Li, X.; Zhang, Z.Q. Mechanism of Stress Generation at the Epoxy-Epoxy Interface. *Colloids Surf. A-Physicochem. Eng. Asp.* **2024**, *685*, 133209. [CrossRef]
19. Qu, C.Y.; Zhang, X.; Wang, D.Z.; Fan, X.P.; Li, H.F.; Liu, C.W.; Feng, H.; Wang, R.K.; Guo, K.; Tian, Y.A.; et al. Residual Stress and Thermal Properties of Rubber-Modified Epoxy Systems for Semiconductor Package. *J. Appl. Polym. Sci.* **2022**, *139*, 51786. [CrossRef]
20. Ye, Y.Y.; Zhang, F.; Chen, Q.S.; Xu, Q.; Ke, Y.L. Viscosity-Coupled Curing Simulation Model for L-Shaped Composites Considering Flow and Compaction. *Polym. Compos.* **2024**, *45*, 12932–12947. [CrossRef]
21. Balaji, A.; Sbarufatti, C.; Dumas, D.; Parmentier, A.; Pierard, O.; Cadini, F. Prediction of Shape Distortions in Thermosetting Composite Parts Using Neural Network Interfaced Visco-Elastic Constitutive Model. *J. Compos. Mater.* **2024**, *58*, 1439–1459. [CrossRef]
22. Wu, R.; Zhou, W.; Fan, Z.H.; Zhou, W.X.; Xiong, Y.J. Analysis of the Curing Deformation of Polyurethane Composite Solar Cell Bezels. *Symmetry* **2024**, *16*, 463. [CrossRef]
23. Yu, X.; Wen, W.; Chunchu, T.; Cuicui, L.; Shaobo, H.; Chen, W. Application of Newly Developed Epoxy-Anhydride VPI resin for High Voltage Motors and Generators. In Proceedings of the 2015 IEEE Electrical Insulation Conference (EIC), Seattle, WA, USA, 7–10 June 2015; pp. 511–514.
24. Han, Y.X.; Shi, X.T.; Yang, X.T.; Guo, Y.Q.; Zhang, J.L.; Kong, J.; Gu, J.W. Enhanced Thermal Conductivities of Epoxy Nanocomposites Via Incorporating In-Situ Fabricated Hetero-Structured SiC-BNNS Fillers. *Compos. Sci. Technol.* **2020**, *187*, 107944. [CrossRef]
25. Guo, R.; Gu, Y.; Zhou, Y.; Wang, S.; Li, M. On-Line Cure Monitoring of Phenol Formaldehyde Resin Using Embedded Fiber Bragg Grating Sensor. *Mater. Today Commun.* **2024**, *39*, 109114. [CrossRef]

26. Shen, X.Y. Cure Monitoring of Composites Based on Embedded Fiber Bragg Gratings. In Proceedings of the International Conference on Mechatronics and Intelligent Materials, Lijiang, China, 21–22 May 2011.
27. Case, S.L.; O'Brien, E.P.; Ward, T.C. Cure Profiles, Crosslink Density, Residual Stresses, and Adhesion in a Model Epoxy. *Polymer* **2005**, *46*, 10831–10840. [CrossRef]
28. Ortega, A.M.; Kasprzak, S.E.; Yakacki, C.M.; Diani, J.; Greenberg, A.R.; Gall, K. Structure-Property Relationships in Photopolymerizable Polymer Networks: Effect of Composition on the Crosslinked Structure and Resulting Thermomechanical Properties of a (meth)acrylate-Based System. *J. Appl. Polym. Sci.* **2008**, *110*, 1559–1572. [CrossRef]
29. Wang, X.X. Numerical Simulation of Curing Deformation of Thermosetting Resin Matrix Composites. Ph.D. Thesis, Shandong University, Jinan, China, 2012.
30. Hardis, R.; Jessop, J.L.P.; Peters, F.E.; Kessler, M.R. Cure Kinetics Characterization and Monitoring of an Epoxy Resin Using DSC, Raman Spectroscopy, and DEA. *Compos. Part A-Appl. Sci. Manuf.* **2013**, *49*, 100–108. [CrossRef]
31. Vyazovkin, S.; Sbirrazzuoli, N. Isoconversional Kinetic Analysis of Thermally Stimulated Processes in Polymers. *Macromol. Rapid Commun.* **2006**, *27*, 1515–1532. [CrossRef]
32. Yang, C.; Jiang, X.L.; Sun, K. Isothermal Cure Kinetics of Epoxy Resin/Montmorillionite Composites Cured by Acid Anhydride. *Acta Phys.-Chim. Sin.* **2005**, *21*, 681–685. [CrossRef]
33. Schafran, B. Chemorheology of Bismaleimide Resins. Master's Thesis, Polytechnic University, New York, NY, USA, 1990.
34. Sultania, M.; Rai, J.S.P.; Srivastava, D. Kinetic Modeling of Esterification of Cardanol-Based Epoxy Resin in the Presence of Triphenylphosphine for Producing Vinyl Ester Resin: Mechanistic Rate Equation. *J. Appl. Polym. Sci.* **2010**, *118*, 1979–1989. [CrossRef]
35. Rosu, D.; Cascaval, C.N.; Mustata, F.; Ciobanu, C. Cure Kinetics of Epoxy Resins Studied by Non-Isothermal DSC Data. *Thermochim. Acta* **2002**, *383*, 119–127. [CrossRef]
36. Wang, J.W.; Laborie, M.P.G.; Wolcott, M.P. Comparison of Model-Fitting Kinetics for Predicting the Cure Behavior of Commercial Phenol-Formaldehyde Resins. *J. Appl. Polym. Sci.* **2007**, *105*, 1289–1296. [CrossRef]
37. Phansalkar, S.P.; Han, B.T. Deterministic Approach to Obtain Autocatalytic Cure Kinetics Model Constants by Normal Equations of Least-Squares Method. *IEEE Trans. Compon. Packag. Manuf. Technol.* **2024**, *14*, 743–746. [CrossRef]
38. Wang, C.S.; Kwag, C. Cure Kinetics of an Epoxy-Anhydride-Imidazole Resin System by Isothermal DSC. *Polym. Polym. Compos.* **2006**, *14*, 445–454. [CrossRef]
39. Jankovic, B. The Kinetic Analysis Of Isothermal Curing Reaction of an Unsaturated Polyester Resin: Estimation of the Density Distribution Function of the Apparent Activation Energy. *Chem. Eng. J.* **2010**, *162*, 331–340. [CrossRef]
40. Málek, J. Kinetic Analysis of Crystallization Processes in Amorphous Materials. *Thermochim. Acta* **2000**, *355*, 239–253. [CrossRef]
41. Starink, M.J. A New Method for the Derivation of Activation Energies from Experiments Performed at Constant Heating Rate. *Thermochim. Acta* **1996**, *288*, 97–104. [CrossRef]
42. Chen, D.; Gao, X.; Dollimore, D. A Generalized form of the Kissinger equation. *Thermochim. Acta* **1993**, *215*, 109–117. [CrossRef]
43. Ozawa, T. Kinetic Analysis of Derivative Curves in Thermal Analysis. *J. Therm. Anal.* **1970**, *2*, 301–324. [CrossRef]
44. Ozawa, T. Estimation of Activation Energy by Isoconversion Methods. *Thermochim. Acta* **1992**, *203*, 159–165. [CrossRef]
45. Starink, M.J. The Determination of Activation Energy from Linear Heating Rate Experiments: A Comparison of the Accuracy of Isoconversion Methods. *Thermochim. Acta* **2003**, *404*, 163–176. [CrossRef]
46. Kandelbauer, A.; Wuzella, G.; Mahendran, A.; Taudes, I.; Widsten, P. Model-Free Kinetic Analysis of Melamine–Formaldehyde Resin Cure. *Chem. Eng. J.* **2009**, *152*, 556–565. [CrossRef]
47. Kada-Benameur, H.; Wirquin, E.; Duthoit, B. Determination of Apparent Activation Energy of Concrete by Isothermal Calorimetry. *Cem. Concr. Res.* **2000**, *30*, 301–305. [CrossRef]
48. Cai, H.; Li, P.; Sui, G.; Yu, Y.; Li, G.; Yang, X.; Ryu, S. Curing Kinetics Study of Epoxy Resin/Flexible Amine Toughness Systems by Dynamic and Isothermal DSC. *Thermochim. Acta* **2008**, *473*, 101–105. [CrossRef]
49. Kamal, M.R. Thermoset characterization for moldability analysis. *Polym. Eng. Sci.* **1974**, *14*, 231–239. [CrossRef]
50. Robertson, I.D.; Yourdkhani, M.; Centellas, P.J.; Aw, J.E.; Ivanoff, D.G.; Goli, E.; Lloyd, E.M.; Dean, L.M.; Sottos, N.R.; Geubelle, P.H.; et al. Rapid Energy-Efficient Manufacturing of Polymers and Composites via Frontal Polymerization. *Nature* **2018**, *557*, 223–227. [CrossRef] [PubMed]
51. Goli, E.; Parikh, N.A.; Yourdkhani, M.; Hibbard, N.G.; Moore, J.S.; Sottos, N.R.; Geubelle, P.H. Frontal Polymerization of Unidirectional Carbon-Fiber-Reinforced Composites. *Compos. Part A Appl. Sci. Manuf.* **2020**, *130*, 105689. [CrossRef]
52. Tarafdar, A.; Jia, C.; Hu, W.; Hosein, I.D.; Fu, K.; Wang, Y. Three-Dimensional Modeling of Frontal Polymerization for Rapid, Efficient, and Uniform Thermoset Composites Manufacturing. *Compos. Part B Eng.* **2023**, *266*, 111029. [CrossRef]
53. Wang, Y.Q. Modeling the Through-Thickness Frontal Polymerization of Unidirectional Carbon Fiber Thermoset Composites: Effect of Microstructures. *J. Appl. Polym. Sci.* **2022**, *139*, e52735. [CrossRef]
54. Šesták, J.; Berggren, G. Study of the Kinetics of the Mechanism of Solid-State Reactions at Increasing Temperatures. *Thermochim. Acta* **1971**, *3*, 1–12. [CrossRef]
55. Wang, C.; Zhou, G.; Sun, Q.; Chen, C.; Zhang, Z.; Li, H.; Peng, Z. Numerical Simulation of the Interfacial Stress Between Epoxy Resin and Metal Conductor of Power Equipment During Epoxy Curing. *High Volt.* **2022**, *7*, 903–915. [CrossRef]
56. Chern, B.-C.; Moon, T.J.; Howell, J.R. Measurement of the Temperature and Cure Dependence of the Thermal Conductivity of Epoxy Resin. *Exp. Heat Transf.* **1993**, *6*, 157–174. [CrossRef]

57. Li, X.; Wang, J.; Li, S.; Ding, A. Cure-Induced Temperature Gradient in Laminated Composite Plate: Numerical Simulation and Experimental Measurement. *Compos. Struct.* **2020**, *253*, 112822. [CrossRef]
58. Bogetti, T.A.; Gillespie, J.W. Process-Induced Stress and Deformation in Thick-Section Thermoset Composite Laminates. *J. Compos. Mater.* **1992**, *26*, 626–660. [CrossRef]
59. Feng, Y.C.; Zheng, W.K.; Zhao, Z.Z.; Liang, B.; Ye, H.T.; Ge, Q.; Zhang, X.Y.; Zhang, W.Z. Characterization and Finite Element Modeling for Thermoset Resin of Carbon Fiber Prepregs During Curing. *J. Manuf. Sci. Eng.-Trans. ASME* **2022**, *144*, 081007. [CrossRef]
60. O'Brien, D.J.; Mather, P.T.; White, S.R. Viscoelastic Properties of an Epoxy Resin During Cure. *J. Compos. Mater.* **2001**, *35*, 883–904. [CrossRef]
61. Kim, Y.K.; White, S.R. Stress Relaxation Behavior of 3501-6 Epoxy Resin During Cure. *Polym. Eng. Sci.* **1996**, *36*, 2852–2862. [CrossRef]
62. Dai, J.F.; Xi, S.B.; Li, D.N. Numerical Analysis of Curing Residual Stress and Deformation in Thermosetting Composite Laminates with Comparison between Different Constitutive Models. *Materials* **2019**, *12*, 572. [CrossRef] [PubMed]

polymers

Article

Constructing a Broad-Pore-Domain Structure of Adsorbents for Acteoside Adsorption

Weibo Ru, Jiaxing Liu, Feng Xiong, Yu Sun, Yong Zhang, Yipei Li, Yin Lv * and Xueqin Li *

School of Chemistry and Chemical Engineering/State Key Laboratory Incubation Base for Green Processing of Chemical Engineering, Shihezi University, Shihezi 832003, China
* Correspondence: ag_125@163.com (Y.L.); lixueqin861003@163.com (X.L.)

Abstract: Acteoside (ACT) is an important medicinal component, but its content is scarce. To obtain higher purity of ACT, the adsorption method was used to purify it. In this study, a broad-pore-domain hyper-crosslinked polymer (BHP-Kae) was prepared to adsorb ACT from *Cistanche tubulosa*, which is a medicinal plant. BHP-Kae-3 possessed a unique broad-pore-domain structure. This structure reduced the transfer resistance of ACT and facilitated the rapid diffusion of ACT into BHP-Kae-3, increasing the adsorption capacity. In addition, the surface and pore channels of BHP-Kae-3 contained abundant functional groups (-OH, C=O), which provided a large number of adsorption sites and facilitated ACT adsorption, thereby improving selectivity. The experimental results showed that BHP-Kae-3 exhibited a good adsorption capacity for ACT; the adsorption capacity was 105.12 mg/g, and the selectivity was 3.41. This study demonstrates the potential for efficient separation of natural products using broad-pore-domain adsorbents.

Keywords: hyper-crosslinked polymer; mesopore; separation; bioactive component

1. Introduction

Cistanche tubulosa is a perennial parasitic plant, which is widely distributed in Inner Mongolia, Ningxia, Gansu, Qinghai, and Xinjiang [1–5]. It is known for its distinctive flowers and unique method of obtaining nutrients by parasitizing other plants. The stems of *Cistanche tubulosa* are highly valued for their medicinal properties and were recorded in ancient Chinese medical literature such as *Shengnong*'s Classic of Materia Medica and *Compendium of Materia Medica* [6]. In recent times, the medicinal value of *Cistanche tubulosa* has been further explored through pharmacological studies [7]. *Acteoside* (ACT) (Figure S1) in *Cistanche tubulosa* has been found to possess a wide range of therapeutic effects [8]. These include neuroprotective properties [9,10], immune system modulation, anti-aging effects, prevention of osteoporosis, intestinal laxative properties, and liver protection [11–16]. There are various methods for extracting the active ingredients of ACT from *Cistanche tubulosa*. Conventional methods like chromatographic separation and extraction have been used [17,18], but they have shown limitations in terms of yield. For example, previous studies by Li et al. using high-speed counter-current chromatography obtained low yields of ACT [19]. Similarly, Nie et al. used deep eutectic solvent to extract ACT, and the yield was only 2.13 mg/g [20]. However, a more promising method for extracting bioactive components from natural products is adsorption [21]. This technique has been proven to be efficient, energy-saving, easy to operate, and environmentally friendly. Research using this method has the potential to improve ACT yield and purity.

The adsorption performance of an adsorbent primarily hinges on its physical and chemical structure [22–24]. Porous organic polymers are a novel type of porous material

 https://doi.org/10.3390/polym17010079

formed through covalent bonding of organic monomers [25,26]. Thanks to their excellent chemical and thermal stability, high specific surface area, and customizable pore structure, they have garnered significant attention in recent times [27]. In particular, hyper-crosslinked polymers possess the advantages of being highly stable [28,29], having a high specific surface area and low density [30], making them an incredibly desirable adsorbent [31,32]. Their stability ensures that they do not degrade in harsh environments or with prolonged use. Their high specific surface area allows for more adsorption sites, thereby increasing their adsorption capacity. Hyper-crosslinked polymers are easy to handle and transport due to their low density, which makes them lightweight [33]. From a synthesis standpoint, hyper-crosslinked polymers can be rapidly produced by a one-step Friedel–Crafts alkylation reaction, and the reaction uses inexpensive reagents and mild reaction conditions [34]. The production of hyper-crosslinked polymers not only reduces the production costs, but also minimizes the environmental effect associated with the use of hazardous reagents and energy-intensive processes [35]. Overall, the unique physical and chemical properties of hyper-crosslinked polymers and their ease of synthesis make them a highly attractive adsorbent for various applications, such as gas separation [36], water purification [37,38], and environmental remediation [39–41]. Researchers and industries are increasingly focusing on exploring the potential of hyper-crosslinked polymers and developing new techniques to further enhance their adsorption properties [28].

Kaempferol is a naturally occurring flavonol that contains high concentrations of phenolic hydroxyl groups [42,43], which contribute to its antioxidant properties [44]. Due to its polyphenol structure, it is an ideal monomer for the synthesis of hydrophilic hydroxyl hyper-crosslinked polymers (BHP-Kae) [45,46]. BHP-Kae, synthesized via a Friedel–Crafts alkylation reaction, possesses a broad-pore-domain structure. This unique structure is conducive to reducing the transfer resistance of ACT into BHP-Kae, facilitating rapid inward diffusion of ACT and improving the adsorption efficiency of BHP-Kae. Furthermore, the surface of BHP-Kae is characterized by the presence of hydroxyl groups and benzene rings. These functional groups have a significant effect on enhancing the interaction forces between BHP-Kae and ACT. This study also explored the effects of different adsorption conditions on the adsorption performance of BHP-Kae. The effects of pH [47], temperature [48], and adsorption time on the adsorption process were investigated [49]. The findings of this study provide valuable insights and guidance for future research in this field.

2. Experimental Section

BHP-Kae was synthesized using the Friedel–Crafts method. First, 0.18 g of kaempferol and 0.48 g of 4,4′-bis(chloromethyl)-1,1′-biphenyl were mixed with 40 mL of 1,2-dichloroethane and sonicated to dissolve completely. Next, 0.30 g of iron (III) chloride was added to the mixture and then heated at 80 °C for a duration of 20 h while being agitated to complete polymerization. After completion of the reaction, it was cooled to room temperature and the solid product was collected through centrifugation and washed three times with ultra-pure water. Finally, methanol was employed as the extraction solvent for Soxhlet extraction. The resulting product was dried under vacuum at 50 °C, resulting in the formation of a dark green solid powder, which was designated as BHP-Kae-3 (Figure 1).

To examine the effect of reaction time on the adsorption capability of the adsorbent, BHP-Kae-1, BHP-Kae-2, and BHP-Kae-4 were also synthesized following the same procedure as BHP-Kae-3, but with different reaction times of 4 h, 12 h, and 28 h, respectively.

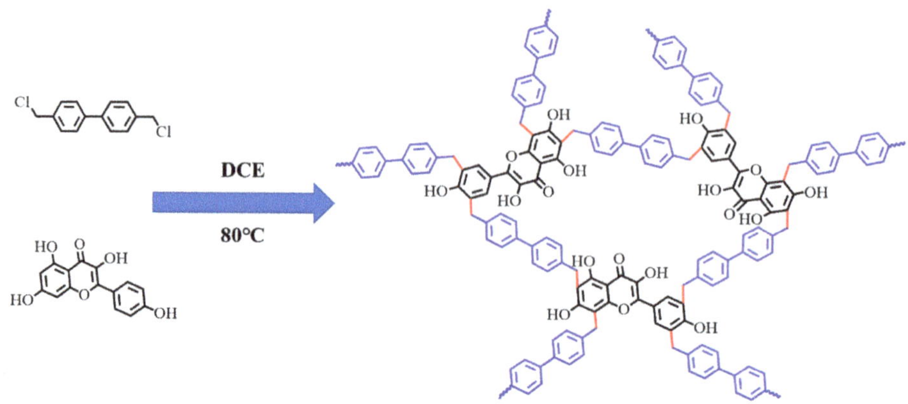

Figure 1. Illustration of BHP-Kae synthesis.

3. Results and Discussion

3.1. Characterization of BHP-Kae

Figure 2 displays scanning electron microscope (SEM) and transmission electron microscope (TEM) images of BHP-Kae at different reaction times. The images revealed that BHP-Kae possessed a rough microsphere structure, with spheres ranging in diameter from 300 nm to 1 μm. Figure 2a,b,e,f show rough microsphere structures, which are caused by short reaction times and incomplete crosslinking. Moreover, the surfaces of these microspheres became smoother as the reaction time increased, indicating that the microstructure of BHP-Kae could be adjusted by manipulating the reaction time. Figure 2c,g provide evidence that when the reaction time reached 20 h, BHP-Kae-3 exhibited a grape-like structure. Figure 2d,h show that BHP-Kae-4 displayed irregular clusters, which were attributed to a prolonged reaction time and higher degree of crosslinking. It could be concluded that controlling the reaction time can adjust the structure of BHP-Kae.

The adsorption–desorption curve and pore size distribution of the adsorbent BHP-Kae were observed, as shown in Figure 3. In Table S1, the specific surface area and pore structure parameters of BHP-Kae are given. From the observation of Figure 3a, the adsorption–desorption curve for BHP-Kae was classified as Type IV, and it showed distinct H3 hysteresis loops. From these characteristics, it could be inferred that BHP-Kae was a mesoporous material. The specific surface area and pore size of BHP-Kae varied depending on the duration of the reaction. Figure 3b shows the broad-pore-domain structure of BHP-Kae-3, and this structure can effectively reduce the transfer resistance when ACT enters BHP-Kae. The molecules inside BHP-Kae could be rapidly and inwardly diffused as well, which would help improve adsorption efficiency.

Based on the parameters shown in Table S1, it was observed that the specific surface area and pore volume of BHP-Kae initially increased and then decreased as the reaction time increased. Among the samples, BHP-Kae-3 exhibited the highest specific surface area of 82.36 m^2/g and the largest pore size of 17.07 nm. June et al. [50] reported that HCP-Fe was 64 m^2/g, and this difference may be related to the sample treatment method in the experiment. The high specific surface area and large pore size of BHP-Kae-3 could be attributed to a couple of reasons. On the one hand, when the crosslinking time was either too long or too short, it hindered the formation of microspheres, leading to a decrease in the specific surface area. On the other hand, excessive crosslinking led to pore clogging, which reduced the specific surface area and pore volume. Theoretically, the high specific surface area and large pore volume provided a significant number of adsorption sites. This facilitated transfer and improved the adsorption performance of BHP-Kae from *Cistanche tubulosa*.

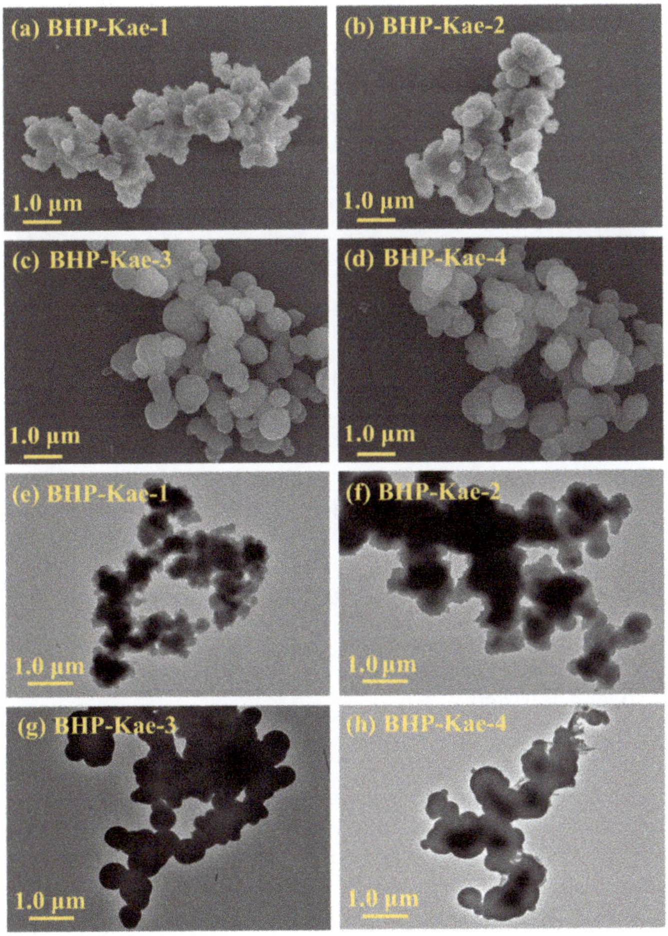

Figure 2. SEM (**a–d**) and TEM (**e–h**) images of BHP-Kae.

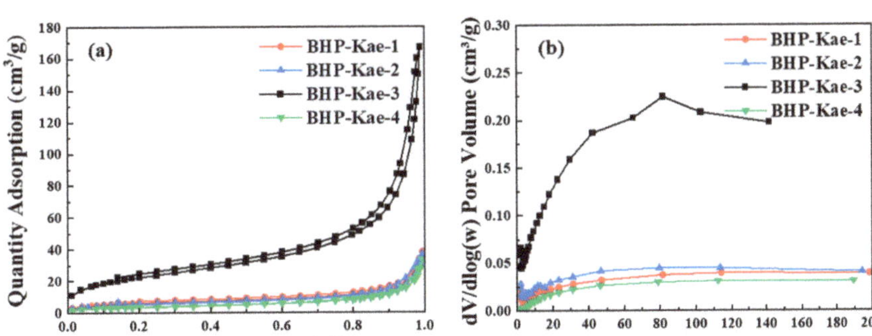

Figure 3. N_2 adsorption and desorption isotherms (**a**), and pore size distributions (**b**) of BHP-Kae.

To investigate the effect of different reaction times on the hydrophilicity of BHP-Kae, water contact angle tests were carried out on the adsorbent. The test results are displayed in Figure 4. It was observed that the water contact angle of BHP-Kae gradually decreased from 40.1° to 23.2° as the reaction time increased from 4 h to 28 h. Significantly, when the reaction time reached 28 h, the water contact angle of BHP-Kae-4 hit its lowest point. This decline in the water contact angle could be attributed to the presence of abundant phenolic hydroxyl groups on the surface of BHP-Kae. These phenolic hydroxyl groups greatly contributed to the hydrophilicity of BHP-Kae, making it exhibit a strong affinity towards water. These findings highlight the importance of reaction time in determining the hydrophilicity of the BHP-Kae adsorbent. Consequently, the adsorbent became more and more hydrophilic, as evidenced by the progressive decrease in the water contact angle.

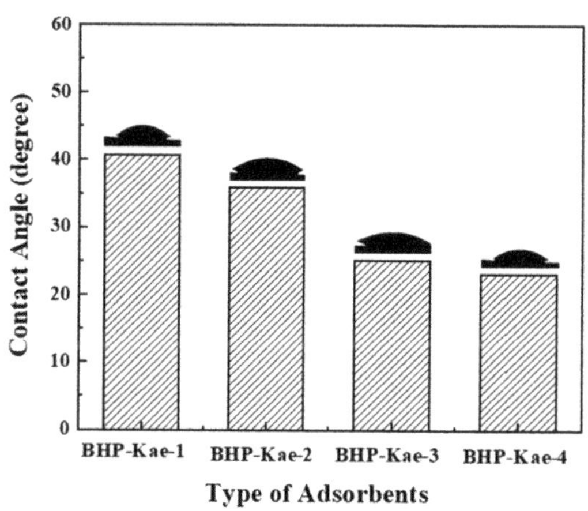

Figure 4. Water contact angle of BHP-Kae.

In Figure 5, the Fourier transform infrared spectroscopy (FTIR) of BHP-Kae is shown. The prepared BHP-Kae exhibited a distinctive peak at 3445 cm^{-1}, indicating the presence of the -OH group on kaempferol. This peak confirmed that the -OH group was not consumed during the crosslinking process. Additionally, a new peak appeared at 2920 cm^{-1}, which was attributed to the stretching vibration of the -CH- bonds on the benzene ring. This peak indicated that the benzene ring was successfully incorporated into the structure of BHP-Kae. Moreover, the disappearance of the characteristic peak at 725 cm^{-1} indicated that the C-Cl bond in 4,4′-bis(chloromethyl)-1,1′-biphenyl was substituted by the benzene ring, resulting in the embedding of the two aromatic rings into the polymer backbone. These findings confirm the successful occurrence of the crosslinking reaction and the successful preparation of BHP-Kae.

Figure 6 presented the zeta potential, which measured the surface charge of BHP-Kae-3 under different pH conditions. The graph revealed that the zeta potential gradually decreased as the pH increased from 2 to 6. This was the same as the experimental results of Li et al. [51] and Wang et al. [52]; in a neutral environment, the zeta potential was negative. Moreover, the zero charge point (pH_{ZPC}) of BHP-Kae-3 was precisely calculated to be 2.82. This meant that at a pH of 2.82, the surface charge of BHP-Kae-3 was neutral. In general, a zero charge adsorbent was advantageous for the formation of hydrogen bonds between the adsorbent and ACT. This was because a neutral adsorbent could establish a strong interaction through hydrogen bonding. Conversely, negatively charged adsorbents tended to exhibit an electrostatic attraction to positively charged target molecules, and vice versa. Notably, in the context of this study, ACT was an acidic molecule, but the zeta potential of the ACT solution at this time was −1.2 mV. During the adsorption process of BHP-Kae-3, the electrostatic interaction was found to be comparatively weak. Instead, the primary force responsible for adsorption was the formation of hydrogen bonds.

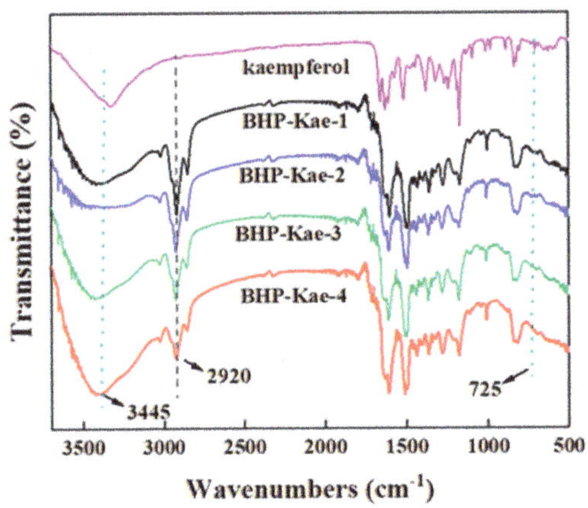

Figure 5. FTIR spectra of BHP-Kae.

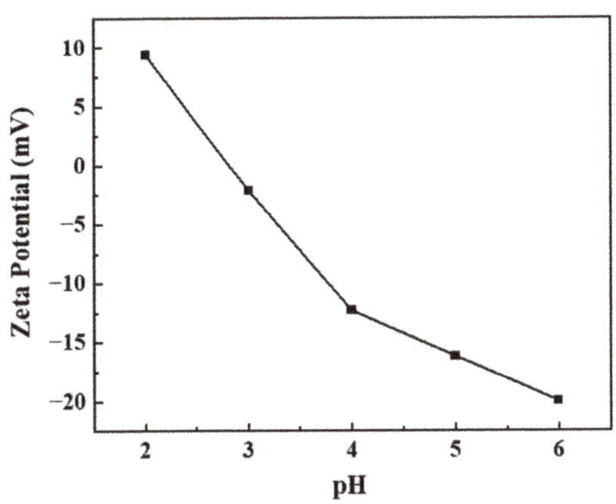

Figure 6. Zeta potential of BHP-Kae-3 at different pH.

3.2. Adsorption Evaluation

3.2.1. Effect of Adsorbent Reaction Time

The results depicted in Figure 7 provide insights into the adsorption capacity and selectivity of BHP-Kae for ACT at different reaction times (BHP-Kae-1, BHP-Kae-2, BHP-Kae-3, and BHP-Kae-4 correspond to 4 h, 12 h, 20h, and 28 h respectively). (High performance liquid chromatography (HPLC) was used for analysis, and HPLC was used for gradient elution. The elution conditions were set as showed in Table S2. The fitting standard curves of ECH and ACT obtained by HPLC are shown in Table S3. The HPLC diagrams of echinacoside (ECH) and ACT were showed in Figure S2.) They clearly demonstrate that the adsorption capacity and selectivity of BHP-Kae for ACT displayed an initial increment followed by a subsequent decrement as the reaction time increased. It is noteworthy that BHP-Kae-3 exhibited the most favorable adsorption performance for ACT, the adsorption capacity of 77.52 mg/g and a selectivity of 3.41 among the adsorbents examined. This superior performance could be attributed to two main reasons. On the one hand, BHP-Kae-3 possessed a broad-pore-domain structure. This unique pore structure facilitated the fast inward diffusion of molecules, thereby enhancing the monolithic transferred capacity of the adsorbent. On the other hand, BHP-Kae-3 contained hydroxyl and benzene rings, which interacted with ACT through multiple forces of interaction. These multiple interactions further enhanced the selective adsorption of ACT. The hydroxyl groups and benzene rings in BHP-Kae formed various types of interactions (hydrogen bonding, Van der Waals force and π–π interactions) with ACT, resulting in an improvement in the adsorption performance of BHP-Kae-3. Therefore, the following studies were all aimed at the adsorption evaluation of BHP-Kae-3.

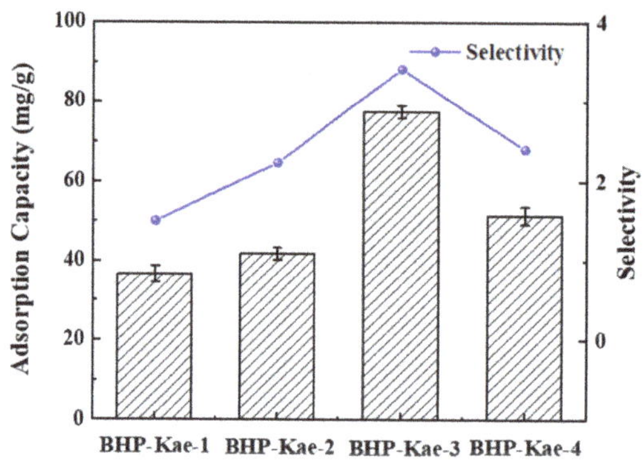

Figure 7. Adsorption performance of the BHP-Kae obtained at different reaction times.

3.2.2. Effect of ACT Concentration

Figure 8 depicts the influence of various initial concentrations on the adsorption performance of BHP-Kae-3. The observed trend indicated that the adsorption capacity of BHP-Kae-3 for ACT consistently rose as the concentration of ACT increased until it reached a point of adsorption equilibrium. As the concentration of ACT in the adsorption solution gradually increased, the number of ACT present also increased. It should be noted that the adsorption sites available on the surface of BHP-Kae-3 were limited. Therefore, the number of ACT occupying the adsorption sites on BHP-Kae-3 also increased as the concentration of ACT continued to rise. This led to a gradual accumulation of ACT on the BHP-Kae-3 until all the adsorption sites became fully occupied.

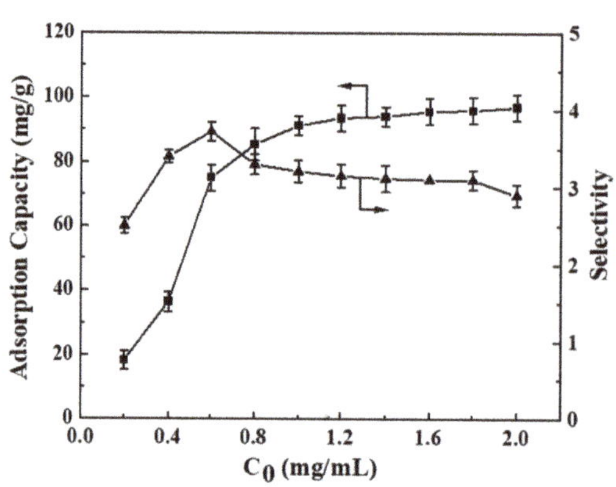

Figure 8. Effect of initial concentrations of ACT on BHP-Kae-3 adsorption performance.

3.2.3. Effect of pH

Figure 9 illustrates the adsorption performance of BHP-Kae-3 for ACT under various pH. The pH of the solution directly influenced the stability of ACT and the formation of hydrogen bonds between the adsorbent and ACT. ACT is an acidic molecule that requires specific pH conditions to maintain its stability in solution. Therefore, it is crucial to control the pH of the solution within the range of 2 to 7. Further analysis of Figure 9 shows that altering the pH had a significant effect on the adsorption process of the adsorbent. The adsorption capacity of BHP-Kae-3 for ACT increased initially and then decreased with the increase in pH. Considering both the adsorption capacity and selectivity of BHP-Kae-3 for ACT, it was found that the optimal pH value of the adsorption solution was 5.

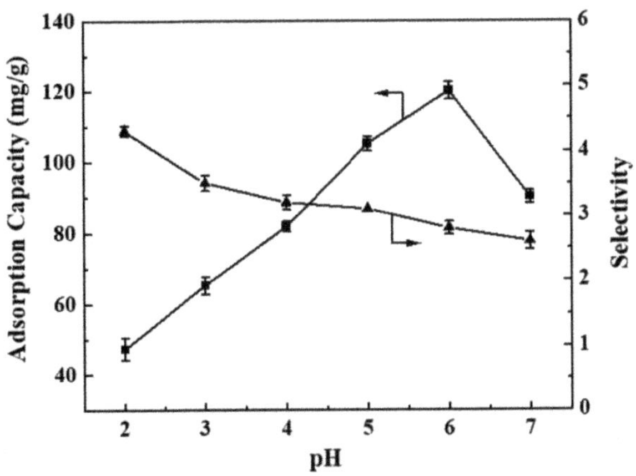

Figure 9. Effect of pH on BHP-Kae-3 adsorption performance.

The reason behind this phenomenon is that the adsorbed solution had a strong polarity at pH = 2. This strong polarity inhibited the formation of hydrogen bonds between BHP-Kae-3 and ACT. In the absence of hydrogen bonding, BHP-Kae-3 had a lower adsorption capacity for ACT. The polarity of the adsorbed solution was weakened as the pH value increased, allowing BHP-Kae-3 and ACT to form stronger interactions. These hydrogen bonds facilitated the adsorption process, which led to a higher adsorption capacity of BHP-Kae-3 for ACT. This suggests that the adsorbent possessed enhanced adsorption capacity towards ACT and effectively captured more ACT at pH = 5.

3.2.4. Influence of Adsorption Temperature

Figure 10 illustrates the adsorption performance of BHP-Kae-3 for ACT at various temperatures. The figure clearly shows that as the temperature increased from 20 °C to 40 °C, the adsorption capacity and selectivity gradually decreased. This decline could be attributed to the fact that the adsorption process of BHP-Kae-3 for ACT was an exothermic reaction, meaning that lower temperatures were more favorable for the adsorption of ACT by BHP-Kae-3. Considering the desired adsorption capacity and selectivity, it could be concluded that the optimal temperature for adsorption was 20 °C.

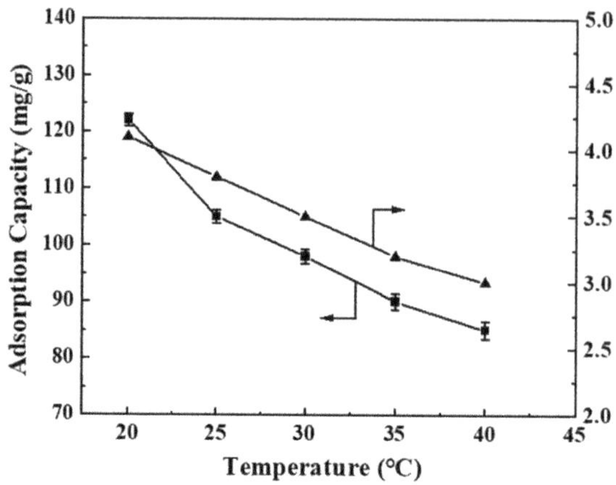

Figure 10. Effect of adsorption temperature on BHP-Kae-3 adsorption performance.

3.3. *Study on Adsorption Kinetics of BHP-Kae-3*

A series of experiments were conducted to further explore the adsorption kinetics of BHP-Kae-3 for ACT. Upon analyzing Figure 11a, it was observed that the adsorption capacity of BHP-Kae-3 for ACT gradually increased with the passage of time until it reached the equilibrium state. Based on the experimental findings, it was found that the adsorption of BHP-Kae-3 for ACT reached equilibrium after approximately 50 min. It could be concluded that the adsorption of BHP-Kae-3 for ACT was a rapid adsorption process.

The results shown in Figure 11b–d and Table S4 clearly illustrate that the theoretical adsorption capacity predicted by the pseudo-second-order kinetic model closely matched the experimental value. Additionally, the correlation coefficient of the pseudo-second-order kinetic model was higher than that of the pseudo-first-order model and Ritchie-second-order [53,54] kinetic model. Based on these findings, it could be concluded that the pseudo-second-order kinetic model was more suitable and reliable for characterizing the adsorption phenomenon of BHP-Kae-3 in ACT.

Figure 11d and Table S5 illustrate the adsorption of BHP-Kae-3 for ACT, depicted by fitting the curve with the intra-particle diffusion model. The curve obtained from the experimental data did not intersect the origin, indicating that there was no singular rate-limiting step in the adsorption process. The adsorption of ACT by BHP-Kae-3 consisted of three distinct stages. The first stage was known as surface diffusion control stage, in which ACT diffused from the adsorbed solution to the BHP-Kae-3 surface to complete the surface adsorption process. The second stage was governed by intra-particle diffusion, in which ACT penetrated from the surface to the interior of BHP-Kae-3 through its broad-pore-domain structure. The third stage represented the equilibrium stage of adsorption and desorption. In this stage, the diffusion rate gradually decreased until it reached zero, signifying the attainment of equilibrium between the adsorption and desorption processes. Based on the adsorption kinetics observed for BHP-Kae-3, it could be concluded that the adsorption of ACT by BHP-Kae-3 was a rapid process.

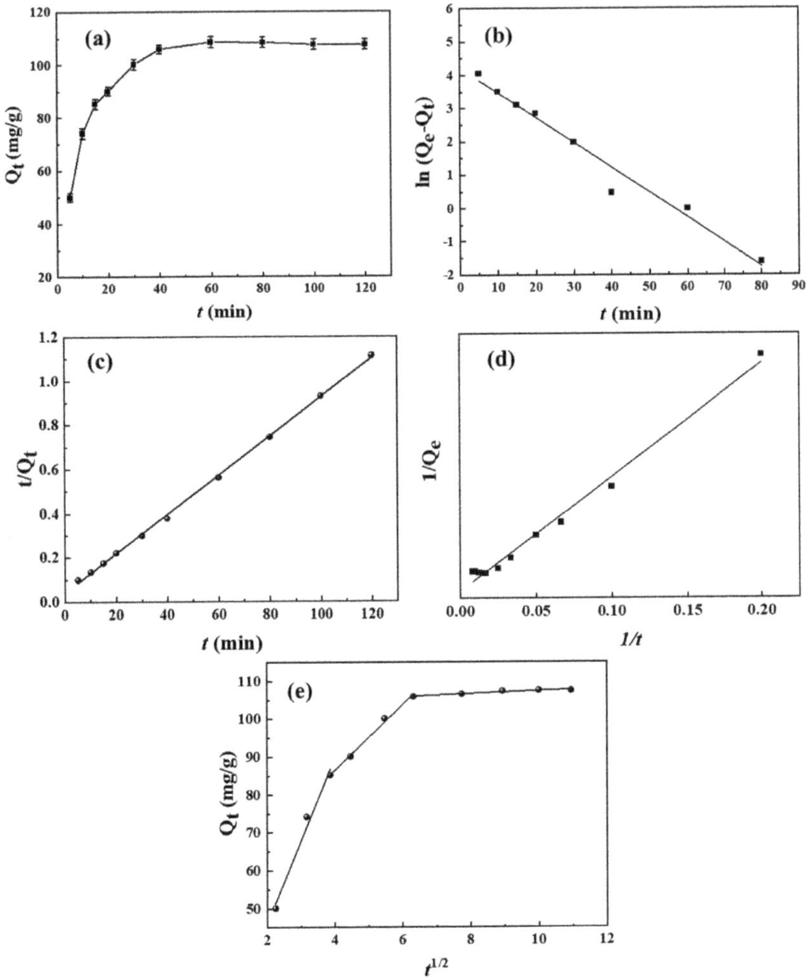

Figure 11. Adsorption kinetics of ACT on BHP-Kae-3: adsorption time (**a**), pseudo-first-order model (**b**), pseudo-second-order model (**c**), Ritchie-second-order model (**d**), and intra-particle diffusion model (**e**). (Q_t (mg/g) and Q_e (mg/g) represent the adsorption capacity at t (min) and the adsorption capacity at equilibrium, respectively).

3.4. Thermodynamic Study on Adsorption of BHP-Kae-3

Figure 12 investigates the adsorption phenomenon of BHP-Kae-3 for ACT through thermodynamic studies. The adsorption capacity of BHP-Kae-3 also gradually increased until it reached a state of equilibrium as the initial concentration of ACT increased, as shown in Figure 12a. Furthermore, it was evident from different adsorption temperatures that the adsorption capacity of BHP-Kae-3 for ACT decreased as the adsorption temperature increased. This suggested that the adsorption process was exothermic during the adsorption of ACT into BHP-Kae-3. The results revealed that the higher initial concentrations and the lower temperature led to increased adsorption capacity.

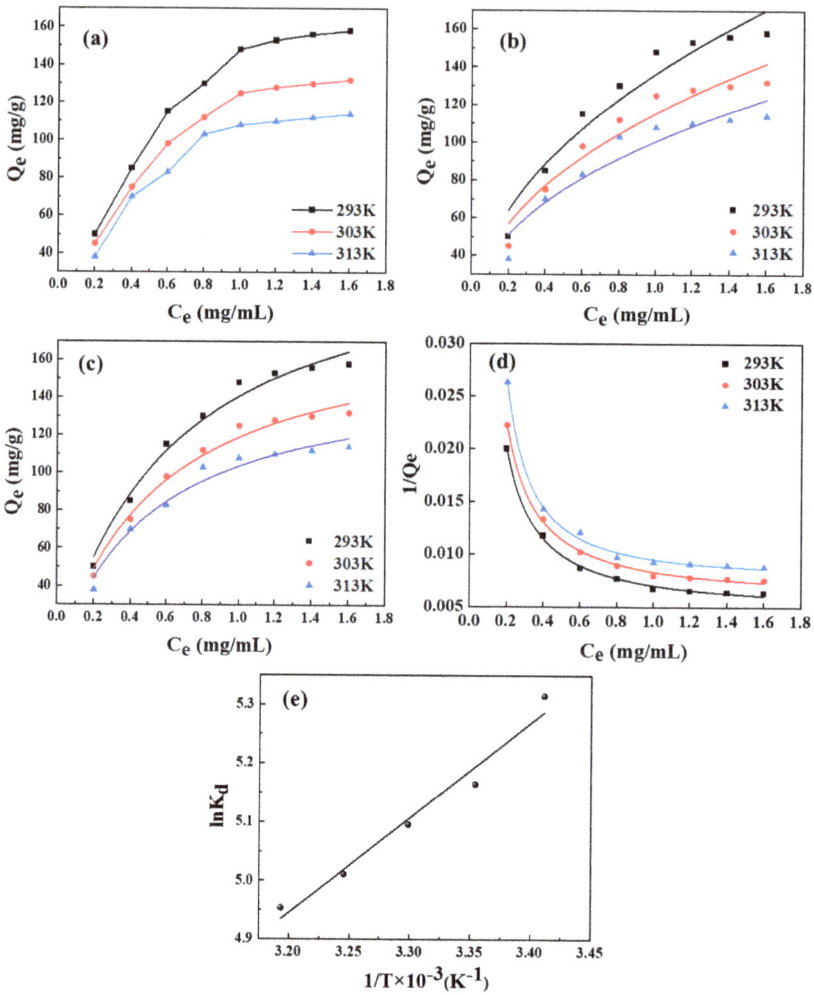

Figure 12. Adsorption isotherms of ACT on BHP-Kae-3: adsorption isotherms (**a**), Freundlich isotherm model (**b**), Langmuir isotherm model (**c**), Liu isotherm model (**d**), and thermodynamics of adsorption (**e**). (Q_e (mg/g) represents the equilibrium adsorption capacity of the adsorbent for ACT. C_e (mg/mL) represents the equilibrium concentration. K_d represents the partition coefficient).

To study the adsorption process of BHP-Kae-3 for ACT in more detail, the Freundlich, Langmuir, and Liu [55] isotherm models were employed to fit the adsorption isotherm data (Figure 12b–d, Table S6). Analyzing the fitted data revealed that the Liu isotherm model yielded a higher correlation coefficient value compared to the Freundlich isotherm and Langmuir isotherm models at each adsorption temperature. Furthermore, the adsorption capacity calculated using the Liu isotherm model exhibited closer agreement with the experimental value. Consequently, the Liu isotherm model could describe the adsorption process of BHP-Kae-3 on ACT.

In order to further study the adsorption thermodynamics of the adsorption process, the change in enthalpy (ΔH), the change in Gibbs free energy (ΔG), and the change in entropy (ΔS) were calculated to explore the adsorption thermodynamics. The magnitude of ΔH was less than 40 kJ/mol, indicating that the adsorption process was predominantly governed by weak intermolecular forces and was a physical adsorption process. The calculated value of ΔG was negative and exhibited a decreasing trend with increasing temperature. This indicated that the surface adsorption process was spontaneous. Furthermore, the positive value of ΔS indicated that the affinity of BHP-Kae-3 for ACT increased during adsorption (Table S7). This further supported the spontaneous and exothermic physical adsorption process for the adsorption of ACT by BHP-Kae-3.

3.5. Adsorption Mechanism

3.5.1. XPS Analysis

Figure 13 presents the XPS spectra of BHP-Kae-3 before and after the ACT adsorption process. In Figure 13a, it is evident that the peak value of the oxygen element in the adsorbent BHP-Kae-3-ACT shifted in energy after ACT adsorption, indicating the formation of hydrogen bonds between BHP-Kae-3 and ACT.

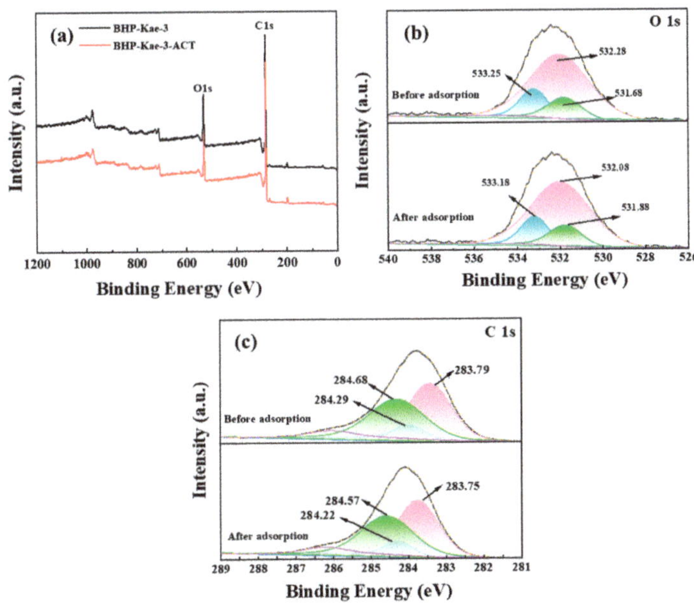

Figure 13. XPS spectra of BHP-Kae-3 before and after adsorption ACT: wide scans (**a**), O 1s (**b**), C 1s (**c**).

In Figure 13b, the XPS spectra of O 1s of BHP-Kae-3 and BHP-Kae-3-ACT are presented. Before the adsorption process, there was a prominent peak observed at 533.25 eV, indicating the presence of hydroxyl (-OH) groups on the BHP-Kae-3 surface. After the adsorption of ACT, the peak shifted slightly to 533.18 eV. This shift suggests that a hydrogen bonding interaction had occurred between the oxygen atom on BHP-Kae-3 and ACT.

Figure 13c presented XPS spectra of C 1s for two samples, BHP-Kae-3 and BHP-Kae-3-ACT. The spectrum of BHP-Kae-3 displayed three distinct peaks at 284.29, 284.68, and 283.79 eV, which corresponded to different carbon–oxygen (C-O), carbon–carbon (C-C), and carbon–carbon double bond (C=C) functionalities present in the benzene ring. The C-O peak shifted to 284.22 eV, the C-C peak shifted to 284.57 eV, and the C=C peak shifted to 283.75 eV. These shifts could be attributed to the formation of hydrogen bonds and π–π interactions between the adsorbent and ACT during the adsorption process. This information is visually represented in Figure S3, illustrating each force involved in the adsorption phenomenon.

3.5.2. Molecular Simulation

In order to fully understand the interaction between BHP-Kae-3 and ACT, DFT calculations (Gaussian 09W) were used to explain the mechanism of their interaction. The optimal binding configuration and interaction patterns of BHP-Kae-3 and ACT were determined by DFT calculations, as shown in Figure 14. A unit on BHP-Kae-3 was chosen for the calculation because it could provide binding sites during the adsorption process and interact with ACT. In addition, the complexes were analyzed by independent gradient modeling (IGM) using Multiwfn software (Multiwfn_3.8_dev_bin_Win64) [56–58]. It was observed that hydrogen bonding and van der Waals interactions occurred simultaneously during the adsorption process. This was consistent with the results of the XPS analysis.

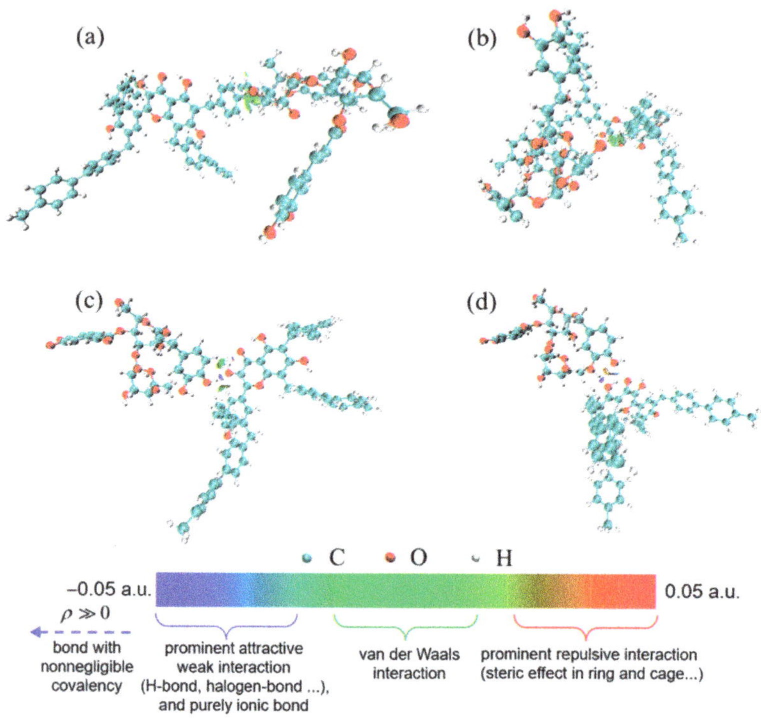

Figure 14. IGM analysis of binding configurations between fragments on BHP-Kae-3 and ACT: (**a**–**d**) four different locations in the BHP-Kae-3 building unit are combined with ACT.

3.6. Study on Regeneration Ability of BHP-Kae-3

The regeneration ability of BHP-Kae-3, a material used for practical applications, was found to be highly important. In order to evaluate its performance, the change in the adsorption capacity of BHP-Kae-3 for ACT was analyzed over the course of eight adsorption–desorption cycles, as depicted in Figure 15. According to the data presented in the figure, it can be observed that after BHP-Kae-3 was subjected to eight consecutive experiments, the adsorption capacity for ACT decreased from an initial value of 105.12 mg/g to 87.66 mg/g. Nonetheless, even after these cycles, the adsorption capacity of BHP-Kae-3 remained above 83% of its original value. Table S8 compared the adsorption capacity of BHP-Kae-3 and other previously reported adsorbents for ACT. Compared with other adsorbents, BHP-Kae-3 showed the maximum adsorption capacity and selectivity for ACT. From these results, one can confidently infer that BHP-Kae-3 possesses excellent regeneration ability and is therefore a suitable material for the adsorption of ACT.

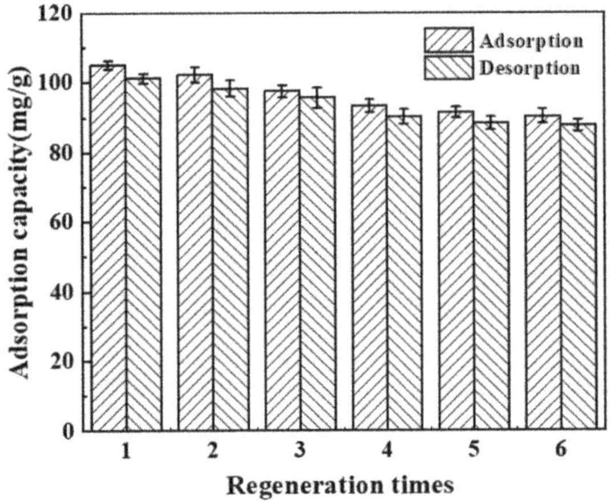

Figure 15. Adsorption–desorption cycle experiments of ACT on BHP-Kae-3.

4. Conclusions

In this study, a series of BHP-Kae adsorbents were successfully synthesized using the Friedel–Crafts reaction method. This method allowed adsorbents to effectively adsorb ACT. The pore structure of the BHP-Kae adsorbents was adjusted by controlling the reaction time. Compared with BHP-Kae-1, BHP-Kae-2, and BHP-Kae-4, the pore size of BHP-Kae-3 increased by 8.2%, 2.3%, and 1.7%, the pore volume increased by 1914.3%, 1310.0%, and 74.1%, and the specific surface area increased by 342.3%, 290.0%, and 150.9%, respectively. Specifically, when the reaction time was 20 h, the resulting BHP-Kae-3 exhibited a broad-pore-domain structure. Adsorption reached equilibrium quickly because the broad-pore-domain structure facilitated fast intra-molecular diffusion by reducing transfer resistance. Moreover, the surface of BHP-Kae-3 was rich in hydroxyl groups and benzene rings. These chemical groups possessed the ability to interact with ACT through multiple forces, including hydrogen bonding and π–π interactions. In addition, BHP-Kae-3 demonstrated excellent regeneration performance. The experimental results showed that, compared with BHP-Kae-1, BHP-Kae-2, and BHP-Kae-4, the adsorption capacity of BHP-Kae-3 increased by 111.7%, 85.8%, and 50.9%, and the selectivity was increased by 127.3%, 52.9%, and 42.1%, respectively. Broad-pore-domain adsorbents, as effective adsorbents, open up new possibilities for the extraction and purification of other bioactive compounds.

Supplementary Materials: The following supporting information can be downloaded at: https://www.mdpi.com/article/10.3390/polym17010079/s1, Figure S1: the structural formula of ACT; Figure S2: HPLC chromatogram of ACT and ECH; Figure S3, the main interaction between BHP-Kae-3 and ACT; Table S1, specific surface area parameters and pore structure of BHP-Kae; Table S2, the gradient elution condition of HPLC; Table S3, the calibration curves of ACT and ECH; Table S4, kinetics parameters of ACT adsorption on BHP-Kae-3; Table S5, intra-molecule diffusion coefficient of ACT on BHP-Kae-3; Table S6, adsorption isotherm parameters for BHP-Kae-3; Table S7, thermodynamic parameters of ACT adsorption; Table S8, comparing the adsorption capacities of various ACT adsorbents [19,20,59–61].

Author Contributions: Conceptualization, data curation, investigation, methodology, writing—original draft, writing—review and editing, W.R.; data curation, writing—review and editing, J.L.; data curation, writing—review and editing; F.X.; data curation, writing—review and editing, Y.S.; data curation, writing—review and editing Y.Z.; data curation, writing—review and editing, Y.L. (Yipei Li); project administration, writing—review and editing, Y.L. (Yin Lv); conceptualization, project administration, writing—review and editing, supervision, X.L. All authors have read and agreed to the published version of the manuscript.

Funding: This work was supported by the Fund for Less Developed Regions of the National Natural Science Foundation of China [grant number 22068032]; the program for Support Plan for Innovation and Development of Key Industries in Xinjiang Province [grant number 2021DB004].

Institutional Review Board Statement: Not applicable for studies not involving humans or animals.

Data Availability Statement: Data are contained within the article and supplementary materials. Further inquiries can be directed to the corresponding author.

Acknowledgments: We wish to thank the Analysis and Testing Center of Shihezi University for the microscopy and microanalysis of our specimens.

Conflicts of Interest: The authors declare that they have no known competing financial interests or personal relationships that could have appeared to influence the work reported in this paper.

References

1. Wang, F.; Zhuo, B.; Wang, S.; Lou, J.; Zhang, Y.; Chen, Q.; Shi, Z.; Song, Y.; Tu, P. *Atriplex canescens*: A new host for *Cistanche deserticola*. *Heliyon* **2021**, *7*, e07368. [CrossRef]
2. Piwowarczyk, R.; Ochmian, I.; Lachowicz, S.; Kapusta, I.; Sotek, Z.; Błaszak, M. Phytochemical parasite-host relations and interactions: A *Cistanche armena* case study. *Sci. Total Environ.* **2020**, *716*, 137071. [CrossRef]
3. Miao, Y.; Zhang, X.; Pei, J.; Liu, C.; Huang, L. Adaptive bacterial and fungal matching between a parasitic plant and its host: A case of *Cistanche deserticola* and *Haloxylon ammodendron*. *Ind. Crops Prod.* **2023**, *191*, 115932. [CrossRef]
4. Jiang, Y.; Tu, P.-F. Analysis of chemical constituents in *Cistanche* species. *J. Chromatogr. A* **2009**, *1216*, 1970–1979. [CrossRef]
5. Dong, J.; Li, J.; Liu, Y.; Cui, L.; Liu, X.; Wang, G.; Wang, Q.; Criddle, D.N.; Tu, P.; Li, C. A comparative study of the anti-fatigue activity of extracts from different parts of *Cistanche tubulosa* (Schenk) Wight. *J. Tradit. Chin. Med. Sci.* **2024**, *11*, 222–231. [CrossRef]
6. Li, Z.; Lin, H.; Gu, L.; Gao, J.; Tzeng, C.-M. *Herba Cistanche* (Rou Cong-Rong): One of the Best Pharmaceutical Gifts of Traditional Chinese Medicine. *Front. Pharmacol.* **2016**, *7*, 41. [CrossRef]
7. Fu, Z.; Fan, X.; Wang, X.; Gao, X. *Cistanches Herba*: An overview of its chemistry, pharmacology, and pharmacokinetics property. *J. Ethnopharmacol.* **2018**, *219*, 233–247. [CrossRef]
8. Zhou, S.; Feng, D.; Zhou, Y.; Duan, H.; Jiang, Y.; Yan, W. Analysis of the active ingredients and health applications of *Cistanche*. *Front. Nutr.* **2023**, *10*, 1101182. [CrossRef]
9. Gu, C.; Yang, X.; Huang, L. *Cistanches Herba*: A Neuropharmacology Review. *Front. Pharmacol.* **2016**, *7*, 289. [CrossRef]
10. Jiang, H.; Ma, R.; Ji, F.; Liu, Y.; Wang, B.; Fu, S.; Ma, L.; Wang, S.; Liu, C.; Guo, Z.; et al. Structure characterization of polysaccharides from *Cistanche deserticola* and their neuroprotective effects against oxidative stress in slow transit constipation mice. *Int. J. Biol. Macromol.* **2024**, *260*, 129527. [CrossRef]
11. Cheng, N.; Wang, H.; Hao, H.; Rahman, F.-U.; Zhang, Y. Research progress on polysaccharide components of *Cistanche deserticola* as potential pharmaceutical agents. *Eur. J. Med. Chem.* **2023**, *245*, 114892. [CrossRef]
12. Li, M.; Tian, X.; Li, X.; Mao, T.; Liu, T. Anti-fatigue activity of gardenia yellow pigment and *Cistanche phenylethanol glycosides* mixture in hypoxia. *Food Biosci.* **2021**, *40*, 100902. [CrossRef]
13. Guo, Y.; Cao, L.; Zhao, Q.; Zhang, L.; Chen, J.; Liu, B.; Zhao, B. Preliminary characterizations, antioxidant and hepatoprotective activity of polysaccharide from *Cistanche deserticola*. *Int. J. Biol. Macromol.* **2016**, *93*, 678–685. [CrossRef]
14. Yan, J.; Wang, H.; Wang, H.; Bian, Y.; Wang, K.; Zhai, X.; Li, Y.; Wu, K.; Wang, W.; Li, J.; et al. Quantitative analysis and hepatoprotective mechanism of *Cistanche deserticola* Y. C. Ma against alcohol-induced liver injury in mice. *Biomed. Pharmacother.* **2023**, *162*, 114719. [CrossRef]
15. Ran, Z.; Ju, B.; Cao, L.; Hou, Q.; Wen, L.; Geng, R.; Liao, Y.; Hu, J.; Yang, J. Microbiome—Metabolomics analysis reveals the potential effect of verbascoside in alleviating cognitive impairment in db/db mice. *Food Funct.* **2023**, *14*, 3488–3508. [CrossRef]
16. Zhang, S.; Gong, F.; Liu, J.; Liu, T.; Yang, J.; Hu, J. A novel PHD2 inhibitor acteoside from *Cistanche tubulosa* induces skeletal muscle mitophagy to improve cancer-related fatigue. *Biomed. Pharmacother.* **2022**, *150*, 113004. [CrossRef] [PubMed]

17. Wu, Y.; Zhang, N.; Luo, K.; Liu, Y.; Bai, Z.; Tang, S. Recent advances of innovative and high-efficiency stationary phases for chromatographic separations. *TrAC Trends Anal. Chem.* **2022**, *153*, 116647. [CrossRef]
18. Sajid, M.; Płotka-Wasylka, J. Combined extraction and microextraction techniques: Recent trends and future perspectives. *TrAC Trends Anal. Chem.* **2018**, *103*, 74–86. [CrossRef]
19. Li, L.; Tsao, R.; Yang, R.; Liu, C.; Young, J.C.; Zhu, H. Isolation and purification of phenylethanoid glycosides from *Cistanche deserticola* by high-speed counter-current chromatography. *Food Chem.* **2008**, *108*, 702–710. [CrossRef]
20. Nie, F.; Feng, C.; Ahmad, N.; Tian, M.; Liu, Q.; Wang, W.; Lin, Z.; Li, C.; Zhao, C. A new green alternative solvent for extracting *echinacoside* and *acteoside* from *Cistanche deserticola* based on ternary natural deep eutectic solvent. *J. Ind. Eng. Chem.* **2023**, *118*, 499–510. [CrossRef]
21. Park, H.I.; Kang, J.; Park, J.-H.; Park, J.C.; Park, J.; Lee, K.B.; Lee, C.H. One-pot synthesis of novel porous carbon adsorbents derived from poly vinyl chloride for high methane adsorption uptake. *Chem. Eng. J.* **2022**, *440*, 135867. [CrossRef]
22. Ioni, Y.; Sapkov, I.; Kirsanova, M.; Dimiev, A.M. Flame modified graphene oxide: Structure and sorption properties. *Carbon* **2023**, *212*, 118122. [CrossRef]
23. Kayan, A. Inorganic-organic hybrid materials and their adsorbent properties. *Adv. Compos. Hybrid Mater.* **2019**, *2*, 34–45. [CrossRef]
24. Abdul, G.; Zhu, X.; Chen, B. Structural characteristics of biochar-graphene nanosheet composites and their adsorption performance for phthalic acid esters. *Chem. Eng. J.* **2017**, *319*, 9–20. [CrossRef]
25. Yang, D.-H.; Tao, Y.; Ding, X.; Han, B.-H. Porous organic polymers for electrocatalysis. *Chem. Soc. Rev.* **2022**, *51*, 761–791. [CrossRef]
26. Shi, Y.; Wang, Y.; Meng, N.; Liao, Y. Photothermal Conversion Porous Organic Polymers: Design, Synthesis, and Applications. *Small Methods* **2024**, *8*, 2301554. [CrossRef] [PubMed]
27. Li, Z.; Yang, Y.-W. Creation and bioapplications of porous organic polymer materials. *J. Mater. Chem. B* **2017**, *5*, 9278–9290. [CrossRef]
28. Huang, J.; Turner, S.R. Hypercrosslinked Polymers: A Review. *Polym. Rev.* **2018**, *58*, 1–41. [CrossRef]
29. Yang, Y.; Tan, B.; Wood, C.D. Solution-processable hypercrosslinked polymers by low cost strategies: A promising platform for gas storage and separation. *J. Mater. Chem. A* **2016**, *4*, 15072–15080. [CrossRef]
30. Wang, X.; Mu, P.; Zhang, C.; Chen, Y.; Zeng, J.; Wang, F.; Jiang, J.-X. Control Synthesis of Tubular Hyper-Cross-Linked Polymers for Highly Porous Carbon Nanotubes. *ACS Appl. Mater. Interfaces* **2017**, *9*, 20779–20786. [CrossRef]
31. Long, C.; Li, Y.; Yu, W.; Li, A. Adsorption characteristics of water vapor on the hypercrosslinked polymeric adsorbent. *Chem. Eng. J.* **2012**, *180*, 106–112. [CrossRef]
32. Wang, R.; Lei, Y.; Ke, L.; Bai, L.; Zheng, Y.; Xu, Z.; Li, T.; Chen, S.; Zhang, D. Construction of self-healing multifunctional hyper-crosslinked hyperbranched polymer@metal-organic polyhedra (HHMOP) membranes. *Eur. Polym. J.* **2024**, *211*, 112966. [CrossRef]
33. Lavine, M.S. Processable cross-linked polymers. *Science* **2017**, *356*, 37–38. [CrossRef] [PubMed]
34. Leveson-Gower, R.B.; Roelfes, G. Biocatalytic Friedel-Crafts Reactions. *ChemCatChem* **2022**, *14*, e202200636. [CrossRef]
35. Gu, Y.; Son, S.U.; Li, T.; Tan, B. Hypercrosslinked Polymers: Low-Cost Hypercrosslinked Polymers by Direct Knitting Strategy for Catalytic Applications. *Adv. Funct. Mater.* **2021**, *31*, 2170082. [CrossRef]
36. Li, M.; Shi, L.; Liu, Y.; Li, S.; Cui, W.; Li, W.; Zhi, Y.; Shan, S.; Miao, Y. A dual-ionic hyper-crosslinked polymer for efficient CO_2 fixation and conversion. *Chem. Eng. J.* **2024**, *481*, 148550. [CrossRef]
37. Sun, L.; Xu, G.; Tu, Y.; Zhang, W.; Hu, X.; Yang, P.; Wu, D.; Liang, Y.; Wei, D.; Li, A.; et al. Multifunctional porous β-cyclodextrin polymer for water purification. *Water Res.* **2022**, *222*, 118917. [CrossRef] [PubMed]
38. Wang, J.; Wu, T.; Wang, X.; Chen, J.; Fan, M.; Shi, Z.; Liu, J.; Xu, L.; Zang, Y. Construction of hydroxyl-functionalized hyper-crosslinked networks from polyimide for highly efficient iodine adsorption. *iScience* **2024**, *27*, 108993. [CrossRef]
39. James, A.M.; Harding, S.; Robshaw, T.; Bramall, N.; Ogden, M.D.; Dawson, R. Selective Environmental Remediation of Strontium and Cesium Using Sulfonated Hyper-Cross-Linked Polymers (SHCPs). *ACS Appl. Mater. Interfaces* **2019**, *11*, 22464–22473. [CrossRef]
40. Yuan, D.; Zhang, S.; Tan, J.; Dai, Y.; Wang, Y.; He, Y.; Liu, Y.; Zhao, X.; Zhang, M.; Zhang, Q. Highly efficacious entrapment of Th (IV) and U (VI) from rare earth elements in concentrated nitric acid solution using a phosphonic acid functionalized porous organic polymer adsorbent. *Sep. Purif. Technol.* **2020**, *237*, 116379. [CrossRef]
41. Yuan, D.; Zhang, S.; Xiang, Z.; He, Y.; Wang, Y.; Liu, Y.; Zhao, X.; Zhou, X.; Zhang, Q. Highly Efficient Removal of Thorium in Strong HNO_3 Media Using a Novel Polymer Adsorbent Bearing a Phosphonic Acid Ligand: A Combined Experimental and Density Functional Theory Study. *ACS Appl. Mater. Interfaces* **2019**, *11*, 24512–24522. [CrossRef]
42. Chen, M.; Xiao, J.; El-Seedi, H.R.; Woźniak, K.S.; Daglia, M.; Little, P.J.; Weng, J.; Xu, S. Kaempferol and atherosclerosis: From mechanism to medicine. *Crit. Rev. Food Sci. Nutr.* **2022**, *64*, 2157–2175. [CrossRef] [PubMed]

43. Barreca, D.; Trombetta, D.; Smeriglio, A.; Mandalari, G.; Romeo, O.; Felice, M.R.; Gattuso, G.; Nabavi, S.M. Food flavonols: Nutraceuticals with complex health benefits and functionalities. *Trends Food Sci. Technol.* **2021**, *117*, 194–204. [CrossRef]
44. Yang, L.; Gao, Y.; Bajpai, V.K.; El-Kammar, H.A.; Simal-Gandara, J.; Cao, H.; Cheng, K.-W.; Wang, M.; Arroo, R.R.J.; Zou, L.; et al. Advance toward isolation, extraction, metabolism and health benefits of kaempferol, a major dietary flavonoid with future perspectives. *Crit. Rev. Food Sci. Nutr.* **2023**, *63*, 2773–2789. [CrossRef]
45. Xu, M.; Wang, J.; Zhang, L.; Wang, Q.; Liu, W.; An, Y.; Hao, L.; Wang, C.; Wang, Z.; Wu, Q. Construction of hydrophilic hypercrosslinked polymer based on natural kaempferol for highly effective extraction of 5-nitroimidazoles in environmental water, honey and fish samples. *J. Hazard. Mater.* **2022**, *429*, 128288. [CrossRef] [PubMed]
46. Kim, N.; Casareto, M.; Mowbray, M.; Henry, R.; Hayden, J.; Rubloff, G.; Lee, S.B.; Gregorczyk, K.E. On-Wafer Wide-Pore Anodic Aluminum Oxide. *J. Electrochem. Soc.* **2023**, *170*, 063507. [CrossRef]
47. Ai, Y.; Liu, Y.; Lan, W.; Jin, J.; Xing, J.; Zou, Y.; Zhao, C.; Wang, X. The effect of pH on the U (VI) sorption on graphene oxide (GO): A theoretical study. *Chem. Eng. J.* **2018**, *343*, 460–466. [CrossRef]
48. Cao, S.; Yin, W.; Yang, B.; Zhu, Z.; Sun, H.; Sheng, Q.; Chen, K. Insights into the influence of temperature on the adsorption behavior of sodium oleate and its response to flotation of quartz. *Int. J. Min. Sci. Technol.* **2022**, *32*, 399–409. [CrossRef]
49. Yi, M.; Cheng, Y.; Wang, Z.; Wang, C.; Hu, B.; He, X. Effect of particle size and adsorption equilibrium time on pore structure characterization in low pressure N_2 adsorption of coal: An experimental study. *Adv. Powder Technol.* **2020**, *31*, 4275–4281. [CrossRef]
50. Jang, J.Y.; Kim, Y.K.; Ko, Y.-J.; Son, S.U. Hollow Microporous Organic Polymer@Hypercrosslinked Polymer Catalysts Bearing *N*-Heterocyclic Carbene-Iron Species for the Synthesis of Biomass-Derived Polymer Platforms. *ACS Sustain. Chem. Eng.* **2024**, *12*, 8880–8889. [CrossRef]
51. Li, Q.; Li, S.; Yu, C.; Zhan, Z.; Cheng, G.; Tan, B.; Ren, S. Hypercrosslinked polymer membranes via interfacial polymerization for organic dye separations. *Commun. Mater.* **2024**, *5*, 255. [CrossRef]
52. Wang, Y.; Zhang, L.; Wang, S.; Feng, H.; Pang, F.; Li, H.; Liu, B.; Zang, X.; Zhang, S.; Wang, Z. Construction of a sulfonic acid functionalized hypercrosslinked polymer for the adsorption of rubidium and cesium. *J. Water Process Eng.* **2024**, *68*, 106321. [CrossRef]
53. Alahabadi, A.; Hosseini-Bandegharaei, A.; Moussavi, G.; Amin, B.; Rastegar, A.; Karimi-Sani, H.; Fattahi, M.; Miri, M. Comparing adsorption properties of NH_4Cl-modified activated carbon towards chlortetracycline antibiotic with those of commercial activated carbon. *J. Mol. Liq.* **2017**, *232*, 367–381. [CrossRef]
54. Yousatit, S.; Rungruangwattanachot, W.; Yuwawanitchakorn, N.; Nuntang, S.; Punyapalakul, P.; Ngamcharussrivichai, C. Amine-Functionalized Natural Rubber/Mesostructured Silica Nanocomposites for Adsorptive Removal of Clofibric Acid in Aqueous Phase. *Molecules* **2023**, *28*, 2330. [CrossRef]
55. Jabar, J.M.; Adebayo, M.A.; Taleat, T.A.A.; Yılmaz, M.; Rangabhashiyam, S. Ipoma batatas (sweet potato) leaf and leaf-based biochar as potential adsorbents for procion orange MX-2R removal from aqueous solution. *J. Anal. Appl. Pyrolysis* **2025**, *185*, 106876. [CrossRef]
56. Lu, T.; Chen, F. Multiwfn: A multifunctional wavefunction analyzer. *J. Comput. Chem.* **2012**, *33*, 580–592. [CrossRef] [PubMed]
57. Lu, T.; Chen, Q. Independent gradient model based on Hirshfeld partition: A new method for visual study of interactions in chemical systems. *J. Comput. Chem.* **2022**, *43*, 539–555. [CrossRef]
58. Lefebvre, C.; Rubez, G.; Khartabil, H.; Boisson, J.-C.; Contreras-García, J.; Hénon, E. Accurately extracting the signature of intermolecular interactions present in the NCI plot of the reduced density gradient versus electron density. *Phys. Chem. Chem. Phys.* **2017**, *19*, 17928–17936. [CrossRef] [PubMed]
59. Li, C.; Nie, F.; Feng, C.; Tian, M.; Yu, M.; Zhao, C.; Fu, Y. Magnetic dual-template molecularly imprinted polymers for separation and enrichment of echinacoside and acteoside from *Cistanche deserticola* Y. C. Ma. *Chem. Eng. Res. Des.* **2022**, *182*, 719–732. [CrossRef]
60. Liu, B.; Ouyang, J.; Yuan, X.; Wang, L.; Zhao, B. Adsorption properties and preparative separation of phenylethanoid glycosides from *Cistanche deserticola* by use of macroporous resins. *J. Chromatogr. B* **2013**, *937*, 84–90. [CrossRef]
61. Fan, Y.; Xu, C.; Li, J.; Zhang, L.; Yang, L.; Zhou, Z.; Zhu, Y.; Zhao, D. Ionic liquid-based microwave-assisted extraction of verbascoside from *Rehmannia* root. *Ind. Crop Prod.* **2018**, *124*, 59–65. [CrossRef]

polymers

MDPI

Article

Application of a New Carbon Black Filler in SBR Composites

Junan Zhou [1], Bainan Tang [1], Liangchen Yu [1], Junping Song [2] and Zepeng Wang [1,*]

[1] College of Mechanical and Electrical Engineering, Qingdao University of Science and Technology, Qingdao 266061, China; 18266629639@163.com (J.Z.); t17661241685@163.com (B.T.); y1047236506@163.com (L.Y.)
[2] Sino-German Institute of Science and Technology, Qingdao University of Science and Technology, Qingdao 266061, China; 02165@qust.edu.cn
* Correspondence: 02546@qust.edu.cn

Abstract: The microstructure and properties of a new type of carbon black produced by a domestic company through a new process were systematically characterized by scanning electron microscopy (SEM), energy dispersive spectroscopy (EDS), Raman spectroscopy, and X-ray diffraction (XRD). The vulcanization properties, mechanical properties, and electrical conductivity of the new carbon black with different filler amounts were investigated in styrene butadiene rubber (SBR), using the traditional reinforcing filler N660 carbon black as a control. The experimental results demonstrate that the new carbon black exhibits a stratified structure with a specific surface area of 345.96 m^2/g, and its particle size distribution is primarily concentrated within the 0.1–1 μm range. When the filling ratio was 30 phr/100 phr, the tensile strength of SBR composites filled with the new carbon black increased by 12.3% and the tear strength increased by 9.6% compared with those filled with N660 carbon black. In summary, the new carbon black can significantly improve the comprehensive performance of SBR composites and reduce the production cost. This provides a new type of material for the rubber industry that takes into account both economy and performance while also providing reference data for basic research in the field of SBR.

Keywords: SBR; new carbon black; N660 carbon black; reinforcement materials

1. Introduction

Carbon black is mainly composed of carbon with small amounts of nitrogen, hydrogen, and oxygen and has an amorphous carbon structure [1]. With its excellent reinforcing ability, carbon black is widely used in industrial production [2]. However, the preparation process of traditional carbon black has high energy consumption and serious pollution emission, so the development of high-performance rubber-reinforcing materials with low cost and low pollution characteristics has become a research focus of academic and industrial attention [3]. SBR is a general-purpose synthetic rubber made from butadiene and styrene through polymerization, which plays an irreplaceable role in various fields due to its excellent electrical conductivity, tear resistance, environmental friendliness, and low cost. However, pure SBR itself has low strength and modulus, and in practical applications, it is often necessary to enhance its comprehensive performance by adding reinforcing materials [4].

Chengyu Tang et al. [5] found that the new homemade high-structure carbon black with high specific surface area and high structural characteristics can significantly improve the tensile and tear strength of natural rubber while optimizing the vulcanization characteristics and showing excellent reinforcing ability. Lili et al. [6] explored the synthesis of graphene oxides and reduced graphene by using graphite foil waste as a carbon precursor, which provided a new way for the preparation of low-cost graphene materials. Yanchen Fang et al. [7] used plasma-modified pyrolytic carbon black and filled it into natural rubber, and found that the modified carbon black had smaller particle size and more uniform distribution and bonded more tightly with the rubber matrix. They also found that the fatigue life of the composite material was extended by 107%. Lei Gao et al. [8] showed that lignin-reinforced emulsion-polymerized SBR could form a good interfacial bond and significantly improve material properties. Tilun et al. [9] improved the filler dispersion and mechanical properties of SBR by using a waste catalyst as filler and utilizing its micro porous structure and active components.

A new carbon black prepared by a domestic company using a new process, while reducing the emission of pollutants in the production process, retains and improves the reinforcing properties of carbon black, and the cost is only two-thirds of the traditional N660 carbon black. As a classic medium reinforcing material, N660 carbon black is widely used in industrial fields [10]. Compared with the traditional high-reinforcing carbon black, the comparative study of N660 carbon black and the new carbon black can better reflect the potential advantages of the new carbon black. By systematically analyzing the effect of the new carbon black produced by the new process on the performance of SBR, this study provides important theoretical support and practical reference for the new carbon black to partially replace the N660 carbon black as a rubber-reinforcing material.

2. Materials and Methods

2.1. Experimental Raw Materials and Formulations

The raw materials and formulations (mass/phr) for the SBR experiments were as follows: SBR1502 100, zinc oxide (ZnO) 5, stearic acid (SA) 1, Si 69 2, accelerator CZ 1.5, rubber accelerator TMTD 1.5, sulphur 0.5, N660 carbon black, and new carbon black (variable).

2.2. Instruments and Equipment

BL-6175-BL double-roller opener (Baolun Precision Testing Instrument Co., Ltd., Dongguan, China); GT-7016-AR pneumatic automatic slicer (LX-A Shore's Rubber Hardness Tester (Shanghai Liuling Instrument Factory, Shanghai, China); MDR2000 Rotorless Vulcanizing Instrument (Alpha Corporation, Dulles, VA, USA); HS-100T-2 Rubber Vulcanizing Machine (Shenzhen Jiaxin Electronics Co., Ltd., Shenzhen, China); GT-AI-7000M Tensile Tester (High Speed Rail Technology Co., Ltd., Beijing, China); ZGS-170808T-722 Wide-Frequency Dielectric Impedance Spectrometer (Beijing Huidexin Technology Co., Ltd., Beijing, China).

2.3. Experimental Scheme

SBR composites and specimens were prepared to study the effect of new carbon black on SBR vulcanization characteristics, mechanical properties, and electrical conductivity. The specific experimental steps are as follows:

Before beginning, clean the rollers of the mixing machine. Next, weigh the SBR and place it into the mixer to soften it until the rubber surface becomes smooth and uniform, followed by mixing. Gradually add the temperature-sensitive raw materials, ensuring they are well mixed, then add the reinforcing materials. Finally, add sulfur, mix thoroughly, adjust the roller pitch, and perform the triangular package operation 8 times until the rubber is in sheet form and set aside. After letting the rubber rest for 24 h, measure the vulcanization curve using a rotorless vulcanizer to determine the optimal vulcanization time (t_{90}). Then, carry out vulcanization at 160 °C for (t_{90} + 2 min) under a pressure of

10 MPa and prepare standard samples. Test the prepared SBR composites and analyze them for vulcanization properties, mechanical properties, and electrical conductivity.

2.4. Performance Test

A scanning electron microscope (SEM) was utilized to observe the microscopic morphology of the new carbon black as well as the cross-section structure of the rubber composite with a resolution of 1 nm (accelerating voltage 15 kV). The vulcanization properties were tested according to the GB/T 9869-2014 standard, the tensile properties were tested according to the GB/T 528-2009 standard, and the tear strength was tested according to the GB/T 529-2008 standard. A broadband dielectric impedance spectrometer was used to determine the conductivity of the composites. The specific surface area of the new carbon black was tested using the Brunauer–Emmett–Teller method (BET). Graphitization degree and structural defects of new carbon black were analyzed by Raman spectroscopy at a fixed collection angle of 90° and by the ratio of the intensity of the D peak to that of the G peak (ID/IG). X-ray diffraction (XRD) was used to analyze the physical phase of the new carbon black, and the compacted samples were placed on blank slides and sent to the sample chamber, using Cu target Kα-rays with a scanning range of 0–90°. Differential scanning calorimetry (DSC) was used to characterize the glass transition temperature (T_g) of the composites.

3. Results and Analysis

3.1. Specific Surface Area and SEM

BET test results show that the specific surface area of the new carbon black is 345.96 m^2/g. Using SEM at 20,000 times magnification to visualize the new carbon black material, as shown in Figure 1a, it can be seen that the new carbon black has a stratified structure and is irregular in shape. The stratified structures of different sizes are closely packed together to form carbon black aggregates, and there is a certain amount of void between the aggregates, contributing to the overall roughness of the structure. Using SEM at 50,000 times magnification to visualize the new carbon black material, as shown in Figure 1b, it can be seen that the new carbon black is mainly in the form of flakes or lumps, with a rough surface and irregular edges, and the voids between particles are more obvious and have a larger aspect ratio.

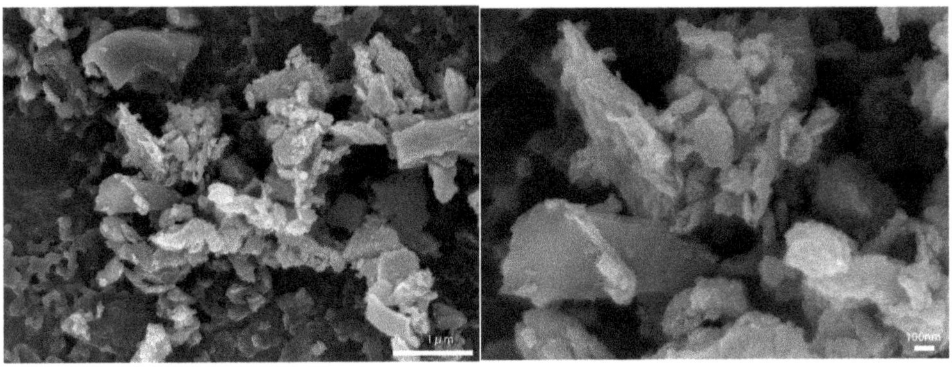

(a) (b)

Figure 1. SEM of new carbon black: (**a**) magnification 20,000 times; (**b**) magnification 50,000 times.

3.2. Laser Particle Size

Figure 2 shows that the particle size of the new carbon black is more concentrated, and its peak corresponds to a smaller particle size, mainly in the range of 0.1–1 μm, with micrometer and nanometer fine particle characteristics, showing good dispersion and filling uniformity. This particle size distribution helps the new carbon black form a uniform dispersion network within the rubber matrix, preventing the issue of stress concentration caused by particle aggregation [11].

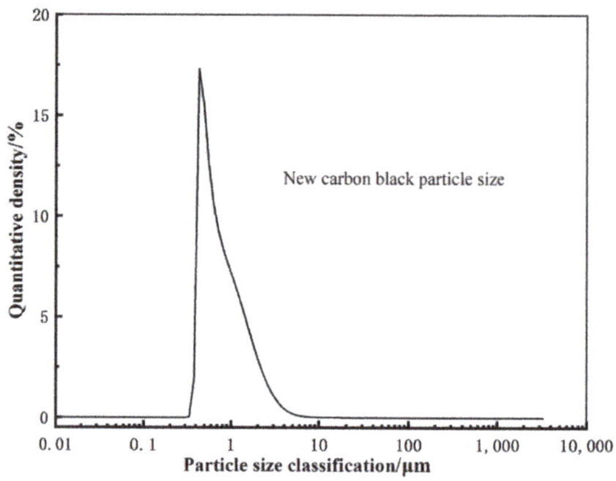

Figure 2. Particle size of new carbon black.

3.3. EDS

As shown in Figure 3, through energy dispersive spectrometer (EDS) analysis of the new type of carbon black, the carbon mass fraction is 87.13%, and the atomic fraction is 90.95%; the oxygen mass fraction is 10.2%, and the atomic fraction is 7.99%. The presence of oxygen imparts a certain degree of oxidizing properties to the surface of the carbon black particles, which facilitates chemical reactions with the rubber matrix and enhances interfacial bonding. Additionally, the presence of impurities such as aluminum, silicon, and calcium provides the potential for performance improvement of the composite material. These impurities create a synergistic effect with the reinforcing role of carbon black, making it a multifunctional and cost-effective rubber filler.

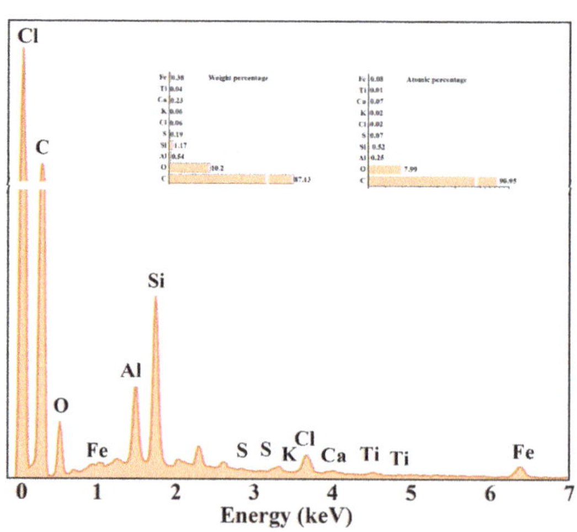

Figure 3. EDS of the new carbon black.

3.4. Raman Spectra

The D peak and G peak in Raman spectroscopy are key peaks used to characterize the properties of carbon materials. The D peak is closely related to the defects and amorphous carbon in the material, reflecting the degree of disorder, while the G peak is associated with the C-C stretching vibration in graphite crystals, reflecting the degree of graphitization or structural order of the material [12].

As shown in the Raman spectra in Figure 4, the D peak of the new carbon black appears at 1570 cm^{-1}, and the G peak is located at 1485 cm^{-1}. Both peaks are pronounced, with the ID/IG ratio greater than 1, suggesting a higher content of disordered or defective carbon, along with some degree of graphitization. The high intensity of the D peak suggests a greater presence of defective structures and amorphous carbon, while the moderate intensity of the G peak indicates that the new carbon black possesses both a disordered structure and the ability to form a conductive network while still maintaining some level of graphitization. Increased structural disorder and defects in the new carbon black enhance the number of active sites on its surface, providing more opportunities for bonding with the rubber matrix or other components. This can strengthen the filler–matrix interactions and, potentially, improve the properties of the composites.

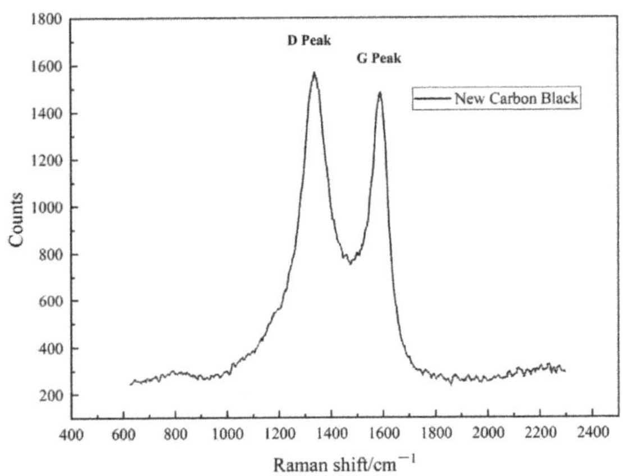

Figure 4. Raman spectra of the new carbon black.

3.5. XRD

Figure 5 shows the XRD refinement of the new carbon black. The cell parameters were calculated by performing Rietveld refinement on the XRD patterns using High-score plus (ver4.8) software with an initial refinement model of C (space group of R-3m, ICDD = 199,068). The back bottom was fitted using the polynomial function and the peak profile was fitted using the pseudo-Voigt function. Refinement confidence criteria references: $R_{wp} < 10\%$, $R_p < 10\%$ and the goodness of fit (χ^2) < 2. The black computed spectral lines in the refined plots have a better fit to the red measured curves, indicating the high reliability of the refined results, which is further supported by the lower confidence factors (R_{wp}, R_p, and χ^2) [13].

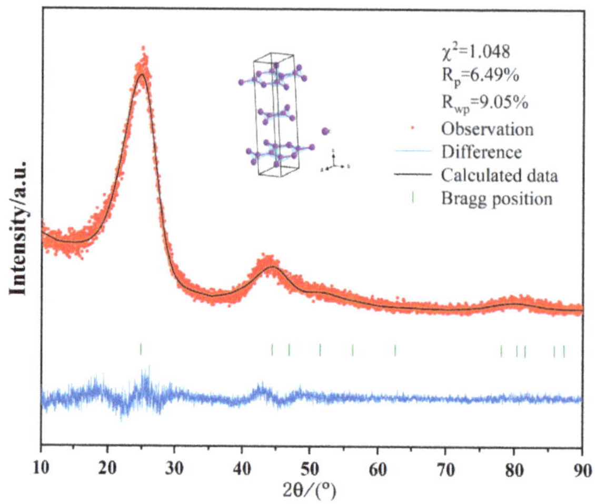

Figure 5. XRD refinement of the new carbon black.

From the match between the Bragg position and the diffraction peaks, it can be seen that the new carbon black exhibits strong crystalline features with a certain degree of long-range ordering within its crystal structure. The main diffraction peaks are concentrated between $2\theta = 20°\sim30°$, with cell parameters a = b = 2.38416, c = 10.60064, suggesting that the crystals may have hexagonal symmetry. The lattice angles are $\alpha = \beta = 90°$, $\gamma = 120°$, the cell volume = 52.18346^3, and the diffraction peak width is moderate, indicating that the grain size of the new carbon black is small, with the potential for microscopic strain on its surface.

4. Discussion of Results

4.1. Dispersibility

Good dispersibility ensures that stress is evenly transferred within the rubber material under external forces, preventing stress concentration and improving mechanical performance [14].

After the NBR composites were subjected to low-temperature embrittlement, the dispersion of the new carbon black particles in the rubber matrix was observed by SEM. As shown in Figure 6a, in the 20 phr new carbon black-filled SBR composites, the carbon black particles were uniformly distributed with moderate particle spacing, indicating good dispersion and some reinforcing effect. Despite the irregular shape and large size of the new carbon black particles, the overall interfacial bonding was good. As shown in Figure 6b, when the filling amount of the new carbon black was increased to 40 phr, the particle density increased significantly, and significant agglomeration also occurred. The bright areas show larger agglomerates, which may lead to localized stress concentration. As shown in Figure 6c, in the 20 phr N660 carbon black-filled SBR composites, the N660 carbon black particles were more uniformly distributed, with finer and denser distribution of bright spot areas. As shown in Figure 6d, when the filling amount of N660 carbon black was increased to 40 phr, although the particle density was increased and some agglomeration phenomenon appeared, the size of the agglomerates was smaller, and the overall dispersion was better than that of the new carbon black.

From the above analysis, it can be seen that both the new carbon black and N660 carbon black exhibit strong molecular chain adsorption and entanglement in SBR.

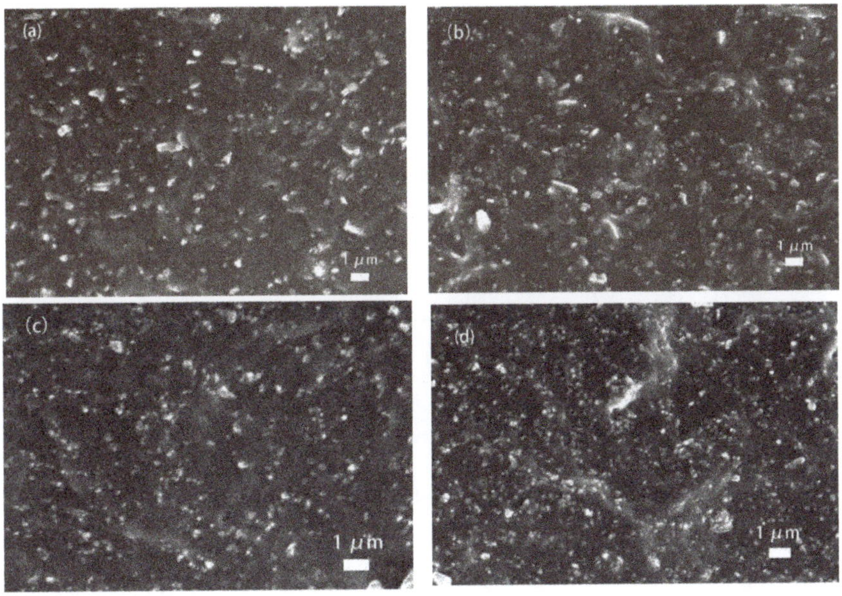

Figure 6. SEM images of different amounts of new carbon black- and N660 carbon black-filled SBR: (**a**) filled with 20 phr new carbon black; (**b**) filled with 40 phr new carbon black; (**c**) filled with 20 phr N660 carbon black; (**d**) filled with 40 phr N660 carbon black.

4.2. DSC

As shown in Figure 7, the glass transition behavior of the SBR composites was investigated by DSC. It was found that the median T_g of the new carbon black-filled SBR was −48.9 °C, which is slightly lower than that of the conventional N660 carbon black-filled SBR (T_g = −47.8 °C) by about 1.1 °C. This suggests that the new carbon black exerts a weaker restriction on the movement of the rubber molecular chain segments.

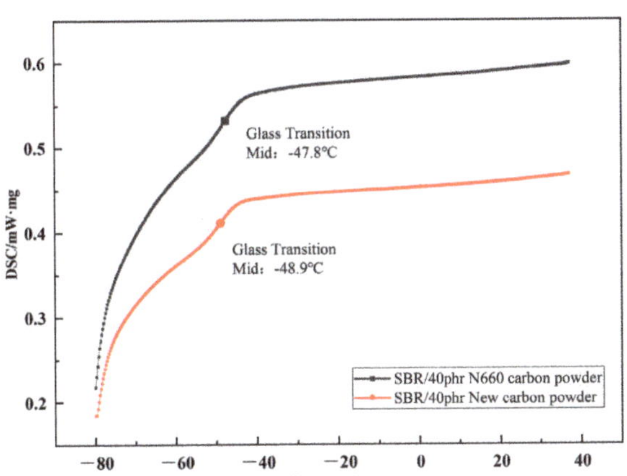

Figure 7. The glass transition temperature of 40 phr N660 carbon black- and new carbon black-filled SBR.

This phenomenon is closely related to the microstructure of the new carbon black and its interfacial interaction with the matrix due to its high specific surface area and uniform particle distribution. The new carbon black can form a more homogeneous filler network in the matrix. This network helps reduce the excessive binding of the local chain segments, allowing the rubber molecular chains greater freedom of movement [15]. Additionally, the lower surface energy of the new carbon black may reduce the physical forces between it and the SBR molecules, thus lowering the stiffness of the composite. The lower T_g indicates that the material exhibits superior flexibility and cold resistance at low temperatures, providing a potential advantage for applications involving the new carbon black in low-temperature conditions.

4.3. Vulcanization Properties

The effects of the dosage of new carbon black and N660 carbon black on the vulcanization characteristics of SBR composites are shown in Tables 1 and 2. The maximum torque (M_H) of the SBR composites increased from 16.123 dN·m to 20.754 dN·m with the increase in the dosage of new carbon black, which reflects the shear modulus of the vulcanized rubber and represents the strength of interaction between the rubber and the filler. This change indicates that the incorporation of the new carbon black significantly increased the cross-link density of the SBR composites. Meanwhile, the minimum torque (M_L) increased from 1.057 dN·m to 2.122 dN·m, indicating that the mobility of the composites gradually weakened with the increase in the content of the new carbon black. In addition, the difference between the maximum and minimum torque ($M_H - M_L$) increased from 15.066 dN·m to 18.628 dN·m, which is an important indicator of the degree of vulcanization cross-linking [16]. Compared with the S1-S4 samples (filled with N660 carbon black) and the S5-S8 samples (filled with new carbon black), the degree of cross-linking between the two samples was similar at 20–30 phr, and with the further increase in filler, the dispersion of the new carbon black was slightly worse than that of N660 carbon black, resulting in a slightly lower degree of cross-linking in SBR composites filled with new carbon black than N660 carbon black.

Table 1. Variable.

Component	Sample Number							
	S1	S2	S3	S4	S5	S6	S7	S8
New carbon black	0	0	0	0	20	30	40	50
N660 carbon black	20	30	40	50	0	0	0	0

Table 2. Vulcanization properties of N660 carbon black- and new carbon black-filled SBR.

Project	Sample Number							
	S1	S2	S3	S4	S5	S6	S7	S8
M_H/(dN·m)	16.916	19.506	22.384	23.83	16.123	18.275	19.999	20.75
M_L/(dN·m)	0.948	1.135	1.304	1.364	1.057	1.397	1.798	2.122
$M_H - M_L$/(dN·m)	15.968	18.371	21.08	22.466	15.066	16.878	18.201	18.628
t_{90}/min	7:18	6:37	6:05	5:39	5:47	6:05	6:10	6:17

The small change in t_{90} indicates that the addition of the new carbon black has a limited effect on vulcanization efficiency, allowing for the maintenance of high vulcanization efficiency. The vulcanization time of the S5–S8 samples ranges from 6:05 min to 6:17 min, which is significantly shorter compared to the S1–S2 formulations using N660 carbon black. This demonstrates that the new carbon black has a beneficial effect on promoting the vulcanization reaction. This improvement is primarily due to the active functional groups on the surface of the new carbon black. As a result, the cross-linking reaction is accelerated, making the vulcanization system more efficient.

4.4. Mechanical Properties

The effect of the dosage of new carbon black and N660 carbon black on the mechanical properties of SBR composites is shown in Table 3. The tensile strength of SBR composites increased significantly with the increase in carbon black and new carbon black filler. At 20 phr, the tensile strength of the new carbon black-filled SBR was 10.4 MPa, which was 7.2% higher than that of the N660 carbon black-filled SBR of 9.7 MPa, and the two reinforcing materials had similar enhancement effects at low filler levels. When the filling amount increased to 30 phr, the tensile strength of the new carbon black-filled SBR increased significantly to 18.3 MPa, which was 12.3% higher than that of the N660 carbon black-filled SBR at 16.3 MPa, showing a better reinforcing effect. When the filling amount was increased to 40 phr, the tensile strength of the new carbon black-filled SBR was 22.2 MPa, which was 4.4% higher than that of the N660 carbon black-filled SBR at 21.27 MPa. When further increased to 50 phr, the tensile strength of the new carbon black-filled SBR was 23.6 MPa, which was 5.2% higher than that of the N660 carbon black-filled SBR of 22.43 MPa. It is true that the new carbon black is slightly less dispersed compared to N660 carbon black, but the higher tensile strength observed in the new carbon black-filled SBR composites can be attributed to its unique lamellar structure as well as a degree of oxidizing properties on its surface, which facilitates its chemical reaction with the rubber matrix and enhances the interaction between the carbon black particles and the SBR matrix. In summary, the tensile strength of the new carbon black-filled SBR was always higher than that of the N660 carbon black-filled SBR at the same filling amount, and the reinforcing effect was the most significant at 30 phr, but the two reinforcing effects were gradually close to each other with the increase in the filling amount.

Table 3. Mechanical properties of N660 carbon black- and new carbon black-filled SBR.

Project	Sample Number							
	S1	S2	S3	S4	S4	S6	S7	S7
Shore A hardness/A	55	59	64	67	54	57	61	64
Tensile strength/MPa	9.7	16.3	21.27	22.43	10.4	18.3	22.2	23.6
Elongation at break/%	214.3	258.7	259.82	232.02	211.8	222.2	279.1	255.4
Tear strength/kN·m^{-1}	22.4	33.46	40.02	47.67	24.46	35.53	43.86	50.87

The tear strength of SBR composites increased with the increase in both carbon black and new carbon black filler amount. Compared with N660 carbon black, the new carbon black showed more significant reinforcement effects at different filler levels. The tear strength of SBR filled with new carbon black was always higher than that of SBR filled with N660 carbon black under the same filling amount, and the tear strength of SBR filled with new carbon black reached 43.86 MPa at 30 phr, while that of SBR filled with N660 carbon black reached 40.02 MPa, with an increase in tear strength of 9.6%. The network structure of the molecular chains of the composites was enhanced with the increase in the filler amount, and the elongation at break increased significantly. However, when the filler amount exceeded 40 phr, the elongation at break tended to stabilize or even decrease due to the decrease in filler dispersion and the appearance of agglomeration phenomenon. At the same time, the distribution density of filler particles in the matrix increased, forming a more compact network structure and leading to a further increase in the hardness of SBR.

4.5. Electrical Properties

Figure 8 demonstrates the effect of filling different amounts of N660 carbon black (S1-S4) and new carbon black (S5-S8), respectively, in a 100 phr SBR matrix on the electrical conductivity of vulcanized rubber. The conductivity of both carbon blacks increases with the increase in the filling level. In the N660 carbon black-filled formulation, the conductivity is gradually improved with the increase in the filling amount. Among them, S1 has the lowest conductivity of about 5.58485×10^{-8} S/cm, and S4 has the highest conductivity of about 5.82370×10^{-7} S/cm. In the new carbon black-filled formulation, the conductivity also increases with the increase in the filling amount. S5 has a conductivity of 7.64468×10^{-8} S/cm, which is about 4% higher than that of the N660 carbon black-filled SBR composite with the same filling amount. The conductivity of S5 was 7.64468×10^{-8} S/cm, which was about 4% higher than that of N660 carbon black-filled SBR composites with the same filling amount, and the conductivity of the S8 formula reached 6.55498×10^{-7} S/cm, which was about 13% higher than that of N660 carbon black-filled SBR composites with the same filling amount.

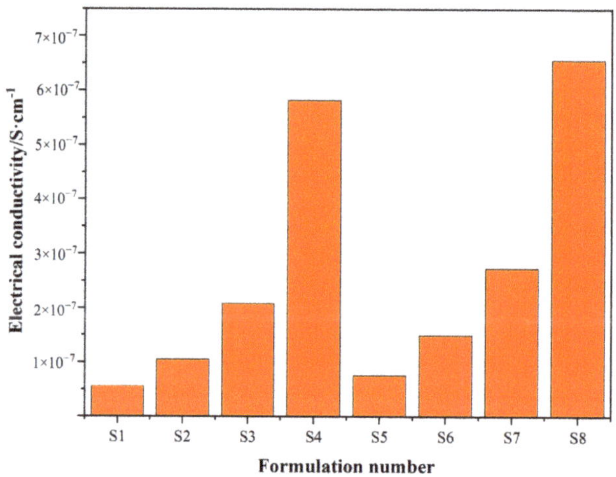

Figure 8. Electrical conductivity of N660 carbon black- and new carbon black-filled SBR.

Comparative results show that the conductivity of the new carbon black is overall better than that of N660 carbon black at the same filler level, and this difference is more significant at high filler levels. This could be attributed to the microstructure and interfacial properties of the new carbon black, which allows it to form a more efficient conductive network within the composites. This performance advantage highlights the potential of new carbon black for applications in conductive materials [17].

5. Conclusions

By analyzing the properties of the new carbon black, the following conclusions were drawn: the new carbon black can partially replace N660 carbon black as a reinforcing filler for SBR. When the filling amount of the new carbon black is 30 phr per 100 phr of SBR, the tensile strength is increased by 12.3% and the tear strength is increased by 9.6% compared with N660 carbon black-filled SBR composites, while the hardness is reduced. The new carbon black helps improve the vulcanization efficiency of SBR, which can shorten the vulcanization time t_{90} of SBR and ensure the uniformity and stability of the vulcanization reaction. In addition, the new carbon black has a stronger conductive reinforcement effect due to its lamellar structure, larger specific surface area, and multi-defect structure. Finally, the new carbon black has economic and environmental advantages—its price is only two-thirds of N660 carbon black and environmental protection is higher—to meet the green low-carbon production requirements.

Author Contributions: Conceptualization, Z.W.; methodology, J.Z. and J.S.; investigation, J.Z., L.Y. and B.T.; resources, Z.W.; data curation, J.Z.; writing—original draft preparation, J.Z.; writing—review and editing, Z.W. and J.S. All authors have read and agreed to the published version of the manuscript.

Funding: Natural Science Foundation of Shandong Province ZR2019MEM050.

Data Availability Statement: Data are contained within the article.

Conflicts of Interest: Author Junan Zhou is studying at the College of Mechanical and Electrical Engineering, Qingdao University of Science and Technology. The remaining authors declare that the research was conducted in the absence of any commercial or financial relationships that could be construed as a potential conflict of interest.

References

1. Ferraiuolo, R.; Alfe, M.; Gargiulo, V.; Pepe, G.P.; Tafuri, F.; Pezzella, A.; Ausanio, G.; Montemurro, D. Insights into the Electrical Characterization of Graphene-like Materials from Carbon Black. *Coatings* **2022**, *12*, 1788. [CrossRef]
2. Balasooriya, W.; Schrittesser, B.; Pinter, G.; Schwarz, T.; Conzatti, L. The Effect of the Surface Area of Carbon Black Grades on HNBR in Harsh Environments. *Polymers* **2019**, *11*, 61. [CrossRef] [PubMed]
3. Mohamad Aini, N.A.; Othman, N.; Hussin, M.H.; Sahakaro, K.; Hayeemasae, N. Lignin as Alternative Reinforcing Filler in the Rubber Industry: A Review. *Front. Mater.* **2020**, *6*, 329. [CrossRef]
4. Zhang, Z.; Yang, X.; Wang, X. Preparation and Performance of Graphene/Three-Component Solid Lubricants/Nitrile Rubber Composites. *Polym. Mater. Sci. Eng.* **2021**, *37*, 94–100.
5. Tang, C.; Ye, Y.; Liu, P.; Li, L.; Tan, S.; Gu, Q.; Chen, J. Application of a new high structure carbon black packing in natural rubber. *Carbon Technol.* **2020**, *39*, 51–55+71.
6. Nadaraia, L.; Dundua, T.; Gamkrelidze, N.; Tsitsishvili, V.; Barbakadze, N.; Chedia, R. Graphite Foil Waste to Graphene: New Carbon Precursors for Synthesis of Graphene and its Oxides. *Key Eng. Mater.* **2021**, *891*, 68–74. [CrossRef]
7. Fang, Y.; Dong, H.; Hao, X.; Liu, Y.; Tang, D.; Zhao, H.; Zhou, W.; Sun, C.; Zhang, L. Enhanced fatigue resistance of plasma modified pyrolysis carbon black filled natural rubber composites. *Appl. Surf. Sci.* **2024**, *643*, 158707. [CrossRef]
8. Gao, L.; Wei, X.; Gong, G.; Liang, T.; Wang, W. Progress of lignin reinforced emulsion polystyrene butadiene rubber research. *Polym. Bull.* **2014**, *11*, 116–121.
9. Shan, T.; Bian, H.; Zhu, D.; Wang, K.; Wang, C.; Tian, X. Study on the Mechanism and Experiment of Styrene Butadiene Rubber Reinforcement by Spent Fluid Catalytic Cracking Catalyst. *Polymers* **2023**, *15*, 1000. [CrossRef] [PubMed]
10. Berki, P.; Kocsis, K.J. Comparative Properties of Styrene-Butadiene Rubbers (SBR) Containing Pyrolytic Carbon Black, Conventional Carbon Black, and Organoclay. *J. Macromol. Sci. Part B* **2016**, *55*, 749–763. [CrossRef]
11. Pöschl, M.; Vašina, M.; Zádrapa, P.; Měřínská, D.; Žaludek, M. Study of Carbon Black Types in SBR Rubber: Mechanical and Vibration Damping Properties. *Materials* **2020**, *13*, 2394. [CrossRef] [PubMed]
12. He, J.; Fernández, C.; Primo, A.; Garcia, H. One-Step Preparation of Large Area Films of Oriented MoS_2 Nanoparticles on Multilayer Graphene and Its Electrocatalytic Activity for Hydrogen Evolution. *Materials* **2018**, *11*, 168. [CrossRef] [PubMed]
13. Pattinson, R.; Trialonis-Suthakharan, N.; Pickles, T.; Austin, J.; FitzGerald, A.; Augustin, M.; Bundy, C. Further refinement of the Patient-Reported Impact of Dermatological Diseases (PRIDD) measure using classical test theory and item response theory. *Br. J. Dermatol.* **2023**, *190*, 718–728. [CrossRef] [PubMed]
14. Fu, Y.; Wu, X.; Zhang, L.; Wang, W.; An, L.; Wang, H. Research progress of carbon black dispersion technology in rubber. *China Rubber Ind.* **2022**, *69*, 228–233.
15. Jiang, S.; Yong, Z. Modulation of Mechanical Properties of Silica-Filled Silicone Rubber by Cross-Linked Network Structure. *Polymers* **2024**, *16*, 2304. [CrossRef] [PubMed]
16. A, N.R.; Jan, M.N.; Hasan, A.D.; Lahmod, N.W.; Abbas, M.M.K. Research Journal of Applied Sciences, Engineering and Technology. *Res. J. Appl. Sci. Eng. Technol.* **2017**, *14*, 324–333.
17. Nan, X.; Zhang, Y.; Shen, J.; Liang, R.; Wang, J.; Jia, L.; Yang, X.; Yu, W.; Zhang, Z. A Review of the Establishment of Effective Conductive Pathways of Conductive Polymer Composites and Advances in Electromagnetic Shielding. *Polymers* **2024**, *16*, 2539. [CrossRef]

polymers

MDPI

Article

The Effect of Fiber Weight Fraction on Tribological Behavior for Glass Fiber Reinforced Polymer

Corina Birleanu [1], Razvan Udroiu [2,*], Mircea Cioaza [1,*], Marius Pustan [1], Bere Paul [3,*] and Cristian Vilau [4]

1 MicroNano Systems Laboratory, Mechanical Systems Engineering Department, Technical University from Cluj-Napoca, Blv. Muncii nr. 103-105, 400641 Cluj-Napoca, Romania; corina.barleanu@omt.utcluj.ro (C.B.); marius.pustan@omt.utcluj.ro (M.P.)

2 Manufacturing Engineering Department, Transilvania University of Brasov, Blv. Eroilor Nr. 29, 500036 Brașov, Romania

3 Manufacturing Engineering Department, Technical University from Cluj-Napoca, 400641 Cluj-Napoca, Romania

4 Mechanical Engineering Department, Technical University from Cluj-Napoca, 400641 Cluj-Napoca, Romania; cristian.vilau@tcm.utcluj.ro

* Correspondence: udroiu.r@unitbv.ro (R.U.); mircea.cioaza@staff.utcluj.ro (M.C.); paul.bere@tcm.utcluj.ro (B.P.)

Abstract: The tribological performance of Glass Fiber Reinforced Polymer (GFRP) composites is essential for applications in automotive, aerospace, and industrial sectors. This study investigates the effect of fiber weight fraction ratio (wf.) (50%, 65%, and 70%), applied load, and sliding speed on the tribological behavior of twill-woven GFRP using a pin-on-disc tribometer. Experimental trials were carried out to assess the impact of control factors on the coefficient of friction, specific wear rate, and contact temperature. Statistical analyses based on generalized linear models (GLM) method or multi-factor ANOVA, identified the most significant factors and their contributions. Results indicate that sliding speed contributes the highest to COF (46.51%), while fiber wf. primarily influences wear rate (34.15%). The applied load was found to have the strongest impact on contact temperature (39.08%). Furthermore, SEM and EDS analyses reveal dominant wear mechanisms, including abrasive wear and transfer layer formation. This study introduces the novelty of using statistical modeling to optimize GFRP for high-performance tribological applications, providing a more precise and efficient approach to enhancing their properties.

Keywords: glass fiber; tensile strength; friction coefficient; abrasive wear; pin-on-disc tribometer; design of experiments; statistical analysis; GLM method

1. Introduction

In recent years, the development of next-generation materials has progressed significantly. Traditional materials are increasingly being replaced by composite materials due to their superior properties, including high tensile strength, excellent strength to wf., and low thermal expansion [1]. Moreover, they exhibit full biodegradability through composting processes and do not emit toxic or harmful components [2].

FRP composites have been extensively employed in diverse engineering fields because of their outstanding combination of stiffness and strength relative to weight. Among them, GFRP composites have gained widespread adoption across multiple industries due to their exceptional strength to wf., corrosion resistance, and adaptability to diverse environmental conditions. These materials are extensively used in the automotive, aerospace, and marine sectors, where friction and wear performance significantly influence component longevity and operational efficiency. Additionally, GFRP composites offer highly flexible design solutions, remarkable durability, and cost-effective production and assembly processes [2].

The mechanical characteristics of GFRP composites are influenced by multiple factors, such as the type of fiber and its volume proportion, distribution, orientation, and void content. Furthermore, interfacial bonding characteristics and load transfer mechanisms at the fiber-matrix interface play a critical role in determining the composite's mechanical behavior [3]. Davallo et al. [4] demonstrated that flexural properties derived from force-deflection data of composites containing 20 wf. and 30 wf. randomly oriented continuous fibers exhibited flexural strength values of 84 MPa and 110 MPa, respectively, with corresponding flexural modules of 7 GPa and 10 GPa.

Experimental studies from the literature suggest that the tensile strength of GFRP composites is greatly influenced by different environmental conditions [3,4]. Abdullah et al. [5] investigated two commercial GFRP composites: chopped glass fiber composites and 0/90 woven fiber-reinforced composites with an unsaturated polyester matrix. Their results revealed that the composite with chopped fibers exhibited higher Young's modulus, maximum stress, and yield strength compared to the 0/90 composite. Likewise, Wazery et al. [3] observed that tensile strength ranged from 28.25 MPa to 78.83 MPa, flexural strength varied between 44.65 MPa and 119.23 MPa, and impact energy at room temperature fluctuated between 3.5 Joules and 6.5 Joules, depending on the glass fiber content (15 wt.% to 60 wt.%). These studies underscore the significant enhancement in mechanical properties, such as tensile and flexural strength, conferred by glass fiber reinforcement in polyester resin matrices [3].

Beyond mechanical performance, FRP composites also represent a vital class of tribomaterials. In applications involving sliding contact, understanding and optimizing the tribomechanical behavior of GFRP composites is essential for improving performance and durability. Tribological performance is influenced by multiple parameters, including operating conditions, fiber and matrix characteristics, manufacturing processes, interfacial properties, and contact conditions [6]. Numerous studies have been carried out in this area to improve the durability and frictional performance of these materials.

Zhao et al. [7,8] and Karsli et al. [9] have investigated various sliding conditions, such as dry adhesion sliding, abrasion, and erosion. Their findings suggest that wear performance under different sliding conditions is not necessarily correlated due to distinct wear mechanisms governing each case. The wear and friction characteristics of GFRP composites are mainly influenced by fiber weight fraction, applied force, and sliding velocity.

These factors influence critical indicators like the coefficient of friction (COF), specific wear rate, and contact temperature. Experimental studies indicate that optimizing these parameters can significantly improve wear resistance and friction efficiency, broadening the application scope of these advanced materials [10–15].

However, optimizing tribomechanical parameters remains challenging due to the complex interplay between material properties and operational conditions. Traditional experimental approaches often fail to fully capture these intricate relationships, necessitating the application of advanced statistical models for comprehensive analysis. Techniques such as analysis of variance (ANOVA) and Grey Relational Analysis (GRA) have been successfully employed to optimize the tribomechanical behavior of composite materials [15–20]. Also, in [10], based on the ELECTRE decision-making method, the tribological parameters have been optimized by assigning weighted factors to the coefficient of friction and specific wear rate. However, the effect of control factors on tribological parameters was not analyzed. The type of reinforcement and the tribological system strongly influence the wear rate, friction coefficient, and wear mechanisms, making them crucial factors in composite selection for specific applications [6]. From the literature survey, few research studies deeply investigated the effect of control factors on tribological properties of different composite materials.

Zaghloul et al. [21] found that glass fiber reinforcement in thermoplastic polymers, particularly at a 33% volume fraction, significantly improved wear resistance. However, increased fiber volume fractions inversely affected wear resistance, highlighting the complex relationship between mechanical properties and tribological performance. GFRP composites exhibit improved tensile strength with increased fiber volume fractions, with optimal performance at 25% to 50% volume fractions [21].

Lower friction coefficients are observed at optimal fiber orientations and volume fractions, enhancing the material's suitability for applications like bearings and gears [22].

Abdurohman et al. [23] suggested that different manufacturing processes (e.g., vacuum infusion, hand lay-up) affect the compressive and shear properties of GFRP, with vacuum infusion yielding superior mechanical performance.

Ali et al. [24,25] focus on the tribological properties of glass fiber reinforced polymer composites (GFRP), highlighting that different thicknesses of GFRP exhibit varying wear rates and sliding characteristics, with thicker samples demonstrating superior performance in mechanical tests and real-world applications.

The number of research papers published in the last two decades focusing on the wear of materials at different fiber volumetric fractions has been limited. The scarcity of studies in previous years has been a key motivation for evaluating the wear performance of reinforced polymers by adjusting fiber volume fractions. Considering the research gaps highlighted in the review article by Zaghloul et al. [21], this study aims to optimize the tribological parameters of a pin-on-disc friction system, where the pin is made of steel, and the disc consists of a GFRP composite. The experimental investigation will analyze the effects of fiber wf. (50%, 65%, and 70% wf.), applied load (10 N, 20 N, and 30 N), and sliding speed (0.1 m/s, 0.25 m/s, and 0.36 m/s) over 120 min. Statistical techniques, based on generalized linear models (GLM) method or multi-factor ANOVA, have been used to identify the most influential factors and their contributions to tribological performance. By optimizing these factors, this study seeks to offer meaningful contributions to the design and advancement of high-performance composite materials for tribological applications. SEM and EDS analyses identify key wear mechanisms, such as abrasive wear and the formation of a transfer layer.

2. Materials and Methods

2.1. Materials and Samples Manufacturing

Three testing samples were manufactured from glass fiber reinforced polymer composite materials at different fiber wf. of 50%, 65%, and 70%. A thermosetting polymer matrix was chosen, specifically the MGS LR135 epoxy resin, which was combined with the MGS LH136 hardener at a mixing ratio of 100:35 $\pm$2 g. Both components were supplied by Hexion (Duisburg, Germany). The epoxy resin MGS LR135, used in manufacturing our composite, is supplied by the manufacturer Hexion, and its material properties, including thermal behavior, are well-documented by the supplier. The heat treatment in the GFRP curing process, according to the tested cycle, includes a curing stage of 3 h at 120 °C. As is known, the resin withstands temperatures up to 10 degrees lower than the temperature it was exposed to during the heat treatment in the curing process, without its properties being affected. According to experimental results, the temperature during the tribological testing process did not reach this value. In this study, we relied on the resin manufacturer's documentation, which specifies the resin's heat treatment and curing procedure for certain applications. The reinforcement of the composite material was achieved using 270 g/m^2 twill-woven glass fiber fabrics to ensure mechanical stability and performance.

The production of GFRP plates with fiber wf. of 50%, 65%, and 70% adhered to a strictly regulated process to ensure uniformity and high-quality test specimens. The primary impregnation method utilized was the hand lay-up technique, wherein fiber layers were manually arranged and impregnated with epoxy resin. A total of six layers of twill-woven glass fiber fabric were used, each measuring 500 mm × 300 mm. The fibers were thoroughly impregnated with the resin matrix to protect them and facilitate effective load transfer. After applying the resin and fabric layers simultaneously, the entire surface was manually rolled to remove any air bubbles trapped between the overlapping layers.

In the subsequent stage, the GFRP material was covered with a release film, and a breather being placed inside a vacuum bag, as shown in Figure 1. For the curing procedure, an autoclave manufactured by Maroso (Maroso SRL, Veneto, Italy) was employed. This equipment enabled precise control of temperature and pressure during the curing process. The vacuum bag was subjected to a vacuum pressure of −0.9 bar during all curing cycle steps. The autoclave parameters were, Step 1, temperature increased at 80 °C, pressure 4 bars 30 min; Step 2, temperature increases at 120 °C pressure at 4 bars, 30 min; Step 3, temperature 120° C, 180 min, pressure 4 Bars; and Step 4, cooling procedure at 25 °C, 120 min.

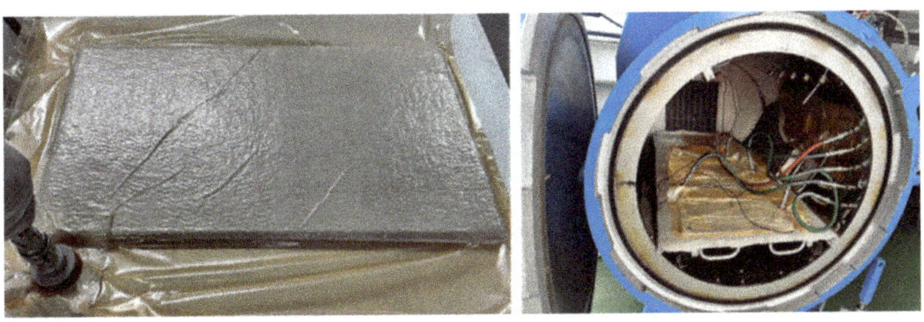

Figure 1. Vacuum bagging technique of the laminated sample and autoclave curing process.

The vacuum bagging process used in this study was carefully controlled to minimize defects and ensure uniform resin impregnation. However, as with any composite manufacturing process, a certain degree of porosity may be present. The wet resin impregnation method was employed, followed by vacuum bagging and autoclave curing. From a practical standpoint, it is impossible to achieve a completely pore-free structure using this method, which is considered one of the most effective in minimizing porosity. Any porosity that does occur is not present within the bulk material structure but manifests as a very slight surface porosity on the composite plates. When using pre-impregnated material, the structures obtained with these technologies may exhibit lower porosity. This is due to the reinforcement weight fraction ratio imposed by manufacturing companies. In this study, the main challenge was to achieve a controlled reinforcement weight fraction ratio at the initially established values while ensuring that the resin was not expelled from the reinforcement material under high pressures (50 tons/m) during the polymerization process.

As a result of this process, multiple composite plates were obtained, each measuring 500 mm × 300 mm × 2 mm, with reinforcement levels of 70%, 65%, and 50% (±0.5%). During the post-processing stage, after curing, the composite parts may require cutting, finishing, or mechanical processing to achieve the final dimensions and specifications. Any defects identified can be rectified through additional application of resin and reinforcing materials.

The prepared test specimens included:

- Tribological test discs: 50 mm in diameter and 2 mm in thickness, used for wear resistance analysis (Figure 2a);
- Tensile test specimens: 250 × 25 × 2 mm, prepared according to ASTM D3039-17 standard [26] (Figure 2b);
- Bending test specimens: 80 × 13 × 2 mm, prepared in accordance with the ASTM D7264D standard (Figure 2c).

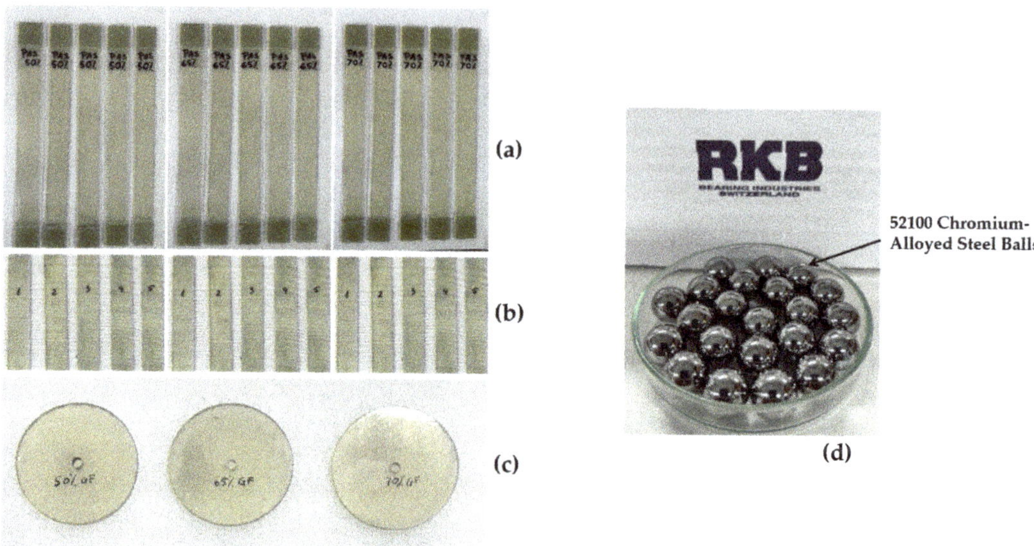

Figure 2. Specimens of GFRP composites (50%, 65%, and 70% wf.) used for: (**a**) mechanical tensile tests; (**b**) mechanical flexure tests; (**c**) tribological tests; and (**d**) 52100-chromium-alloyed steel balls used in tribological tests.

Compliance with these standards ensured consistent sample quality, enabling accurate and reproducible mechanical testing. In the initial phase, the polymeric semi-couples manufactured for experimental testing are presented in Table 1 and Figure 2a,b.

Table 1. Laminated composite—name and composition.

Laminated Composite Symbolization	Fiber (%)	Epoxy Resin (%)
Laminated epoxy/glass fiber composite (wf. 50%)—GFRP_50	50	50
Laminated epoxy/glass fiber composite (wf. 65%)—GFRP_65	65	35
Laminated epoxy/glass fiber composite (wf. 70%)—GFRP_70	70	30

Bearing balls made of 52100-chromium-alloyed steel (RKB Bearing Industries Group, Balerna, Switzerland) measuring 12.7 mm in diameter were utilized during the wear tests based on the recommendations of the ASTM A295 standard, (Figure 2d). Bearing balls were used in the wear tests as counter pieces against the circular composite material specimens (discs).

2.2. Mechanical Tests

The analysis of experimental data related to the tribological and mechanical testing of FRP materials involves several systematic steps, from processing primary data to interpreting and comparing the results. The mechanical testing of GFRP specimens was conducted in accordance with ISO 527-1:2019 (standard for tensile testing) and ISO 178:2003 (standard for flexural testing). The ISO standards correspond to ASTM D638-14 (Standard Test Method for Tensile Properties of Plastics) and ASTM D790-03 (Standard Test Methods for Flexural Properties of both Unreinforced and Reinforced Plastics).

The mechanical testing was conducted in the Accredited Mechanical Testing Laboratory of the Department of Mechanical Engineering, Technical University of Cluj-Napoca, using a mechanical testing machine, specifically the Instron 8801 (Instron, Norwood, MA, USA), with a maximum load capacity of 100 kN. The equipment automatically records the loading process in the form of a stress-strain curve, which represents the applied stress in the specimen and the corresponding strain. Each test involved five specimens. The applied load was controlled at a crosshead speed of 2 mm/min until specimen failure, under ambient conditions of 18 °C temperature and 50% relative humidity. For flexural testing, the specimens (Figure 2b) were evaluated using a three-point bending test with a universal testing machine, Instron 3366-10 kN (Instron, Norwood, MA, USA). The parameters set for the flexural tests were identical to those used for the tensile tests.

2.3. Wear Tests

Tribological testing is crucial for evaluating the behavior of materials under conditions similar to their actual operational environment, enabling the understanding and optimization of performance in industrial applications. In this context, the tests performed on GFRP specimens in contact with steel balls aimed at characterizing the composite material's behavior under specific dry friction conditions. The experimental procedures for these tests follow established standards, including ASTM G99 (Wear Testing with a Pin-on-Disk Apparatus) and ASTM F732 (Wear Testing of Polymeric Materials).

The study of surface interactions and behaviors in relative motion is highly dependent on operating conditions, which play a fundamental role in evaluating the performance of systems and materials. The establishment and control of operating conditions, as detailed in Table 2, are essential for ensuring accurate and relevant results. Tribological variables such as the coefficient of friction and wear rate can be significantly influenced by working parameters and environmental conditions during the experiment.

Table 2. Testing parameters.

Parameters	Operating Conditions
Load (Force)	10, 20, 30 N
Sliding Speed	0.1, 0.25, 0.36 m/s
Rotational Speed (RPM)	Max 215 (±3) rpm
Relative Humidity	40 (±5)%
Initial Temperature	20 (±2) °C
Test Duration	120 min
Conditions	Dry friction
Materials Disc and Ball	GFRP composites and 52100-chromium-alloyed carbon steel balls
Roughness of the Disc and Ball [Ra]	0.32 μm and 0.04 μm

Prior to testing, specimen preparation involved careful cleaning with technical alcohol and weighing using an analytical balance with 0.1 mg precision. After each test, both the steel balls and composite discs were replaced with new ones to ensure consistency and avoid cross-contamination. By establishing and maintaining appropriate and consistent operating conditions, reliable and reproducible data can be obtained, enabling the evaluation and comparison of different materials or systems across various operating scenarios. Moreover, this approach allows for the identification of key factors influencing tribological behavior and the development of effective strategies for improving the performance and durability of studied materials. During the experimental trials, operating conditions were strictly monitored and controlled to guarantee precision and reliability of the obtained data, allowing for meaningful and applicable conclusions to be drawn from the results.

The tribological analysis focused on abrasive wear evaluation, material transfer mechanisms, and thermal stability, with continuous monitoring of humidity, chamber temperature, and local frictional temperature. Additionally, the coefficient of friction was continuously recorded, and wear assessments were performed. Surface wear images were captured, test parameters were documented, and material transfer was analyzed.

Friction and wear tests were conducted using a pin-on-disc tribometer under the conditions outlined in Table 2. To improve the reliability of the results, each experiment was repeated five times, with the reported values representing the average measurements. The dispersion of results was highlighted using error bars, representing the corresponding standard deviation.

For each test, a new ball and disc were utilized, and prior to experimentation, they were cleaned with ethanol and then dried in a controlled environment. Throughout the test, temperature, worn surface, and friction force were continuously monitored. Coefficients of variation (CV) of the tribo-mechanical parameters were calculated based on standard deviation and mean values. The coefficient of variation is a measure of spread that describes the variation in the data relative to the mean. CV coefficients less than 10% were determined, which confirm the repeatability of the experiments.

The test was conducted for 120 min, during which the coefficient of friction was recorded in real time, capturing both the initial run-in phase and the steady-state friction regime. The temperature generated during the experiments was also continuously monitored. After a specified number of cycles, the test was temporarily halted to perform a topographic surface analysis, allowing for wear quantification and assessment of surface roughness evolution.

To achieve a more comprehensive understanding of the wear phenomenon and material removal mechanisms, 3D optical microscopy (OM) was employed to examine the worn surfaces of the ball and the running surface of the disc after each test. Surface modifications were assessed using the 3D Nano Focus optical microscope (NanoFocus AG, Oberhausen, Germany), which incorporates advanced µsurf technology for high-precision 3D surface measurements [16].

Accurate measurement of the wear track is functionally essential. The profilometric measurement method is particularly useful but is susceptible to measurement errors. Factors such as human judgment and measurement techniques significantly influence the accuracy of wear track evaluation. The best practice involves performing a complete surface measurement of the wear track. The number of scans represents the most critical factor in measuring the wear track in the pin-on-disc configuration, with a higher number of scans being required for non-homogeneous wear tracks. In this study, eight scans were performed on each disc at a 45° angle between them, allowing for a more precise estimation of volumetric wear, which led to a reduction in uncertainty. To validate the measurements, in addition to µsurf technology, the discs were also scanned using the wear modulus of the Tribometer TRB^3 (Anton Paar GmbH, Graz, Austria). The modelization and tribometer software share the same InstrumX user interface with modular technology, which enables a complete integration of various data in a single document including tribometer measurements and modelization, image management, statistics, and reporting functions.

The wear rate (K, mm^3/Nm) was determined using Equation (1) [16]:

$$K_{disc} = \frac{V_{disk}}{L_{sliding} \cdot F}; \; K_{ball} = \frac{V_{ball}}{L_{sliding} \cdot F} \quad (1)$$

where it is noted that:

- V_{disk}—wear volume of the disc (mm^3);
- V_{ball}—wear volume of the ball (mm^3);
- F—normal force (N);
- $L_{sliding}$—total sliding distance (m);
- K_{disk}, K_{ball}—wear factors of the disc and ball (mm^3/Nm).

The wear volume of the specimens was calculated by measuring the width and depth of the wear tracks using advanced µsurf technology. These calculations were derived from empirical mathematical equations, assuming the ideal geometric accuracy of the ball as a model for the wear marks.

Additionally, an infrared thermal camera, FLIR E5xt (Teledyne FLIR Company, Wilsonville, OR, USA), equipped with MSX and Wi-Fi technology, within a temperature range of −20 °C to +400 °C, a 160 × 120-pixel resolution, and a 9 Hz refresh rate, was used to monitor temperature variations during the tests.

2.4. Design of the Experiments and Statistical Method

The tribological experimental plan was designed to assess the impact of three control factors—fiber weight fraction (wf.), applied force (F), and sliding speed (v)—on three response variables: coefficient of friction (COF), specific wear rate (K), and temperature (T). Each control factor was evaluated at three distinct levels, as detailed in Table 3.

Table 3. Control factors and their levels for statistical analysis.

Targets	Glass Fiber, wf.		Applied Load, F		Sliding Speed, v	
	Symbol	**Value [%]**	**Symbol**	**Value [N]**	**Symbol**	**Value [m/s]**
Coefficient of friction	1	50	1	10	1	0.10
Specific wear rate	2	65	2	20	2	0.25
Temperature	3	70	3	30	3	0.36

A general full factorial design with 27 factor combinations was implemented to examine the effects of control factors on target variables. Statistical analysis was conducted using Generalized Linear Models (GLM) or multi-factor ANOVA in Minitab 19 software (Coventry, UK). The percentage contribution ratio (PC%) of individual factors and their interactions was determined. Significant control factors were identified from the ANOVA tables based on F-values and p-statistical parameters. Graphical methods, including main effects plots, interaction effects plots, and interval plots of target variables against control factors, were utilized for data visualization. Lastly, ANOVA assumptions were verified to ensure the validity of the analysis.

2.5. Morphological Analysis

The JEOL JSM-5600LV Scanning Electron Microscope (SEM), manufactured by JEOL Ltd., Tokyo, Japan, analyzes samples in both high and low vacuum modes, enabling the investigation of conductive and non-conductive materials. Equipped with a tungsten filament electron gun, secondary electron (SE), and backscattered electron (BSE) detectors, it provides high-resolution imaging and compositional contrast.

For elemental analysis, the ULTIM MAX 65 Energy Dispersive X-ray Spectroscopy (EDX) detector from Oxford Instruments, High Wycombe, UK, is used. Featuring Aztec 4.2 software, it ensures precise spectral acquisition, real-time element identification, and high-sensitivity mapping. The SEM-EDX system supports microstructural and compositional studies across various scientific and industrial applications.

3. Results and Discussion

3.1. Results of Mechanical Testing

For mechanical testing, five specimens were prepared for each type of GFRP and subsequently subjected to tensile and flexural testing using the standardized testing methods described above. The mean values of the mechanical properties, along with their standard deviations for all tested specimens, are summarized in Tables 4 and 5. Specifically, the results of the tensile tests are detailed in Table 4. Flexural tests are used to evaluate the material's resistance to forces acting perpendicular to its length, measuring failure mode and deformation under bending stresses, and the results are shown in Table 5.

Table 4. Tensile properties of GFRP specimens.

Specimen	Tensile Strength [MPa]/ SD [MPa]/CV [%]	Tensile Strain at Maximum Load [%]/ SD [%]/CV [%]	Elastic Modulus E [MPa]/ SD [MPa]/CV [%]
GFRP 70%	480.1/(25.92)/5.39	2.94/(0.03)/1.02	22,181.7/(253.2)/1.14
GFRP 65%	376.57/(24.37)/6.47	2.63/(0.02)/0.76	18,720.6/(316.36)/1.73
GFRP 50%	318.8/(22.45)/7.04	2.41/(0.02)/0.82	14,602/(301.25)/2.06

SD—Standard Deviation; CV—Coefficient of Variation

The increase in glass fiber content significantly improves tensile strength: GFRP 70% exhibits 59% higher tensile strength compared to GFRP 50%. The elastic modulus also increases, indicating higher stiffness. Figure 3 illustrates the variation in tensile strength and elastic modulus for GFRP specimens with different fiber content (70%, 65%, and 50%), highlighting the direct correlation between increased fiber reinforcement and enhanced mechanical performance, particularly in terms of strength and stiffness.

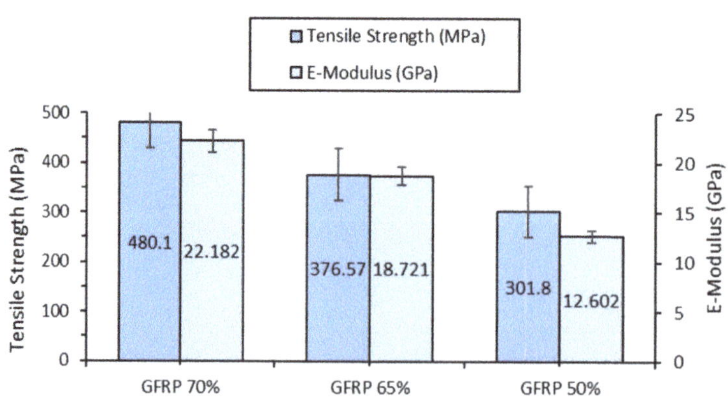

Figure 3. Tensile strength and elastic modulus of GFRP 70%, GFRP 65%, and GFRP 50% specimens.

Table 5. Flexural properties of GFRP specimens.

Specimen	Flexural Strength [MPa]/ SD [MPa]/CV [%]	Flexural Strain [%]/ SD [%]/CV [%]	Elastic Modulus E [MPa]/ SD [MPa]/CV [%]
GFRP 70%	464.8/(11.47)/2.46	3.77/(0.16)/4.24	21,009/(392)/1.52
GFRP 65%	415.5/(21.56)/5.18	3.08/(0.13)/4.22	18,218/(484)/2.65
GFRP 50%	388.6/(36.8)/9.46	2.36/(0.25)/9.59	14,935/(199)/2.94

SD—Standard Deviation; CV—Coefficient of Variation

Flexural tests confirm similar trends observed in tensile testing, demonstrating an increase in resistance to forces acting perpendicular to the specimen's length. GFRP 70% reaches a flexural strength of 464.8 MPa, approximately 19.6% higher than that of GFRP 50%, which exhibits a strength of 388.6 MPa. The flexural elastic modulus follows the same trend (Figure 4 and Table 4), reaching 21,009 MPa for GFRP 70%, a 40.7% increase compared to GFRP 50%. A notable observation is the variation in flexural strain, where GFRP 50% shows the highest deformation (3.8%), while GFRP 65% exhibits the lowest value (2.4%), indicating a possible nonlinear influence of fiber content on flexural behavior.

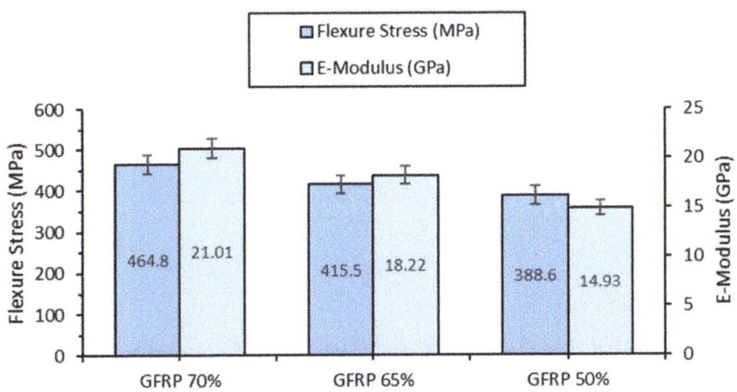

Figure 4. Flexural strength and elastic modulus of GFRP 70%, GFRP 65%, and GFRP 50% specimens.

Overall, increasing the fiber weight fraction significantly enhances the strength and stiffness of the material, enhancing its suitability for applications that demand a high load-bearing capacity. However, as fiber becomes more dominant in the polymer matrix, ductility may decrease, potentially affecting tribological behavior, particularly in terms of wear particle formation and friction coefficient. These findings suggest that optimizing the tribological and mechanical performance of GFRP composites requires balancing strength, stiffness, and deformation capacity to achieve the best overall performance.

3.2. Tribological Testing Results

Tribological testing results provide essential data for determining the coefficient of friction, wear rate, and understanding the degradation mechanisms of the fiber-reinforced composite material in relation to the contact material of the ball.

Throughout the 135 experimentally validated tests conducted for the friction pair, special attention was given to strict control and monitoring of the operating conditions. This rigorous approach ensured the precision and significance of the acquired data, allowing meaningful and applicable conclusions to be drawn from the experimental results. The results of the tribological analysis are summarized in Table 6, highlighting the control factors (fiber weight fraction, applied force, and sliding speed) and the target parameters (coefficient of friction, specific wear rate, and temperature).

Table 6. Experimental results of dry sliding wear.

	Experimental Parameters			Mean of Measured Parameters		
Exp.nr.	Applied Load F [N]	Sliding Speed v [m/s]	Glass Fiber wf. [%]	Specific Wear Rate K [10^{-5} mm^3 × (Nm)$^{-1}$]	Coefficient of Friction [μ] Average of the Last 60 min	Temperature Average of the Last 60 min [°C]
1	10	0.1	50	5.758	0.54	32.7
2	10	0.1	65	3.42	0.5	29.4
3	10	0.1	70	1.5488	0.43	35.1
4	10	0.25	50	20.985	0.4	35
5	10	0.25	65	12.267	0.39	36.2
6	10	0.25	70	6.4267	0.38	31
7	10	0.36	50	14.907	0.39	40.2
8	10	0.36	65	16.022	0.38	41.4
9	10	0.36	70	9.59	0.34	25.1
10	20	0.1	50	15.208	0.54	32
11	20	0.1	65	14.189	0.48	29.2
12	20	0.1	70	11.456	0.41	35.9
13	20	0.25	50	24.906	0.45	47.5
14	20	0.25	65	32.737	0.43	45.7
15	20	0.25	70	6.12	0.42	37.8
16	20	0.36	50	26.019	0.37	48.7
17	20	0.36	65	23.291	0.36	51.7
18	20	0.36	70	4.42	0.35	46.3
19	30	0.1	50	21.671	0.51	40.8
20	30	0.1	65	18.256	0.49	34.9
21	30	0.1	70	14.426	0.48	62.7
22	30	0.25	50	32.385	0.37	52.4
23	30	0.25	65	23.727	0.36	52.7
24	30	0.25	70	14.341	0.33	47.1
25	30	0.36	50	14.488	0.49	61.3
26	30	0.36	65	19.191	0.48	62.7
27	30	0.36	70	10.492	0.47	34.2

The mass reduction of the chromium-alloyed steel balls was negligible, whereas for the disc, the measured weight loss across all experimental tests ranged from 0.001 to 0.030 g, increasing with both sliding speed and applied force. Consequently, the wear rate (K, mm^3/Nm) was determined using the results obtained through 2D Optical Scanning.

Figure 5a–c shows the 2D optical images of the wear track profiles obtained on the GFRP discs in contact with the 52100-steel ball. Based on the experimental profilometry findings, Table 6 presents the wear rate range of the composite disc for three different sliding speeds and corresponding normal values of the applied force.

The minimal wear (Figure 5) observed on the 52100-chromium-alloyed steel balls when in contact with GFRP discs of varying fiber weight fractions (50%, 65%, and 70%) can be attributed to several key factors related to the material properties and tribological interactions. The 52100-steel is a high-carbon, chromium-alloyed bearing steel with excellent wear resistance and high hardness (60–64 HRC), which significantly limits material removal due to abrasive contact. Given that GFRP is significantly softer than steel, the interaction results in minimal wear on the steel balls, as the primary material loss occurs within the polymeric matrix of the composite rather than on the metallic counter face.

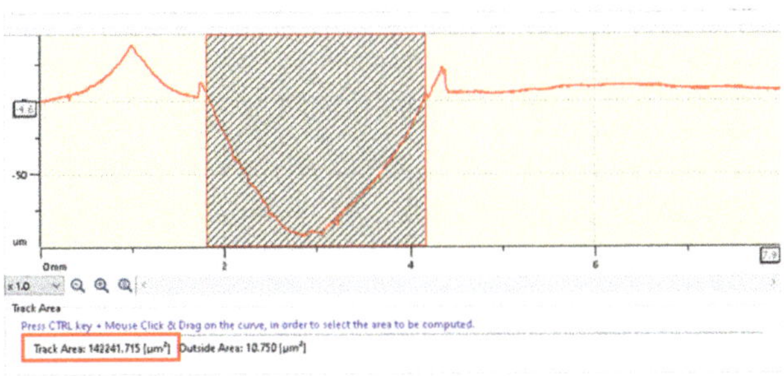

(**a**)

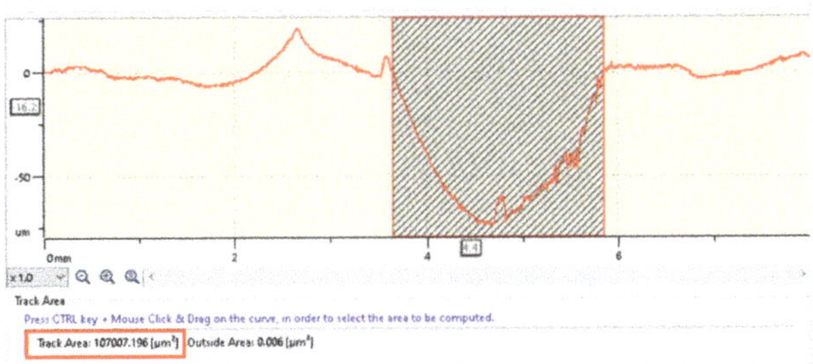

(**b**)

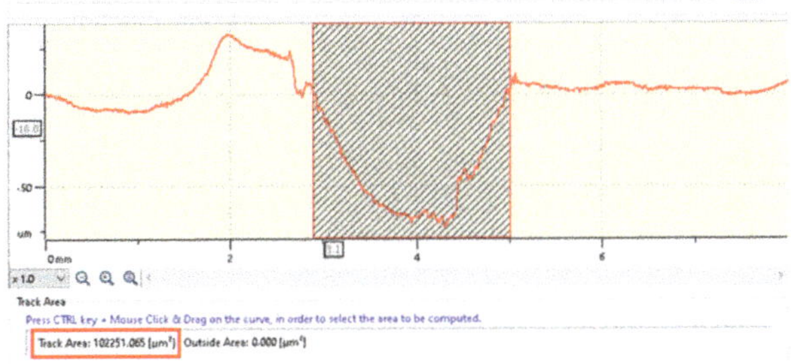

(**c**)

Figure 5. (**a**) A 2D profilometry measurement of wear tracks of the GFRP50 wf. specimen after the dry friction test with F = 30 N, v = 0.25 m/s, duration of 120 min, in contact with a chromium-alloyed steel ball (**left**) and 3D optical profilometry images (**right**); (**b**) similar images for GFRP65 wf.; (**c**) similar images for GFRP70 wf.

3.3. Results of Statistical Analysis

The key findings of the statistical analysis for the coefficient of friction, specific wear rate, and temperature are presented in Table 7. A control factor was considered to have a significant influence on the target parameters when the p-value was lower than the significance threshold of 0.05.

The most significant percentage contribution ratios on the coefficient of friction were 46.51% obtained for sliding speed, 32.51% for applied force to sliding speed interaction, and 11% for glass fiber wf. ratio, as is shown in Table 7. The applied load factor and the other two interactions between factors had lower percentage contributions to the coefficient of friction.

This behavior can be explained by the complex interplay of tribological mechanisms that govern friction behavior in glass fiber-reinforced polymer (GFRP) composites under different operating conditions.

The effect of applied load on COF is primarily related to the contact pressure at the interface between the 52100-chromium-alloyed steel ball and the GFRP composite. At lower loads, the contact pressure remains moderate, resulting in stable friction behavior. However, as the load increases, localized deformation at the interface intensifies, leading to fiber-matrix detachment, increased material transfer, and potential modifications in the surface topography. These effects can either increase or decrease COF, depending on the extent of the deformation and wear mechanisms involved.

Sliding speed also plays a critical role in COF variation. At lower speeds, friction remains relatively high due to limited thermal effects. As the sliding speed increases, frictional heating causes localized softening of the polymer matrix, reducing adhesion and lowering the COF. However, at very high speeds, excessive heat accumulation may degrade the polymer surface, potentially increasing COF due to surface roughening or instability in the wear mechanism.

The interaction between load and sliding speed emerges from their combined influence on surface deformation, material transfer, and thermal effects. At high loads and low speeds, stronger adhesive interactions increase COF due to intensified fiber-matrix detachment. As the speed increases under high load conditions, frictional heating facilitates the formation of a polymeric transfer layer, reducing direct contact and lowering COF. However, at high loads and very high speeds, excessive heat buildup can degrade the composite surface, leading to unstable friction behavior. In contrast, low loads combined with high speeds result in lower contact pressure, reducing heat generation and stabilizing COF.

These findings align with our SEM and EDS analyses, which indicate that increased load and sliding speed alter wear mechanisms by affecting fiber exposure, polymer matrix softening, and transfer layer formation. The observed trends further support the explanation of this significant interaction effect.

The glass fiber weight fraction (wf.) exhibited the most significant impact on the specific wear rate, contributing 34.15%, followed by the applied load at 21.87% and the sliding speed at 14.11%, as detailed in Table 7. The interactions among these factors were not statistically significant, as their p-values exceeded the 0.05 significance threshold, as shown in Table 7.

Table 7. The statistical results for coefficient of friction, specific wear rate, and temperature.

	COF			K			T		
Source	**F-Value**	***p*-Value**	**PC [%]**	**F-Value**	***p*-Value**	**PC [%]**	**F-Value**	***p*-Value**	**PC [%]**
wf	18.17	0.001	11	17.08	0.001	34.15	0.84	0.466	2.71
F	5.07	0.038	3.07	10.94	0.005	21.87	12.13	0.004	39.08
v	76.85	<0.001	46.51	7.05	0.017	14.11	3.85	0.068	12.39
wf*F	0.81	0.551	0.98	1.47	0.296	5.89	0.06	0.992	0.4
wf*v	2.90	0.094	3.51	2.07	0.177	8.28	4.23	0.039	27.24
F*v	26.86	<0.001	32.51	1.92	0.200	7.7	0.82	0.546	5.3
Error	-	-	2.42	-	-	8	-	-	12.88
Total	-	-	100	-	-	100	-	-	100

Note: "*" signifies the interaction between factors.

The applied load and the interaction between glass fiber weight fraction (wf.) and sliding speed were the most influential factors affecting temperature, contributing 39.08% and 27.24%, respectively. In contrast, the individual effects of wf. and sliding speed, along with other factor interactions, had lower contributions to temperature, as shown in Table 7.

The graphical results of statistical analysis for all three targets (COF, K, and T) are explained based on the main effects plot, interaction plot and interval plot.

The maximum average values of COF were obtained for wf. 50%, force at level 3 (30 N) and sliding velocity at level 1 (0.1 m/s), as shown in Figure 6a. It was observed that there was a decreasing trend for COF with increasing wf., an increasing trend for COF with increasing F, and a large decrease in COF from level 1 to level 2 of v, followed by a small increase. The lowest mean values of COF were obtained for wf. = 70%, F = 10 N, and v = 0.25 m/s.

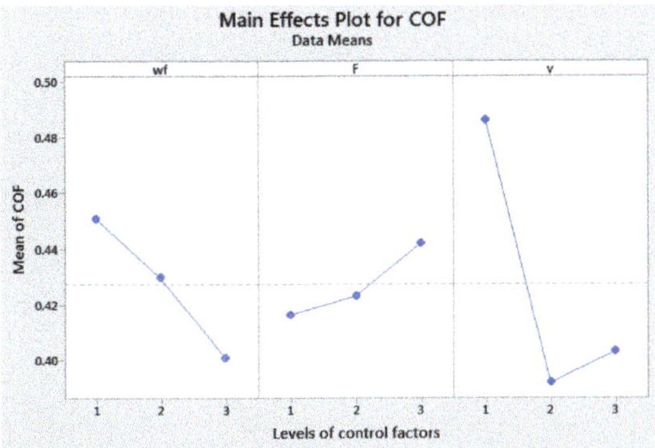

(a)

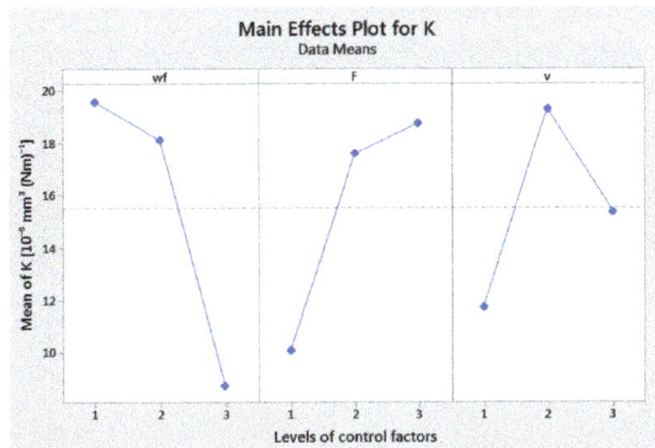

(b)

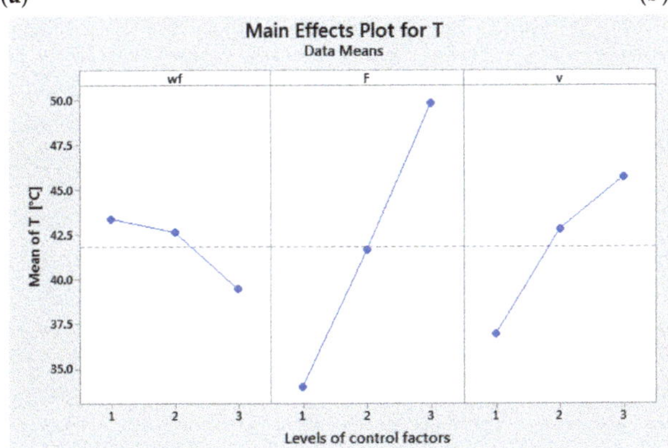

(c)

Figure 6. Main effects plot for: (**a**) coefficient of friction; (**b**) specific wear rate; and (**c**) temperature.

The main effects plot results indicate that the highest mean values of K were observed for a glass fiber weight fraction (wf.) at level 1 (50%), an applied force at level 3 (30 N), and a sliding speed at level 2 (0.25 m/s), as illustrated in Figure 6b. The specific wear rate decreases with an increase in glass weight fraction. The specific wear rate raises with increasing load applied. It was observed that there was a large increase in K from level 1 (0.1 m/s) to level 2 (0.25 m/s) of v, followed by a small decrease to level 3 (0.36 m/s).

The primary effects on temperature were identified for the glass fiber weight fraction (wf.) at level 1 (50%), the applied force at level 3 (30 N), and the sliding speed at level 3 (0.36 m/s), as illustrated in Figure 6c. Additionally, it was observed that temperature decreased as the glass fiber weight fraction increased.

The interaction between applied force (F) and sliding speed (v) had a significant effect on the coefficient of friction, as illustrated in Figure 7a and confirmed in Table 7. Minimal interactions between factors were observed for the specific wear rate, as shown in Figure 7b. Additionally, the interaction between glass fiber weight fraction (wf.) and sliding speed (v) had a notable impact on temperature, as depicted in Figure 7c.

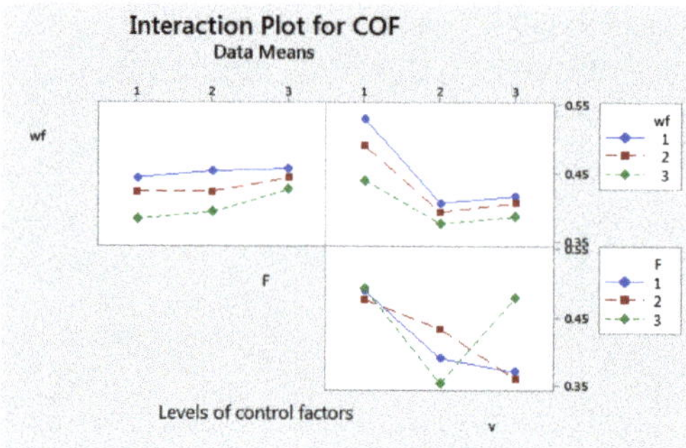

(a)

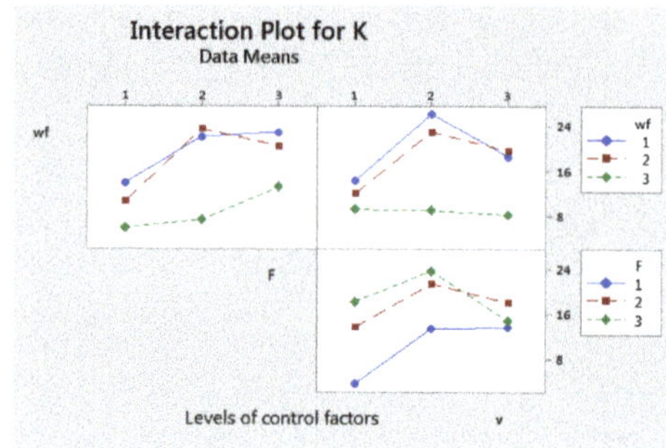

(b)

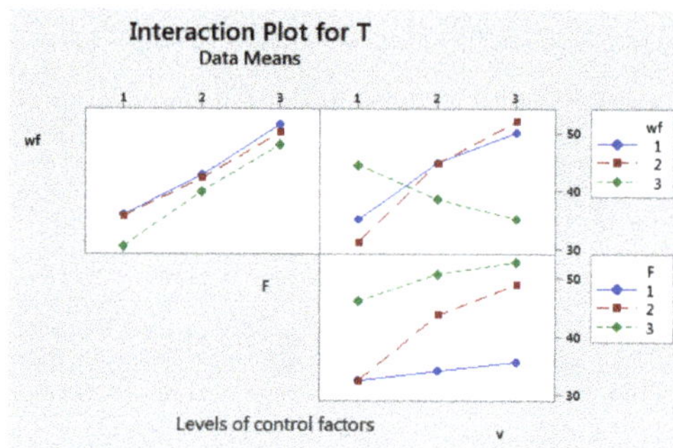

(c)

Figure 7. Interaction effects plot for: (**a**) coefficient of friction; (**b**) specific wear rate; and (**c**) temperature.

The interval plots with standard error bars for each factor in relation to the coefficient of friction are presented in Figure 8. The difference between the means for coefficient of friction in v, were significant because the interval bar for level 1 (v = 0.1 m/s) and the interval bars for level 2 (v = 0.25 m/s) and 3 (v = 0.36 m/s) did not overlap (Figure 8c). The differences for glass weight fraction and applied force were probably not significant because all the interval bars have over-lapped (Figure 8a,b).

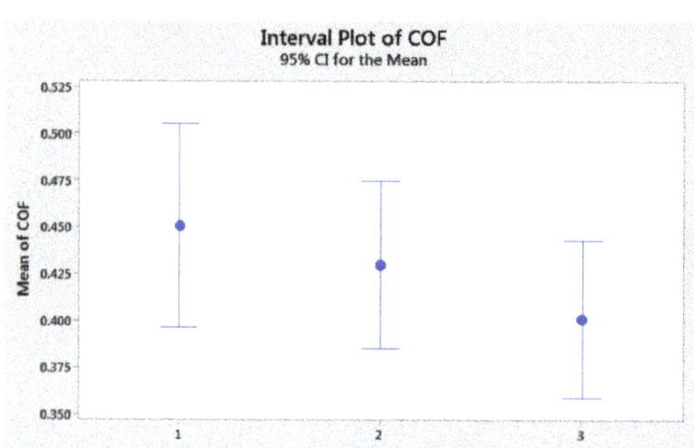

(a)

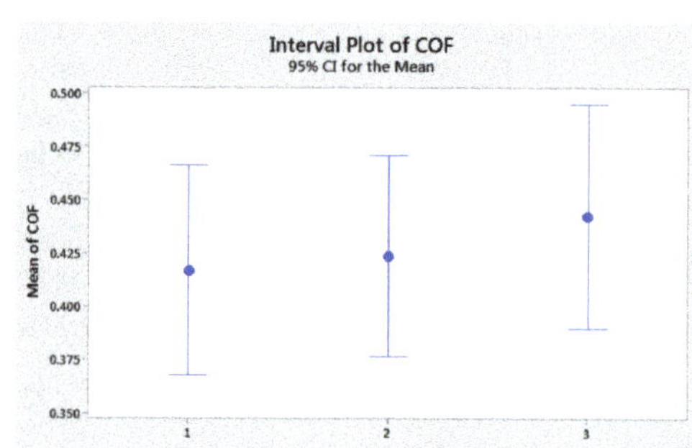

(b)

Figure 8. *Cont.*

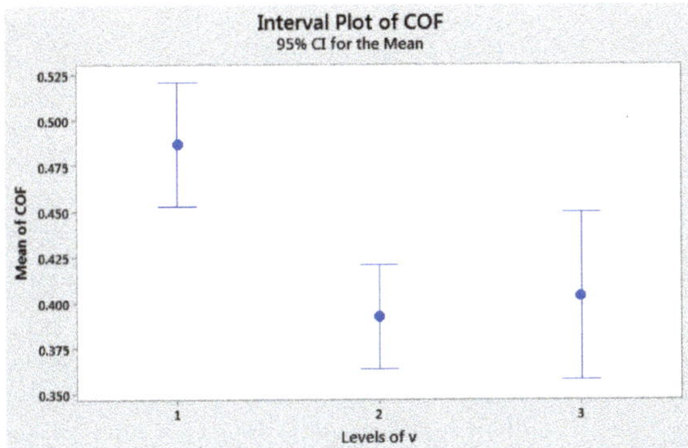

(c)

Figure 8. Interval plot of the coefficient of friction (COF) factor with: (**a**) wf.; (**b**) F; and (**c**) v. Individual standard deviations were used to calculate the interval plot. Bars are standard errors of the mean.

The interval plots of factors versus specific wear rate are shown in Figure 9. The lowest mean values of K were recorded for wf. = 70% (level 3), F = 10 N (level 1), and v = 0.1 m/s (level 2). Similarly, the interval plots of each factor versus temperature are illustrated in Figure 10. The lowest mean temperature values were observed for wf. = 70%, F = 10 N, and v = 0.1 m/s. The difference in temperature means for sliding speed (v) was statistically significant, as indicated by the non-overlapping interval bars between levels 1 and 3.

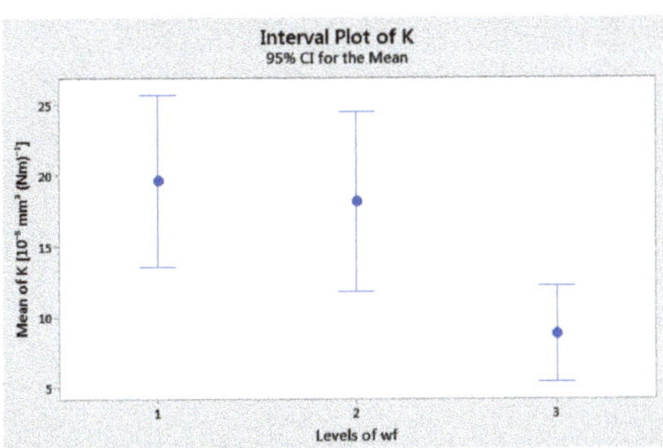

(a)

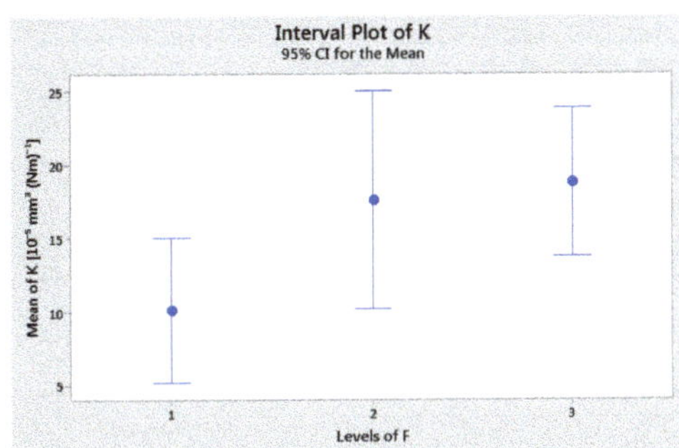

(b)

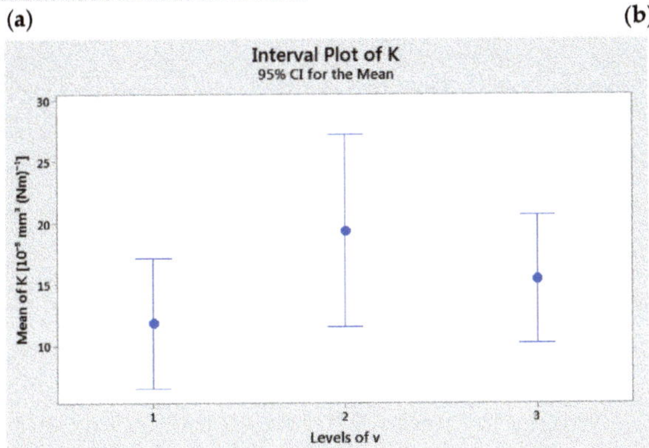

(c)

Figure 9. Interval plot of the specific wear rate (K) factor with: (**a**) wf.; (**b**) F; and (**c**) v. Individual standard deviations were used to calculate the interval plot. Bars are standard errors of the mean.

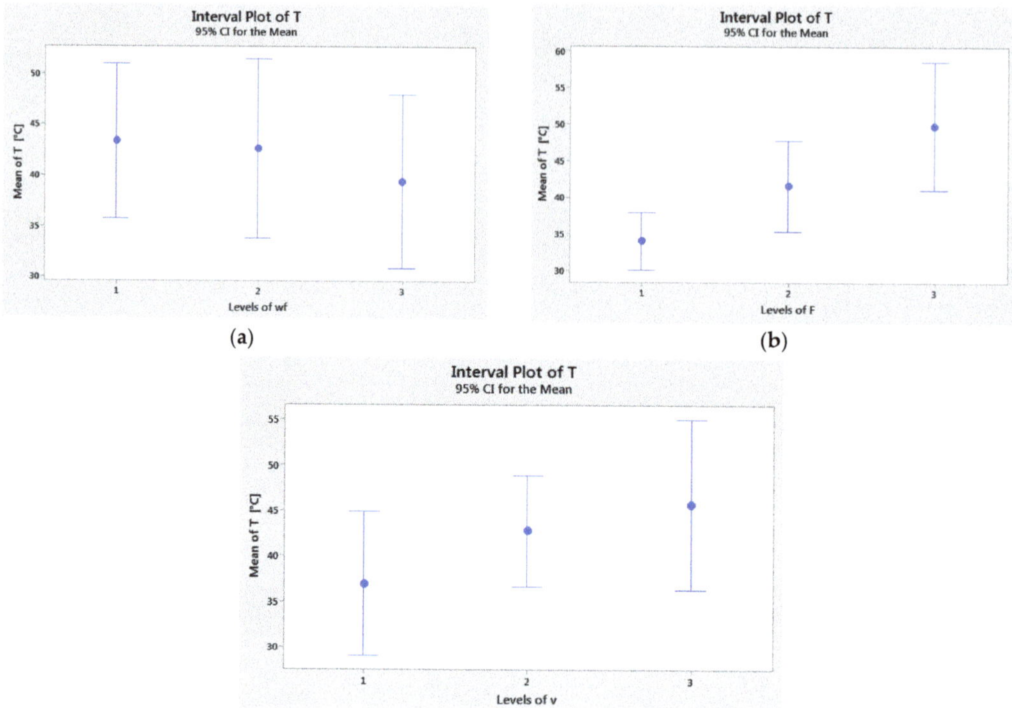

(a) (b)

(c)

Figure 10. Interval plot of the temperature (T) factor with: (**a**) wf.; (**b**) F; and (**c**) v. Individual standard deviations were used to calculate the interval plot. Bars are standard errors of the mean.

Normal probability plots were obtained for residuals of COF, K, and T, that confirmed the generalized linear models were adequacy [27–30]. Figure 11 shows the normal probability plot for COF. Similar distributions were obtained for K and T.

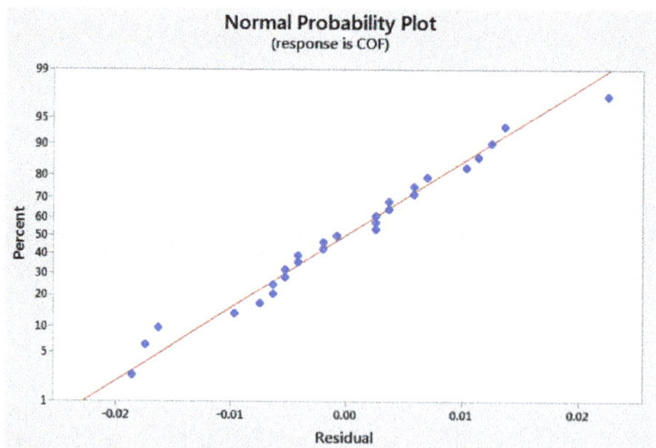

Figure 11. Normal probability plots of residuals for coefficient of friction (COF).

3.4. Results of the Morphological Analysis

Based on the results obtained from the statistical analysis, a thorough investigation of the worn surface and an SEM analysis were necessary for a complete characterization of the wear mechanism. Thus, after 120 min, the following investigations were performed:

optical microscopic investigations of the appearance of the GFRP disc and the steel ball used in the test, as well as SEM and EDS investigations of the worn surfaces of the GFRP samples (50%, 65%, and 70% wf.).

After a 120 min test interval, the GFRP disc and the steel ball subjected to the test parameters look as in Figure 12. Around the contact area of the ball with the disc, material removed by abrasion from the glass fiber reinforced polymer composite disc can be observed.

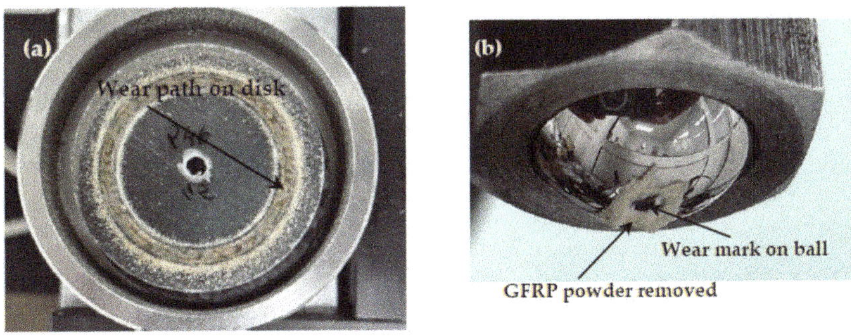

Figure 12. Aspects of wear marks on (**a**) the GFRP disc and (**b**) the steel ball, 52100.

Various factors, including sliding speed, significantly influence the wear rate (K) in the GFRP/chromium-alloyed steel tribo-system. In general, an overall decrease in the wear rate is observed as sliding speed increases. A possible explanation for this trend is the effect of sliding speed on the development of a transfer layer. At low sliding speeds, the interaction between the ball and the disc results in the accumulation of wear particles, promoting abrasive wear and increasing the wear rate. However, as sliding speed increases, wear particles are partially displaced from the contact area and partially redeposited onto the counterface surface, aiding in the formation of a transfer layer that helps reduce the wear rate.

The wear rate shows a direct increase with sliding duration for all tested specimens. Additionally, higher glass fiber content correlates with a lower wear rate, while composites with lower fiber content experience an increased wear rate. These findings are consistent with the conclusions of Kukureka et al. [19,20], who studied rolling-sliding contact in a similar material system.

The GFRP 50 wt.% specimen demonstrated higher sensitivity to sliding speed compared to composites with greater fiber content, especially under a normal load of 10 N. This wear pattern is attributed to surface property modifications as the glass fiber content decreases in the polymer matrix. A notable difference emerges when the normal load increases to 30 N and the sliding speed reaches approximately 0.25 m/s, especially for the GFRP 50 wf. specimen. At this higher load, the rougher metallic surface asperities of the steel ball deform the polymeric surface, leading to brinelling effects, micro-cutting, and the development of abrasive wear tracks. The observed wear is attributed to the abrasive action of metallic asperities, which is further intensified by the interaction with fragmented glass fiber debris.

Glass fibers demonstrate abrasive and lubricating effects on the sliding surface, significantly contributing to friction reduction, especially during the running-in period. This dual behavior arises from their hard and rigid nature. In the early sliding stages, glass fibers may protrude from the polymer matrix, causing mechanical abrasion on the steel ball, as illustrated in Figure 13.

Although glass fibers are harder than steel, they are embedded in an epoxy polymer matrix, this serves as a protective layer, minimizing direct contact between the fibers and the steel surface. This prevents the aggressive abrasive interactions typically responsible for severe metallic wear. Additionally, a thin transfer layer consisting of polymeric material from the GFRP forms on the steel ball's surface during the test, further reducing friction and shielding the steel from direct abrasive action. This transfer layer minimizes the plowing effect of the fibers and helps stabilize the coefficient of friction (COF), which remains within a low range (~0.35–0.54), indicating the absence of severe adhesion or three-body abrasion.

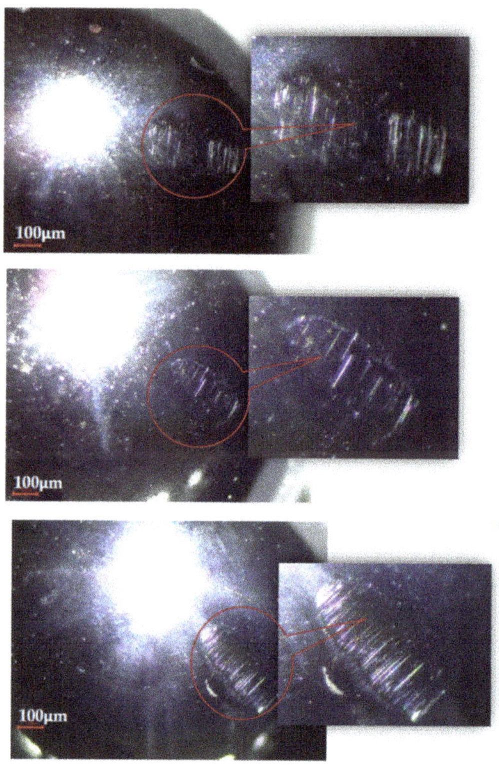

(a)

(b)

(c)

Figure 13. Wear marks on the chromium-alloyed steel ball at a sliding speed of 0.25 m/s under a 30 N load, after sliding on: (**a**) GFRP50 wf. specimen; (**b**) GFRP65 wf. specimen and (**c**) GFRP70 wf.

Moreover, the variation in fiber weight fraction (50%, 65%, and 70%) within the GFRP does not significantly impact the wear behavior of the steel balls. Since the primary wear mechanisms in this system involve polymer matrix degradation and fiber pull-out, rather than direct fiber-to-steel abrasion, the steel balls remain largely unaffected in all tested cases. While increasing the fiber content strengthens the composite, the absence of direct fiber contact with the steel means that no notable increase in steel ball wear occurs, regardless of the fiber weight fraction.

In conclusion, the minimal wear of the 52100-steel balls across all test conditions is due to their intrinsic hardness and wear resistance, the cushioning effect of the polymer matrix, the development of a protective transfer layer, and the absence of severe wear mechanisms such as adhesion or deep abrasive plowing. These findings confirm that 52100-steel maintains its integrity in GFRP tribological applications, making it a suitable material for prolonged frictional use without significant degradation.

The SEM images presented in Figure 14 illustrate the microstructural modifications of the worn surfaces of GFRP samples (50%, 65%, and 70% wf.) after 120 min of testing under a 30 N load and a sliding speed of 0.25 m/s. These images highlight the differences in wear behavior of the composites, emphasizing the dominant mechanisms that contribute to material degradation.

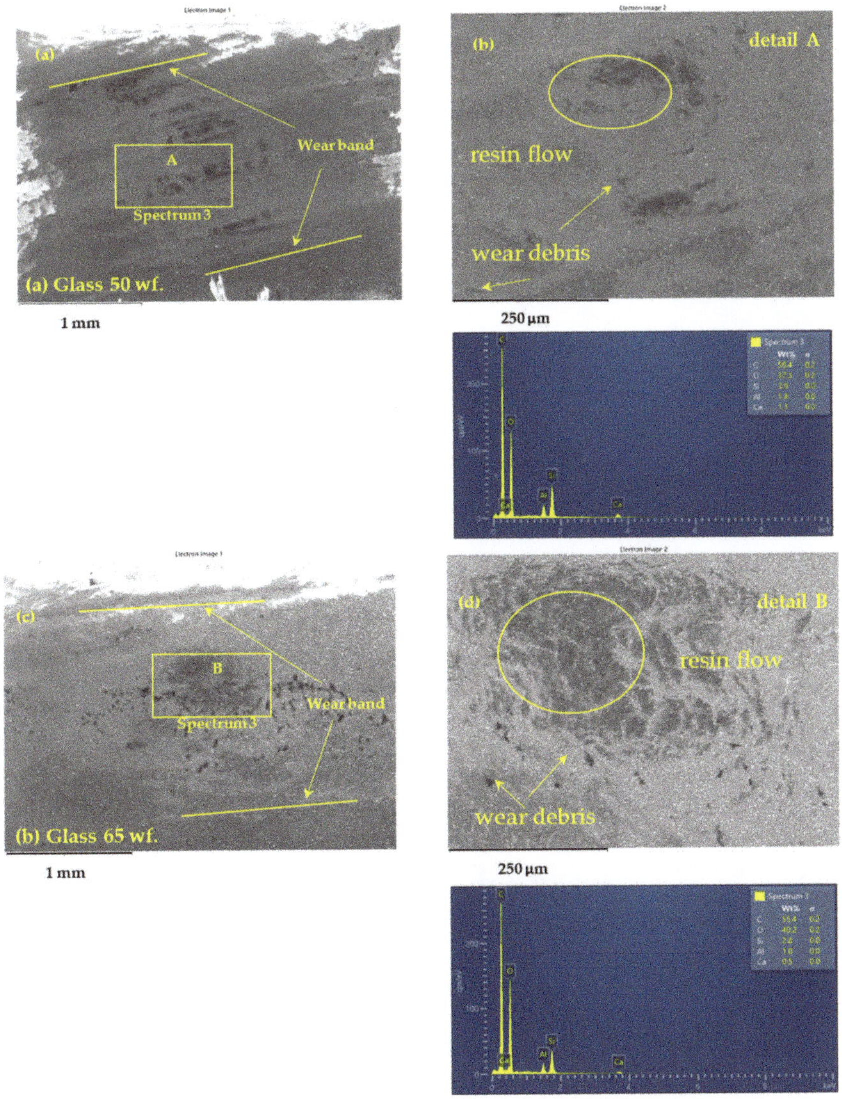

Figure 14. *Cont.*

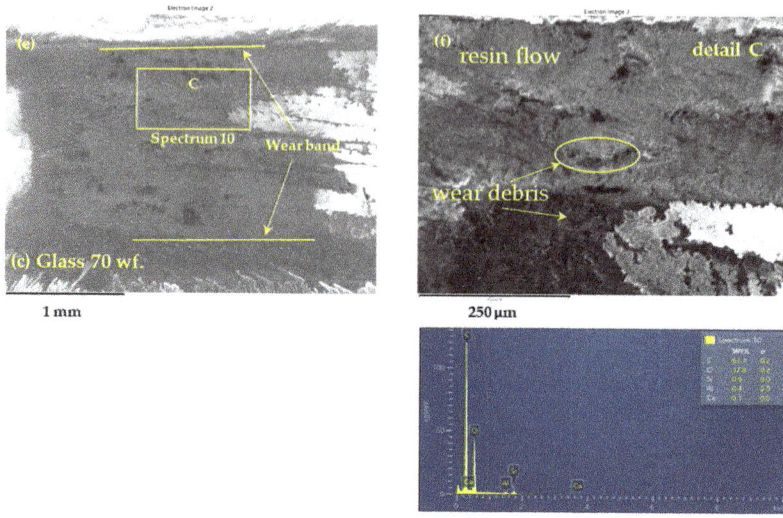

Figure 14. SEM images of wear tracks on GFRP samples after 120 min of dry sliding against a 52100-steel ball under a 30 N load and a 0.25 m/s sliding speed: (**a**,**b**) GFRP 50% overview and detailed wear track and EDS spectra; (**c**,**d**) GFRP 65% overview and detailed wear track and EDS spectra; and (**e**,**f**) GFRP 70% overview and detailed wear track and EDS spectra.

In the case of the GFRP 50%, sample, prominent abrasive wear marks are observed, characterized by significant removal of the polymer matrix and exposure of the glass fibers. This sample exhibits the deepest wear marks, indicating lower friction resistance and accelerated degradation under load. In contrast, the GFRP 65% sample shows uniform wear marks with reduced material removal, suggesting a more stable interface between the epoxy matrix and glass fibers. The GFRP 70% sample presents a relatively smoother surface, indicating improved wear resistance and greater protection against abrasion.

The main identified wear mechanisms include micro cutting and the formation of fine grooves, which are more pronounced in the GFRP 50% sample, suggesting a higher material fragility. Delamination wear is evident in samples with a lower fiber content, where the composite's surface layer separates more easily under frictional forces. In the GFRP 65% and 70% samples, the formation of a transfer layer is observed, which contributes to surface protection and reduces the coefficient of friction.

The local temperature in the contact area increases with the applied load, leading to localized softening of the polymer matrix under the influence of the high temperature generated by friction. As the matrix softens, it loses rigidity, which can result in reduced adhesion between the glass fibers and the epoxy matrix, promoting delamination wear and fiber extraction. However, in samples with a higher fiber content (70% wf.), the contact temperature is lower compared to the 50% and 65% samples due to more efficient heat dissipation through the fiber network. This observation correlates with a reduction in the coefficient of friction and the volume of material removed during the tribological tests.

Energy-dispersive X-ray spectroscopy (EDS) detected the presence of aluminum and silicon elements on the worn surfaces, with an increase in their concentration as the fiber content increases. In the case of the GFRP 50% sample, the surface layer contains smaller amounts of silicon and aluminum, suggesting faster wear of the polymer matrix. In contrast, for the GFRP 65% and 70% samples, the worn layer contains a higher proportion of these elements, indicating a protective effect of the glass fibers on the composite. The silicon detected in the EDS analysis primarily originates from the glass fibers of the GFRP composite, being released because of abrasive wear and tribological phenomena. Part of this silicon plays a crucial role in the formation of a protective layer on the contact surface, which explains the reduction in the friction coefficient at higher fiber contents (65–70%). The source of the aluminum detected in the wear layer is its natural presence in the glass fibers in the form of aluminosilicates.

The SEM and EDS results confirm that a higher fiber content improves the wear resistance of GFRP composites, reducing the coefficient of friction and dissipating thermal energy more efficiently during friction. Additionally, the microstructural analysis shows that abrasive wear is the dominant mechanism for GFRP 50%, while in GFRP 70%, the formation of a transfer layer enhances surface protection. These findings are crucial for optimizing the use of GFRP composites in tribological applications, where reducing friction and improving wear resistance are critical factors.

3.5. Interactions and Transformations Between Wear Mechanisms

The tribological behavior of GFRP composites in contact with 52100-chromium-alloyed steel balls is governed by multiple wear mechanisms that act concurrently. These mechanisms include abrasive wear, transfer layer formation, and polymer matrix degradation, each influenced by applied load, sliding speed, and fiber weight fraction. The transition between these mechanisms determines the wear performance and surface integrity of the composite material.

At low sliding speeds and high loads, abrasive wear is the dominant mechanism. The mechanical interaction between the steel ball and the composite leads to the removal of polymer matrix material, exposing the reinforcing glass fibers. However, as sliding speed increases, the frictional heat softens the polymer matrix, facilitating the formation of a transfer layer. This layer, consisting of polymeric material adhered to the steel counterface, acts as a protective barrier and reduces direct abrasive interactions, leading to a lower wear rate.

At very high sliding speeds, the accumulated heat can exceed the thermal stability of the polymer matrix, leading to localized softening or degradation. This weakens the interfacial bonding between fibers and the matrix, increasing the likelihood of fiber pull-out and adhesive wear. Additionally, thermal degradation can result in an unstable transfer layer, which may intermittently detach and expose fresh material to wear, creating fluctuations in the wear rate.

The interplay between these wear mechanisms is evident from SEM and EDS analyses, which reveal signs of both abrasion marks and polymer transfer layers. The transformation between these mechanisms is primarily dictated by the tribological parameters. At low loads and low speeds, minimal wear occurs, with limited fiber exposure and negligible polymer degradation. An optimal transfer layer is formed at moderate loads and sliding speeds, balancing wear resistance and reducing COF. At high loads and high speeds, polymer degradation intensifies, leading to increased fiber pull-out and fluctuating wear behavior.

The combination of abrasive, adhesive, and thermal effects explains the observed wear trends in the experimental results. These findings highlight the complex interactions between material properties and tribological conditions, providing valuable insights into the optimization of GFRP composites for frictional applications.

3.6. Correlation Between Structural Changes and Tribological Performance

The SEM and EDS analyses provide critical insights into the microstructural evolution of the worn surfaces and their direct impact on tribological behavior. The morphological differences observed among GFRP composites with varying fiber content reveal that higher fiber weight fractions contribute to lower polymer degradation and improved heat dissipation, ultimately enhancing wear resistance.

The EDS analysis of transfer layers further clarifies the role of material transfer in stabilizing frictional performance. The presence of silicon and aluminum elements from the glass fibers in the worn layers suggests that these components contribute to the formation of a protective surface layer. This helps reduce the coefficient of friction and mitigates excessive material removal.

Additionally, fiber pull-out and polymer degradation observed in SEM images indicate that at high loads and sliding speeds, weak fiber-matrix interfacial bonding accelerates surface damage, leading to increased wear. In contrast, composites with higher fiber content maintain structural integrity for longer durations, delaying wear initiation and improving overall tribological performance.

4. Conclusions

This study investigated the mechanical and tribological performance of twill-woven glass fiber-reinforced polymer (GFRP) composites with different fiber weight fractions (50%, 65%, and 70%). The research evaluated the influence of fiber content, applied load, and sliding speed on the coefficient of friction, specific wear rate, and contact temperature. Experimental results and statistical analysis provided an in-depth understanding of the wear mechanisms and the key factors influencing the tribological behavior of these materials.

The results demonstrate that the coefficient of friction is significantly influenced by sliding speed, which accounts for 46.51% of its variation. Additionally, the interaction between applied load and sliding speed had a considerable impact (32.51%), while the fiber weight fraction had a lower contribution (11%). Samples with a higher fiber content (70%) exhibited the lowest friction coefficient values, indicating improved wear resistance and reduced abrasive interaction at the contact surface. The maximum COF was obtained for a wf. of 50%, force of 30N, and v of 0.1m/s.

The analysis of specific wear indicated that fiber weight fraction is the most influential factor, contributing 34.15%, followed by applied load (21.87%) and sliding speed (14.11%). Wear decreased as the fiber content increased, confirming that a higher proportion of glass fibers enhances the material's resistance to wear. In samples with 50% fibers, abrasive wear was more intense, whereas in those with 70% fibers, a protective effect was observed due to the formation of a transfer layer. The minimum K was obtained for a wf. of 70%, a force of 10 N, and v of 0.1 m/s.

Contact temperature was primarily influenced by applied load (39.08%) and the interaction between fiber fraction and sliding speed (27.24%). Temperature increases favored localized softening of the polymer matrix, leading to reduced adhesion between fibers and the epoxy matrix, thus promoting delamination wear. However, in composites with 70% fiber content, temperature was lower compared to 50% and 65% samples due to more efficient heat dissipation through the fiber network. This observation correlates with a reduction in the friction coefficient and specific wear rate.

SEM and EDS results confirmed that abrasive wear is the dominant mechanism in samples with 50% fibers, where pronounced micro-cutting and delamination marks were observed. In samples with 65% and 70% fibers, analysis revealed the presence of a transfer layer and an increase in aluminum and silicon concentrations on the worn surface, indicating a protective effect of the glass fibers. These observations support the conclusion that a higher fiber content contributes to stabilizing tribological interactions and improving the wear behavior of GFRP composites.

These findings are essential for optimizing the use of GFRP composites in industrial applications where friction and wear are critical factors, such as the aerospace, automotive, and mechanical engineering industries. Future studies should explore the long-term behavior of these materials under varying load and environmental conditions, as well as analyze the impact of other types of reinforcements on their tribological performance. Additionally, using predictive models based on machine learning could help optimize operating parameters and improve the durability of these advanced composites.

Author Contributions: C.B., M.P., R.U. and B.P.: conceptualization, visualization, writing—original draft, designed the experiments; M.C., R.U. and C.V.: writing—review and editing, investigation; M.C., M.P., R.U., C.V. and B.P.: formal analysis. All authors reviewed the manuscript. All authors have read and agreed to the published version of the manuscript.

Funding: This research received no external funding.

Institutional Review Board Statement: Not applicable.

Data Availability Statement: Data are contained within the article.

Conflicts of Interest: The authors declare no conflicts of interest.

References

1. Palanikumar, K.; Ramesh, M.; Hemachandra Reddy, K. Experimental investigation on the mechanical properties of green hybrid sisal and glass fiber reinforced polymer composites. *J. Nat. Fibers* **2016**, *13*, 321–331. [CrossRef]
2. Landesmann, A.; Seruti, C.A.; Batista ED, M. Mechanical properties of glass fiber reinforced polymers members for structural applications. *Mater. Res.* **2015**, *18*, 1372–1383. [CrossRef]
3. El-Wazery, M.S.; El-Elamy, M.I.; Zoalfakar, S.H. Mechanical properties of glass fiber reinforced polyester composites. *Int. J. Appl. Sci. Eng.* **2017**, *14*, 121–131.
4. Davallo, M.; Pasdar, H. Comparison of mechanical properties of glass-polyester composites formed by resin transfer molding and hand lay-up technique. *Int. J. ChemTech Res.* **2009**, *1*, 470–475.
5. Abdullah, E.T. A study of bending properties of unsaturated polyester/glass fiber reinforced composites. *Al-Nahrain J. Sci.* **2013**, *16*, 129–132. [CrossRef]
6. Talib AA, A.; Jumahat, A.; Jawaid, M.; Sapiai, N.; Leao, A.L. Effect of wear conditions, parameters and sliding motions on tribological characteristics of basalt and glass fibre reinforced epoxy composites. *Materials* **2021**, *14*, 701. [CrossRef]
7. Zhao, G.; Hussainova, I.; Antonov, M.; Wang, Q.; Wang, T.; Yung, D.-L. Effect of temperature on sliding and erosive wear of fiber reinforced polyimide hybrids. *Tribol. Int.* **2015**, *82*, 525–533. [CrossRef]
8. Zhao, G.; Hussainova, I.; Antonov, M.; Wang, Q.; Wang, T. Friction and wear of fiber reinforced polyimide composites. *Wear* **2013**, *301*, 122–129. [CrossRef]
9. Karsli, N.G.; Demirkol, S.; Yilmaz, T. Thermal aging and reinforcement type effects on the tribological, thermal, thermomechanical, physical and morphological properties of poly(ether ether ketone) composites. *Compos. Part B Eng.* **2016**, *88*, 253–263. [CrossRef]
10. Birleanu, C.; Cioaza, M.; Serdean, F.; Pustan, M.; Bere, P.; Contiu, G. Tribological investigation of glass fiber reinforced polymer composites against 52100 chrome alloy steel based on ELECTRE decision-making method. *Polymers* **2023**, *16*, 62. [CrossRef]
11. Asyraf, M.R.; Ishak, M.R.; Syamsir, A.; Amir, A.L.; Nurazzi, N.M.; Norrrahim, M.N.F.; Asrofi, M.; Rafidah, M.; Ilyas, R.A.; Rashid, M.Z.A.; et al. Filament-wound glass-fibre reinforced polymer composites: Potential applications for cross arm structure in transmission towers. *Polym. Bull.* **2023**, *80*, 1059–1084. [CrossRef]
12. Saba, N.; Paridah, M.T.; Jawaid, M. Mechanical properties of kenaf fibre reinforced polymer composite: A review. *Constr. Build. Mater.* **2015**, *76*, 87–96. [CrossRef]
13. Sanjay, M.; Yogesha, B. Studies on natural/glass fiber reinforced polymer hybrid composites: An evolution. *Mater. Today Proc.* **2017**, *4*, 2739–2747. [CrossRef]
14. Collar, E.P.; García-Martínez, J.M. Mechanical Behavior of Polymeric Materials: Recent Studies. *Polymers* **2024**, *16*, 2821. [CrossRef]
15. Hosseini, S.A.; Torabizadeh, M.; Eisazadeh, H. Experimental study of the effect of strain rate on the mechanical behavior of assorted thermoplastic polymers. *J. Mater. Eng. Perform.* **2024**, *33*, 6942–6951. [CrossRef]
16. Nurazzi, N.M.; Sabaruddin, F.A.; Harussani, M.M.; Kamarudin, S.H.; Rayung, M.; Asyraf, M.R.M.; Aisyah, H.A.; Norrrahim, M.N.F.; Ilyas, R.A.; Abdullah, N.; et al. Mechanical performance and applications of cnts reinforced polymer composites—A review. *Nanomaterials* **2021**, *11*, 2186. [CrossRef]
17. Bazli, M.; Abolfazli, M. Mechanical properties of fibre reinforced polymers under elevated temperatures: An overview. *Polymers* **2020**, *12*, 2600. [CrossRef]

18. Khieng, T.K.; Debnath, S.; Ting Chaw Liang, E.; Anwar, M.; Pramanik, A.; Basak, A.K. A review on mechanical properties of natural fibre reinforced polymer composites under various strain rates. *J. Compos. Sci.* **2021**, *5*, 130. [CrossRef]
19. Kukureka, S.N.; Hooke, C.J.; Rao, M.; Liao, P.; Chen, Y.K. The effect of fibre reinforcement on the friction and wear of polyamide 66 under dry rolling–sliding contact. *Tribol. Int.* **1999**, *32*, 107–116. [CrossRef]
20. Dearn, K.D.; Kukureka, S.N.; Walton, D. Engineering polymers and composites for machine elements. In *Handbook of Polymer Tribology*; World Scientific: Singapore, 2018; pp. 441–479.
21. Zaghloul MM, Y.; Steel, K.; Veidt, M.; Heitzmann, M.T. Mechanical and tribological performances of thermoplastic polymers reinforced with glass fibres at variable fibre volume fractions. *Polymers* **2023**, *15*, 694. [CrossRef]
22. Yaşar, İ.; Arslan, F. Effects of fiber volume fraction and fiber orientation on the tribological behaviour of unidirectionally oriented glass fiber-polyester composites. *Turk. J. Eng. Environ. Sci.* **2000**, *24*, 181–192.
23. Abdurohman, K.; Pratomo, R.A.; Hidayat, R.; Ramadhan, R.A.; Nurtiasto, T.S.; Ardiansyah, R.; Pribadi Putra Pratama, M.G. A comparison of vacuum infusion, vacuum bagging, and hand lay-up process on the compressive and shear properties of GFRP materials. *Indones. J. Aerosp.* **2023**, *21*, 39–50. [CrossRef]
24. Ali, F.A.; Choudhary, V.; Priya, I.M.I. Experimental study on the tribological properties of polymer composites. In *AIP Conference Proceedings*; AIP Publishing: Mellville, NY, USA, 2022; Volume 2460.
25. Sun, X.; Chi, M.; Yu, Y.; Wu, Y.; Zhang, X.; Chen, Q.; Shi, J.; Lan, D. High temperature resistance oil-paper insulation with polyimide: A study on dielectric properties. *IEEE Trans. Dielectr. Electr. Insul.* **2024**. [CrossRef]
26. Available online: https://www.astm.org/d3039_d3039m-17.html (accessed on 6 March 2025).
27. Udroiu, R. Quality Analysis of Micro-Holes Made by Polymer Jetting Additive Manufacturing. *Polymers* **2023**, *16*, 32. [CrossRef] [PubMed]
28. Birleanu, C.; Cioaza, M.; Udroiu, R.; Pustan, M.; Bere, P.; Lazarescu, L. Enhanced Tribological and Mechanical Properties of Copper-Modified Basalt-Reinforced Epoxy Composites. *Polymers* **2005**, *17*, 91. [CrossRef]
29. Popișter, F.; Dragomir, M.; Ciudin, P.; Goia, H.Ș. Empowering Rehabilitation: Design and Structural Analysis of a Low-Cost 3D-Printed Smart Orthosis. *Polymers* **2024**, *16*, 1303. [CrossRef]
30. Cioaza, M.; Birleanu, C.; Pustan, M.; Bere, P.; Contiu, G. Evaluation the Tribological Properties of Different Fiber Reinforced Polymers. In *International Tribology Symposium of IFToMM*; Springer Nature: Cham, Switzerland, 2024; pp. 85–95.

polymers

MDPI

Article

Improvement of Bonding Strength Between Polyphenylene Sulfide/Glass Fiber Composites and Epoxy via Atmospheric-Pressure Plasma Surface Treatment

Hwan-Gi Do [1], Pyoung-Chan Lee [2,*] and Beom-Gon Cho [1,*]

[1] Department of Polymer Science and Engineering, Kumoh National Institute of Technology, 61 Daehak-ro, Gumi 39177, Gyeongbuk, Republic of Korea; 2025210703@kumoh.ac.kr

[2] Chassis & Materials Research Laboratory, Korea Automotive Technology Institute, 303 Pungse-ro, Pungse-myeon, Dongnam-gu, Cheonan-si 31214, Chungnam, Republic of Korea

* Correspondence: pclee@katech.re.kr (P.-C.L.); bgcho@kumoh.ac.kr (B.-G.C.); Tel.: +82-41-559-3144 (P.-C.L.); +82-54-478-7684 (B.-G.C.)

Abstract: Polyphenylene sulfide (PPS) is becoming increasingly valuable in the electrical, electronic, and automotive industries. In particular, PPS composites reinforced with glass fiber (GF) have better dimensional stability and mechanical properties than conventional PPS materials and can be used in applications like electric vehicle capacitor housing. In the electric vehicle industry, the epoxy-molding process is essential for manufacturing capacitor housings, where the bonding strength between the PPS/GF composites and epoxy significantly affects the durability of the product. However, the inert surface characteristics of polymers like PPS limit their interaction with epoxy, decreasing the bonding strength. This study was aimed at enhancing the bonding strength between PPS/GF composites and epoxy by modifying the PPS surface using atmospheric-pressure plasma treatment. The surface modification resulted in increased surface roughness and the introduction of polar functional groups, which improved both mechanical interlocking and chemical affinity to the epoxy. Surface changes were analyzed using atomic force microscopy and scanning electron microscopy, and chemical characterization was conducted using X-ray photoelectron spectroscopy and Fourier-transform infrared spectroscopy. Surface energy was determined via contact angle measurements, and bonding strength was evaluated through single-lap shear tests. The results showed a 55% increase in surface energy and a 24.8% improvement in bonding strength due to the surface modification.

Keywords: polyphenylene sulfide; glass fiber; composites; surface modification; atmospheric-pressure plasma; bonding strength

1. Introduction

Polyphenylene sulfide (PPS) is a super-engineered plastic known for its outstanding thermal stability, chemical resistance, and mechanical properties. In particular, PPS exhibits excellent dimensional stability and electrical insulation, making it widely used in electronics, aerospace, and automotive applications [1–3]. Compared with other thermoplastics like polyethylene terephthalate (PET) or polyamide (PA), PPS offers higher temperature resistance, flame retardancy, and long-term mechanical stability. In addition, compared with conventional thermosetting polymers such as epoxy and rigid polyurethane (RPU), PPS exhibits distinct advantages in terms of long-term performance and environmental resistance. Epoxy resins are prone to cracking and stress concentration under complex loading conditions, and their hydroxyl-rich molecular structure readily absorbs moisture, leading to hydrolysis, plasticization, and fiber/matrix debonding [4]. RPU is inherently brittle, and despite various reinforcement strategies, its mechanical limitations continue to hinder broader application [5]. In contrast, PPS offers excellent thermal resistance, chemical stability, mechanical performance, and recyclability due to its thermoplastic nature. Recent studies have demonstrated its potential as a reliable alternative in high-performance composite systems [1].

To improve its mechanical strength, PPS is often reinforced with fibers such as carbon fiber (CF) or glass fiber (GF). PPS-based composites reinforced with carbon fiber (CF) or glass fiber (GF) are lightweight and possess enhanced mechanical strength and dimensional stability compared with PPS [6–8]. Compared with CF, GF offers better cost-effectiveness and electrical insulation, which are desirable properties for electrical components such as capacitor housings. Therefore, GF-reinforced PPS (PPS/GF) is a promising material for applications such as capacitor housings in electric vehicles [9].

In electric vehicles, the fabrication of capacitor housings requires epoxy molding, where the bonding strength between the PPS/GF composites and the epoxy resin is a crucial factor influencing the final product's performance. However, polymers like PPS inherently possess inert surfaces; therefore, formation of chemical bonds with adhesives is difficult. This issue arises owing to their non-polar nature, low surface energy, and lack of polar functional groups, all of which contribute to their weak bonding strength [10,11]. To overcome these limitations, surface treatments are essential for improving adhesion performance [12–14]. Several surface modification strategies have been explored, including chemical grafting, silane coupling, nanoparticle coating, laser pattering, and plasma treatment. Tutunchi et al. introduced a two-step chemical method combining acid oxidation and epichlorohydrin grafting, which significantly enhanced bonding strength with epoxy resin through strong chemical interactions. This method offers robust surface functionalization but involves multi-step wet chemistry and environmental concerns due to acid usage [15]. Nurazzi et al. reviewed silane-based treatments, showing that they are cost-effective and widely applicable for natural fibers, although they are sensitive to moisture and require precise control of reaction conditions [16]. Hwang et al. employed laser patterning micro-patterned CFRP surfaces, which improved adhesion via enhanced mechanical interlocking without significant damage to the fibers. However, the requirement for expensive equipment and the limited scalability remain key drawbacks [17]. Du et al. utilized silane coupling agents and carbon nanotube coatings to reinforce carbon fiber surfaces, achieving increased mechanical strength of the composites [18]. While effective, this hybrid approach involves complex surface preparation and material costs. In contrast, atmospheric-pressure plasma (APP) treatment is a promising surface modification technique that enables mass production via continuous processing, while preserving the intrinsic properties of materials by minimizing excessive energy exposure [19,20]. In addition, APP treatment is a dry, eco-friendly process, unlike wet chemical surface modification, making it suitable for heat-sensitive polymer materials. This plasma treatment was applied to PPS/GF composites and improved the bonding strength with epoxy through several mechanisms: (1) removal of organic surface contaminants; (2) etching-induced surface roughness enhancement, which increases the adhesive contact area; and (3) introduction of polar functional groups such as hydroxyl (–OH) and carboxyl (–COOH) groups, facilitating stronger chemical interactions with epoxy adhesives [21–23].

Although previous studies have examined plasma treatment on various thermoplastics [21–23], systematic analysis focusing on how plasma-treatment parameters (e.g., treatment speed) affect the adhesion of PPS/GF composites has not been sufficiently explored. Furthermore, few studies provide comprehensive correlation among surface chemistry, roughness, and mechanical bonding strength in PPS-based composites. In this study, we investigated the effects of APP treatment on the surface properties of PPS/GF composites and examined the influence of varying plasma-treatment speeds on the surface structure and bonding strength. The aim of this study is to enhance the adhesion performance in the epoxy-molding process of PPS/GF-based capacitor housings for electric vehicles through a simple, rapid, and effective atmospheric-pressure plasma treatment. In addition, the optimal plasma-treatment speed that is needed to maximize bonding strength is investigated. The chemical composition and functional group introduction were analyzed using X-ray photoelectron spectroscopy (XPS) and Fourier-transform infrared (FT-IR) spectroscopy. The changes in surface morphology and roughness were examined using atomic force microscopy (AFM) and scanning electron microscopy (SEM), whereas energy-dispersive spectroscopy (EDS) was used to evaluate the elemental composition of the etched surfaces. Contact angle measurements were conducted to assess the changes in surface energy, and the bonding strength was assessed using a single-lap shear test.

2. Materials and Methods

2.1. Preparation of PPS/GF Composites

The PPS compounds reinforced with 65 wt% short glass fiber were supplied by Toray Advanced Composites Corporation (Tokyo, Japan). The PPS matrix (TORELINA^{TM} series, Toray Advanced Composites Corporation, Tokyo, Japan) exhibits excellent thermal and mechanical properties, including a tensile strength of approximately 150 MPa, a tensile modulus of ~6.0 GPa, a flexural strength of ~496 MPa, and a heat deflection temperature (1.8 MPa) of approximately 237 °C. The glass fiber contributes additional strength, with a tensile strength of ~610 MPa and a tensile modulus of ~33.8 GPa. The resulting PPS/GF composites exhibit a composite density of approximately 1.73 g/cm^3. The plate samples, with dimensions of 100 mm (width) × 100 mm (length) × 3 mm (thickness), were manufactured using an injection molding process at a barrel temperature of 300 °C and a mold temperature of 140 °C, injection pressure of 80 MPa, cooling time of 30 s, and mold holding time of 10 s [9]. Furthermore, the injected plate samples were cut to dimensions of 25 mm × 100 mm × 3 mm (width × length × thickness) for the single-lap shear test. A schematic illustration of the sample preparation is shown in Figure S1.

2.2. Atmospheric-Pressure Plasma Surface Treatment

The surfaces of the composites were treated using an atmospheric-pressure plasma apparatus (3D i REV, Applied Plasma Inc., Gumi, Republic of Korea) at an exposure power of 950 W. The plasma apparatus was supplied with compressed air at a pressure of 0.3 MPa and a frequency of 60 Hz. A nozzle with a diameter of 60 mm was fixed to the plasma apparatus, and the plate on which the composites were placed was moved back and forth at speeds of 1, 2, or 6 m/min. A slower plasma-treatment speed resulted in greater surface plasma irradiation. The area that can be treated at once is mainly determined by the width of the plasma nozzle because this width controls how wide the plasma spreads when it exits [24]. In this study, the nozzle width was 60 mm, while the specimen was only 25 mm wide, meaning that the entire surface could be moved back and forth to treat the entire surface at one time. As a result, all parts of the surface were equally exposed to the plasma, enabling uniform treatment. Figure 1 shows a schematic of the atmospheric-pressure plasma-treatment apparatus. Each specimen was treated with plasma once, and specimens without plasma treatment were used as control samples. Additionally, considering that the specimen length is 100 mm, even at the slowest treatment speed of 1 m/min, the plasma exposure takes only about 6 s. Thus, the total processing time remains under 10 s, demonstrating the high efficiency and speed of the proposed method.

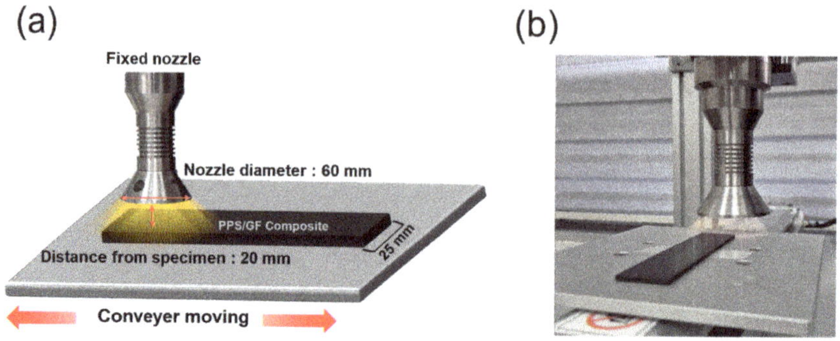

Figure 1. (**a**) Scheme of atmospheric-pressure plasma treatment for PPS/GF composites (Horizontal arrow: Nozzle diameter, Vertical arrow: Distance from specimen), (**b**) Photograph of PPS/GF composites under atmospheric-pressure plasma treatment.

2.3. Characterization

2.3.1. Structural Analysis

The surface chemical composition of the PPS/GF composites was analyzed using XPS (Nexsa G2, Thermo Fisher, Waltham, MA, USA) with a monochromatic Al Kα X-ray source (1486.6 eV). For the XPS analysis, samples were prepared with dimensions of 25 mm $\times$ 25 mm $\times$ 3 mm. The oxygen functional groups introduced onto the PPS/GF composites were examined using attenuated total reflectance (ATR)-FT-IR spectroscopy (IRAffinity-1S, Shimadzu, Kyoto, Japan). The analysis was carried out in transmittance mode over the scanning range of 600–4000 cm^{-1}.

2.3.2. Morphological Analysis

The surface roughness change of the PPS/GF composites was analyzed using AFM (NX20, Park Systems, Suwon, Republic of Korea) in the non-contact mode. The surface morphologies of the PPS/GF composites were observed using field-emission (FE)-SEM (JSM-IT700HR, JEOL, Tokyo, Japan). The surface elemental distribution of the PPS/GF composites was analyzed using EDS data collected from the SEM. The specimens used for SEM analysis were prepared with dimensions of 12.5 mm $\times$ 12.5 mm $\times$ 3 mm using a diamond saw cutting machine.

2.3.3. Wettability Analysis

The surface energies of the PPS/GF composites were determined by measuring the contact angles (Phoenix 300, SEO, Suwon, Republic of Korea) using the Owens–Wendt equation.

$$\gamma_S = \gamma_S^D + \gamma_S^P \tag{1}$$

$$(1 + cos\theta)\gamma_{LV} = 2\left(\gamma_S^D \gamma_{LV}^D\right)^{1/2} + 2\left(\gamma_S^P \gamma_{LV}^P\right)^{1/2} \tag{2}$$

where γ_S is the total surface energy of the PPS/GF composites; γ_{LV} is the surface tension of the liquid; and γ_S^D and γ_S^P represent the dispersive and polar components, respectively. The contact angle (θ) measurements were performed using the sessile drop method with deionized (DI) water and diiodomethane. The surface tensions of the liquids are listed in Table 1 [25].

Table 1. Surface tension values of DI water and diiodomethane.

Liquid	γ_{LV} (mN/m)	γ_{LV}^{D} (mN/m)	γ_{LV}^{P} (mN/m)
DI water	72.8	51.0	21.8
Diiodomethane	50.8	50.8	0

2.3.4. Single-Lap Shear Testing

The adhesive used to evaluate the bonding strength was prepared using epoxy and a hardener provided by ESP Chem (Hwaseong, Republic of Korea). The epoxy and hardener were mixed in a weight ratio of 10:9, followed by degassing in a vacuum oven for 20 min. Subsequently, in accordance with the ISO 4587 standard, the adhesive was applied to the PPS/GF composites over an area of 12.5 mm × 25 mm [26]. Finally, plastic was attached to the grip to prevent specimen slippage and to ensure uniform load application during the bonding strength measurement.

The bonding strength between the PPS/GF composites and the epoxy adhesive was evaluated using a universal testing machine (AG-50kNX, Shimadzu, Kyoto, Japan) with a 50 kN load cell at a crosshead speed of 5 mm/min. The fractured surfaces after the single-lap shear test were observed using optical microscopy (OM; HVM0850D, Hana-vision, Seoul, Republic of Korea) to analyze the failure modes. Figure 2a shows the ISO 4587 standard dimensions of the specimen. Figure 2b,c show a schematic and a photograph of the PPS/GF composite specimen for the single-lap shear test.

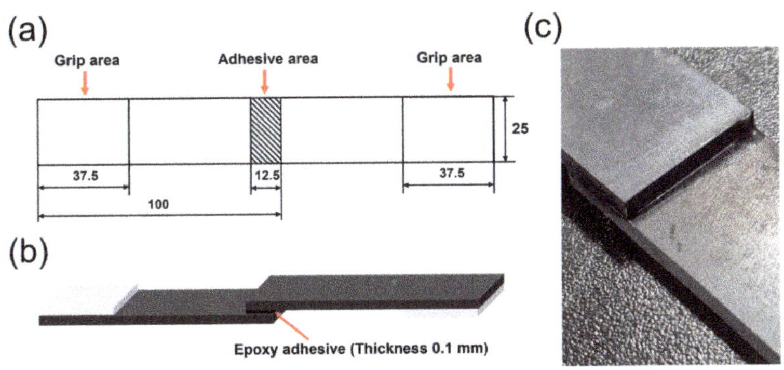

Figure 2. (**a**) Scheme of specimen dimensions in accordance of the ISO 4587 standards, (**b**) Schematic illustration of the single lap shear test specimen, and (**c**) Photograph the bonded region of the PPS/GF composites specimen for the single lap shear test.

3. Results

3.1. Structural Analysis of PPS/GF Composites

XPS is a widely recognized technique for detecting all elements, except hydrogen and helium, within ~10 nm of a polymer surface. The elemental composition of the samples was analyzed by deconvoluting the XPS spectra using CASA XPS software (version 2.3.26). Additionally, the chemical composition of the PPS/GF composite surface before and after plasma treatment was quantitatively characterized using XPS.

Figure S2 shows the XPS profiles of the PPS/GF composite surfaces obtained at varying plasma-treatment speeds. Table 2 shows the percentage of C, O, and S (at.%) and the oxygen ratio to carbon on the composites. The chemical changes were observed to be closely related to the plasma-treatment speed. As the treatment time increased, the concentration of oxygen atoms increased, corresponding to a lower treatment speed, whereas the carbon atom content decreased. The carbon content decreased from 86.51% in the untreated sample to 73.32% in the sample treated with plasma at a speed of 6 m/min. This reduction is attributed to plasma treatment, which removes organic pollutants from the surface and breaks or oxidizes carbon-based bonds. Conversely, the oxygen (O) ratio increased from 11.73% in the untreated specimen to 18.93% and 20.49% in the samples treated with plasma at speeds of 6 m/min and 1 m/min, respectively. As the plasma-treatment speed decreased, the plasma irradiation time on the surface increased, leading to greater incorporation of oxygen-containing functional groups. Meanwhile, the sulfur (S) ratio increased from 1.04% in the untreated specimen to 5.61% and 5.48% at 6 m/min and 2 m/min, respectively. This increase may be due to the removal of superficial low-molecular-weight contaminants or surface oxidation products during mild plasma exposure, thereby revealing more sulfur-containing groups of the PPS matrix. In addition, short-term plasma exposure may induce surface chain re-orientation, resulting in greater surface accessibility of sulfur atoms. However, at a treatment speed of 1 m/min, the sulfur content decreased to 2.11% owing to oxidation by the plasma treatment. At this lower speed, the plasma exposure was more intense and prolonged, potentially leading to excessive oxidation and surface etching. This could result in the formation of volatile sulfur compounds such as SO_2 and SO_3, which were subsequently desorbed and removed from the surface [27,28].

Table 2. Percentage of elements on surface of PPS/GF composites at varying plasma-treatment speeds.

Plasma-Treatment Speed	Surface Elements (at.%)			
	C	O	S	O/C
No plasma	86.51	11.73	1.04	13.55
6 m/min	73.32	18.93	5.61	25.82
2 m/min	72.13	20.25	5.48	28.08
1 m/min	75.38	20.49	2.11	27.18

The XPS C1s spectra of the PPS/GF composites at varying plasma-treatment speeds are shown in Figure 3 and summarized in Table 3. C1s peaks were deconvoluted into C–C (285 eV), C–OH (286.2 eV), C=O (286.9 eV), and COOH (288.7 eV) [29]. Comparing the control PPS/GF composites and the plasma-treated PPS/GF composites, the C–C/C–H ratio in the untreated sample was 89.67%, which decreased to 78.85%, 79.5%, and 77.45% at 6, 2, and 1 m/min, respectively. This reduction was likely due to the removal of organic contaminants from the surface and oxidation caused by the plasma treatment. On the contrary, the C–OH and COOH ratios increased from 2.05% (no plasma) to 4.02% (1 m/min plasma) and from 1.53% (no plasma) to 5.22% (1 m/min plasma), respectively. However, the C=O ratio, which increased from 6.74% to 11.91%, decreased to 8.11% at a speed of 2 m/min and to 3.46% at a speed of 1 m/min. This phenomenon is presumed to be due to the conversion of C=O to COOH, and this tendency becomes more pronounced as plasma irradiation increases.

Table 3. Percentage of carbon bonds on surface of PPS/GF composites at varying plasma-treatment speeds.

Plasma-Treatment Speed	Carbon Bonds (at.%)			
	C–C/C–H	C–OH	C=O	COOH
No plasma	89.67	2.05	6.74	1.53
6 m/min	78.85	4.02	11.91	5.22
2 m/min	79.57	6.48	8.11	5.83
1 m/min	77.45	13.14	3.46	5.93

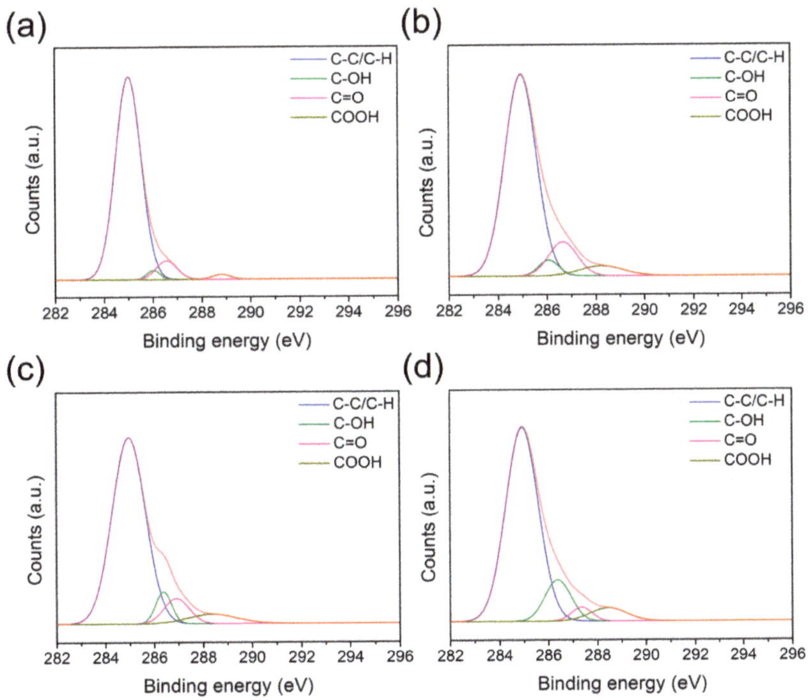

Figure 3. C1s spectra of PPS/GF composites at varying plasma-treatment speeds: (**a**) No plasma, (**b**) 6 m/min, (**c**) 2 m/min, and (**d**) 1 m/min. The orange curve represents the original C1s spectra before peak deconvolution.

The XPS S2p spectra of the PPS/GF composites at varying plasma-treatment speeds are shown in Figure 4 and summarized in Table 4. The S2p peaks were deconvoluted into –S– (163.9 eV), S=O (165.0 eV), and O=S=O (168.4 eV) peaks [27,30]. The –S– ratio of the control specimen, initially at 68.99%, decreased to 63.83%, 62.11%, and 56.28% at plasma-treatment speeds of 6, 2, and 1 m/min, respectively. This reduction was attributed to the etching effect induced by the plasma treatment and the oxidation of sulfur, leading to the formation of S=O and O=S=O bonds. At a speed of 6 m/min, the S=O and O=S=O ratios slightly increased from 29.09% (no plasma) to 33.29% and from 1.91% (no plasma) to 2.86%, respectively. However, at speeds of 2 and 1 m/min, the S=O ratios decreased to 30.01% and 30.07%, whereas the O=S=O ratios increased to 7.86% and 13.64%. These results indicate that the S=O bonds are gradually converted into O=S=O bonds, which are more stable [27].

Table 4. Percentage of sulfur bonds on surface of PPS/GF composites at varying plasma-treatment speeds.

Plasma-Treatment Speed	Sulfur Bonds (at.%)		
	–S–	S=O	O=S=O
No plasma	68.99	29.09	1.91
6 m/min	63.83	33.29	2.86
2 m/min	62.11	30.01	7.86
1 m/min	56.28	30.07	13.64

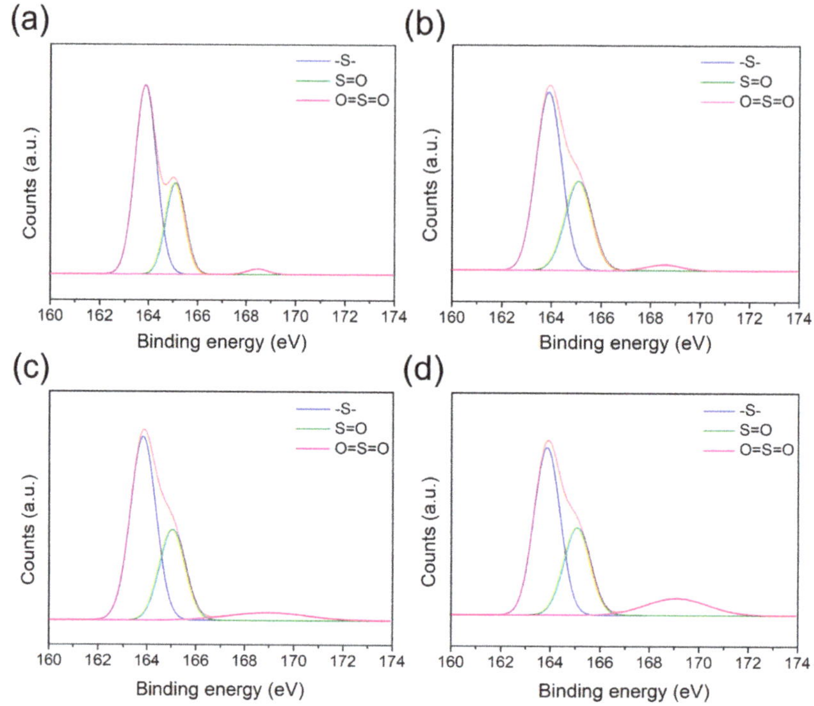

Figure 4. S2p spectra of PPS/GF composites at varying plasma-treatment speeds: (**a**) No plasma, (**b**) 6 m/min, (**c**) 2 m/min, and (**d**) 1 m/min. The orange curve represents the original S2p spectra before peak deconvolution.

FT-IR spectroscopy is a widely used analytical technique for identifying functional groups and characterizing chemical bonds on polymer surfaces. ATR-FT-IR spectroscopy allows the direct analysis of the composite surface by measuring the IR absorption of the evanescent wave that penetrates a few micrometers into the material. The acquired spectra were analyzed to investigate the chemical modifications induced by the plasma treatment. The FT-IR spectra of the surface of the PPS/GF composites at varying plasma-treatment speeds are shown in Figure 5 and summarized in Table 5. Figure 5a shows the entire wavenumber range of 600–4000 cm^{-1}. As shown in the 2750–4000 cm^{-1} region (Figure 5b), the plasma-treated specimens exhibit distinct peaks in the ranges of 3433–4000 cm^{-1}, 3190–3417 cm^{-1}, and 2750–3086 cm^{-1}, indicating the formation of O–H and N–H groups. Specifically, the peaks in the 3190–3417 cm^{-1} range were attributed to C–OH, whereas those in the 2750–3086 cm^{-1} range were assigned to O–H and COOH groups. The intensities of these peaks increased as the plasma-treatment speed decreased, indicating enhanced surface functionalization. Figure 5c shows a zoom-in of the 1200–1800 cm^{-1} region. The peaks at 1682 cm^{-1} and 1751 cm^{-1} correspond to the C=O stretching vibrations of the conjugated aldehyde and carboxylic acids, respectively. Additionally, the peak at 1631 cm^{-1} is attributed to the N–H bending vibration. The presence of O–H, N–H, COOH, and C=O peaks indicate surface modification induced by plasma treatment. The peaks at 1470 cm^{-1} and 1516 cm^{-1} are related to the aromatic ring in the PPS molecule structure. The peak at 1516 cm^{-1} corresponds to the C=C stretching vibration of the aromatic ring, and its intensity increased as the surface organic contaminants were removed by plasma treatment. The peak at 1470 cm^{-1} exhibited weaker intensity as the plasma-treatment speed decreased, indicating the breaking of C–H bonds. The sulfone group exhibits two S=O stretching peaks, one of which appears at 1240 cm^{-1}. Figure 5d presents a magnified view of the wavenumber range of 600–1200 cm^{-1}, where the other sulfone peak can be observed at ~1087 cm^{-1}. The peaks at 875 cm^{-1} and 817 cm^{-1} correspond to the out-of-plane C–H bending vibrations of the aromatic ring within the PPS structure, while the peak at 702 cm^{-1} represents the stretching vibration of the C–S bond [31,32].

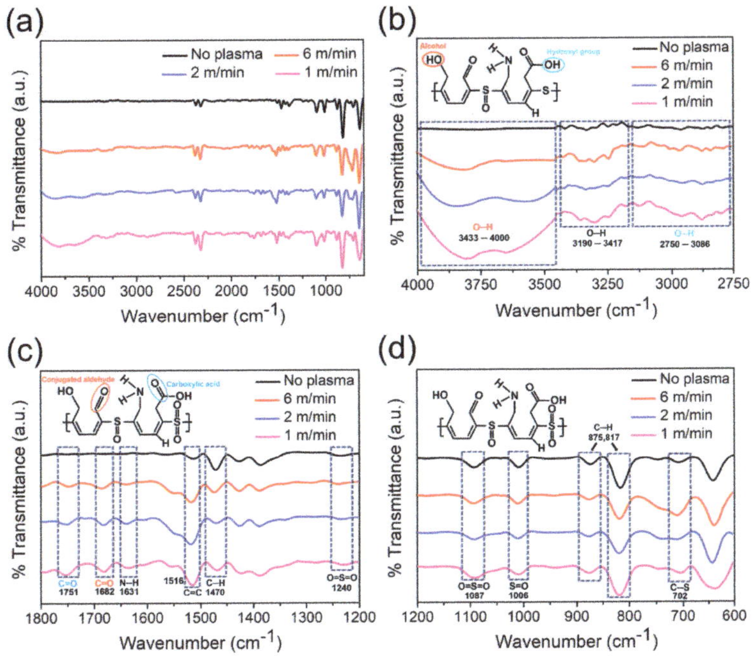

Figure 5. FT-IR spectra of the PPS/GF composites with various plasma treatment speeds (No plasma, 6 m/min, 2 m/min, 1 m/min), highlighting the formation and evolution of oxygen-containing functional groups: (**a**) Full range spectra (600–4000 cm^{-1}), (**b**) O–H stretching region (2750–4000 cm^{-1}), (**c**) carbonyl and aromatic C=C region (1200–1800 cm^{-1}) and (**d**) Fingerprint region (600–1200 cm^{-1}).

Table 5. FT-IR spectral peak assignments of PPS/GF composites treated with plasma at atmospheric pressure.

Wavenumber (cm^{-1})	Assignment
3433–4000	O–H stretching of alcohol
3190–3417	O–H, N–H stretching
2750–3086	O–H stretching of hydroxyl group
1751	C=O stretching of carboxylic acid
1682	C=O stretching of conjugated aldehyde
1631	N–H bending of amine
1516	C=C stretching of aromatic ring
1470	C–H (in-plane) bending of aromatic ring
1240, 1087	S=O stretching of sulfone
1006	S=O stretching of sulfoxide
875, 817	C–H (out-of-plane) bending of aromatic ring
702	C–S–C stretching of PPS

3.2. Morphology of PPS/GF Composites

The surface roughness profile and 10-point average roughness (R_Z) of the PPS/GF composites at varying plasma-treatment speeds are shown in Figure 6. At a plasma-treatment speed of 6 m/min, R_Z increased by 64%, from 42.74 nm to 70.02 nm. This increase can be explained by the partial removal of surface contaminants and mild surface etching caused by reactive species in the plasma. At lower plasma-treatment speeds of 2 m/min and 1 m/min, the R_Z significantly increased to 206.52 nm (+383%) and 392.21 nm (+817%), respectively, compared with the control sample. At lower treatment speeds, the longer exposure time allows the plasma to more strongly react with the surface. This causes more noticeable etching, forming uneven microstructures. These structures increase the surface area, improving mechanical interlocking with the epoxy adhesive.

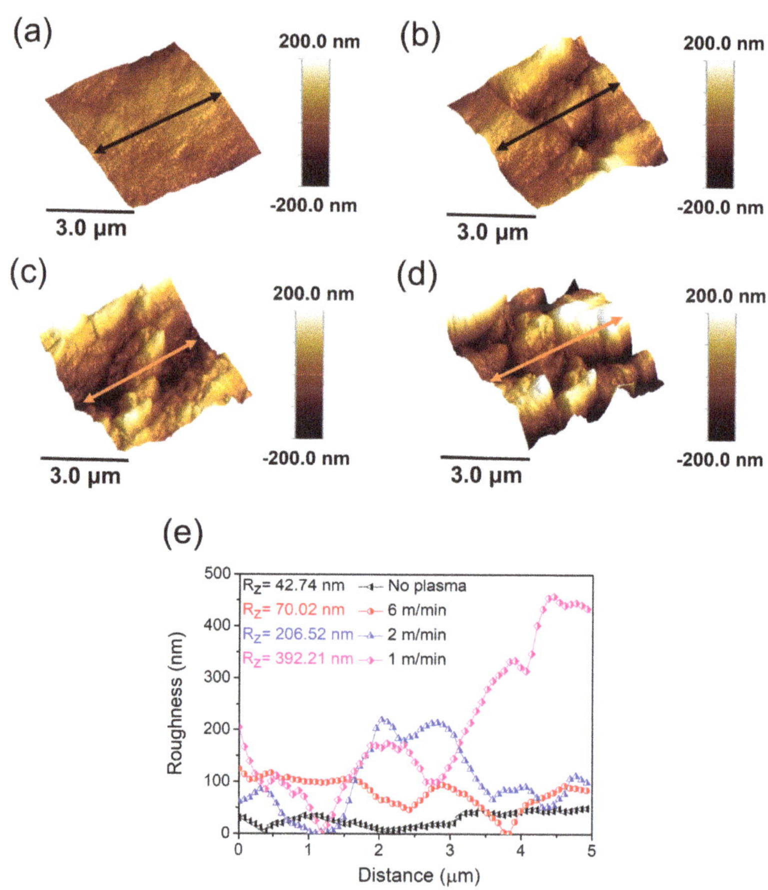

Figure 6. Surface roughness profile of PPS/GF composites at varying plasma-treatment speeds: (**a**) No plasma, (**b**) 6 m/min, (**c**) 2 m/min, and (**d**) 1 m/min. (**e**) Roughness profile of the arrow section along the longitudinal direction of the 5 μm scale.

SEM images and EDS mapping of the PPS/GF composites at varying plasma-treatment speeds are shown in Figure 7. These images clearly illustrate the etching effects on the composite surface as a function of the plasma-treatment speed, confirming the trend of increasing surface roughness with prolonged plasma irradiation. Additionally, EDS analysis revealed that oxygen was concentrated in the plasma-treated regions, whereas the carbon and sulfur contents were partially reduced. This suggests that surface etching and oxidation occurred simultaneously, removing surface impurities while activating the surface chemically.

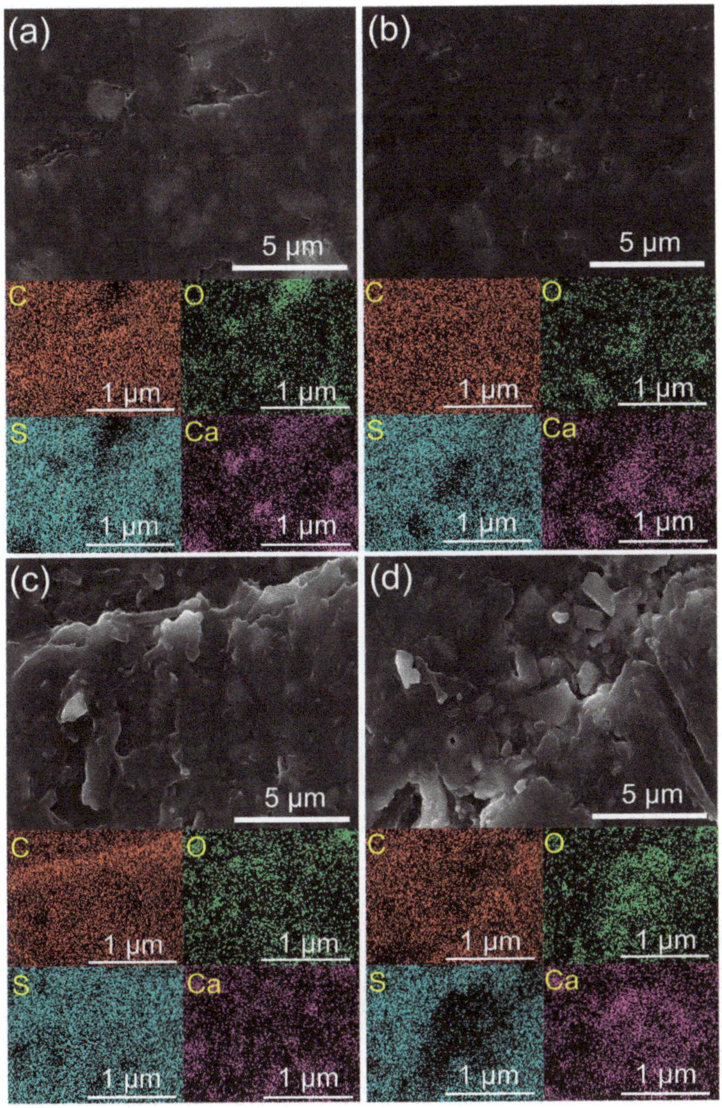

Figure 7. SEM images and EDS mapping of PPS/GF composites at varying plasma-treatment speeds: (**a**) No plasma, (**b**) 6 m/min, (**c**) 2 m/min, and (**d**) 1 m/min.

3.3. Wettability Analysis of PPS/GF Composites

Wettability describes how well a liquid spreads on a solid surface and plays a crucial role in bonding strength. Higher wettability increases the contact area between the adhesive and substrate, leading to stronger bonding. Wettability is typically evaluated using contact angle measurements, where a lower contact angle corresponds to better wettability. The contact angles—measured using diiodomethane (the dispersive component) and DI water (the polar component)—are shown in Figure 8. The water contact angle in the control sample was 78.9°, whereas at a plasma-treatment speed of 6 m/min, it decreased significantly to 31.5°, indicating a sharp increase in wettability. At speeds of 2 m/min and 1 m/min, the contact angles decreased further to 28.5° and 17.5°, respectively. This trend indicates that lower plasma-treatment speeds introduce a greater number of oxygen-containing functional groups onto the surface, thereby enhancing wettability. For diiodomethane, the contact angle of the control sample was 38.8°, which decreased to 28.0°, 21.0°, and 15.0° at 6, 2, and 1 m/min, respectively. Because diiodomethane contains only dispersive components, the reduction in contact angle can be attributed to the increase in surface roughness caused by the plasma treatment, which likely enhances wettability.

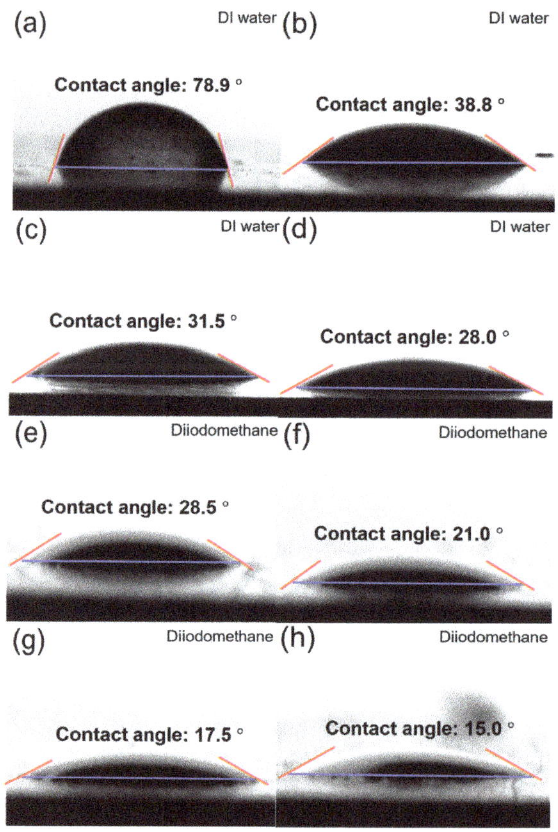

Figure 8. Contact angles measured using DI water and diiodomethane on the surface of the PPS/GF composites at varying plasma-treatment speeds: (**a**,**e**) No plasma, (**b**,**f**) 6 m/min, (**c**,**g**) 2 m/min, and (**d**,**h**) 1 m/min. (**a**–**d**) show DI water droplets, and (**e**–**h**) show diiodomethane droplets. Red lines indicate the fitted tangent at the droplet-solid interface used to calculate the contact angle, and blue lines represent the baseline of the solid surface.

The surface energy components calculated by applying the Owens–Wendt equation are shown in Figure 9 and are summarized in Table 6 [33]. The total surface energy of the control sample was 47.7 mN/m, with 44.9 mN/m from the dispersive component and 2.8 mN/m from the polar component, indicating that the surface was predominantly dispersive and relatively inert. After plasma treatment at 6 m/min, the total surface energy increased by 43% to 68.2 mN/m, with 47.3 mN/m from the dispersive component and 20.9 mN/m from the polar component. While the dispersive component showed a slight increase, the polar component clearly increased, leading to an increase in the total surface energy. Similarly, at a speed of 2 m/min, the total surface energy reached 72.2 mN/m (48.3 mN/m dispersive, 23.9 mN/m polar), and at 1 m/min, it further increased to 74.0 mN/m (48.8 mN/m dispersive, 25.2 mN/m polar), representing increases of 51% and 55%, respectively. These results indicate that lower plasma-treatment speeds introduce a greater number of polar functional groups on the surface, leading to a significant increase in surface energy. This enhancement in surface energy is expected to improve the interfacial interactions between the PPS/GF composites and the epoxy adhesive, ultimately contributing to the improved bonding strength.

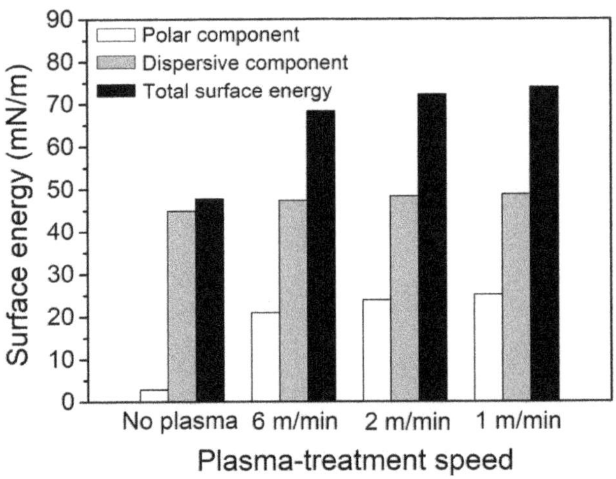

Figure 9. Variation in surface energy with plasma-treatment speed of the PPS/GF composites.

Table 6. Polar, dispersive, and surface energies of the PPS/GF composites at varying plasma-treatment speeds.

Plasma-Treatment Speed	γ_S^D (mN/m)	γ_S^P (mN/m)	γ_S
No plasma	44.9	2.8	47.7
6 m/min	47.3	20.9	68.2
2 m/min	48.3	23.9	72.2
1 m/min	48.8	25.2	74.0

3.4. Adhesive Behavior of PPS/GF Composites

Figure 10 illustrates the failure modes observed after single-lap shear testing and the trends in failure behavior according to the speed of plasma treatment. Figure 10a shows a partial adhesive failure mode, in which the epoxy adhesive delaminates from the composite surface due to insufficient interfacial bonding between the PPS/GF composites and the epoxy. The presence of residual epoxy on the composite surface is clearly visible in Figure S3, an optical microscope (OM) image. In contrast, Figure 10b presents a stock-break failure mode, where the failure occurs within the PPS/GF composite itself, indicating that the interfacial bonding strength exceeded the cohesive strength of the composite. Figure 10c summarizes the overall failure mode trends with respect to the plasma-treatment speed. All control samples exhibited partial adhesive failure. At a treatment speed of 6 m/min, some stock-break failures were observed. However, the failure mode was still predominantly adhesive. In contrast, most samples treated at 2 m/min and 1 m/min exhibited stock-break failure, indicating significantly improved adhesion. This change in failure mode is attributed to the increased incorporation of hydrophilic functional groups and the enhanced surface roughness caused by longer plasma exposure times, both of which contribute to stronger bonding with the epoxy [34].

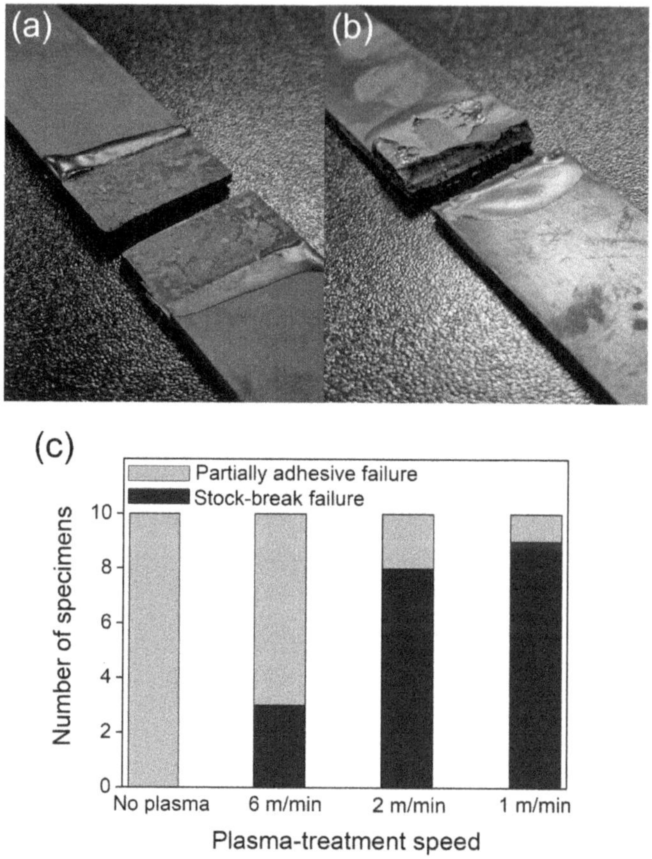

Figure 10. Representative failure morphologies and failure mode distribution according to plasma-treatment speed: (**a**) partial adhesive failure showing interfacial debonding, (**b**) stock-break failure indicating cohesive fracture within the substrate, and (**c**) failure mode distribution (partial adhesive vs. stock-break) across different plasma-treatment speeds.

The effect of plasma-treatment speed on mechanical performance is further demonstrated in Figure 11, which shows the stress–strain curve and bonding strengths, and in Table 7, which summarizes the numerical values. Compared with the control sample, a treatment speed of 6 m/min resulted in a 4.5% increase in bonding strength (from 3.76 MPa to 3.93 MPa). At speeds of 2 m/min and 1 m/min, the bonding strength further increased to 12.8% (4.24 MPa) and 24.8% (4.69 MPa), respectively, indicating a trend of lower plasma-treatment speeds leading to higher bonding strengths. Similarly, the elastic modulus increased compared with the control sample (113.76 MPa), reaching 146.52 MPa (+ 12.87%) at 6 m/min, 156.51 MPa (+13.76%) at 2 m/min, and 170.60 MPa (+ 15%) at 1 m/min. As discussed earlier, samples treated at speeds of 2 m/min and 1 m/min predominantly exhibited stock-break failure, which corresponds to their higher bonding strength and elastic modulus values. A lower speed of plasma treatment leads to improved interfacial adhesion, as evidenced by the transition from partial adhesive failure to stock-break failure and the enhancement in mechanical properties. Furthermore, as shown in Figure S4, there is a clear positive correlation between surface energy and bonding strength. The increase in surface energy, attributed to the introduction of polar functional groups, directly contributed to enhanced wettability and chemical interaction with the epoxy adhesive. Among all tested conditions, a plasma-treatment speed of 1 m/min was found to be the optimal, yielding the highest bonding strength.

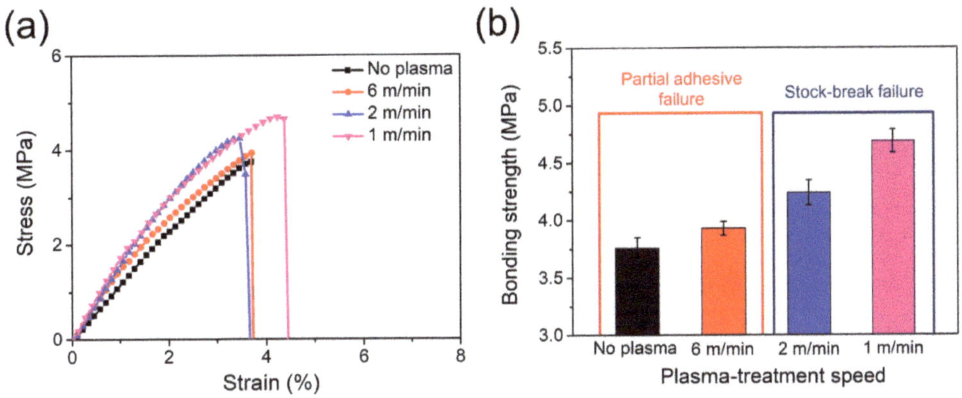

Figure 11. (**a**) Stress–strain curves and (**b**) bonding strength between PPS/GF composites and epoxy adhesives at varying plasma-treatment speeds.

Table 7. Bonding strength and failure mode of PPS/GF composites at varying plasma-treatment speeds.

Plasma-Treatment Speed	Bonding Strength (MPa)	Elastic Modulus (MPa)	Failure Mode
No plasma	3.76 ± 0.09	113.76 ± 6.80	Partially adhesive
6 m/min	3.93 ± 0.06	146.52 ± 5.62	Partially adhesive
2 m/min	4.24 ± 0.11	156.51 ± 7.88	Stock-break
1 m/min	4.69 ± 0.10	170.60 ± 4.21	Stock-break

4. Conclusions

In this study, the bonding strength between PPS/GF composites and epoxy was successfully enhanced through atmospheric-pressure plasma surface treatment. As the treatment speed decreased, the bonding strength increased, accompanied by a distinct transition in failure mode from partial adhesive failure to stock-break failure. This improvement is attributed to the increased introduction of oxygen-containing functional groups, such as hydroxyl (–OH) and carboxyl (–COOH) groups on the surface of the composites during prolonged plasma exposure, which promoted stronger chemical bonding with the epoxy adhesive. Additionally, the plasma treatment induced the surface etching of the PPS/GF composites, increasing surface roughness. This enhanced roughness expanded the contact area with the epoxy adhesive, thereby improving mechanical interlocking and further contributing to the enhanced bonding strength. At the optimized treatment speed of 1 m/min, the plasma treatment resulted in an 817% increase in surface roughness, a 55% increase in surface energy, and ultimately a 24.8% improvement in bonding strength compared with control samples. The findings of this study demonstrate that a simple and facile plasma treatment significantly improves adhesion performance, which plays a crucial role in the fabrication of capacitor housings for electric vehicles.

Supplementary Materials: The following supporting information can be downloaded at: https://www.mdpi.com/article/10.3390/polym17101344/s1. Figure S1: A schematic illustration of PPS/GF composites specimen preparation for single lap test. Figure S2: XPS spectra of PPS/GF composites with various plasma treatment speeds: (a) No plasma, (b) 6 m/min, (c) 2 m/min, and (d) 1 m/min. Figure S3: OM image of fractured surface showing partially adhesive failure at a plasma treatment speed of 6 m/min. Figure S4: Correlation between surface energy and bonding strength of PPS/GF composites treated by plasma irradiation with different treatment speeds. The red regression line illustrates the positive trend between increasing surface energy and corresponding bonding strength.

Author Contributions: Conceptualization, B.-G.C.; methodology, B.-G.C.; validation, B.-G.C.; formal analysis, B.-G.C. and H.-G.D.; investigation, B.-G.C. and H.-G.D.; data curation, B.-G.C. and H.-G.D.; writing—original draft preparation, H.-G.D.; writing—review and editing, B.-G.C.; supervision, B.-G.C. and P.-C.L.; funding acquisition, P.-C.L. All authors have read and agreed to the published version of the manuscript.

Funding: This research was funded by the Ministry of Trade, Industry, and Energy (MOTIE) of the Republic of Korea (grant number 20019105).

Institutional Review Board Statement: Not applicable.

Data Availability Statement: All data used in this study appear in the submitted article.

Acknowledgments: This work was supported by the Industrial Technology Innovation Program (20019105) and (20024220) funded by the Ministry of Trade, Industry, and Energy (MOTIE) of the Republic of Korea.

Conflicts of Interest: The authors declare no conflicts of interest. The funders had no role in the study design, collection, analyses, interpretation of data, writing of the manuscript, or decision to publish the results.

References

1. Chen, G.; Mohanty, A.K.; Misra, M. Progress in research and applications of Polyphenylene Sulfide blends and composites with carbons. *Compos. B Eng.* **2021**, *209*, 108553. [CrossRef]
2. Zhao, L.; Yu, Y.; Huang, H.; Yin, X.; Peng, J.; Sun, J.; Huang, L.; Tang, Y.; Wang, L. High-performance polyphenylene sulfide composites with ultra-high content of glass fiber fabrics. *Compos. B Eng.* **2019**, *174*, 106790. [CrossRef]
3. Gao, Y.; Zhou, X.; Zhang, M.; Lyu, L.; Li, Z. Polyphenylene sulfide-based membranes: Recent progress and future perspectives. *Membranes* **2022**, *12*, 924. [CrossRef] [PubMed]
4. Xian, G.; Zhou, P.; Li, C.; Dong, S.; Du, H.; Tian, J.; Guo, R.; Peng, Z.; Zhang, Z.; He, T. Mechanical properties evaluation of glass fiber reinforced thermoplastic composite plate under combined bending loading and water immersion. *Constr. Build. Mater.* **2024**, *440*, 137470. [CrossRef]
5. Li, W.; Yang, J.; Liu, X. Long carbon fibers reinforced rigid polyurethane composites: An investigation in strength and toughness. *Fibers Polymers* **2020**, *21*, 1605–1610. [CrossRef]
6. Shao, L.; Huang, J.; Feng, X.; Sun, Z.; Qiu, Y.; Tian, W.; Zhu, C. Study on preparation and properties of glass fibre fabric reinforced polyphenylene sulphide composites. *Materials* **2022**, *15*, 9036. [CrossRef]
7. Sun, Z.; Sun, L.; Zhu, C.; Tian, W.; Shao, L.; Feng, X.; Huang, K. Effect of polyphenylene sulphide particles and films on the properties of polyphenylene sulphide composites. *Materials* **2022**, *15*, 7616. [CrossRef]
8. Batista, N.L.; Anagnostopoulos, K.; Botelho, E.C.; Kim, H. Influence of crystallinity on interlaminar fracture toughness and impact properties of polyphenylene sulfide/carbon fiber laminates. *Eng. Fail. Anal.* **2021**, *119*, 104976. [CrossRef]
9. Hong, S.; Cho, B. Effect of Cooling Temperature on Crystalline Behavior of Polyphenylene Sulfide/Glass Fiber Composites. *Polymers* **2023**, *15*, 3179. [CrossRef]
10. Yu, L.; Wang, K.; Guan, Y.; Liu, Z.; Sun, M.; Zhao, Y. Effect of carbon fiber surface properties on carbon fiber/polyphenylene sulfide composite interfacial property. *Polym. Compos.* **2023**, *44*, 2005–2015. [CrossRef]
11. Astrouski, I.; Kudelova, T.; Kalivoda, J.; Raudensky, M. Shear strength of adhesive bonding of plastics intended for high temperature plastic radiators. *Processes* **2022**, *10*, 806. [CrossRef]
12. Aydemir, C.; Altay, B.N.; Akyol, M. Surface analysis of polymer films for wettability and ink adhesion. *Color Res. Appl.* **2021**, *46*, 489–499. [CrossRef]

13. Gupta, U.S.; Dhamarikar, M.; Dharkar, A.; Chaturvedi, S.; Kumrawat, A.; Giri, N.; Tiwari, S.; Namdeo, R. Plasma modification of natural fiber: A review. *Mater. Today Proc.* **2021**, *43*, 451–457. [CrossRef]
14. Mrsic, I.; Baeuerle, T.; Ulitzsch, S.; Lorenz, G.; Rebner, K.; Kandelbauer, A.; Chasse, T. Oxygen plasma surface treatment of polymer films—Pellethane 55DE and EPR-g-VTMS. *Appl. Surf. Sci.* **2021**, *536*, 147782. [CrossRef]
15. Tutunchi, A.; Ghodrati, T.; Tabrizi, A.T.; Osouli-Bostanabad, K. Enhancing the mechanical properties of CF-reinforced epoxy composites through chemically surface modification of carbon fibers via novel two-step approach by addition of epichlorohydrin. *Funct. Compos. Struct.* **2024**, *6*, 035005. [CrossRef]
16. Nurazzi, N.M.; Shazleen, S.S.; Aisyah, H.A.; Asyraf, M.R.M.; Sabaruddin, F.A.; Mohidem, N.A.; Norrahim, M.N.F.; Kamarudin, S.H.; Ilyas, R.A.; Ishak, M.R. Effect of silane treatments on mechanical performance of kenaf fibre reinforced polymer composites: A review. *Fuct. Compos. Struct.* **2021**, *3*, 045003. [CrossRef]
17. Hwang, M.-Y.; Kang, L.-H.; Huh, M. Application of laser surface treatment technique for adhesive bonding of carbon fiber reinforced composites. *Compos. Res.* **2020**, *33*, 371–376. [CrossRef]
18. Du, H.; Xian, G.; Tian, J.; Ma, Z. Effect of fiber surface treatment with silane coupling agents and carbon nanotubes on mechanical properties of carbon fiber reinforced polyamide 6 composites. *Polym. Compos.* **2024**, *46*, 1267–1283. [CrossRef]
19. Domonkos, M.; Tichá, P.; Trejbal, J.; Demo, P. Applications of cold atmospheric pressure plasma technology in medicine, agriculture and food industry. *Appl. Sci.* **2021**, *11*, 4809. [CrossRef]
20. Ghobeira, R.; Tabaei, P.S.E.; Morent, R.; De Geyter, N. Chemical characterization of plasma-activated polymeric surfaces via XPS analyses: A review. *Surf. Interfaces* **2022**, *31*, 102087. [CrossRef]
21. Turkoglu Sasmazel, H.; Alazzawi, M.; Kadim Abid Alsahib, N. Atmospheric pressure plasma surface treatment of polymers and influence on cell cultivation. *Molecules* **2021**, *26*, 1665. [CrossRef]
22. Cho, B.; Hwang, S.; Park, M.; Park, J.K.; Park, Y.; Chae, H.G. The effects of plasma surface treatment on the mechanical properties of polycarbonate/carbon nanotube/carbon fiber composites. *Compos. B Eng.* **2019**, *160*, 436–445. [CrossRef]
23. Mohammed, M.; Rahman, R.; Mohammed, A.M.; Adam, T.; Betar, B.O.; Osman, A.F.; Dahham, O.S. Surface treatment to improve water repellence and compatibility of natural fiber with polymer matrix: Recent advancement. *Polym. Test.* **2022**, *115*, 107707. [CrossRef]
24. Kodaira, F.V.P.; Almeida, A.C.P.L.; Tavares, T.F.; Quade, A.; Hein, L.R.O.; Kostov, K.G. Study of a conical plasma jet with a cloth-covered nozzle for polymer treatment. *Polymers* **2023**, *15*, 3344. [CrossRef]
25. Zhou, H.; Wang, F.; Wang, Y.; Li, C.; Shi, C.; Liu, Y.; Ling, Z. Study on contact angles and surface energy of MXene films. *RSC Adv.* **2021**, *11*, 5512–5520. [CrossRef] [PubMed]
26. *ISO 4587:2003*; Adhesives—Determination of Tensile Lap-Shear Strength of Rigid-to-Rigid Bonded Assemblies. ISO: Geneva, Switzerland, 2003.
27. Zhang, S.; Huang, G.; Wang, X.; Huang, Y.; Yang, J.; Li, G. Effect of air plasma treatment on the mechanical properties of polyphenylene sulfide/glass fiber cloth composites. *J. Reinf. Plast. Compos.* **2013**, *32*, 786–793. [CrossRef]
28. Cvelbar, U.; Mozetič, M.; Junkar, I.; Vesel, A.; Kovač, J.; Drenik, A.; Vrlinič, T.; Hauptman, N.; Klanjšek-Gunde, M.; Markoli, B. Oxygen plasma functionalization of poly (p-phenilene sulphide). *Appl. Surf. Sci.* **2007**, *253*, 8669–8673. [CrossRef]
29. Hellgren, N.; Haasch, R.T.; Schmidt, S.; Hultman, L.; Petrov, I. Interpretation of X-ray photoelectron spectra of carbon-nitride thin films: New insights from in situ XPS. *Carbon* **2016**, *108*, 242–252. [CrossRef]
30. de Castro, E.M.; Bozorgmehrian, F.; Carrola, M.; Koerner, H.; Samouei, H.; Asadi, A. Sulfur-Driven Reactive Processing of Multiscale Graphene/Carbon Fiber-Polyether Ether Ketone (PEEK) Composites with Tailored Crystallinity and Enhanced Mechanical Performance. *Compos. B Eng.* **2025**, *295*, 112180. [CrossRef]
31. Batista, N.L.; de Faria, M.C.M.; Iha, K.; de Oliveira, P.C.; Botelho, E.C. Influence of water immersion and ultraviolet weathering on mechanical and viscoelastic properties of polyphenylene sulfide–carbon fiber composites. *J. Thermoplast. Compos. Mater.* **2015**, *28*, 340–356. [CrossRef]
32. Li, Q.; Qin, Y.; Ren, S. Structural characterization analysis and macromolecular model construction of coal from Qinggangping coal mine. *Sci. Rep.* **2023**, *13*, 14365. [CrossRef] [PubMed]
33. Liber-Kneć, A.; Łagan, S. Surface testing of dental biomaterials—Determination of contact angle and surface free energy. *Materials* **2021**, *14*, 2716. [CrossRef] [PubMed]
34. Primc, G.; Mozetič, M. Surface modification of polymers by plasma treatment for appropriate adhesion of coatings. *Materials* **2024**, *17*, 1494. [CrossRef] [PubMed]

polymers

MDPI

Article

The Influence of the Dispersion and Covalent Functionalization of CNTs on Electrical Conduction Under an Electric Field in LDPE/CNT Composites

Xiaoli Wu [1], Ting Yin [1], Yi Yang [1], Wenyan Liu [1], Danping Wang [2,*], Libo Wan [2] and Yijun Liao [1,*]

1 School of Materials and Environmental Engineering, Chengdu Technological University, Chengdu 611730, China; wxli1@cdtu.edu.cn (X.W.)
2 School of Humanities and Design, Chengdu Technological University, Chengdu 611730, China; wanlibo@cdtu.edu.cn
* Correspondence: danpingwang@cdtu.edu.cn (D.W.); liaoyijun@cdtu.edu.cn (Y.L.)

Abstract: In this study, we comprehensively compare electrical conduction behavior under an applied electric field and electrical conductivity variation with temperature in low-density polyethylene (LDPE)/CNT composites with different dispersions and covalent functionalizations. Composites with different dispersions were prepared using solution and melt mixing processes. The solution-mixed composites exhibited better dispersion and higher electrical conductivity compared to the melt-mixed composites. At a high critical content (beyond the percolation threshold), the current–voltage (I–V) curve of the solution-mixed composites exhibited linear conduction behavior due to the formation of a continuous conductive network. In contrast, the melt-mixed composites exhibited nonlinear conduction behavior, with the conductive mechanism attributed to the field emission effect caused by poor interfacial contact between the CNTs. Additionally, LDPE/CNT-carboxyl (LDPE/CNT-COOH) and LDPE/CNT-hydroxy (LDPE/CNT-OH) composites demonstrated better dispersion but displayed lower electrical conductivity and similar nonlinear conduction behavior when compared to unmodified ones. This is attributed to the surface defects caused by the modification process, which lead to an increased energy barrier and a decreased transition frequency in the field emission effect. Furthermore, the temperature-dependent electrical conductivity results indicate that the variation trend in current with temperature differed among LDPE/CNT composites with different dispersions and covalent functionalizations. These differences were mainly influenced by the gap width between CNTs (mainly affected by dispersion and aspect ratio of CNTs), as well as the electrical conductivity of CNTs (mainly influenced by surface modification and intrinsic electrical conductivity of CNTs).

Keywords: dispersion; covalent functionalization; CNT-based composites; electrical conduction; temperature; field emission

1. Introduction

Carbon nanotubes (CNTs) are one-dimensional nanotubular materials with a structure in which carbon atoms are predominantly sp^2-hybridized [1,2]. In recent years, carbon nanotubes have attracted much attention due to their large aspect ratio, excellent electrical conductivity, and outstanding mechanical properties [3]. They are often incorporated as conductive particles into polymers to prepare conductive polymer composites, which are widely used in the fields of electromagnetic shielding [4,5], electric heating [6], antistaticity [7,8], etc. Due to the high aspect ratio and pronounced entanglement characteristics of CNTs, CNTs can easily agglomerate within a matrix. Thus, the dispersion of CNTs is a critical factor that significantly influences the electrical conductivity and mechanical properties of CNT–polymer composites [9,10].

The dispersion of conductive fillers in polymer composites is influenced by various factors, including the preparation method (including shear force) [11,12], surface modification, the length of CNTs [13], the viscosity of polymers [14], etc. For instance, Bai et al. [13] investigated the effect of nanotube length and aggregate size on the mechanical and electrical properties of composites and reported that the length and aggregate size had a very clear impact on the dispersion and dielectric properties of the composites. Sui et al. [15] found that the effect of preparation methods on dispersion and electrical conductivity is related to the content of CNTs. Faraguna et al. [16] prepared differently functionalized-CNT–polystyrene (PS) nanocomposites by melt and solution mixing, revealing that the solution-mixed nanocomposites exhibited lower percolation thresholds due to better dispersion compared to the melt-mixed composites. These studies have provided important guidance for the effect of CNT dispersion on the electrical conductivity and percolation threshold of CNT–polymer composites. However, the investigation of the influence of CNT dispersion on electrical conduction behavior under electric fields remains sluggish.

Most previous studies reported that the electrical conductivity of polymer composites increased with the increase in electric field when the filler content approached the percolation threshold [17–19]. That is, the intensity current–voltage (I–V) curve shows a nonlinear characteristic [20]. Previous studies have suggested that the nonlinearity of each element may be responsible for its intrinsically nonlinear characteristic [21]. Alternatively, it has been proposed that the transition from insulating to conducting channels under strong local fields could account for nonlinear electrical conductivity [22]. This transition is likely due to the nonlinear hopping or tunneling of charge carriers. He et al. [23] have demonstrated that a combination of nonlinear conduction induced by internal field emission and linear conduction contributed by a conducting network governs the overall electrical conduction process in percolating systems. Although these studies have described nonlinear electrical conduction for composites with different contents near the percolation threshold, none of them have focused on the investigation of electrical conduction under an electric field for composites with different dispersion states.

In most cases, it is necessary to chemically modify the surface of CNTs in order to achieve good dispersion in CNT–polymer composites and strong interfacial adhesion between CNTs and polymer matrices [24–27]. However, chemical modification processes may destroy the surface structure of CNTs. For example, Sui et al. [15] reported that carboxyl–CNT (COOH-CNT) had better dispersion than unmodified CNT, while COOH-CNT/thermoplastic polyether urethanes (TPU) composites had lower conductivity and higher percolation thresholds, which can be attributed to the structure defects and the reduced length of CNTs caused by the acidification process. Similarly, Trinh et al. [28] also demonstrated that COOH-CNT nanofluids showed lower electrical conductivity due to structure defects in CNTs, in comparison to unmodified CNT nanofluids, which even revealed a higher agglomerated size. These investigations have focused on the effect of surface chemical modification on dispersion, mechanical properties, and thermal and electrical conductivity [26,27,29]. However, the influence of surface chemical modification on electrical conduction behavior under electric fields is not very clear. Furthermore, there is a lack of research exploring how the covalent functionalization method affects the electrical conductivity variation as a function temperature for CNT based composites.

In order to systematically investigate the effect of CNT dispersion on electrical conduction behavior under the applied electric field and the temperature-dependent electrical conductivity variation, the low-density polyethylene (LDPE)/CNT composites with different dispersion states were prepared by the melt and solution mixing methods. The reason for using LDPE as the matrix is due to its ability to melt at high temperatures and dissolve in the solvent xylene [30], and its excellent mechanical properties [31]. What is more, COOH-CNT and OH-CNT were employed as the conductive fillers of the composites to systematically investigate the influence of surface chemical modification on CNT dispersion, electrical conduction behavior under an applied electric field, and temperature-dependent electrical conductivity variation.

2. Experimental Section

2.1. Materials

Low-density polyethylene (LDPE) was provided by Dongguan Shanhe Plastic Technology Co., Ltd., Dongguan, China. Multi-walled CNT (TNM1), the corresponding carboxyl-CNT (COOH-CNT, TNMC5) and hydroxy-CNT (OH-CNT, TNMH5) with 5–15 nm diameter and 98% purity were provided by the Chengdu Organic Chemicals Pty Ltd., Chengdu, China. Xylene (AR grade) was purchased from Chengdu Jinshan Chemical Co., LTD., Chengdu, China. Anhydrous ethanol (AR grade) was provided by Chengdu Colon Chemical Co., LTD., Chengdu, China.

2.2. Preparation of Composites

For the melt mixing process, the dried unmodified CNT and LDPE were mixed in a Torque Rheometer (Rheoccord System 40, Haake Technik GmbH, Vreden, Germany) at 200 °C. The mixing speed was set at 60 and 120 rpm, respectively. Then the LDPE/CNT composites were hot-pressed at 190 °C and 10 MPa for 5 min into a sheet with a size of $50 \times 50 \times 2$ mm^3 for different characterizations. A series of LDPE/CNT composites with different volume fractions of unmodified CNT (0.7, 1.4, 2, 2.8, 3.5, 4.5, 5, and 6 vol.%) were prepared. The LDPE/CNT-COOH and LDPE/CNT-OH composites with the same volume fraction were fabricated using the above process with a mixing speed of 120 rpm.

For the solution mixing process, a certain amount of CNTs was suspended in ethanol by sonication. The LDPE was dissolved in the xylene at 130 °C by magnetic stirring. When the mixture became transparent, the CNT/ethanol suspension was slowly dropped into the LDPE/xylene solution, and the mixture was stirred for 1 h. Subsequently, the mixture was filtered and vacuum dried at 60 °C for 72 h to remove the xylene. By measuring the weight, we confirmed that the xylene had been completely evaporated. Finally, the composite was put into the mold and hot pressed at 190 °C and 10 MPa for 20 min to obtain a plate-like sample with a size of $50 \times 50 \times 2$ mm^3. The volume fraction of CNT in the LDPE/CNT composites prepared by the solution mixing method is 0.7, 1.4, 2, 2.8, 3.5, 4.5, 5, and 6 vol.%.

The obtained LDPE/CNT composites prepared by different methods were marked as CNT-60-x, CNT-120-x, and CNT-solution-x, respectively, where x stands for the CNT content. The composites with and without grafting groups were named as CNT-x, CNT-COOH-x, and CNT-OH-x, respectively, where x represented the CNT content.

2.3. Material Characterization

The cross-sectional morphology of the LDPE/CNT composites was characterized by scanning electron microscopy (SEM, Helios 5 CX, Thermo Scientific, Waltham, MA, USA). The cross-section was obtained by brittle fracture in liquid nitrogen. The tensile and flexural tests were conducted using an electronic universal testing machine (68TM-10, Instron Corporation, Boston, MA, USA). The specimens have a size of $50 \times 10 \times 2$ mm^3. The loading rate was 50 mm/min and the gauge length was 25 mm. The flexural test was carried out with a loading rate of 10 mm/min, and the flexural displacement was set to 8 mm, at which point the test stopped.

2.4. Electrical Performance Characterizations

The volume electrical conductivity of the composites was examined using the RTS-11 four-probe resistivity tester (Shenzhen Junda Times Instrument, Shenzhen, China) when the electrical conductivity was above 10^{-6} S/cm and the ZST121 electric resistance meter (Avic Xi'an Aircraft Industry Group Company Ltd., Xi'an, China) when the electrical conductivity was below 10^{-6} S/cm. The electrical conductivity of CNTs was examined by the RTS-11 four-probe resistivity tester after the CNTs were pressed into a sheet at 15 MPa. The electrical conductivity of the sample was calculated using the following equation:

$$\sigma = \frac{t}{R \times S} \tag{1}$$

where σ is the electrical conductivity, t is the thickness, R is the electrical resistance, and S is the area of the samples.

The electrical conduction was carried out using a DC voltage source (DC1203D, Dingce Power Technology, Dongguan, China). The electrical conduction and the volume electrical resistivity at different temperatures (−40–80 °C) were measured in a freezer and a drying oven. The connection area of the equipment was wrapped with insulation foam to ensure the accuracy of the temperature.

The effective aspect ratio of CNTs in the matrix was calculated by substituting the percolation threshold into the following equations, which are given below [32–34]:

$$\psi_c = \frac{18S^2{}_{11} - 9S_{11}}{18S^2{}_{11} - 3S_{11} - 4} \tag{2}$$

$$S_{11} = S_{22} = \begin{cases} \frac{\alpha}{2(1-\alpha^2)^{\frac{3}{2}}}\left[\cos^{-1}\alpha - \alpha\left(1-\alpha^2\right)^{\frac{1}{2}}\right], \alpha < 1 \\ \frac{\alpha}{2(\alpha^2-1)^{\frac{3}{2}}}\left[\alpha\left(1-\alpha^2\right)^{\frac{1}{2}} - cosh^{-1}\alpha\right], \alpha > 1 \end{cases} \tag{3}$$

where ψ_c is the volume fraction of the percolation threshold, α is the aspect ratio of CNT, and S_{11} and S_{22} are the components of the shape-dependent depolarization tensor S of associated phases. For fillers with spherical symmetric shape, $S_{11} = S_{22} = 1/3$ [32]. For fillers with an asymmetric axis 3, such as carbon nanotubes, graphene, and other ellipsoidal fillers, S_{11} can be calculated by the formula mentioned above [33]. Previous research reports that the percolation threshold (ψ_c) is a strictly geometrical parameter [34], which represents the onset of the connective networks through the composite. It is only related to the aspect ratio of the fillers and can be used to calculate ψ_c [35,36]. Therefore, in this study, we inversely used the above equations to calculate the effective aspect ratio of CNTs in the matrix.

3. Results and Discussion

3.1. The Effect of CNT Dispersion on the Electrical Conduction Under an Electric Field for the LDPE/CNT Composites

The LDPE/CNT composites with different dispersion states were fabricated using melt and solution mixing methods. The dispersion of the LDPE/CNT composites was observed by SEM. At a low volume fraction (3.5 vol.%), for the melt-mixed composites, as depicted in Figure 1a, the distribution of CNTs was less uniform, with some agglomerations observed for the composites subjected to a stirring speed of 60 rpm (see the red dashed boxes). When the stirring speed increased to 120 rpm, the amount of the CNT agglomerates reduced, as depicted in Figure 1b, exhibiting a better CNT dispersion primarily due to enhanced shear force. The solution-mixed composites exhibit a more uniform dispersion, without any noticeable agglomerates (Figure 1c,d). As shown in Figure 1c,d, the CNTs were observed to be in contact with each other, thereby establishing a continuous conductive network within the solution-mixed composites. When the volume fraction was increased to 6 vol.% (Figure 1e–h), the agglomeration of CNTs in the melt-mixed composites was more obvious, while the solution-mixed composites still exhibited the best dispersion. For the melt-mixed composites with a stirring speed of 120 rpm, the CNTs formed interconnected conduction paths but failed to establish a continuous network. When the stirring speed decreased to 60 rpm, obviously isolated CNTs were observed (Figure 1e). The aforementioned results demonstrated that the dispersion of CNT in the solution-mixed composite surpassed that in the melt-mixed composite, and that an increase in shear force can enhance CNT dispersion. This result was in agreement with other previous studies [11,16,37]. They have reported that a better dispersion enabled CNTs to form continuous conducting networks more easily.

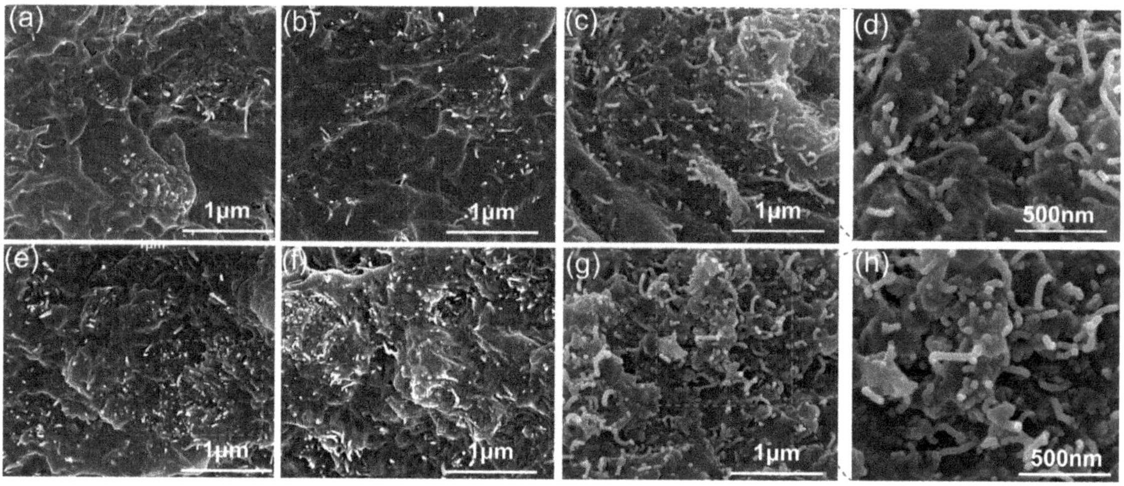

Figure 1. The SEM morphology of the LDPE/CNT composites prepared by different mixing processes: (**a**) CNT-60-3.5 100,000×; (**b**) CNT-120-3.5 100,000×; (**c**) CNT-Solution-3.5 100,000×; (**d**) CNT-Solution-3.5 200,000×; (**e**) CNT-60-6 100,000×; (**f**) CNT-120-6 100,000×; (**g**) CNT-Solution-6 100,000×; (**h**) CNT-Solution-6 200,000×.

In order to further confirm the dispersion state of the LDPE/CNT composites with different preparation methods, the mechanical properties were tested. Figure 2a,b show the tensile modulus and tensile strength of the LDPE/CNT composites with different preparation methods, respectively. It can be clearly observed that the solution-mixed LDPE/CNT composites exhibit the highest tensile modulus and strength at both low and high contents. The melt-mixed composites with a stirring speed of 120 rpm exhibited higher tensile modulus and strength compared to those with a stirring speed of 60 rpm. These results should be related to the homogeneity of CNT. The excellent dispersion of CNT in LDPE resulted in the interconnection between the polymer matrix and CNTs, thereby reducing the slip and enhancing the mechanical strength of the composites [38,39]. In addition, a decline in tensile strength was observed at 6 vol.% CNTs, which was likely attributed to the more agglomeration resulting in poor stress transfer [15]. Figure 2c shows the flexural strength of the LDPE/CNT composites with different preparation methods. The solution-mixed LDPE/CNT composites also exhibited the highest flexural strength at both low and high contents. These results suggest that the solution-mixed LDPE/CNT composite exhibited superior dispersion, while the melt-mixed composite with a stirring speed of 60 rpm showed the poorest dispersion.

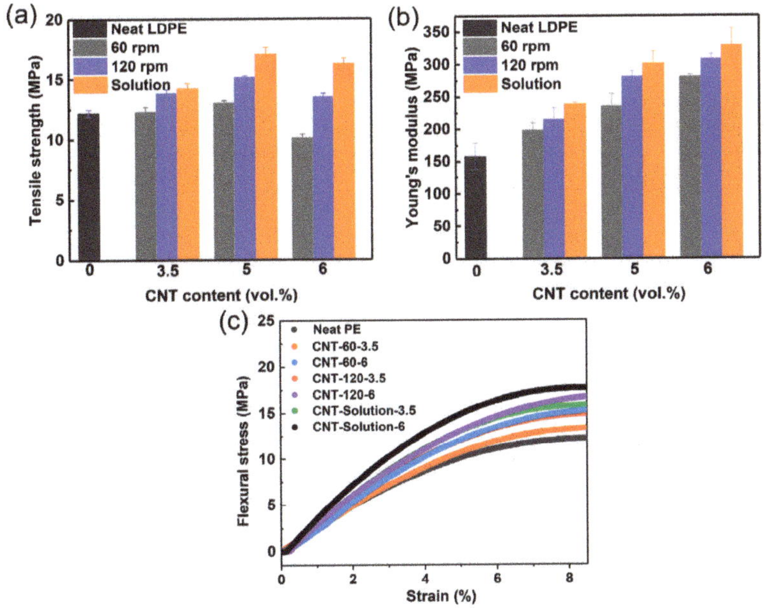

Figure 2. The mechanical strength of the LDPE/CNT composites prepared by different mixing processes at different CNT volume fractions: (**a**) tensile modulus; (**b**) the ultimate tensile strength; (**c**) the flexural strength and strain curves.

Figure 3a shows the volume electrical conductivity of the LDPE/CNT composites prepared by different methods. For the melt-mixed composites, the electrical conductivity of the composite with a stirring speed of 120 rpm was higher than that of the composite with a stirring speed of 60 rpm. The solution-mixed composites demonstrated a higher electrical conductivity than the melt-mixed composites. The percolation threshold was calculated using the classical percolation formula, as delineated below [40]:

$$\Sigma = \sigma_f (\psi - \psi_c)^t \ (\psi > \psi_c) \quad (4)$$

where ψ is the volume fraction of the filler, ψ_c is the percolation threshold, and σ and σ_f are the conductivities of the composite and filler, respectively. In addition, t is a critical power exponent that depends on the dimensionality of the conducting network and typically takes the value of ~1.33 for two-dimensional (2D) systems (e.g., in coatings) [41], For three-dimensional (3D) systems, this is usually taken as ~2 [42]. Log values were taken for each side of the equation, and a linear line fit was used. The fitting results are shown in Figure 3b. The percolation thresholds of the melt-mixed composites with stirring speeds of 60 and 120 rpm were fitted to be 4.2 and 3.2 vol.%, respectively. Moreover, the corresponding t-values were significantly high at 4.17 and 5.12. Combined with the SEM results, the reason for such a high t-value was likely attributed to the formation of tunneling percolation systems [37,43], which were formed by conductive particles located apart from each other and enclosed by the polymer matrix. This system may arise from extensive CNT aggregations in the melt-mixed composites. For the solution-mixed composites, the percolation threshold (2 vol.%) was obviously lower than that of the melt-mixed composites. Moreover, the t-value reached as high as 5.7, which was significantly higher than that of the melt-mixed composites. Such a high t-value was likely due to the formation of a 3D conductive network (see Figure 1h) as well as the presence of small CNT aggregations. These results further confirmed the differences in the dispersion state of the LDPE/CNT composites prepared by different methods.

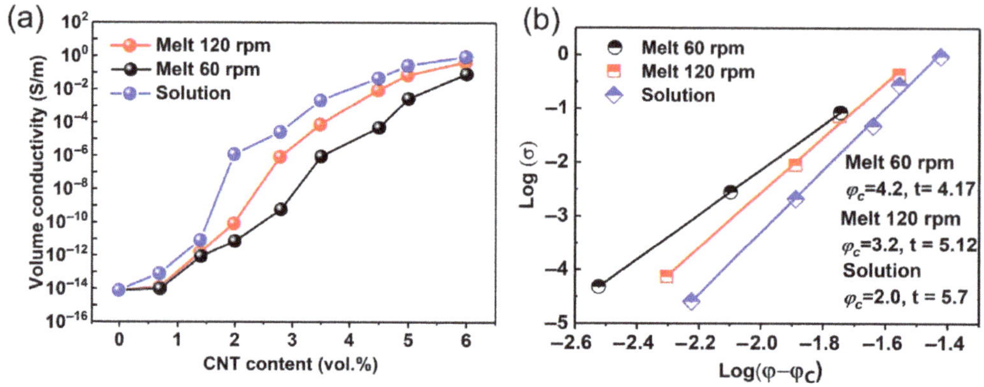

Figure 3. (**a**) The volume conductivity (σ) of the LDPE/CNT composites prepared by different mixing processes; (**b**) the Log plots of σ versus ψ-ψ_c and their fit curves.

The electrical conduction under an electric field of the LDPE/CNT composites with different preparation methods was examined at ambient temperature. The composites with a volume fraction of 6 vol.% were selected for testing due to their content exceeding the percolation threshold and showing conductive behavior ($\sigma < 10^{-4}$ S/m) [44]. The current–voltage (I–V) curves are shown in Figure 4a–c. The solution-mixed composite demonstrated the highest current value at a given voltage among the three groups of composites, owing to its superior electrical conductivity. The current of all three groups of composites exhibited an increasing trend with the applied voltage; however, it is evident that the I–V curves of these three groups have obvious differences. For the solution-mixed composite, the I–V curve showed a linear characteristic, whereas the I–V curves of the melt-mixed composites exhibited a nonlinear variation. Moreover, the resistance (R) of the melt-mixed composites decreased with the applied voltage (Figure 4d). Previous studies have reported that the nonlinear variation in percolated composites primarily arose from electron tunneling induced by the field emission effect occurring between isolated fillers [22]. The internal field emission effect was proposed by Zener in the 1930s [45], through which a large number of isolated conductive particles that are not connected to the whole conductive network can also participate in the conduction process by jumping over the insulating barriers through electron tunneling. The curves were fitted using linear and field emission formulas, respectively. The specific formulas employed are as follows:

$$I = a + bV \quad (5)$$

$$I = AV^n exp\,(-B/V) \quad (6)$$

where I is the current intensity; V is the applied voltage; a, b, A, B, and n are constants; and n is usually between 1 and 3 depending on various corrections included in the theory, such as the effects of image fields or Coulombic forces [46]. A is a function of tunneling frequency, i.e., the number of attempts to cross the barrier per second, and B represents the energy barrier between the insulating matrix and the conducting fillers, which is mainly related to the gap width between the conducting fillers [47]. In this study, the fitted curves and parameters for the two formulas are also shown in Figure 4a–c. The I–V curves of the melt-mixed composites with stirring speeds of 60 and 120 rpm exhibited a closer adherence to the field emission equation as compared to that of the linear one. This result suggests that the dominant conductive mechanism of melt-mixed composites is the field emission effect. This can be attributed to non-uniform dispersion resulting in insufficient interconnection between CNTs, see schematic diagram in Figure 5a. Moreover, it is apparent that the composite stirred at 120 rpm exhibits a higher tunneling frequency (A) and lower potential barrier (B) compared to the one stirred at 60 rpm (Figure 4e). This can be attributed to the better dispersion, resulting in a narrower gap width between CNTs in the former.

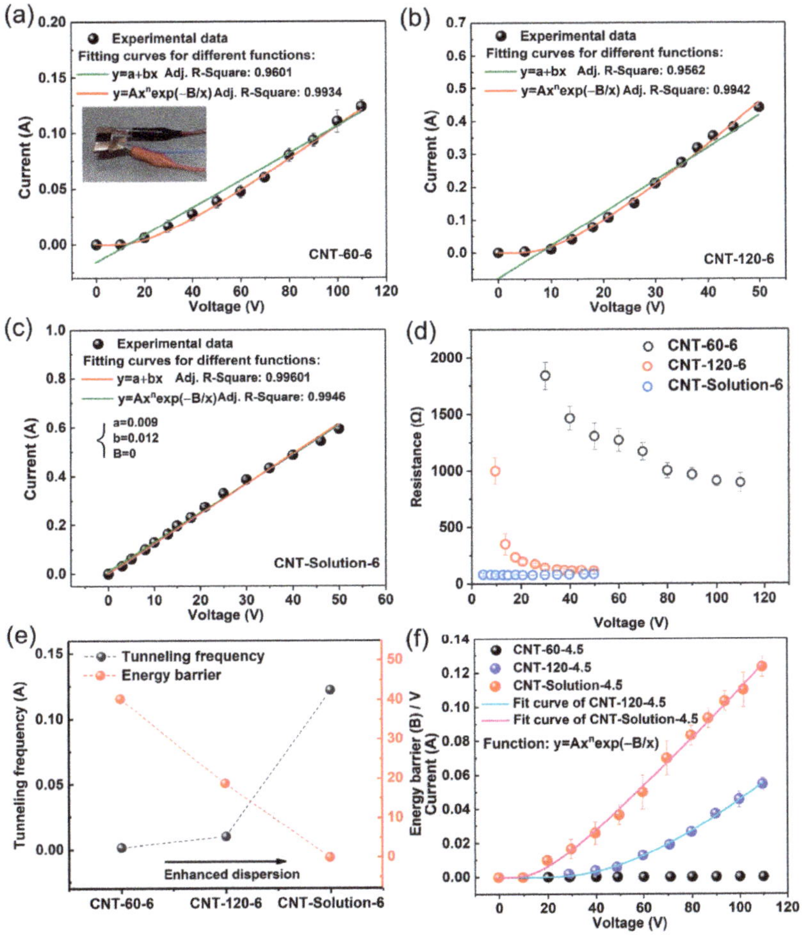

Figure 4. (**a**–**c**) The current intensity variation as a function of applied voltage for the LDPE/CNT composites prepared by different mixing processes: (**a**) CNT-60-6; (**b**) CNT-120-6; (**c**) CNT-Solution-6. The solid lines correspond to the best linear-fitting result and fitted curve from the electric field emission equation, respectively. (**d**) The resistance variation as a function of applied voltage; (**e**) the fitted parameters of transition probability A and the energy barrier B; (**f**) the current intensity variation as a function of applied voltage and fitted curve from the electric field emission equation for the LDPE/CNT composites prepared by different mixing process at a volume fraction of 4.5 vol.%.

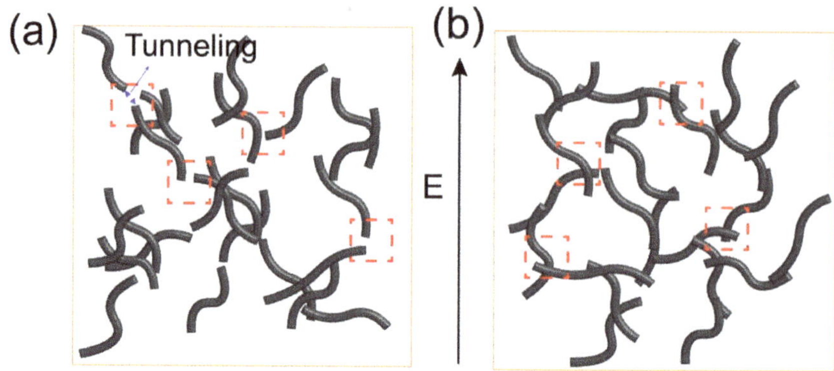

Figure 5. (**a**) The schematic diagram of the LDPE/CNT composites with CNT aggregates; (**b**) the schematic diagram of the LDPE/CNT composites with good dispersion.

The *I–V* curve of the solution-mixed composite demonstrated a linear characteristic, indicating that the conductivity mechanism follows the percolation theory. This finding was different from the melt-mixed composites. It was likely arising from the more effective interconnection between CNTs, resulting in a reduction in gap width between CNTs, see schematic diagram in Figure 5b. This results in a decrease in potential barrier (*B*) derived from the field emission formula, as shown in Table 1. The *B* value became zero, leading to the transformation of the field emission equation into linear equation of $I = 0.012\ V$. When the CNT content decreased to 4.5 vol.%, the conductive mechanism of the solution-mixed composite became the field emission effect, as evidenced by the experimental data and its fitting curve depicted in Figure 4f. *A* higher *B* value was obtained (Table 1), which was likely due to the larger gap width between CNTs. Therefore, the above results indicate that the CNT dispersion appears to exhibit a similar effect as the filler content, influencing not only the electrical conductivity and percolation threshold but also the gap width of CNTs and the electrical conduction behavior under the electric field.

Table 1. The fitted parameters of the field emission equation for the LDPE/CNT composites prepared by different methods.

Specimens	*A*	*B*	*n*	Adj. R-Square
CNT-120-4.5	0.00141	113.86	1.05	0.9964
CNT-Solution-4.5	0.00161	31.68	1.08	0.9977
CNT-60-6	0.00158	40.33	1.06	0.9934
CNT-120-6	0.010	18.79	1.09	0.9942
CNT-Solution-6	0.122	0	—	0.9946

In engineering applications, the variation in electrical conductivity as a function of temperature serves as a crucial indicator. Therefore, in this study, the electrical properties of the LDPE/CNT composites with different dispersion states at different environmental temperatures were investigated. Figure 6a shows the current variation in the composites with content of 6 vol.% prepared by different preparation methods at environment temperatures of −40–80 °C. It can be seen that there is an obvious difference in the current variation trends of the three LDPE/CNT composites at their respective maximum operating voltages. When the environmental temperature decreased from 20 to −20 °C, the current values of these composites exhibited an increasing trend. The melt-mixed composite with a stirring speed of 60 rpm showed a maximum increase value. It can be attributed to its poorer dispersion, resulting in inadequate contact between CNTs. Consequently, when the temperature was reduced, the gap width between CNTs was significantly decreased due to the matrix contracting to some extent as the temperature dropped [48]. As the environmental temperature decreased to −40 °C, the current values of all three groups of composites showed a decreasing trend, which was mainly related to the decrease in the energy of charge carriers at low temperatures [49].

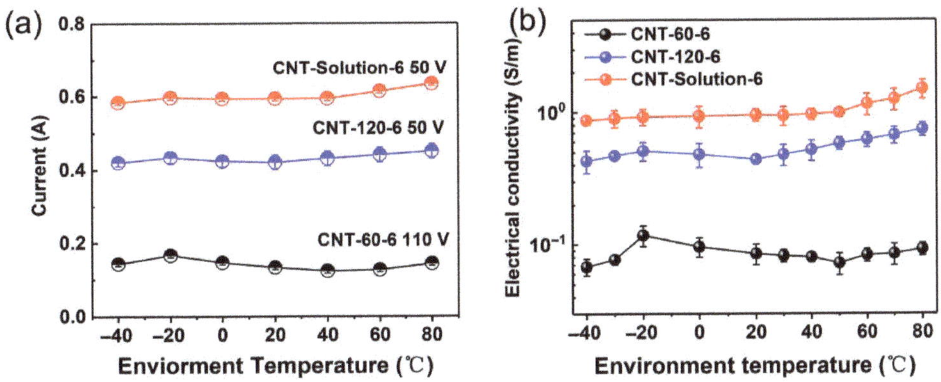

Figure 6. (**a**) The current variation as a function of temperature for the LDPE/CNT composites prepared by different mixing processes; (**b**) the electrical conductivity variation as a function of temperature for the LDPE/CNT composites prepared by different mixing processes.

As the temperature increased from 20 °C to 80 °C, the current value of the melt-mixed composite with a stirring speed of 60 rpm initially exhibited a decreasing trend, followed by a slight increase. The decrease may be attributed to the increase in the gap width between CNTs induced by the thermal expansion of the polymer matrix as the temperature rose [50]. The increase may be due to the fact that the carrier energy was elevated at higher temperatures [51,52]. The increasing trend also occurred in the current values of melt-mixed composites with a stirring speed of 120 rpm and the solution-mixed composites. It is interesting that these two groups did not exhibit a declining trend, potentially due to the better dispersion state (more effective contact) resulting in less variation in gap width between CNTs, see schematic diagram in Figure 5. The above results are further confirmed by the electrical conductivity variation with temperature, see Figure 6b. It can be concluded that the dispersion states of CNTs have a significant impact on electrical conduction and conductivity of the composites with temperature.

3.2. The Effect of Covalent Functionalization on the Dispersion of CNT and the Electrical Conduction Under an Electric Field for the LDPE/CNT Composites

The SEM morphologies of the LDPE/CNT-COOH and LDPE/CNT-OH composites with a volume fraction of 6 vol.% are shown in Figure 7a,b. It can be seen that there is no obvious difference in the dispersion of the LDPE/CNT-COOH and LDPE/CNT-OH composites. When compared to the unmodified LDPE/CNT composites, they exhibited a smaller agglomeration size and more uniform distribution of CNTs within the matrix. However, it is worth noting that the LDPE/CNT-COOH and LDPE/CNT-OH composites have formed conductive paths, while there was a certain gap width between the conductive chains, and there were some isolated CNTs. In addition, the gap width between CNTs seemed to be larger than that of the unmodified LDPE/CNT composites. It could be due to the destruction of the surface structure of the CNT during the acidification modification and the reduction in length, which was determined by the SEM morphology of COOH-CNT and OH-CNT (Figure 7c,d).

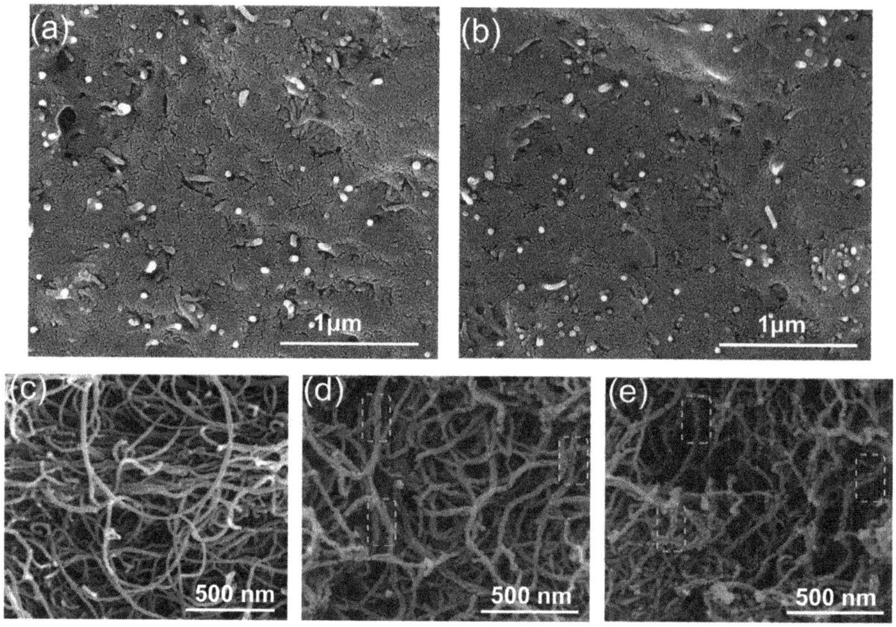

Figure 7. The SEM morphology of (**a**) LDPE/CNT-COOH-6 composite; (**b**) LDPE/CNT-OH-6 composite; (**c**) CNT; (**d**) COOH-CNT; (**e**) OH-CNT.

Figure 8 shows the tensile modulus, tensile strength, and flexural strength–strain curves of the LDPE/CNT-COOH and LDPE/CNT-OH composites, respectively. The result of the unmodified LDPE/CNT composite is also displayed for comparison. It is clearly observed that the LDPE/CNT-COOH and LDPE/CNT-OH composites exhibit lower tensile modulus and tensile strength compared to those of unmodified composites at both low and high contents, see Figure 8a,b. Figure 8c shows the flexural strength of the LDPE/CNT-COOH and LDPE/CNT-OH composites. It can be seen that the flexural strengths of the LDPE/CNT-COOH and LDPE/CNT-OH composites are also lower than the unmodified composites at both low and high contents. The poor reinforcing effect of COOH-CNT and OH-CNT may be attributed to the reduction in CNT length and the many structural defects. This indicates that the uniform dispersion of CNTs does not appear to be a critical factor influencing the enhancement of mechanical properties in the composites as compared to the structural defects induced by the chemical modification process.

Figure 9a,b show the volume electrical resistivity and the percolation threshold of LDPE/CNT-COOH and LDPE/CNT-OH composites. It can be seen that the LDPE/CNT-COOH and LDPE/CNT-OH composites exhibit lower electrical conductivity and higher percolation thresholds compared with the unmodified LDPE/CNT composite. The percolation thresholds of the LDPE/CNT-COOH and LDPE/CNT-OH composites were as high as 4.0 and 4.3 vol.%, respectively. This result could be attributed to two factors. Firstly, the presence of structural defects led to a decrease in electrical conductivity [15,29,53]. This was supported by the results of the electrical conductivity of the modified and unmodified carbon nanotubes (CNTs), as shown in Figure 9c. The data indicates that the electrical conductivity of COOH-CNT (1124.3 S/m) and OH-CNT (813.6 S/m) is significantly lower than that of pristine CNT (2501.7 S/m). Secondly, the existence of structural defects caused a reduction in length, which affected the formation of conductive networks [54]. This was supported by the calculated effective aspect ratios of carbon nanotubes (CNTs) in the matrix, obtained using Equations

(2)–(4). These values were 25.0, 23.2, and 31.7 for LDPE/CNT-COOH, LDPE/CNT-OH, and unmodified LDPE/CNT composites, respectively.

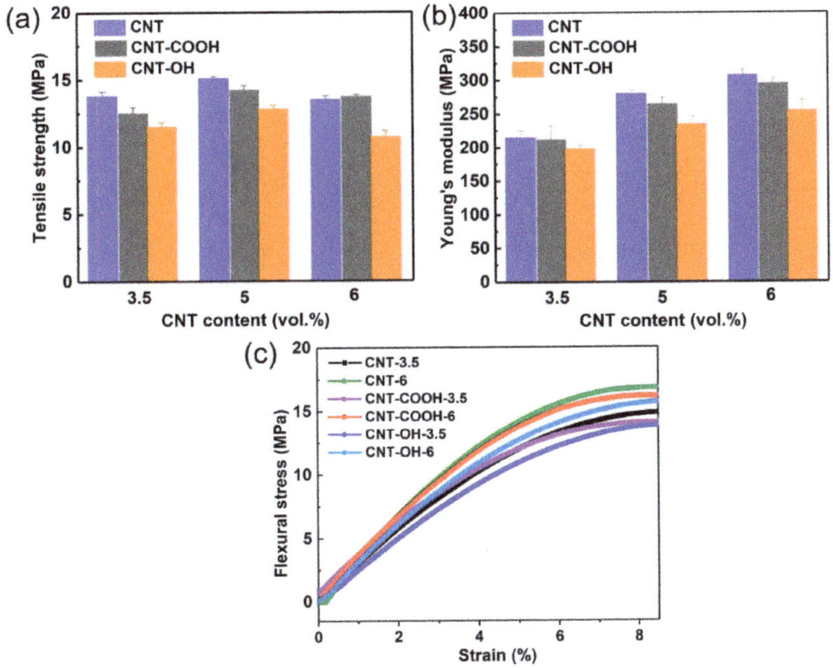

Figure 8. The mechanical strength of the modified CNT/ LDPE composites at different CNT volume fractions: (**a**) tensile modulus; (**b**) the ultimate tensile strength; (**c**) the flexural strength and strain curves.

The current–voltage (I–V) curves of the LDPE/CNT-COOH and LDPE/CNT-OH composites with a fraction volume of 6.0 vol.% are depicted in Figure 9d. It can be observed that the LDPE/CNT-COOH and LDPE/CNT-OH composites show lower current values than the unmodified LDPE/CNT composite. The I–V curves of the two modified LDPE/CNT composites exhibited nonlinear characteristics (Figure 9d,e) and were fitted to the field emission equation, which was similar to that of the LDPE/CNT composite. Differently, the two modified composites exhibit significantly larger values of potential barrier (B) and lower values of tunneling frequency (A) compared with the unmodified composite (Figure 9f). This could be due to their reduced charge carrier concentration and mobility of CNT caused by surface defects, leading to a decrease in tunneling frequency (A). Additionally, the reduction in CNT length increases the gap width between CNTs, leading to an increase in the potential barriers (B). These results indicate that the structural defects induced by the chemical modification process seem to be a more critical factor influencing the enhancement of electrical properties in the composites as compared to that of the uniform dispersion of CNTs.

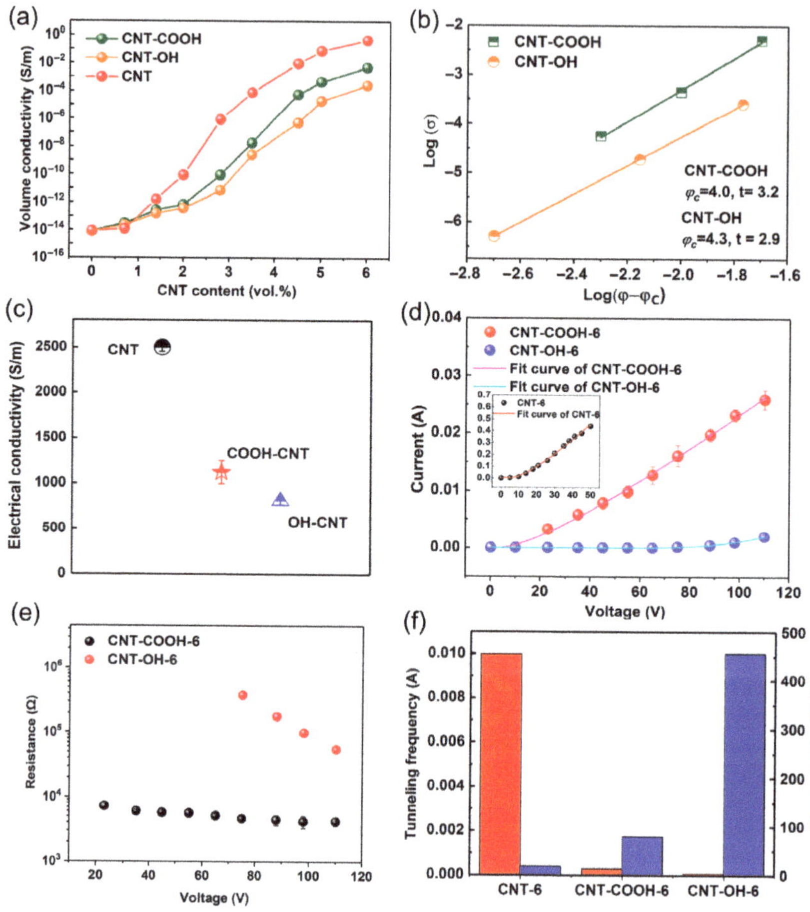

Figure 9. (**a**) The volume conductivity (σ) of the modified LDPE/CNT composites. (**b**) The Log plots of σ versus ψ-ψ_c and their fit curves. (**c**) The volume conductivity of CNT, COOH-CNT, and OH-CNT pressed sheets. (**d**) The current intensity variation as a function of applied voltage for the modified LDPE/CNT composites and unmodified LDPE/CNT composites. The solid lines correspond to the best-fitted curve from the electric field emission equation. (**e**) The resistance variation as a function of applied voltage. (**f**) The fitted parameters of transition probability A and the energy barrier B.

Figure 10a shows the current variation in the LDPE/CNT-COOH and LDPE/CNT-OH composites with a volume fraction of 6 vol.% at environmental temperatures of −40–80 °C. It can be seen that the two modified LDPE/CNT composites exhibit different current variations with environmental temperature at their respective operating voltages. When the environmental temperature changed from 20 °C to −40 °C, the current variations in the two modified LDPE/CNT composites initially increased and subsequently decreased, exhibiting a similar trend to that observed in the unmodified LDPE/CNT composite. It is worth noting that the change in slope of the two modified composites is higher, which is likely attributed to the reduced effective aspect ratio of CNT. The shorter CNTs resulted in a larger gap width between CNTs compared to long CNTs [55]. While the larger gap width between CNTs may lead to more significant changes during heating or cooling, as illustrated in the schematic diagram of Figure 11.

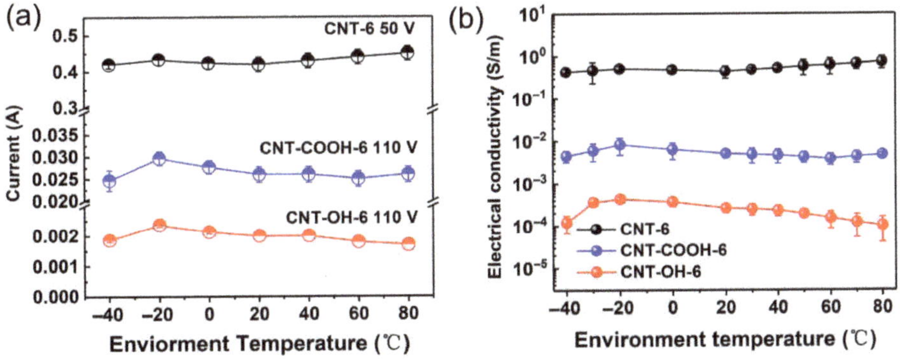

Figure 10. (**a**) The current variation as a function of temperature for the modified LDPE/CNT composites; (**b**) the electrical conductivity variation as a function of temperature for the modified LDPE/CNT composites.

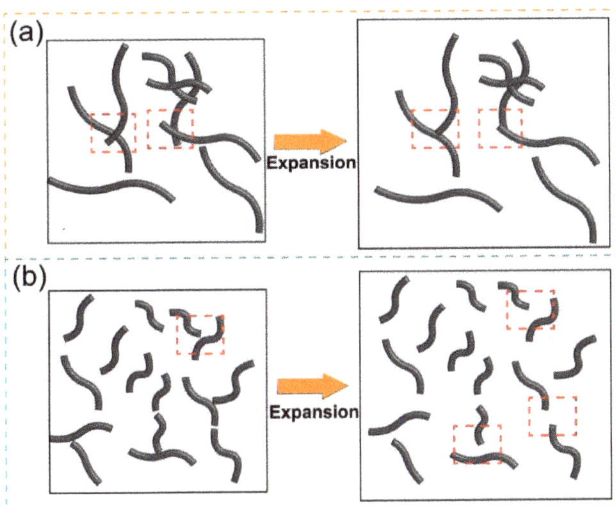

Figure 11. The schematic diagram pre- and post-expansion for (**a**) the unmodified LDPE/CNT composites with long CNT and several aggregates and (**b**) the modified LDPE/CNT composites with good dispersion but short CNT.

When the environmental temperatures increased from 20 °C to 60 °C, the current values of both the modified composites decreased with the increase in temperature, which was different from the results of the unmodified LDPE/CNT composite. It could be attributed to the expansion of the matrix at higher temperatures, leading to an increased gap width between CNTs. As the environmental temperature rose from 60 °C to 80 °C, the LDPE/CNT-COOH composite exhibited a slight increasing trend in current. This behavior was likely due to the effect of the increase in energy of the carriers caused by elevated temperature, surpassing the effect of the gap width variation. However, the LDPE/CNT-OH composite continued to show a decreasing trend in current, primarily due to its more severe structural disruption and lower electrical conductivity of CNTs. The electrical conductivity of both the modified composites exhibited similar variations with temperature (refer to Figure 10b). Consequently, we can conclude that the variations in electrical conductivity with temperature are mainly influenced by the gap width between

CNTs (mainly affected by dispersion and aspect ratio of CNTs) as well as the electrical conductivity of CNTs (mainly influenced by surface modification and intrinsic electrical conductivity of CNTs).

4. Conclusions

In summary, we have studied the effect of CNT dispersion and covalent functionalization on the electrical conduction behavior under the applied electric field for LDPE/CNT composites for the first time. At a high critical content beyond the percolation threshold, the I–V curve showed a linear characteristic for the solution-mixed composite due to its continuous network of CNTs. While the melt-mixed composites exhibited a nonlinear characteristic, the conductive mechanism was found to be the field emission effect, which was likely contributed to the poor contact of the CNTs. On the other hand, the LDPE/CNT-COOH and LDPE/CNT-OH composites demonstrated better dispersion, but lower electrical conductivity and similar nonlinear conduction behavior compared to that of the unmodified ones. This was attributed to the surface defects caused by the covalent functionalization process, which led to an increased energy barrier and a reduced transition frequency in the field emission effect. Furthermore, the results of the electrical conductivity variation with temperature showed that the composites with different dispersions and covalent functionalizations of CNTs exhibited varying trends in current variation with temperature. Such findings might have critical implications in completing the electrical conduction mechanism of polymer composites.

Author Contributions: Writing—original draft, X.W.; writing—review and editing, X.W., D.W. and Y.L.; investigation, X.W., T.Y., Y.Y., W.L. and L.W.; data curation, X.W.; conceptualization, X.W. and Y.L.; methodology, X.W., D.W. and Y.L.; software, D.W. and L.W.; project administration, D.W., Y.L.; supervision, Y.L.; funding acquisition, Y.L. All authors have read and agreed to the published version of the manuscript.

Funding: This research was funded by the Science and Technology Project of Sichuan Province (2022YFG0107).

Institutional Review Board Statement: Not applicable.

Data Availability Statement: The raw/processed data required to reproduce these findings cannot be shared at this time as the data also forms part of an ongoing study.

Conflicts of Interest: The authors declare no conflict of interest.

References

1. Rathinavel, S.; Priyadharshini, K.; Panda, D. A review on carbon nanotube: An overview of synthesis, properties, functionalization, characterization, and the application. *Mater. Sci. Eng. B* **2021**, *268*, 115095. [CrossRef]
2. Bhatia, R.; Kumari, K.; Rani, R.; Suri, A.; Pahuja, U.; Singh, D. A critical review of experimental results on low temperature charge transport in carbon nanotubes based composites. *Rev. Phys.* **2018**, *3*, 15–25. [CrossRef]
3. Khan, T.; Irfan, M.S.; Ali, M.; Dong, Y.; Ramakrishna, S.; Umer, R. Insights to low electrical percolation thresholds of carbon-based polypropylene nanocomposites. *Carbon* **2021**, *176*, 602–631. [CrossRef]
4. Soares, B.G.; Cordeiro, E.; Maia, J.; Pereira, E.; Silva, A. The effect of the noncovalent functionalization of CNT by ionic liquid on electrical conductivity and electromagnetic interference shielding effectiveness of semi-biodegradable polypropylene/poly(lactic acid) composites. *Polym. Compos.* **2019**, *41*, 82–93. [CrossRef]
5. Thi, T.B.N.; Ata, S.; Morimoto, T.; Kato, Y.; Horibe, M.; Yamada, T.; Okazaki, T.; Hata, K. Annealing-induced enhancement of electrical conductivity and electromagnetic interference shielding in injection-molded CNT polymer composites. *Polymer* **2022**, *245*, 124680. [CrossRef]
6. Zheng, B.; Wang, H.; Wu, X.; Yang, K.; Yu, Y.; Cui, H.; Gao, F.; Qian, K.; Yao, H.; Li, J.; et al. Flexible nanocomposite electrothermal films based on carbon nanotubes and waterborne polyurethane with high reliability, stretchability and low-temperature performance for wind turbine blade deicing. *Compos. Part A Appl. Sci. Manuf.* **2022**, *158*, 106979. [CrossRef]

7. Yadav, R.; Tirumali, M.; Wang, X.; Naebe, M.; Kandasubramanian, B. Polymer composite for antistatic application in aerospace. *Def. Technol.* **2020**, *16*, 107–118. [CrossRef]
8. Tian, Y.; Zhong, J.; Hu, L.; Zheng, X.; Cheng, J.; Pu, Z. Preparation of carbon nanotubes/polyethersulfone antistatic composite materials by a mixing process. *Polym. Compos.* **2019**, *41*, 556–563. [CrossRef]
9. Ma, P.C.; Siddiqui, N.A.; Marom, G.; Kim, J.K. Dispersion and functionalization of carbon nanotubes for polymer-based nanocomposites: A review. *Compos. Part A Appl. Sci. Manuf.* **2010**, *41*, 1345–1367. [CrossRef]
10. Siddiqui, N.A.; Li, E.L.; Sham, M.L.; Tang, B.Z.; Munir, A.; Kim, J.K. Tensile strength of glass fibres with carbon nanotube–epoxy nanocomposite coating: Effects of CNT morphology and dispersion state. *Compos. Part A Appl. Sci. Manuf.* **2010**, *41*, 539–548. [CrossRef]
11. Xu, H.; Gong, L.X.; Wang, X.; Zhao, L.; Pei, Y.B.; Wang, G.; Liu, Y.J.; Wu, L.B.; Jiang, J.X.; Tang, L.C. Influence of processing conditions on dispersion, electrical and mechanical properties of graphene-filled-silicone rubber composites. *Compos. Part A Appl. Sci. Manuf.* **2016**, *91*, 53–64. [CrossRef]
12. Pötschke, P.; Villmow, T.; Krause, B. Melt mixed PCL/MWCNT composites prepared at different rotation speeds: Characterization of rheological, thermal, and electrical properties, molecular weight, MWCNT macrodispersion, and MWCNT length distribution. *Polymer* **2013**, *54*, 3071–3078. [CrossRef]
13. Bai, J.B.; Allaoui, A. Effect of the length and the aggregate size of MWNTs on the improvement efficiency of the mechanical and electrical properties of nanocomposites-experimental investigation. *Compos. Part A Appl. Sci. Manuf.* **2003**, *34*, 689–694. [CrossRef]
14. Zhang, W.; Dehghani-Sanij, A.A.; Blackburn, R.S. Carbon based conductive polymer composites. *J. Mater. Sci.* **2007**, *42*, 3408–3418. [CrossRef]
15. Sui, G.; Liu, D.; Liu, Y.; Ji, W.; Zhang, Q.; Fu, Q. The dispersion of CNT in TPU matrix with different preparation methods: Solution mixing vs melt mixing. *Polymer* **2019**, *182*, 121838. [CrossRef]
16. Faraguna, F.; Pötschke, P.; Pionteck, J. Preparation of polystyrene nanocomposites with functionalized carbon nanotubes by melt and solution mixing: Investigation of dispersion, melt rheology, electrical and thermal properties. *Polymer* **2017**, *132*, 325–341. [CrossRef]
17. Liu, C.H.; Fan, S.S. Nonlinear electrical conducting behavior of carbon nanotube networks in silicone elastomer. *Appl. Phys. Lett.* **2007**, *90*, 041905. [CrossRef]
18. Lin, H.; Lu, W.; Chen, G. Nonlinear DC conduction behavior in epoxy resin/graphite nanosheets composites. *Phys. Rev. B Condens.* **2007**, *400*, 229–236. [CrossRef]
19. Bardhan, K.K. Nonlinear conduction in composites above percolation threshold-beyond the backbone. *Phys. A Stat. Mech. Its Appl.* **1997**, *241*, 267–277. [CrossRef]
20. Lu, W.; Wu, D.; Wu, C.; Chen, G. Nonlinear DC response in high-density polyethylene/graphite nanosheets composites. *J. Mater. Sci.* **2006**, *41*, 1785–1790. [CrossRef]
21. Kenkel, S.W.; Straley, J.P. Percolation theory of nonlinear circuit elements. *Phys. Rev. Lett.* **1982**, *49*, 767–770. [CrossRef]
22. Zheng, Q.; Song, Y.; Wu, G.; Yi, X. Reversible Nonlinear Conduction Behavior for High-Density Polyethylene/Graphite Powder Composites near the Percolation Threshold. *J. Polym. Sci. Part B* **2001**, *39*, 2833–2842. [CrossRef]
23. He, L.; Tjiong, S.C. Nonlinear electrical conduction in percolating systems induced by internal field emission. *Synth. Met.* **2011**, *161*, 540–543. [CrossRef]
24. Liu, X.L.; Li, M.; Geng, W.H.; Cao, W.; Tian, Y.H.; Li, T.Y.; Bin, P.S.; Qian, P.F.; Geng, H.Z. Modified carbon nanotubes/polyvinyl alcohol composite electrothermal films. *Surf. Interfaces* **2023**, *36*, 102540. [CrossRef]
25. Yanmaz, E.; Doğan, M.; Turhan, Y. Effect of sodium dodecyl sulfate on thermal properties of polyvinyl alcohol (PVA)/modified single-walled carbon nanotube (SWCNT) nanocomposites. *Diam. Relat. Mater.* **2021**, *115*, 108359. [CrossRef]
26. Barick, A.K.; Tripathy, D.K. Preparation, characterization and properties of acid functionalized multi-walled carbon nanotube reinforced thermoplastic polyurethane nanocomposites. *Mater. Sci. Eng. B* **2011**, *176*, 1435–1447. [CrossRef]
27. Wongtimnoi, K.; Guiffard, B.; Moortèle, A.; Seveyrat, L.; Cavaillé, J.Y. Electrostrictive thermoplastic polyurethane-based nanocomposites filled with carboxyl-functionalized multi-walled carbon nanotubes (MWCNT-COOH): Properties and improvement of electromechanical activity. *Compos. Sci. Technol.* **2013**, *85*, 23–28. [CrossRef]
28. Trinh, P.V.; Anh, N.N.; Tam, N.T.; Hong, N.T.; Hong, P.N.; Minh, P.N.; Thang, B.H. Influence of defects induced by chemical treatment on the electrical and thermal conductivity of nanofluids containing carboxyl-functionalized multi-walled carbon nanotubes. *RSC Adv.* **2017**, *7*, 49937–49946. [CrossRef]
29. Guadagno, L.; De Vivo, B.; Di Bartolomeo, A.; Lamberti, P.; Sorrentino, A.; Tucci, V.; Vertuccio, L.; Vittoria, V. Effect of functionalization on the thermo-mechanical and electrical behavior of multi-wall carbon nanotube/epoxy composites. *Carbon* **2011**, *49*, 1919–1930. [CrossRef]
30. Hay, J.N.; Zhou, X.Q. The effect of mixing on the properties of polyethylene blends. *Polymer* **1993**, *34*, 2282–2288. [CrossRef]
31. Aljoumaa, K.; Ajji, Z. Mechanical and electrical properties of gamma-irradiated silane crosslinked polyethylene (Si-XLPE). *J. Radioanal. Nucl. Chem.* **2015**, *307*, 1391–1399. [CrossRef]

32. Wang, Y.; Weng, G.J.; Meguid, S.A.; Hamouda, A.M. A continuum model with a percolation threshold and tunneling-assisted interfacial conductivity for carbon nanotube-based nanocomposites. *J. Appl. Phys* **2014**, *115*, 193706. [CrossRef]
33. Souri, H.; Yu, J.; Jeon, H.; Kim, J.W.; Yang, C.-M.; You, N.-H.; Yang, B.J. A theoretical study on the piezoresistive response of carbon nanotubes embedded in polymer nanocomposites in an elastic region. *Carbon* **2017**, *120*, 427–437. [CrossRef]
34. Gao, L.; Li, Z. Effective medium approximation for two-component nonlinear composites with shape distribution. *J. Phys. Condens. Matter* **2003**, *15*, 4397. [CrossRef]
35. Hashemi, R.; Weng, G.J. A theoretical treatment of graphene nanocomposites with percolation threshold, tunneling-assisted conductivity and microcapacitor effect in AC and DC electrical settings. *Carbon* **2016**, *96*, 474–490. [CrossRef]
36. Pan, Y.; Weng, G.J.; Meguid, S.A.; Bao, W.S.; Zhu, Z.H.; Hamouda, A.M.S. Percolation threshold and electrical conductivity of a two-phase composite containing randomly oriented ellipsoidal inclusions. *J. Appl. Phys.* **2011**, *110*, 123715. [CrossRef]
37. Li, W.; Liu, Z.Y.; Yang, M.B. Preparation of carbon black/polypropylene nanocomposite with low percolation threshold using mild blending method. *J. Appl. Polym. Sci.* **2009**, *115*, 2629–2634. [CrossRef]
38. Tanabi, H.; Erdal, M. Effect of CNTs dispersion on electrical, mechanical and strain sensing properties of CNT/epoxy nanocomposites. *Results Phys.* **2019**, *12*, 486–503. [CrossRef]
39. Guo, W.; Liu, C.; Sun, X.; Yang, Z.; Kia, H.G.; Peng, H. Aligned carbon nanotube/polymer composite fibers with improved mechanical strength and electrical conductivity. *J. Mater. Chem.* **2012**, *22*, 903–908. [CrossRef]
40. Russ, M.; Rahatekar, S.S.; Koziol, K.; Farmer, B.; Peng, H.X. Length-dependent electrical and thermal properties of carbon nanotube-loaded epoxy nanocomposites. *Compos. Sci. Technol.* **2013**, *81*, 42–47. [CrossRef]
41. Gao, J.F.; Li, Z.M.; Meng, Q.J.; Yang, Q. CNTs/ UHMWPE composites with a two-dimensional conductive network. *Mater. Lett.* **2008**, *62*, 3530–3532. [CrossRef]
42. Bauhofer, W.; Kovacs, J. A review and analysis of electrical percolation in carbon nanotube polymer composites. *Compos. Sci. Technol.* **2009**, *6*, 1486–1498. [CrossRef]
43. Vionnet-Menot, S.; Grimaldi, C.; Maeder, T.; Strassler, S.; Ryser, P. Tunnelingpercolation origin of nonuniversality: Theory and experiments. *Phys. Rev. B (Condens. Matter Mater. Phys.)* **2005**, *71*, 064201. [CrossRef]
44. Feng, C.P.; Wan, S.S.; Wu, W.C.; Bai, L.; Bao, R.Y.; Liu, Z.Y.; Yang, M.B.; Chen, J.; Yang, W. Electrically insulating, layer structured SiR/GNPs/BN thermal management materials with enhanced thermal conductivity and breakdown voltage. *Compos. Sci. Technol.* **2018**, *167*, 456–462. [CrossRef]
45. Zener, C. A theory of the electrical breakdown of solid dielectrics. *Proc. R. Soc. Lond. A* **1934**, *145*, 523–529. [CrossRef]
46. Franz, W. *Handbook der Physik*; Springer: Berlin/Heidelberg, Germany, 1956.
47. Beek, L.K.H.V.; Pu, B.I.C.F.V. Internal field emission in carbon black-loaded natural rubber vulcanizates. *Rubber Chem. Technol.* **1964**, *37*, 348–354. [CrossRef]
48. He, L.; Tjiong, S.C. Electrical behavior and positive temperature coefficient effect of graphene/polyvinylidene fluoride composites containing silver nanowires. *Nanoscale Res. Lett.* **2014**, *9*, 375. [CrossRef]
49. Muchharla, B.; Narayanan, T.N.; Balakrishnan, K.; Ajayan, P.M.; Talapatra, S. Temperature dependent electrical transport of disordered reduced graphene oxide. *2D Mater.* **2014**, *1*, 011008. [CrossRef]
50. Shen, J.T.; Buschhorn, S.T.; Hosson, J.T.M.D.; Schulte, K.; Fiedler, B. Pressure and temperature induced electrical resistance change in nano-carbon/epoxy composites. *Compos. Sci. Technol.* **2015**, *115*, 1–8. [CrossRef]
51. Trushin, M. Thermally activated conductivity in gapped bilayer graphene. *Europhys. Lett.* **2012**, *98*, 47007. [CrossRef]
52. Eda, G.; Mattevi, C.; Yamaguchi, H.; Kim, H.; Chhowalla, M. Insulator to Semimetal Transition in Graphene Oxide. *J. Phys. Chem. C* **2009**, *35*, 15768–15771. [CrossRef]
53. Rajendran, D.; Bautista-Quijano, J.R.; Bouhamed, A.; Kanoun, O. Influence of Carbon nanotubes (CNT) Functionalization on CNT/PDMS Nanocomposite Pressure Sensor. In Proceedings of the 2019 5th International Conference on NanofIM, Sfax, Tunisia, 30–31 October 2019. [CrossRef]
54. Disfani, M.N.; Jafari, S.H. Assessment of intertube interactions in different functionalized multiwalled carbon nanotubes incorporated in a phenoxy resin. *Polym. Eng. Sci.* **2013**, *53*, 168–175. [CrossRef]
55. Li, Z.I.; Lou, C.W.; Pan, Y.J.; Hsieh, C.T.; Huang, C.L.; Huang, C.H.; Chen, Y.S.; Lin, J.H. The effects of MWCNT length on the mechanical, crystallization and electromagnetic interference shielding effectiveness of PP/MWCNT composites. *J. Polym. Res.* **2017**, *24*, 32. [CrossRef]

MDPI AG
Grosspeteranlage 5
4052 Basel
Switzerland
Tel.: +41 61 683 77 34

Polymers Editorial Office
E-mail: polymers@mdpi.com
www.mdpi.com/journal/polymers

www.ingramcontent.com/pod-product-compliance
Lightning Source LLC
Chambersburg PA
CBHW061818210926
51850CB00006B/746
9783725861750